Principles and Techniques of Radiological Protection

Online at: https://doi.org/10.1088/978-0-7503-2424-3

IPEM–IOP Series in Physics and Engineering in Medicine and Biology

About the Series

The series in Physics and Engineering in Medicine and Biology will allow the Institute of Physics and Engineering in Medicine (IPEM) to enhance its mission to 'advance physics and engineering applied to medicine and biology for the public good'.

It is focused on key areas including, but not limited to:

- clinical engineering
- diagnostic radiology
- informatics and computing
- magnetic resonance imaging
- nuclear medicine
- physiological measurement
- radiation protection
- radiotherapy
- rehabilitation engineering
- ultrasound and non-ionising radiation.

A number of IPEM–IOP titles are being published as part of the EUTEMPE Network Series for Medical Physics Experts.

A list of recently published titles published in this series can be found here: https://iopscience.iop.org/bookListInfo/physics-engineering-medicine-biology-series.

Principles and Techniques of Radiological Protection

Edited by
Michael Charles Thorne
Mike Thorne and Associates Limited, Durham, UK

IOP Publishing, Bristol, UK

ISBN 978-0-7503-2424-3 (ebook)
ISBN 978-0-7503-2422-9 (print)
ISBN 978-0-7503-2425-0 (myPrint)
ISBN 978-0-7503-2423-6 (mobi)

DOI 10.1088/978-0-7503-2424-3

Version: 20260301

IOP ebooks

British Library Cataloguing-in-Publication Data: A catalogue record for this book is available from the British Library.

Published by IOP Publishing, wholly owned by The Institute of Physics, London

IOP Publishing, No.2 The Distillery, Glassfields, Avon Street, Bristol, BS2 0GR, UK

US Office: IOP Publishing, Inc., 190 North Independence Mall West, Suite 601, Philadelphia, PA 19106, USA

In memory of Professor N A Beresford.

Contents

Preface xvii

Editor biography xix

List of contributors xx

1 Introduction and overview **1-1**

M C Thorne

1.1 Introduction 1-1

1.2 A brief history of radiological protection 1-4

References 1-8

2 The physics of ionising radiations and radioactive decay **2-1**

M C Thorne

2.1 Introduction 2-1

2.2 Types of radioactive emission 2-1

2.3 Radioactive decay 2-3

2.4 Sources of ionising radiation 2-6

2.5 The interactions of ionising radiations with matter 2-8

2.6 Dosimetric quantities 2-9

References 2-12

3 Radiation interactions with matter **3-1**

P Burgess

3.1 X- and gamma-radiation 3-1

3.2 Electron interactions 3-11

3.3 Alpha particles 3-15

3.4 Neutron radiation 3-18

References 3-21

4 Radiation monitoring and instruments **4-1**

P Burgess

4.1 Introduction 4-1

4.2 Fundamental considerations in instrument choice 4-2

4.3 External x- and gamma-dose-rate monitoring 4-4

4.4 External beta dose rate monitoring 4-5

4.5	Neutron dose equivalent rate monitoring	4-6
4.6	Detector characteristics	4-7
	4.6.1 Ionisation chambers	4-7
	4.6.2 The proportional counter	4-10
	4.6.3 Geiger-Müller (GM) detectors	4-13
	4.6.4 Scintillation detectors	4-22
	4.6.5 Other inorganic scintillators	4-27
	4.6.6 The photomultiplier tube	4-27
	4.6.7 The silicon photomultiplier	4-30
	4.6.8 Solid state—semiconductor—detectors	4-31
	4.6.9 Alpha and beta in air monitors	4-31
	4.6.10 Installed gamma monitors and personal dosemeters	4-31
	4.6.11 Gamma spectrometry detectors	4-33
4.7	Management of monitoring equipment	4-34
4.8	Design of monitoring programmes	4-36
	4.8.1 Use of historical data	4-37
	4.8.2 Planning	4-38
	4.8.3 Predicting the radiation fields	4-39
	4.8.4 Lowest levels of interest	4-40
	4.8.5 Feedback	4-40
5	**The natural radiation environment and human modifications to that environment** *N Mitchell and M C Thorne*	**5-1**
5.1	Introduction	5-1
5.2	Natural radiation	5-1
	5.2.1 Primordial radionuclides	5-2
	5.2.2 Cosmic rays	5-7
	5.2.3 Cosmogenic radionuclides	5-9
5.3	Human modifications of the natural radiation environment	5-9
	5.3.1 Nuclear weapons testing	5-10
	5.3.2 Radiation exposure away from Earth's surface	5-15
	References	5-20
6	**Principles of internal dosimetry** *M C Thorne*	**6-1**
6.1	Introduction	6-1
6.2	Dosimetric quantities	6-2

6.3 Biokinetic modelling 6-4
6.3.1 The respiratory tract 6-5
6.3.2 The gastrointestinal tract 6-7
6.3.3 Element-specific models for radionuclides entering the systemic circulation 6-9
6.4 Radiation transport modelling 6-11
6.5 Monitoring for internal contamination 6-13
References 6-16

7 Physical and mathematical phantoms for human dosimetry 7-1
M C Thorne
7.1 Introduction 7-1
7.2 Models used in estimating operational quantities 7-2
7.3 Mathematical models of human anatomy 7-5
7.4 Physical phantoms 7-10
7.5 Considerations in monitoring to determine operational dose quantities 7-12
7.5.1 Ambient dose equivalent rate 7-12
7.5.2 Directional dose equivalent rate 7-14
7.5.3 Individual dose equivalents 7-14
7.5.4 The influence of dosemeter position 7-14
References 7-15

8 General principles of mathematical models for the transport of radionuclides 8-1
M C Thorne
8.1 Introduction 8-1
8.2 Developing and justifying a conceptual model 8-1
8.3 Common types of mathematical model 8-4
8.3.1 Equilibrium models 8-4
8.3.2 Compartmental models 8-5
8.3.3 Advective-diffusive 8-7
8.4 Model verification and validation 8-10
8.5 Modelling under uncertainty 8-12
References 8-14

9 Biokinetic models for humans 9-1
M C Thorne
9.1 Introduction 9-1

9.2 Behaviour of inhaled radionuclides in the respiratory system 9-1
9.3 Behaviour of ingested radionuclides in the gastrointestinal tract 9-7
9.4 Behaviour of radionuclides present in wounds 9-11
9.5 Retention after entry into the systemic circulation 9-13
9.6 Special issues with H-3 and C-14 9-15
9.7 Loss of radon-222 from the body 9-16
9.8 The interpretation of monitoring data 9-17
References 9-20

10 Models for the environmental transport of radionuclides 10-1
R Klos and M C Thorne

10.1 Introduction 10-1
10.2 The role of models in radiological impact assessment 10-2
10.3 Types of model 10-3
10.4 Atmospheric dispersion models 10-6
10.5 Modelling of surface-water catchments 10-13
10.6 Modelling of estuaries 10-20
10.7 Modelling of the marine environment 10-21
10.8 Radiological impact assessment 10-27
10.8.1 Pathways of exposure 10-28
10.8.2 Individual and collective doses 10-29
10.8.3 Handling uncertainties 10-30
10.8.4 Defining human habits 10-31
10.8.5 Assessing radiological impacts on non-human biota 10-33
References 10-34

11 Radiochemical analysis 11-1
D Read and B Russell

11.1 Introduction 11-1
11.2 Radionuclides in the environment 11-1
11.3 Sampling and sample preparation 11-2
11.4 Analytical techniques 11-5
11.4.1 General principles 11-5
11.4.2 Non-destructive analysis and its limitations 11-5
11.4.3 Radiometric methods 11-6
11.4.4 Mass spectrometry 11-8
11.5 Quality control 11-11
References 11-12

12 Effects of ionising radiations on the environment **12-1**
N Horemans and N A Beresford

12.1 Introduction 12-1
12.2 Overview of approaches used to assess risk to wildlife from ionising radiations 12-2
12.2.1 Simplification of ecosystems and biodiversity 12-2
12.2.2 Radionuclide transfer 12-3
12.2.3 Dosimetry 12-5
12.2.4 Estimating risk 12-6
12.3 Ionising radiation effects on different organisms 12-8
12.3.1 Learning from radiologically contaminated areas 12-9
12.3.2 Studying species sensitivities 12-17
12.3.3 Adaptation and epigenetic regulation of radiation responses 12-19
12.3.4 The way forward: towards an ecosystem approach 12-21
References 12-21

13 International organisations influencing radiological protection and recommendations on protection against ionising radiation **13-1**
C Clement

13.1 History of radiological protection 13-1
13.2 The international framework of science, recommendations, guidance, and standards 13-4
13.2.1 United Nations Scientific Committee on the Effects of Atomic Radiation (UNSCEAR) 13-7
13.2.2 International Commission on Radiological Protection (ICRP) 13-7
13.2.3 International Atomic Energy Agency (IAEA) 13-8
13.2.4 International Radiation Protection Association (IRPA) 13-9
13.2.5 European Commission (EC) 13-9
13.2.6 European Nuclear Installations Safety Standards Initiative (ENISS) 13-10
13.2.7 Heads of the European Radiological Protection Competent Authorities (HERCA) 13-11
13.2.8 International Electrotechnical Commission (IEC) 13-12
13.2.9 International Labour Organization (ILO) 13-12
13.2.10 OECD Nuclear Energy Agency (NEA) 13-14
13.2.11 World Health Organization (WHO) 13-14
13.3 An overview of the system of radiological protection developed by ICRP 13-15
13.3.1 Science, ethics, and experience 13-15

13.3.2 Scope 13-16
13.3.3 Scientific basis 13-16
13.3.4 Ethical basis 13-17
13.3.5 Aims and objectives 13-18
13.3.6 Fundamental principles 13-19
13.3.7 Concepts, tools, and requisites 13-20
13.4 Evolution of the system of radiological protection 13-22
References 13-23

14 Radiological protection in the nuclear industry 14-1
M C Thorne and J K Humphries

14.1 Introduction 14-1
14.2 The mining and milling of uranium 14-1
14.2.1 The mining and milling process and the history of uranium exploitation 14-1
14.2.2 Main routes of occupational exposure 14-3
14.3 Reprocessing 14-6
14.3.1 Stages of reprocessing and their radiological protection issues 14-6
14.3.2 Radiological hazards from nuclear reprocessing 14-7
14.4 Waste disposal 14-8
14.4.1 Preferred approaches to the disposal of solid radioactive wastes 14-8
14.4.2 Radiological impact assessments for geological disposal facilities for solid radioactive wastes 14-13
References 14-17

15 Radiological protection in radiology 15-1
C J Martin

15.1 The early use of radiation in medicine 15-1
15.2 X-ray imaging equipment 15-2
15.2.1 Radiography 15-2
15.2.2 Fluoroscopy and interventional radiology 15-5
15.2.3 Computed tomography scanners 15-7
15.3 Measurements of patient doses 15-8
15.3.1 Radiography and fluoroscopy 15-8
15.3.2 CT dosimetry 15-9
15.3.3 Patient dose surveys 15-9

15.4 Radiotherapy 15-10
15.4.1 Planning radiotherapy treatments 15-11
15.4.2 Linear accelerator treatments 15-12
15.4.3 Image guided radiation therapy (IGRT) 15-13
15.4.4 Particle therapy 15-13
15.4.5 Other external beam radiotherapy devices 15-14
15.4.6 Brachytherapy 15-15
15.5 Protection of staff and members of the public 15-16
15.5.1 General principles of x-ray installation design 15-16
15.5.2 Design of diagnostic and interventional radiology rooms 15-17
15.5.3 Evaluation of protective barrier requirements for radiology 15-19
15.5.4 Design of radiotherapy bunkers 15-19
15.5.5 Personal protective equipment and personnel monitoring 15-21
References 15-23

16 Radiological protection in nuclear medicine 16-1
C J Martin

16.1 Introduction 16-1
16.2 Radionuclide imaging 16-1
16.2.1 Radiopharmaceuticals with radionuclides emitting photon radiation 16-1
16.2.2 Gamma cameras 16-3
16.2.3 Single-photon emission computed tomography (SPECT) 16-5
16.2.4 Positron-emitting radiopharmaceuticals 16-5
16.2.5 Positron emission tomography (PET) 16-6
16.3 Preparation and administration of radiopharmaceuticals 16-8
16.3.1 The radiopharmacy 16-8
16.3.2 PET radionuclide production with a cyclotron 16-9
16.3.3 Design of nuclear medicine facilities 16-10
16.4 Medical diagnostic procedures 16-11
16.4.1 Single-photon emission imaging 16-11
16.4.2 PET procedures 16-12
16.4.3 *In vitro* nuclear medicine 16-13
16.4.4 Optimization of radiation protection for the patient 16-13
16.5 Radionuclide therapy 16-15
16.5.1 Planning treatments 16-15
16.5.2 Iodine therapies 16-16

16.5.3 Other therapy radionuclides 16-17
16.6 Radiation protection of nuclear medicine staff 16-18
16.6.1 Handling of radiopharmaceuticals 16-19
16.6.2 Protection from contamination 16-20
16.6.3 The patient journey though the imaging department 16-20
16.6.4 Issues relating to radionuclide therapy 16-21
16.6.5 Personnel dosimetry 16-22
References 16-24

17 Radiological protection and naturally occurring radioactive materials **17-1**
D Read and M C Thorne

17.1 Introduction 17-1
17.2 Mineral processing and metal smelting 17-3
17.3 Phosphate extraction 17-6
17.4 Mineral processing 17-7
References 17-7

18 The transport of radioactive materials **18-1**
M C Thorne

18.1 Introduction 18-1
18.2 Classes of radioactive materials and their packaging 18-2
18.3 Package designs 18-6
18.4 Dose rates in the vicinity of packages, transport indices and criticality indices 18-7
18.5 Conditions of transport 18-10
18.6 Security considerations 18-11
18.7 Accidents and incidents 18-12
References 18-12

19 Radiation accidents leading to large-scale releases of radionuclides to the environment **19-1**
M C Thorne and R Wakeford

19.1 Introduction 19-1
19.2 Summaries of the five major accidents 19-4
19.2.1 The Windscale Reactor Fire (INES level 5) 19-4
19.2.2 The Kyshtym Waste Tank Explosion (INES level 6) 19-6
19.2.3 The Three Mile Island accident (INES level 5) 19-7

19.2.4 The Chornobyl accident 19-9
19.2.5 The Fukushima Dai-ichi accident (INES level 7) 19-11
19.3 Radiological and health effects impacts on workers 19-14
19.3.1 The Windscale reactor fire 19-14
19.3.2 The Kyshtym accident at Mayak, Russia 19-15
19.3.3 The Three Mile Island accident 19-15
19.3.4 The Chornobyl accident 19-15
19.3.5 The Fukushima Dai-ichi accident 19-16
19.4 Radiological and health effects impacts on members of the public 19-16
19.4.1 The Windscale Reactor fire 19-16
19.4.2 The Kyshtym accident at Mayak, Russia 19-17
19.4.3 The Three Mile Island accident 19-17
19.4.4 The Chornobyl accident 19-17
19.4.5 The Fukushima Dai-ichi accident 19-19
References 19-20

20 Radiation incidents and accidents involving localised sources of exposure 20-1
M C Thorne

20.1 Introduction 20-1
20.2 Industrial uses of sealed radioactive sources 20-1
20.3 Incidents associated with radiotherapy and nuclear medicine 20-19
20.4 Criticality incidents and accidents 20-21
20.5 Larger-scale accidents 20-23
20.5.1 Loss or abandonment of radiotherapy sources 20-23
20.5.2 Theft of sources 20-24
20.5.3 Accidental discovery of sources 20-25
20.5.4 Entry into high-dose-rate areas 20-25
20.6 Malicious activities 20-26
References 20-26

21 Emergency preparedness, planning and response 21-1
K I Pearce

21.1 Introduction 21-1
21.2 Learning from experience 21-1
21.3 The protection of the public in the event of an atmospheric release 21-4
21.4 International advice 21-10
21.4.1 ICRP 21-10

21.4.2 IAEA 21-10
21.4.3 EU basic safety standards 21-12
21.5 National guidance 21-13
21.6 The development of an off-site emergency plan 21-14
21.6.1 Purpose 21-14
21.6.2 Contents of an on-site plan 21-14
21.6.3 Contents of an off-site plan 21-15
21.6.4 Emergency planning zones 21-15
21.6.5 Risk assessment 21-17
21.6.6 Functional specification 21-18
21.6.7 Concept of operations 21-21
21.6.8 Consultation and plan writing 21-23
21.6.9 Reviewing and testing of emergency plans 21-23
21.6.10 Prior and real-time information to the public 21-24
21.7 Emergency response 21-24
21.7.1 Event initiation 21-24
21.7.2 Decision making 21-25
21.7.3 Strategic, tactical and operational considerations 21-26
21.7.4 Command and control 21-27
21.7.5 Radiological advice 21-28
21.7.6 Detection of oversea accidents 21-31
21.7.7 Integration of weather forecast information into dose assessments 21-31
21.7.8 Protecting the environment 21-31
21.7.9 Public information 21-32
21.7.10 Care for the public 21-33
21.7.11 Recovery 21-35
21.8 Non-reactor faults 21-37
21.8.1 Criticality accidents 21-37
21.8.2 Industrial accidents 21-39
21.8.3 Orphan sources and malicious source placement 21-39
21.8.4 Dirty bombs 21-40
References 21-41

Preface

This compilation of information on various aspects of radiological protection has been some five years in the making, largely because of the inactivity of the editor for long periods, and his distraction on other projects. However, I am pleased to be able to provide what I hope will be a useful resource to the radiological protection community. The reader will not find here everything that they need to know, for example radiobiology and epidemiology are under-represented, but I trust that they will find much that is relevant to their professional work and some things that are irrelevant, but fascinating. Also, such is the multi-disciplinary and ever-changing nature of our profession that material that is of only marginal relevance today may become central tomorrow.

I should stress that this is not an elementary text. The individual chapters often plunge the reader into specialist areas. But we have tried to highlight broadly applicable concepts, techniques and tools of the trade with the intent that this work will not become outdated too rapidly. However, developments in science and technology (e.g. in the use of AI), changes in policy (e.g. in respect of nuclear power programmes), accidents and incidents (including those due to military conflicts), and advances in medical practice will necessarily mean that updating of this text is appropriate. To that end, comments from readers on the scope and limitations of this first edition will be welcome, as will information on new developments that should be considered. Detailed comments pointing out both technical and typographical faults will also be welcomed. The editor takes full responsibility for these and promises to do better next time, if he has the privilege of participating in the preparation of a new edition in the future.

Overall, I should like to thank the authors of the various chapters who have gone beyond the call of duty in providing detailed accounts of various specialised topics. If this work has any particular overall merit, it is largely due to their efforts. The various chapter authors and their affiliations are listed below.

Author	Contributions	Affiliation
N A Beresford (Deceased)	Chapter 12	UK Centre for Ecology and Hydrology, Lancaster Environment Centre, Library Av., Bailrigg, Lancaster, LA1 4AP, United Kingdom
P Burgess	Chapters 3 and 4	Radiation Metrology Ltd, North Craigearn, Kemnay, AB51 5 LN
C Clement	Chapter 13	International Commission on Radiological Protection, Ottawa, ON, Canada
N Horemans	Chapter 12	Belgian Nuclear Research Centre (SCK CEN), Biosphere Impact Studies, Boeretang 200, 2400 Mol, Belgium
J K Humphries	Chapter 14	EHS Valuestream Lead Spent Fuel Management, Sellafield Sites Ltd, UK

(*Continued*)

(*Continued*)

Author	Contributions	Affiliation
R Klos	Chapter 10	Aleksandria Sciences Limited, 37 Coverdale Road, Sheffield, S7 2DD
C J Martin	Chapters 15 and 16	Department of Clinical Physics and Bio-Engineering, University of Glasgow, Gartnavel Royal Hospital, Glasgow, UK
N D Mitchell	Chapter 5	Eden Nuclear and Environment Ltd, Unit 1 Mereside, Greenbank Road, Eden Business Park, Penrith, Cumbria, CA11 9FB
K I Pearce	Chapter 21	Katmal Limited, 1 Draycott, Cam, Dursley, Gloucestershire, GL11 5NF
B Russell	Chapter 11	National Physical Laboratory, Teddington, TW11 0LW
D Read	Chapters 11 and 17	University of Surrey, Guildford, GU2 7XH
R Wakeford	Chapter 19	Centre for Occupational and Environmental Health, Institute of Population Health, Faculty of Biology, Medicine and Health, The University of Manchester, Ellen Wilkinson Building, Oxford Road, Manchester, M13 9PL.

I should also like to thank the staff at IoPP for their forbearance and for finally shepherding this volume into print. Finally, I should like to thank my long-suffering wife, who has put up with not only my efforts on this book, but also my duties as Editor-in-Chief of the *Journal of Radiological Protection* and my ongoing consultancy work.

Editor biography

Michael Charles Thorne

Dr Mike Thorne is a Chartered Radiation Protection Professional with 50 years of experience in radiological protection, He has been involved in various projects related to the radiological impacts of nuclear power stations, constraints on land use in the vicinity of nuclear and other hazardous facilities, and site characterisation and assessment studies for solid radioactive waste disposal in the UK, Scandinavia and the United States.

Mike has a PhD in experimental high energy physics. Having completed this in 1974, he spent five years with the Medical Research Council working on the biokinetics and radiotoxicity of radionuclides. Subsequently, Mike has had over forty years of consultancy experience in modelling the transport of radionuclides in the environment, and he is also a former Scientific Secretary to the International Commission on Radiological Protection. He maintains a wide interest in all aspects of radiological protection and is a staunch supporter of the Society for Radiological Protection (SRP). He is an Honorary Fellow of the SRP and a Fellow of the Institute of Physics.

List of contributors

Professor N A Beresford (deceased)
UK Centre for Ecology and Hydrology, Bailrigg, Lancaster, UK

Dr P Burgess
Radiation Metrology Ltd, North Craigearn, Kemnay, AB51 5 LN

Dr C Clement
International Commission on Radiological Protection, Ottawa, ON, Canada

Dr N Horemans
Belgian Nuclear Research Centre (SCK CEN), Biosphere Impact Studies, Boeretang, Belgium

Dr J K Humphries
EHS Valuestream Lead Spent Fuel Management, Sellafield Sites Ltd, UK

Dr R Klos
Aleksandria Sciences Limited, Sheffield, UK

Dr C J Martin
Department of Clinical Physics and Bio-Engineering, University of Glasgow, Gartnavel Royal Hospital, Glasgow, UK

Dr N D Mitchell
Unit 1 Mereside, Greenbank Road, Eden Business Park, Penrith, Cumbria, UK

Dr K I Pearce
Katmal Limited, Dursley, Gloucestershire, UK

Dr B Russell
National Physical Laboratory, Teddington, UK

Professor D Read
University of Surrey, Guildford, UK

Professor R Wakeford
6 Bourne Street, Wilmslow, Cheshire, UK

Chapter 1

Introduction and overview

M C Thorne

1.1 Introduction

Protection against the adverse effects of ionising radiations has been required ever since the discovery of x-rays and radioactivity in the late 19th century. To achieve this, a complex international regulatory system has developed since the 1920s and this is supported and implemented through a wide variety of organisational and technical measures. Protection against the adverse effects of ionising radiations is required in nuclear and non-nuclear industries, in medicine, in defence-related activities and in research and development contexts. This protection is required for workers, members of the public and the environment, both in respect of routine and accidental releases. Although there are some introductory texts on radiological protection and various advanced texts on specific aspects, there seemed to be the need for an advanced text covering the whole subject area that could be used by postgraduate students and new entrants to the field to obtain an overview of the issues and current state of the art. This book is intended to satisfy this need and to serve as a reference work for professionals, such as Radiation Protection Advisors (RPAs), needing information across a wide range of radiological protection topics on a day-to-day basis. Throughout, the emphasis is on broad principles and widely applicable techniques, because it is considered that an individual who gains a clear understanding of these principles and techniques will be readily able to apply that understanding to the diverse and changing set of challenges that arise.

Radiological protection is, necessarily, a quantitative discipline. Therefore, it requires the application of various mathematical procedures relating to topics such as kinetic modelling, simulation of radiation transport through media and the statistical interpretation of data. These considerations mean that some parts of this text make use of mathematical formulae. However, the mathematical techniques applied will, in general, be familiar to graduates in science, technology, engineering and medicine. Also, every effort has been made to provide word pictures and figures to illustrate the implications of the mathematical material.

doi:10.1088/978-0-7503-2424-3ch1

The book is structured to first deal with physical aspects of radiological protection. Thus, it begins by addressing basic nuclear physics, the sources of radiation (both from generators such as x-ray sets and from radioactive materials), and various dosimetric quantities (chapter 2). This leads on naturally to a consideration of how such radiations are transported through matter and deposit their energy by ionising and exciting atoms (chapter 3). This is of relevance both to protection against the different types of radiations and their effects. However, detailed consideration of the physical mechanisms by which adverse effects are generated is deferred to later chapters.

The transport of radiation through matter and the associated deposition of energy forms the basis for the various types of instruments that are used to measure fluxes of particles and the energy deposited per unit mass. Such detectors use a wide variety of techniques and involve energy deposition in solids, liquids and gases. These matters are addressed in chapter 4, which covers not only the initial detection process, but also signal amplification, data recording and interpretation in terms of radiological protection quantities. Ancillary issues such as instrument calibration, robustness and fitness for purpose are also addressed.

Having introduced the different types of radiation and radioactive materials, an account is provided in chapter 5 of the natural radiation environment to which we are all exposed. This distinguishes between the various sources of external and internal exposure and quantifies their magnitude and variability.

The regulation of radiation exposures depends strongly on the concept of dose. Various dosimetric quantities must be distinguished (e.g. absorbed dose, equivalent dose, effective dose, committed dose). A major distinction in techniques of dose assessment occurs between external exposure to radiation fields and internal exposure to radioactive materials. Chapter 6 focuses on the principles of internal dosimetry and shows that the two main considerations are the biokinetics of radionuclides in the body and the transport of radiation within and between various tissues and organs. Radiation transport studies for both external and internal dosimetry require either physical or mathematical models of the human body (typically known as phantoms). Therefore, chapter 7 discusses the characteristics of such phantoms, which range considerably in complexity and adaptability to different exposure contexts. Although the first generation of such phantoms exhibited simple geometries, complex, voxel-based phantoms are now being developed for research purposes and these are also reviewed.

Chapter 8 addresses the general principles of mathematical modelling of radionuclide transport. Relevant techniques range from simple equilibrium and empirically based kinetic approaches through to complex 3D, physically based representations. Methods of solution, computational techniques, and model verification and validation are all addressed. This overview provides a basis for the following two chapters. The first of these (chapter 9) expands upon material included in chapter 6 by describing the types of biokinetic models used for humans. It includes examples of models used for representing retention in the organs of intake (lungs and gastrointestinal tract) and transport and retention in other tissues and organs for key radionuclides. Chapter 10 then addresses the types of models used to characterise radionuclide transport in the environment and how these models are used in radiological impact assessment studies.

Once radionuclides are distributed throughout the environment or within the human body, it becomes important to be able to determine the changing concentrations that are present. This requires both external radiation survey and radiochemical analysis techniques. As radiation survey techniques are covered in the earlier chapter on instrumentation, chapter 11 focuses on radiochemical analyses. It discusses alpha and gamma spectroscopy, liquid scintillation counting, and a wide variety of other techniques, including some, such as ICP-MS (inductively coupled plasma mass spectroscopy) that detect the mass of material present rather than its radioactivity. Both radiochemical preparation procedures and detection techniques are addressed and preferred methods for key radionuclides are identified.

This text does not address the biological basis of induction of adverse effects of radiation in humans or the epidemiological techniques used to quantify those effects. These are fully addressed in textbooks on radiobiology and epidemiology, and in the various reports of the United Nations Scientific Committee on the Effects of Atomic Radiation (UNSCEAR). However, a major topic of current interest in radiological protection is the induction of adverse effects in non-human populations. This is discussed in chapter 12 in relation to both laboratory and field evidence, but also in relation to the responses of individuals, populations, communities and ecosystems to multiple stressors. Current approaches to ensuring that adverse effects on the environment are avoided or kept to acceptable levels are addressed.

The chapters outlined above are intended to provide a comprehensive understanding of scientific issues relating to radiological protection. This understanding is needed to appreciate the basis for the regulation of exposures to ionising radiations and radioactive materials. A brief history of such standards is provided later in this introductory chapter. However, details of the current regulatory context are set out in chapter 13, which focuses on international guidance and recommendations. This sets out the roles and responsibilities of the various organisations involved before describing in greater detail the recommendations of the International Commission on Radiological Protection (ICRP) that form the basis of legislative and regulatory regimes throughout the world.

The next few chapters of the book then show how the information provided in earlier chapters is used in various practical contexts including the nuclear industry (chapter 14), medicine (chapter 15 on radiology and chapter 16 on nuclear medicine), and industries handling naturally occurring radioactive materials (chapter 17). The transport of radioactive materials is addressed in chapter 18. It is discussed separately because it is regulated by specific international conventions and relates to a wide range of materials from small, sealed sources to high-level waste and spent nuclear fuel.

Over the last few decades, there has been a great deal of interest and concern over the potential for, and consequences of, radiation accidents. These include large-scale accidents resulting in widespread environmental contamination and giving rise to requirements for evacuation, sheltering and prophylactic measures such as stable iodine administration. However, of equal interest are smaller-scale accidents that may result in external or internal exposures to one or a few individuals. Thus, chapters 19 and 20 are devoted to large-scale and small-scale accidents, respectively. These chapters lead naturally to chapter 21 on emergency planning and response, addressing, for example, roles and responsibilities of different organisations.

References are given separately for each chapter. Because this is intended to be a textbook rather than a comprehensive literature review, the lists of references have generally been kept relatively short, focusing on key texts that are definitive on specific topics or draw together an extensive range of underlying information. Exceptions are made for areas that are currently developing rapidly, where reference has been made to the primary literature, as appropriate.

The remainder of this introductory chapter provides a brief history of radiological protection because this historical development continues to influence various aspects of the discipline today, e.g. in terms of the overall philosophy of approach and the quantities used.

1.2 A brief history of radiological protection

The discovery of x-rays by Wilhelm Conrad Röntgen in November 1895 led to an explosion of interest. Thus, in 1896 alone some 50 books and pamphlets and nearly 1000 papers were published on their properties and uses (Meggitt 2008). This is not surprising, because nearly every laboratory of consequence already had the apparatus necessary for studying cathode rays and could, therefore, produce x-rays. With this intense interest, deleterious effects on health began to be identified. For example, Drury (1896) described radiation damage to the hands and fingers of early investigators in the UK and similar observations were made in Germany by Leppin (1896). Subsequently, the discovery and application of radioactive elements also resulted in adverse tissue reactions, and the ICRP (Clarke and Valentin 2009) comments that it was soon established that the deleterious effects induced by x-rays and radium could be both gruesome and lethal. Specifically, the monument to 'x-ray and radium martyrs' in Hamburg erected in 1936 names several hundred medical workers of many nationalities who died from radiation damage (Molineus *et al* 1992).

In the early 1900s, a few practitioners recognised the need to take precautions when working with x-rays or radium. Thus, for example, William Rollins, a Boston dentist proposed that physicians should wear glasses that keep out x-rays when using fluoroscopes, that x-ray tubes should be kept in shielded boxes with a small window to give a cone of radiation no larger than needed, and that patients should be shielded except where necessary for examination or treatment (Meggitt 2008). Overall, Meggitt (2008) concludes that by about 1910 'the dangers of acute and disastrous tissue damage were widely recognised and there were some straightforward protection measures being adopted. The means of measuring larger doses were available and were used for control of radiation exposures. Together these things could, if sensibly applied, reduce and perhaps eliminate the dreadful acute effects… However, many of the early workers were to die because of their injuries and even more were to suffer and die from unsuspected long-term effects that were a long way from being understood.'

It is also worth noting that this early period was also characterised by a widely held belief that small amounts of radiation were beneficial either cosmetically or medicinally. This led to a proliferation of consumer products for external irradiation of various parts of the body or containing radioactive material to be inhaled or ingested. In addition, the use of radium in luminous paint resulted in high levels of exposure, and a high incidence of cancer, in workers who used these paints, notably in the production of luminous watches, clocks and various gauges and instruments. A useful history of these various uses and consumer products is provided by Santos (2020).

Although, the sizes of large exposures could be estimated by 1910, it is important to recognise that units for measuring radiation exposures were not agreed internationally until the 1920s. Thus, at the First International Congress of Radiology in 1925, the need for quantifying radiation exposure was recognised and, as a result, in 1928 the röntgen was adopted by the International Committee on x-ray and Radium Protection {the forerunner of the International Commission on Radiological Protection (ICRP)} as a measure of exposure to x-rays and gamma rays (Smith and Thorne 1987). However, at that time, no quantitative standards of protection were proposed. Such standards were first introduced in the 1934 Recommendations (Clarke and Valentin 2009) which claimed that 'the evidence at present available appears to suggest that under satisfactory working conditions a person in normal health can tolerate an exposure to x-rays to an extent of about 0.2 international röntgens (r) per day…The protective values given in these recommendations are in general harmony with this figure under average conditions.' This standard is approximately 25 times larger than the present annual limit for average occupational dose and approximately 10 times the present limit for occupational dose in one year (Clarke and Valentin 2009).

There were only limited updates to the 1934 Recommendations before the Second World War and the first post-war meeting of what was now the ICRP took place in 1950. Over the intervening period, considerations in radiological protection had changed fundamentally. Whereas before the war the focus had been on the control of exposures of workers and patients in a medical context, the development of nuclear weapons and prospects of commercial applications of nuclear energy meant that potential occupational and public exposures had to be considered in a wider diversity of contexts. Furthermore, there had been a substantial increase in knowledge of the effects of radiation and on the behaviour of radionuclides in the human body. The ICRP issued an eight-page report in 1951 (ICRP 1951) that has been described as quite comprehensive (Clarke and Valentin 2009), but the 1954 Recommendations (ICRP 1955) were much more extensive, incorporating both the recommendations from the Commission and those of its various sub-committees. In both the 1951 and 1954 recommendations, it was recommended that every effort be made to reduce exposures to the lowest possible level, reflecting the consideration that there was no reason to assume a threshold exposure for the induction of cancer and genetic effects. It was in the 1954 Recommendations that the ICRP first proposed restrictions on exposures of the general public, suggesting that 'in the case of prolonged exposure of a large population, the maximum permissible levels should be reduced by a factor of ten below those accepted for occupational exposure.' These recommendations also introduced the dosimetric quantities the rad and the rem, which were replaced in the 1970s by the SI equivalents the gray (Gy) and the sievert (Sv) (1 Gy = 100 rad; 1 Sv = 100 rem; but see chapter 2 for more details on absorbed, equivalent and effective dose). The 1954 Recommendations also introduced the concept of the critical organ and set Maximum Permissible Concentration (MPC) values in air and water for occupational exposure and for some 90 radionuclides based on a weekly dose of 0.3 rem (3 mSv) to the organ that was critical in each case (Clarke and Valentin 2009).

Notwithstanding the substantial nature of the recommendations developed in the early to mid-1950s, the modern era in radiological protection is probably properly dated from the publication of the 1958 Recommendations in what came to be known as ICRP Publication 1 (ICRP 1959a) complemented by a second report that set MPC values for a large number of radionuclides (ICRP 1959b). These recommendations included a limit on accumulated dose equivalent of 5(N-18) rem, where N is the age in years of the exposed individual. This can be thought of as similar, in current terms, to a limit on annual effective dose of 5 rem or 50 mSv. This is the limit on annual effective dose in a single year that applies today. However, at that time the system of dose limitation was not yet risk-based, as it is today, and MPC values continued to be set on a critical organ basis. However, by the time the 1958 Recommendations were revised in ICRP Publication 9 (ICRP 1966) the Commission had become increasingly doubtful of the existence of a threshold dose for the induction of cancer and saw 'no practical alternative, for the purposes of radiological protection, to assuming a linear relationship between dose and effect, and that doses act cumulatively. Thus, the issue had become one of defining acceptable levels of risk and establishing an approach to protection that ensured risks were controlled to acceptable levels (Clarke and Valentin 2009). Furthermore, increasing knowledge of cancer risks in individual tissues and organs led to the concept of calculating a weighted, whole-body dose that was a measure of the risk to health incurred when the body was irradiated non-uniformly (Jacobi 1975). In order to control risks to acceptable levels, the recommendations in ICRP Publication 9 included the first clear formulation of the principle of optimisation recommending that 'all doses be kept as low as is readily achievable, economic and social consequences being taken into account.'

These various ideas came together in the 1977 Recommendations set out in ICRP Publication 26 (ICRP 1977), which formally introduced the principles of justification, optimisation and the application of dose limits. At that time, optimisation was closely tied to the use of cost-benefit analysis, but it has subsequently been recognised that alternative approaches, such as multi-attribute utility analysis, may be more useful in addressing complex issues. Also, the concept of collective dose was introduced at the time partly because it facilitated the undertaking of cost-benefit analyses. However, it has proved to be of limited usefulness and now features much less in decision-aiding procedures than when it was first introduced. The 1977 recommendations also introduced a formal procedure for computing the quantity now known as effective dose (see chapter 2). This meant that limits on annual intakes of radionuclides could be set using a coherent risk-based approach. However, this necessitated the development of a suite of models that permitted calculation of a wide range of tissue and organ doses for unit intakes of all radionuclides of interest by ingestion and inhalation. This required models for the distribution and retention of those radionuclides in the body plus models relating the concentrations of each radionuclide in the various source tissues and organs of the body to dose rates in the various target tissues and organs. These models were developed and applied in an occupational context as described in ICRP Publication 30 (ICRP 1979, 1980, 1982).

ICRP Publication 26 (ICRP 1977) also included, for the first time, remarks on the protection of organisms other than humans. It stated that: 'Although the principal objective of radiation protection is the achievement and maintenance of appropriately safe conditions for activities involving human exposure, the level of safety required for the protection of all human individuals is thought likely to protect other species, although not necessarily individual members of those species. The Commission therefore believes that if man is adequately protected then other living things are also likely to be sufficiently protected.' This statement has given rise to considerable debate, and it has been recognised that there is a need to demonstrate that it is true in a variety of specific assessment contexts. Therefore, the ICRP has subsequently developed recommendations on the protection of non-human biota (ICRP 2003; 2008; 2009). This topic is discussed further in chapter 12.

A comprehensive compilation of documentation relating to radiological protection over the period 1928–74 is provided by Taylor (1979). A detailed history of the subject from the late 19th Century to the early 21st Century is provided by Lindell (2019a; 2019b; 2020a; 2020b).

During the 1970s and 1980s, epidemiological data from Hiroshima and Nagasaki became increasingly recognised as the 'gold standard' upon which risk estimates for use in radiological protection should be based, though the relevance of other exposed populations was also taken into account and the issues involved in interpreting these data in the context of chronic exposures of populations of different composition were well recognised. However, during the 1980s, there were re-evaluations of the data from Hiroshima and Nagasaki, partly due to revisions in the dosimetry, and by 1989 the ICRP had revised upward its estimates of the risks of carcinogenesis from exposure to ionising radiation. This upward revision was addressed in its 1990 Recommendations (ICRP 1991), which also included a fuller statement of the principles of justification, optimisation and dose and risk limitation than had been given previously. These principles were defined as follows (Clarke and Valentin 2009).

- No practice involving exposures to radiation should be adopted unless it produces sufficient benefit to the exposed individuals or to society to offset the radiation detriment it causes. (The justification of a practice.)
- In relation to any particular source within a practice, the magnitude of individual doses, the number of people exposed, and the likelihood of incurring exposures where these are not certain to be received should all be kept as low as reasonably achievable, economic and social factors being taken into account. This procedure should be constrained by restrictions on the doses to individuals

(dose constraints) or the risks to individuals in the case of potential exposures (risk constraints) so as to limit the inequity likely to result from the inherent economic and social judgements. (The optimisation of protection.)
- The exposures of individuals resulting from the combination of all the relevant practices should be subject to dose limits, or to some control of risk in the case of potential exposures. These are aimed at ensuring that no individual is exposed to radiation risks that are judged to be unacceptable from these practices in any normal circumstances. (Individual dose and risk limits.)

Since ICRP Publication 60 (ICRP 1991), the ICRP has issued a substantial number of publications on diverse topics in radiological protection and a new set of basic recommendations (ICRP 2007). These publications form the basis of the present-day system of radiological protection that is adopted throughout the world. This present-day system of recommendations that is translated into more specific guidance and regulations by international organisations, such as the International Atomic Energy Agency (IAEA) and the European Union (EU), and by national authorities, is the topic of chapter 13.

Although the focus of this brief historical account has been the development of the recommendations of the ICRP, it should be appreciated that these recommendations are based upon an enormous body of primary research that is, in turn, reviewed and summarised by organisations such as the IAEA, the Nuclear Energy Agency (NEA) of the Organisation for Economic Co-operation and Development (OECD) and, in particular, the United Nations Scientific Committee on the Effects of Atomic Radiation (UNSCEAR). The UNSCEAR reports from 1958 onward are available at https://www.unscear.org/unscear/en/publications.html and provide an unrivalled source of detailed information. Relevant reports from the IAEA are available at https://www.iaea.org/publications. National organisations also undertake extensive studies and provide reports of broad international relevance, e.g., those from the US National Council on Radiation Protection and Measurements (NCRP) (see https://ncrponline.org/publications/).

References

Clarke R and Valentin J 2009 The history of ICRP and the evolution of its policies, ICRP Publication 109 *Ann. ICRP* **39** 75–110

Drury H C 1896 Dermatitis caused by Roentgen x-rays *Br. J. Med.* **2** 1377

ICRP 1951 International recommendations on radiological protection, revised by the International Commission on Radiological Protection and the 6th International Congress of Radiology, London, 1950 *Br. J. Radiol.* **24** 46–53

ICRP 1955 Recommendations of the International Commission on Radiological Protection *Br. J. Radiol.* **Suppl. 6** (Oxford: Pergamon)

ICRP 1959a Recommendations of the International Commission on Radiological Protection *ICRP Publication 1* (Oxford: Pergamon)

ICRP 1959b Recommendations of the International Commission on Radiological Protection *ICRP Publication 2: Report of Committee II on Permissible Dose for Internal Radiation* (Oxford: Pergamon)

ICRP 1966 Recommendations of the International Commission on Radiological Protection *ICRP Publication 9* (Oxford: Pergamon)

ICRP 1977 Recommendations of the International Commission on Radiological Protection *ICRP Publication 26* **1(3)** Ann. ICRP

ICRP 1979 Limits for intakes of radionuclides by workers: Part 1, ICRP Publication 30 *Ann. ICRP* **2(3/4)**

ICRP 1980 Limits for intakes of radionuclides by workers: Part 2, ICRP Publication 30 *Ann. ICRP* **4(3/4)**

ICRP 1982 Limits for intakes of radionuclides by workers: Part 3, ICRP Publication 30 *Ann. ICRP* **6(2/3)**

ICRP 1991 1990 Recommendations of the International Commission on Radiological Protection, ICRP Publication 60 *Ann. ICRP* **21(1–3)**

ICRP 2003 A framework for assessing the impact of ionising radiation on non-human species, ICRP Publication 91 *Ann. ICRP* **33(3)**

ICRP 2007 The 2007 recommendations of the international commission on radiological protection: user's edition, ICRP Publication 103 *Ann. ICRP* **37(2–4)**

ICRP 2008 Environmental protection: the concept and use of reference animals and plants, ICRP Publication 108 *Ann. ICRP* **38(4–6)**

ICRP 2009 Environmental protection: transfer parameters for reference animals and plants, ICRP Publication 114 *Ann. ICRP* **39(6)**

Jacobi W 1975 The concept of the effective dose—a proposal for the combination of organ doses *Rad. Environ. Biophys.* **12** 101–9

Leppin O 1896 Aus kleine Mitteilungen. Wirkung der Röntgenstrahlen auf die Haut *Dtsch. Med. Wschr.* **28** 454

Lindell B 2019a *Pandora's Box, The History of Radiation, Radioactivity, and Radiological Protection: Part 1. The Time Before World War II* (Nordic Society for Radiation Protection) English Edition—first published in Swedish in 1996

Lindell B 2019b *The Sword of Damocles, The History of Radiation, Radioactivity, and Radiological Protection: Part 2. The 1940s* (Nordic Society for Radiation Protection) English Edition—first published in Swedish in 1999

Lindell B 2020a *The Labours of Hercules, The History of Radiation, Radioactivity, and Radiological Protection: Part 3, 1950–1966* (Nordic Society for Radiation Protection) English Edition—first published in Swedish in 2003

Lindell B 2020b *The Toil of Sisyphus, The History of Radiation, Radioactivity, and Radiological Protection: Part 4, 1967–1999+* (Nordic Society for Radiation Protection) English Edition—first published in Swedish in 2011

Meggitt G 2008 *Taming the Rays: A History of Radiation and Protection* (Lulu.com)

Molineus W, Holthusen H and Meyer H 1992 *Ehrenbuch der Radiologen aller Nationen* 3rd edn (Berlin: Blackwell Wissenschaft)

Santos L J 2020 *Half Lives: The Unlikely History of Radium* (London: Icon Books Ltd)

Smith H and Thorne M C 1987 The origins and work of the international commission on radiological protection *Invest. Radiol.* **22** 918–21

Taylor L S 1979 *Organization for Radiation Protection: The Operations of the ICRP and NCRP 1928–1974* (Office of Technical Information, US Department of Energy) DOE/TIC-10124

Chapter 2

The physics of ionising radiations and radioactive decay

M C Thorne

2.1 Introduction

In assessing radiation doses to humans and other organisms, determining radionuclide concentrations in various media and measuring radiation fields in the workplace and the wider environment, it is often necessary to have an appreciation for the physics of ionising radiations and of radioactive decay. This does not imply an in-depth understanding of nuclear and elementary particle physics, but it does require a general understanding of these topics, insofar as they impact upon day-to-day aspects of radiological protection. It is the aim of this chapter to provide such a general understanding. Thus, section 2.2 describes the various types of radioactive emissions that can arise from radioactive decay. Following this, section 2.3 sets out the mathematics governing radioactive decay, including chain decay, and gives details of the decay characteristics of some radiologically important radionuclides and decay chains. Section 2.4 then outlines how ionising radiations are produced from various sources, including x-ray generators, accelerators and nuclear reactors. Finally, section 2.5 briefly outlines how ionising radiations interact with matter and section 2.6 introduces some fundamental concepts relating to radiation dosimetry.

2.2 Types of radioactive emission

Radioactive emissions arise from the transformation of the nucleus of an atom from one state to another. However, in some types of radioactive decay, the nuclear transformation is associated with changes in the number and distribution of the electrons that surround the nucleus. The nucleus itself is composed of protons and neutrons. The number of protons present defines the atomic number (Z), whereas the number of neutrons plus protons defines the atomic mass (A), because the proton and neutron both have very similar masses. In an un-ionised atom, the number of protons is equal to the number of electrons, rendering the atom neutral overall.

doi:10.1088/978-0-7503-2424-3ch2

The nucleus is made stable by the strong nuclear force. This force is attractive and acts between all nucleons (protons and neutrons). However, it is very short range, with a range that is less than the diameter of the larger nuclei. Because protons have a positive charge, they repel each other electrostatically with a force that varies as the inverse square of the distance between them. Thus, the stability of the nucleus is determined by a competition between the attractive, but short-range, strong nuclear force acting between all nucleons and the repulsive, long-range force acting between protons. With increasing nuclear size, the effect of the strong force saturates, whereas the effect of the electrostatic repulsion between protons continues to increase. To achieve stability, as the atomic number increases, the ratio of neutrons to protons increases (diluting the density of protons in the nucleus). Thus, light nuclei have approximately equal numbers of protons and neutrons, whereas the longest-lived isotope of uranium has approximately 1.6 neutrons for every proton.

Most, but not all, elements have one or more stable isotopes, where an isotope is characterised by a specific number of protons and neutrons. Individual isotopes are generally specified by the chemical symbol for the element and the atomic mass number of the isotope. Thus, for example, ^{60}Co or Co-60 represents an isotope of cobalt with a nucleus comprising 60 nucleons. As the atomic number of cobalt is 27, this means that the nucleus of ^{60}Co comprises 27 protons and 33 neutrons. It is generally helpful to distinguish the terms 'nuclide' and 'radionuclide' from 'isotope' and 'radioisotope'. Nuclides comprise isotopes of any element, whereas 'isotope' relates to a specific element. Thus, it is appropriate to refer to the radionuclides produced in nuclear fission, but the radioisotopes of caesium produced in nuclear fission.

Because nuclei can contain either an excess or a deficit of neutrons, they can move closer to stability by either converting a neutron to a proton, or, more unusually, by converting a proton to a neutron. In the first case, the change from a neutral to a positively charged particle is compensated by the emission of an electron, which is much lighter than a nucleon and carries a single unit of negative charge. In the second case, the compensation is by emission of a positron, the positively charged antiparticle of an electron. These processes are referred to as beta-minus ($\beta-$) and beta-plus ($\beta+$) decay, respectively. In general, positrons are not observed directly because soon after emission they interact with atomic electrons in the medium through which they are travelling. The interacting positron and electron annihilate each other producing two gamma-ray photons (quanta of electromagnetic radiation), which are emitted in opposite directions to each other to conserve momentum.

Nuclei of large atomic number can decay by a different type of emission. Because the helium-4 (He-4) nucleus is particularly strongly bound and because nuclear stability decreases with increasing atomic number for elements with atomic number higher than that of iron, an increase in stability can be achieved by emitting a He-4 nucleus (generally known as an alpha particle) to decrease the mass of the resulting nucleus. Some nuclides, primarily specific isotopes of the heaviest actinides, can decay by spontaneous fission, with the nucleus splitting into two (or occasionally three) fragments plus some individual neutrons. Spontaneous fission is seldom of much significance in radiological protection, but neutron-induced fission in nuclear reactors is of much greater importance. It is addressed in section 2.4 on sources of ionising radiation.

Atomic nuclei can be produced in an excited state. This typically occurs in radioactive decay in which one nucleus is transformed into another. The resulting nucleus may have rotational or vibrational energy that must be released in order that the nucleus can achieve its most stable ('ground') state. As no change in the charge of the nucleus is required, the de-excitation process usually occurs through the emission of one or more energetic photons that are termed gamma rays.

In addition to these primary modes of emission, various other types of emission can occur for some radionuclides. For example, rather than a nucleus emitting a positron, it can achieve the same result by absorbing one of the electrons from the cloud around the nucleus. This is known as electron capture. Also, photons emitted from the nucleus may interact with and eject atomic electrons, leaving vacancies in the electron orbitals. These vacancies can then be filled by other electrons falling from higher energy levels and emitting the difference in energy as characteristic x-ray photons. Alternatively, electrons falling from higher energy levels can transfer the difference in energy to other orbital electrons, ejecting then from the atom. Where multiple electrons are emitted by this process, it is often referred to as an Auger cascade.

All the different types of emission described above are governed by quantum mechanical considerations. They are, therefore, characterised probabilistically. This means that where a radionuclide can exhibit different modes of decay, the specific mode that occurs will be determined by the relative probabilities of the different modes and the rate of decay will be determined by the sum of the frequencies of occurrence of the various modes. However, a mode of decay does not necessarily result in a single emission, e.g. a beta decay event may be associated with several gamma emissions arising from de-excitation of the resulting nucleus. Therefore, compilations of radioactive decay data typically give average yields per transformation for all identified emissions from a radionuclide, together with the radioactive decay constant (or radioactive half-life) for that radionuclide. Thus, for example, table 2.1 gives a simplified version of the radioactive decay scheme for I-131.

.Note that I-131 can decay by emitting beta-particles of several different energies or by capture of atomic electrons from either the K shell or the L shell (ce-K and ce-L). There are also some transformations involving emissions of gamma rays and/or emissions of characteristic K shell and L shell x-rays. Energies of the emissions are given in MeV, where 1 eV is equivalent to 1.602×10^{-19} J. Unique energies are given for photon emissions. However, in beta decay both a beta particle and a neutrino are emitted. This means that the energy of the transformation is partitioned between them. Consequently, there is a spectrum of beta energies. As the neutrino does not deposit energy locally, it is the average energy of the beta particle that is the primary determinant of the radiation dose that is received.

2.3 Radioactive decay

As described in section 2.2, radioactive decay is determined by quantum mechanical considerations. Thus, the timing of individual nuclear transformations is random,

Table 2.1. Decay characteristics of I-131.

Radiation	Yield per transformation	Average or unique energy (MeV)
β^-	2.13E−2	6.935E−2
β^-	6.20E−3	8.693E−2
β^-	7.36E−2	9.660E−2
β^-	8.94E−1	1.915E−1
β^-	4.20E−3	2.832E−1
γ	2.62E−2	8.018E−2
ce-K	3.63E−2	4.562E−2
ce-L	4.30E−3	7.473E−2
γ	2.65E−3	1.772E−1
γ	6.06E−2	2.843E−1
ce-K	2.48E−3	2.497E−1
γ	2.51E−3	3.258E−1
γ	8.12E−1	3.645E−1
ce-K	1.55E−2	3.299E−1
ce-L	1.71E−3	3.590E−1
γ	3.61E−3	5.030E−1
γ	7.27E−2	6.370E−1
γ	2.20E−3	6.427E−1
γ	1.80E−2	7.229E−1
K x-ray	2.59E−2	2.978E−2
K x-ray	1.40E−2	2.946E−2

time-independent and independent of most external factors. This means that for a radionuclide R, with a decay constant $\lambda_R(s^{-1})$ and an amount at any time, t(s), of $Q_R(t)$ atoms, the rate of change in the number of atoms with time is given by:

$$dQ_R(t)/dt = -\lambda_R Q_R(t) \tag{2.1}$$

For an initial number of atoms, $Q_R(0)$, this gives:

$$Q_R(t) = Q_R(0)\exp\{-\lambda_R t\} \tag{2.2}$$

Although equation (2.2) is written in terms of numbers of atoms, multiplying each side by λ_R would leave the expression unchanged. Thus, the same form applies both for mass and for activity.

If $t_{1/2} = \ln 2/\lambda_R$, it follows from equation (2.2), that:

$$Q_R(t_{1/2}) = Q_R(0)/2 \tag{2.3}$$

Thus, $t_{1/2}$ is known as the radioactive half-life.

Although many radionuclides decay directly to stable nuclides, there are many others that decay via one or more generations of radioactive progeny. In this case, there is a governing equation for each member of the chain. Here, the situation is illustrated for an unbranched chain in which the parent radionuclide decays only to one progeny radionuclide, in contrast to a branched chain in which various alternative progeny radionuclides may be produced in one or more of the transformations. The governing equations are:

$$dQ_1(t)/dt = -\lambda_1 Q_1(t) \tag{2.4a}$$

$$dQ_2(t)/dt = -\lambda_2 Q_2(t) + \lambda_1 Q_1(t) \tag{2.4b}$$

$$dQ_J(t)/dt = -\lambda_J Q_J(t) + \lambda_{J-1} Q_{J-1}(t) \tag{2.4c}$$

For the first and second members of such a chain and for $Q_1(0) \neq 0$, $Q_2(t) = 0$, the solution is:

$$Q_1(t) = Q_1(0)\exp\{-\lambda_1 t\} \tag{2.5a}$$

$$Q_2(t) = \left[\lambda_1 Q_1(0)/(\lambda_2 - \lambda_1)\right][\exp\{-\lambda_1 t\} - \exp\{-\lambda_2 t\}] \tag{2.5b}$$

Figure 2.1 illustrates the case for $Q_1(0) = 1.0$, $\lambda_1 = 1.0$ and $\lambda_2 = 0.1$.

As the number of atoms of type 1 declines, so the number of atoms of type 2 increases. The total number of atoms stays approximately constant at first, as each atom of type 1 becomes an atom of type 2. However, in the longer term there is a decrease in the total number of atoms due to the radioactive decay of type 2.

Note that equations (2.4) and (2.5) are appropriate only when expressed in mass terms, e.g. numbers of atoms or moles of each radionuclide. The results can be converted to activity by multiplying by the appropriate decay constant, i.e. giving activities of $\lambda_i Q_i(t)$ for $i = 1, J$. Also, where branched chains are to be addressed,

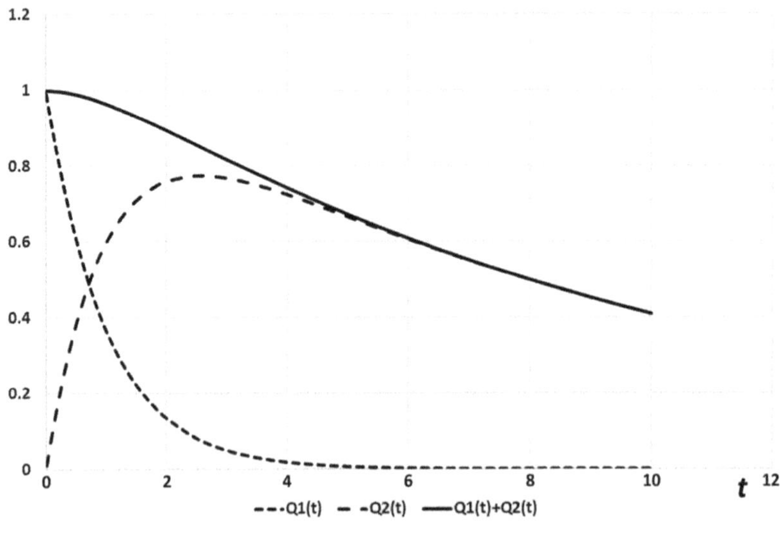

Figure 2.1. Solution of equation (2.5) for $Q_1(0) = 1.0$, $\lambda_1 = 1.0$ and $\lambda_2 = 0.1$.

equation (2.4) is modified by the inclusion of branching ratios from one generation to the next.

The mathematical model of a decay chain given by equation (2.4) is a specific example of a multi-compartmental model with constant rate coefficients. Approaches that can be used for obtaining solutions of the equations for generalised multi-compartmental models are described in chapter 8. In practice, the modelling of radioactive decay is often integrated with modelling the environmental transport of radionuclides or with modelling their biokinetics in the human body. These matters are addressed in chapters 9 and 10. Complexities arise because the radionuclides in a decay chain may have very different radioactive half lives and the individual members of the chain may differ considerably in their biogeochemical behaviour and in their uptake and retention in the human body. Thus, for example, the U-238 decay chain includes isotopes of the two actinides uranium and thorium, the alkaline earth radium, the inert gas radon, and the metals lead and polonium. Half lives of these various radionuclides vary from 3.8 days up to 4.5 10^9 years. More information on the long decay series that occur in Nature is given in chapter 5.

In other contexts, decay chains can be of importance in providing an assured supply of short-lived radionuclides, e.g. for nuclear medicine. For example, Tc-99m, which is a metastable state of technetium that decays with a half-life of 6.02 h is provided in the form of a generator comprising a source of Mo-99. The Mo-99 has a radioactive half-life of 66.02 h. Tc-99m can be extracted from this generator as required for nuclear medicine examinations, providing a useful source for a period of up to about a week.

Some radioactive progeny have half lives that are very short compared with those of their parents and compared with the characteristic timescales by which they would be transported away from their places of production. Thus, for example, Sr-90, with a half-life of 29 years decays to Y-90 with a half-life of 64 h. In many environmental contexts, the short half-life of Y-90 means that it decays very close to where it was produced. Where this is the case, it can be assumed that the short-lived progeny will be in secular equilibrium with its longer-lived parent, i.e. the activity or activity concentration of the progeny radionuclide will be identical to that of its parent (because each transformation of the parent will be associated with one transformation of its progeny at the same location). Where a branching ratio of less than 1.0 exists in the decay chain, this must be addressed. Thus, for example, Cs-137 with a half-life of 30.0 years decays to Ba-137m (with a half-life of 2.55 min) in 94.6% of its decays. Thus, when considering Ba-137m to be in secular equilibrium with Cs-137, its activity should be taken as 0.946 of that of the Cs-137.

2.4 Sources of ionising radiation

In broad terms, sources of ionising radiation can be distinguished into accelerators, radioactive materials and nuclear reactors. The initial discovery of ionising radiations arose from an accelerator experiment. On 8 November 1895, Wilhelm Conrad Roentgen was repeating some experiments on cathode rays (which had not yet been identified as electrons) that had been undertaken by Heinrich Hertz and his student Philipp Lenard. It had been found possible to pass these cathode rays a short distance outside the generating vacuum tube. What Roentgen observed was that 'if one covers the tube with a fairly close fitting mantle…one observes in the completely darkened room that a paper screen painted with barium platinocyanide…glows brightly and becomes fluorescent with each discharge…at a distance of two metres from the apparatus' (cited at pages 37 and 38 of Pais [1986]). As discussed below, the scattering of the electrons at the anode had given rise to energetic photons, or x-rays.

In an accelerator, a beam of particles is accelerated to a high velocity. This beam may either be used directly as a source of ionising radiation or may be used to generate a secondary source. For example, in an electron synchrotron, a beam of high-energy electrons is forced to travel in a circular orbit by the application of a magnetic field perpendicular to the direction of motion. Because circular motion involves acceleration along the radius of the circle, the electron loses energy by radiating photons along the direction of travel. This is an example of bremsstrahlung (or 'braking radiation'). The energy lost in such radiation can be replaced by accelerating the electrons in radio-frequency cavities included around the orbital path. Bremsstrahlung also occurs at an atomic scale when beta-particles emitted in radioactive decay are scattered, and hence accelerated, in the surrounding medium. Scattering processes are considered further in section 2.5.

Accelerated electrons are used to produce secondary sources of photons in both x-ray tubes and linear accelerators. In either case, the fast-moving electrons impinge on a solid target and two different types of x-rays are produced. These are bremsstrahlung, arising from acceleration of the impinging electrons by the nuclei of atoms in the target and characteristic x-rays that arise because the incoming electron displaces an orbital electron from an atom in the target leaving a vacancy. That vacancy is filled by an electron from a less tightly bound orbit, with the difference in energy being emitted as a photon with an energy equal to the difference in energy between the two orbits involved. Because heavy atoms have several filled orbitals from which electrons can be ejected, x-rays of various characteristic energies can be produced. However, in practice, those arising from the ejection of inner (K-shell) electrons are of greatest practical importance.

Accelerators can produce beams of electrons, protons and heavier charged ions directly. In addition, if proton and ion beams are incident on suitable targets, fluxes of neutrons can be produced. For example, if heavy hydrogen (deuterium) ions are accelerated and interact with tritium nuclei a neutron with an energy of 14.1 MeV and a recoil He-4 nucleus are produced. A similar reaction involves a collision between two deuterium nuclei, resulting in a neutron with an energy of 2.5 MeV and a recoil He-3 nucleus.

Neutrons are also copiously produced in nuclear reactors. At the point of production, these neutrons have energies of about 2 MeV on average. However, they can be slowed down (moderated) to thermal energies of about 0.025 eV by repeated collisions with a material such as graphite that scatters the neutrons but does not absorb them to a significant degree. Thus, thermal neutrons can be extracted from a reactor using a graphite column that extends into the reactor core.

Finally, consideration needs to be given to the production of ionising radiations from radioactive materials. These include materials that incorporate naturally occurring radionuclides and sources that incorporate either naturally occurring or artificially produced radionuclides. Artificial radionuclides may be produced using cyclotrons or in nuclear reactors. They vary immensely in lifetime and decay characteristics, and this facilitates the selection of a specific radionuclide that has the optimum characteristics for a specific task. Thus, for example, various physicochemical forms of the positron emitters C-11 (half-life 20 min) and F-18 (half-life 110 min) are extensively used in medical applications of positron-emission tomography, whereas Co-60 (half-life 5.27 years) emits energetic gamma photons (1.17 and 1.33 MeV) and has been useful in applications such as industrial radiology (though Ir-192, half-life 73.8 days, is increasingly used for this purpose). An important distinction is between sealed sources that are used for external irradiation and unsealed sources, such as those used in diagnostic tests in nuclear medicine. Sealed sources can also be used to produce neutrons, e.g. a mixture of the alpha-emitter Am-241 with beryllium produces neutrons through the reaction Be-9$\{\alpha,n\}$C-12.

2.5 The interactions of ionising radiations with matter

The interactions of photons (x-rays and gamma rays) with matter at energies from a few eV up to tens of MeV are dominated by three processes. These are the photoelectric effect, which dominates at the lower end of the energy range, Compton scattering, which dominates at higher energies, and pair-production, which becomes of progressively greater importance as the energy of the photon rises above the pair-production threshold of 1.022 MeV, which is the minimum energy required to create an electron–positron pair.

In the photoelectric effect, the incoming photon ejects an electron from an atom. Some of the energy of the photon is used to supply the potential energy needed by the electron to escape from the atom, whereas the remainder is transferred to the electron as kinetic energy of motion. Thus, in the photoelectric effect, the incoming photon is annihilated.

In contrast, in the Compton effect, the incoming photon is scattered inelastically by an electron in the medium. Therefore, the photon both loses energy and is scattered from its original direction, with both energy and momentum being carried away by the recoiling electron.

As the photon energy increases above the sum of the rest mass energies of an electron and a positron, it becomes possible for the incoming photon to be converted into an electron–positron pair. However, there is also the requirement that this process should conserve both energy and momentum, and this is only possible if another particle is involved. In general, this is the nucleus of an atom in the medium. When pair-production occurs in the electric field of the nucleus, energy and momentum are carried away both by the electron and positron, and by the recoiling nucleus.

The electron and positron arising from pair-production can be very energetic. This means that they can be scattered in the bremsstrahlung process leading to photons that may, in turn, have sufficient energy to give rise to electron–positron pairs. This electromagnetic cascade effect is of great importance in the context of cosmic ray particles impinging on the atmosphere of the Earth. These can result in an extensive shower of electrons, positrons and photons reaching the surface of the Earth.

The various photon-scattering processes outlined above result in the production of secondary charged particles that typically have sufficient energy to result in ionisation of atoms of the medium through which they pass. The degree of ionisation per unit length in the medium depends strongly on the type of particle produced by the photon interactions or originating by other mechanisms, such as radioactive decay. Low mass particles, such as the electron and positron deposit only small amounts of energy per unit length of path and are known as radiations of low Linear Energy Transfer (low-LET radiations). In contrast, high-mass particles, such as alpha particles, heavy ions and recoil nuclei are high-LET radiations. This distinction is important because the sparsely ionising tracks of low-LET radiations are much less effective at inducing biological effects than are the densely ionising tracks of high-LET radiations. Neutrons do not cause ionisation directly, but rather through scattering processes that result in secondary charged particles. The LET of neutrons depends strongly upon their energy, which means in turn that their effectiveness at inducing biological effects depends strongly on their energy.

To quantify the amount of ionising radiation received by a tissue or organ, various dosimetric quantities have been defined. These quantities are described below. In addition, other quantities have been defined related to the monitoring of external exposure to various types of radiation field. These quantities are addressed in chapters 3 and 4.

2.6 Dosimetric quantities

The fundamental quantity used in radiological protection is the absorbed dose. This is defined as the amount of energy, ΔE (J), deposited in an increment of mass, Δm (kg), in the form of ionisations. Absorbed dose is often described as being defined at a point, but this is not quite accurate, because ionisations are spatially discrete events and it requires a finite-size increment of mass to include enough ionisations to give a well-defined absorbed dose. In micro-dosimetric studies, other quantities are used to describe the amount of energy absorbed in a small mass and fluctuations in that amount from one location to another.

However, in radiological protection, the focus is on relating the absorbed dose to adverse biological effects observed at the tissue or organ level, e.g. cancer or hereditary disease in humans. Therefore, the values of absorbed dose of relevance are averages over individual tissues or organs, defined as the total energy absorbed in a tissue or organ divided by the mass of that tissue or organ.

The units of absorbed dose are J kg^{-1}. This is given the special name gray (Gy). Thus, 1 Gy = 1 J kg^{-1}.

As mentioned in section 2.5, the various types of ionising radiation give rise to different spatial densities of ionisation in tissues and organs, and these different densities of ionisation in turn result in different levels of induction of adverse biological effects for the same amount of deposited energy. In fundamental radiobiology, this distinction between radiations of different types is quantified using the concept of Relative Biological Effectiveness (RBE). The RBE of one radiation relative to another is defined as the ratio of the absorbed doses at which they give the same level of a specific effect, e.g. the same degree of cell sterilisation in a cell-culture experiment. Thus:

$$RBE_{\rm i} = D_{\rm ref}/D_{\rm i} \qquad (2.6)$$

where $RBE_{\rm i}$ is the RBE of radiation type i relative to a reference radiation, and $D_{\rm i}$ is the absorbed dose of radiation type i that gives the same level of effect as a dose of the reference radiation $D_{\rm ref}$. The reference radiation adopted is typically x-rays with a specific spectrum or gamma rays.

Values of RBE differ depending upon the effect analysed and as a function of the size of the absorbed dose. This latter dependence arises because various types of radiation typically exhibit different shapes of dose–effect relationship. This is illustrated in figure 2.2. The linear dose–effect relationship tends to be characteristic of high-LET radiations, whereas the curvilinear relationship tends to be characteristic of low-LET radiations. It is often characterised as a linear-quadratic relationship, i.e.

$$R(D) = \alpha D + \beta D^2 \qquad (2.7)$$

where $R(D)$ is the response at absorbed dose D, and α and β are calibration coefficients.

This expression can be used to represent the probability of cell sterilisation. In that case, the probability of a cell being sterilised is $R(D)$, so the probability of a cell

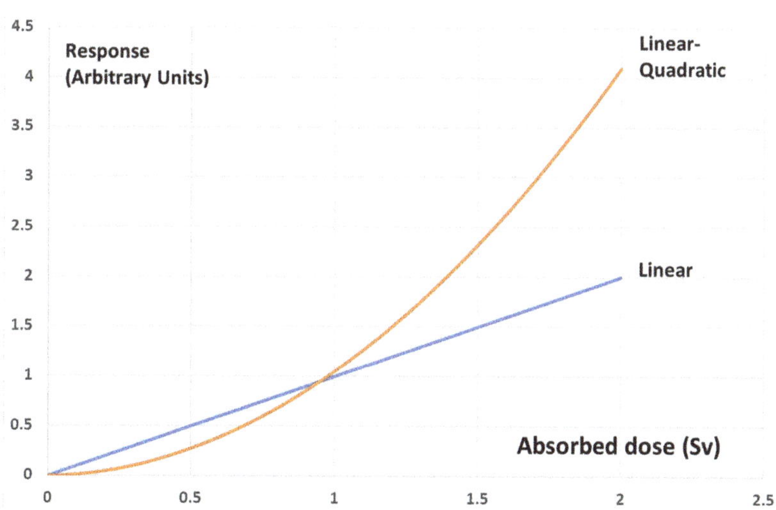

Figure 2.2. Linear and linear-quadratic dose–response relationships.

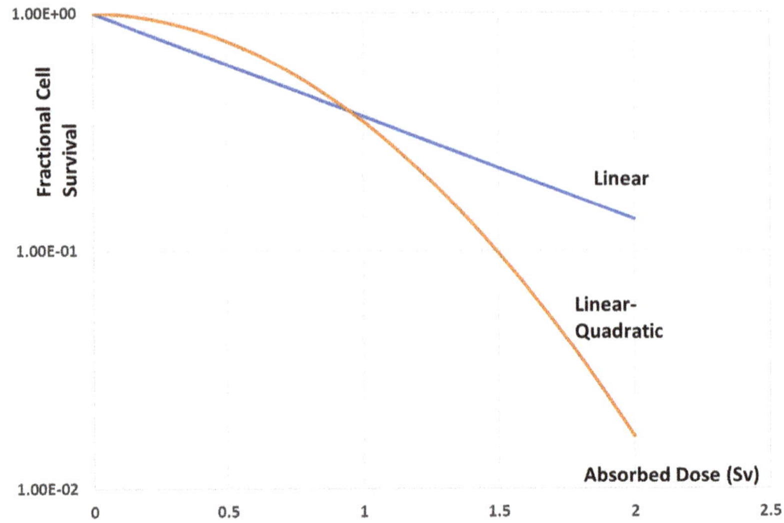

Figure 2.3. Cell survival curves determined by linear and linear-quadratic response relationships. Note that fractional cell survival is given on a logarithmic scale.

surviving an absorbed dose D without being sterilised is $\exp[-R(D)]$. Thus, fractional cell survival, $S(D)$, can be expressed as:

$$S(D) = \exp[-(\alpha D + \beta D^2)] \tag{2.8}$$

This is illustrated in figure 2.3. In both figures 2.2 and 2.3, $\alpha = 1.0$ for the linear response, and $\alpha = 0.05$, $\beta = 1.0$ for the linear-quadratic response.

In considering cell mutation or cancer induction, these two forms are often combined, because a cell mutation can only be expressed if the cell survives and is capable of reproduction. Thus, the general form for induction of an effect, $I(D)$, is often written:

$$I(D) = (\alpha_1 D + \beta_1 D^2) \times \exp[-(\alpha_2 D + \beta_2 D^2)] \tag{2.9}$$

In these expressions, the linear term is often attributed to the passage of a single track of ionising radiation through a cell, with the quadratic term attributed to the combined effect of two tracks. If a radiation dose is protracted in time, there exists the possibility that damage arising from the first track is repaired before the second track is induced. Thus, as a radiation dose is protracted, or fractionated, the β coefficient tends to decrease and dose–response relationships become more nearly linear. Also, because high-LET radiations give rise to much higher ionisation densities along their tracks than low-LET radiations, single-track effects dominate. Thus, linear dose–response relationships are characteristic of high-LET radiations and curvilinear dose–response relationships are characteristic of low-LET radiations.

In radiological protection, rather than using RBE, absorbed doses are averaged over the tissues and organs at risk and radiation weighting factors, w_R, are applied to those average doses to take account of differences in the efficiency with which various types of radiation induce the deleterious effects of greatest significance to human health, i.e. cancer and serious hereditary disease. Thus, w_R is taken as 1.0 for x-rays and gamma rays and is taken as 20 for α particles (ICRP 2007). Because these weighting factors have been set to be appropriate at low doses and dose rates, they cannot automatically be applied in other contexts, e.g. where the induction of tissue reactions is being addressed or where the implications of exposures of non-human biota are to be evaluated.

The product of w_R and the absorbed dose to a tissue or organ is known as the equivalent dose to that tissue or organ, $H_T = w_R D_T$. Because w_R is dimensionless, equivalent dose has the fundamental units of J kg^{-1}, but it has the special name of sievert (Sv). Equivalent doses to a specific tissue or organ from different types of radiation can be summed to give the total equivalent dose.

The various tissues or organs of the body differ in their sensitivity to the induction of deleterious effects, and this is addressed by using a tissue weighting factor, w_T. Currently adopted values of w_T are 0.12 for red bone marrow, colon, lung, stomach, breast and remainder tissues, 0.08 for the gonads, 0.04 for bladder, oesophagus, liver and thyroid, and 0.01 for bone surface, brain, salivary glands and skin. The sum of the w_T values is 1.0. Effective dose, E, is defined as:

$$E = \sum_T w_T H_T \tag{2.10}$$

The effective dose also has fundamental units of J kg^{-1}. Like equivalent dose, it has the special name sievert (Sv). Because of the normalisation adopted, for uniform whole-body irradiation, the effective dose is numerically equal to the equivalent dose in any tissue or organ.

When radionuclides are taken into the body, additional considerations arise because the dose received is protracted in time after the intake with which it is associated. This is addressed by defining the quantity committed dose, as discussed in chapter 6.

References

ICRP 2007 The 2007 Recommendations of the International Commission on Radiological Protection: User's Edition, ICRP Publication 103 *Ann. ICRP* **37(2–4)**

Pais A 1986 *Inward Bound: Of Matter and Forces in the Physical World* (Oxford: Oxford University Press)

Van de Casteele E 2004 Model-based approach for beam hardening correction and resolution measurements in microtomography *PhD Thesis* University of Antwerp

Chapter 3

Radiation interactions with matter

P Burgess

3.1 X- and gamma-radiation

Exposure to photon fields is, operationally, often the main contributor to effective dose. People who work routinely in such fields get used to their presence and can develop a good intuitive feel for how those fields behave. This section covers the basic physics of interactions of photons with matter. The treatment does not go into detail but explores some aspects that are generally ignored in textbooks, despite their importance operationally.

The text below does not distinguish gamma and x-radiation as there is no difference between them in terms of interactions. It uses the more general term 'photon radiation'. What differentiates x- and gamma-radiation is where they are generated. Gamma radiation is generated in the nucleus and is basically the nucleus shedding energy which has resulted from a particulate emission, generally a beta particle. This can leave the nucleus in an excited state which will eventually result in the emission of a gamma photon. There is a simple rule, which is not perfect, which says that the more the energy, the shorter the half-life, i.e. the Geiger–Nutall law. Looking at common sources, the highest one encountered by workers is often the 6 and 7 MeV radiation generated by the decay of N-16, an activation product of oxygen, found in CO_2 and water-cooled reactors. Its half-life is 7 s. Its production is interesting—it starts with an n,p reaction on O-16, which generates N-16. This decays by beta emission, with main branches with maximum energies of 10.4 MeV at a yield per transformation of 28% and 4.3 MeV with a yield of 66%. The 'residual' energy left after the 4.3 MeV decay is then emitted at 6 MeV. The product is O-16. Effectively, the process has turned the energy of the original neutron into beta radiation and gamma radiation, starting and finishing with O-16. This reaction has a neutron energy threshold of over 10 MeV. In natural radiation, the most energetic photon of significance is the 2.6 MeV gamma from Tl-208, the major final radionuclide in the Th-232 chain. This has a half-life of 3 min.

doi:10.1088/978-0-7503-2424-3ch3

X-radiation, in contrast, is generated in the electron shell. Its characteristics are influenced by the relevant nucleus, but generation is in the shell. Since the dawn of time, the photon spectrum has been illustrated by a split into energy bands covering radio, microwaves, infra-red, visible light, UV, x-rays and gamma radiation. This last split is misleading. The highest energies of x-radiation encountered routinely in the workplace can be 18 MeV, in radiotherapy, far higher than the maximum gamma energies.

A good thing to keep in mind is that the way photon radiation interacts is very different from alpha and beta radiation. A useful way to visualise this is to think of a source that generates alpha, beta and gamma radiation, such as a disc of uranium which is commonly used as a check source (figure 3.1).

If we could visualise the energy loss process in air, there would be a series of short, straight, intense lines emerging from the source with virtually the same length of a few cm. These are the alphas. The beta radiation would be a series of very much less intense, much longer (up to metres) lines, also emerging from the source, with erratic and unpredictable shapes and widely varying total lengths. The consequences of gamma radiation would be lines rather like the betas, but generally shorter and with very few starting at the source. These lines can 'pop into existence' at any distance from the source. The 'popping into existence' will follow the inverse square law, if you observe a very large number of them. If you only observe a small number, say 10, then there may be no real observable pattern. It is this capacity to travel into a material without interacting that means that we cannot ascribe a range to gamma radiation. In theory, the lowest energy photon radiation can penetrate the thickest material.

The effectiveness of materials for interacting with photons is expressed in terms of a stopping power with usual units of $cm^2\ g^{-1}$. If this was expressed in a correct SI form, then the units would be $m^2\ kg^{-1}$ but this form is very rarely used. This may seem a strange unit but when multiplied by the mass per unit area of the absorber in $g\ cm^{-2}$ (i.e. thickness in cm multiplied by density in $g\ cm^{-3}$) this gives a dimensionless number, which is what is needed in calculation. One way of thinking about this is to consider it as a measure of the target area represented by 1 g of the material, i.e. the target area of each atom multiplied by the number of atoms in each

Figure 3.1. A uranium check source.

gram. It is important to remember that the target represented by an individual atom is very small indeed. For example, at 1 MeV for lead, the stopping power value is 0.07 cm^2 g^{-1}. In 1 g of lead there are approximately 3×10^{21} atoms which means that each atom represents a target area of about 2.4×10^{-23} cm^2. Thinking of this target area as a disc, the disc has a diameter of 5.5×10^{-12} cm. If we consider the lead atoms as each occupying a cube then the corresponding dimension is 3×10^{-7} cm, i.e. about 60 000 times bigger.

There are three important photon interaction mechanisms in radiation protection: photoelectric, Compton and pair production. The less important interaction considered is photoneutron production.

The photoelectric effect dominates for high Z materials and low photon energies. In this process, the photon is completely absorbed, and an electron is emitted that has the energy of the incoming photon minus the biding energy of the electron. If this is with one of the inner shell electrons, e.g. from the K shell, then the binding energy can be significant. Immediately after emission of the electron, the vacancy is filled by an electron from a further-out shell, either in one step or several. This process can generate characteristic x-rays, i.e. having energies equal to the difference in binding energies between the two shells involved, or Auger electrons. In shielding, for example, these x-rays are often absorbed close to the point of generation. In spectrometry, this will result in a full energy peak. However, these x-rays may escape from the spectrometry detector and the result is a count with an energy lower than the full energy peak. The photoelectric interaction probability for an atom is proportional to $Z^n/E^{3.5}$ where n varies from 4 to 5 over the energy range considered here. This is per atom so the bulk probability will go down as the atomic number A, which, as Z increases, increases faster. The result for a bulk material approximates to $Z^3/E^{3.5}$. There are significant discontinuities. When the incoming photon energy is just above that of the binding energies of the electrons in a particular shell then the interaction is relatively very likely. When the energy drops below that of the electron shell then electron removal becomes impossible—the target represented by the atom gets suddenly smaller. This process is obvious with high Z elements. The example in figure 3.2 is for lead. Notice the factor of 7 drop at about 83 keV (the K edge) and also the L, M and N edges at much lower energies.

Note that the variation of the mass attenuation coefficient with photon energy is very rapid. This has major implications for shielding—the lead thickness which is sufficient at 100 keV may be very ineffective at 200 keV. Also, note how, in this case, Compton scattering is dominant between 800 keV and 3.5 MeV with photoelectric scattering dominant below 800 keV and pair production above 3.5 MeV.

A further consideration apart from mass attenuation is the difference between the dose rate in shielding materials and the dose rate in air at the same location. Indeed, a possible definition of a material useful as wall shielding is one where the dose rate in the material is very much higher than the dose rate in air or tissue for the same photon spectrum. The consequence of the very high photoelectric interaction with lead at 100 keV is that the dose rate to lead in lead at the output side of any lead shielding is very much higher than the dose rate to air in air at the same position. What is occurring is that the density of emitted electrons with energies up to 100 keV

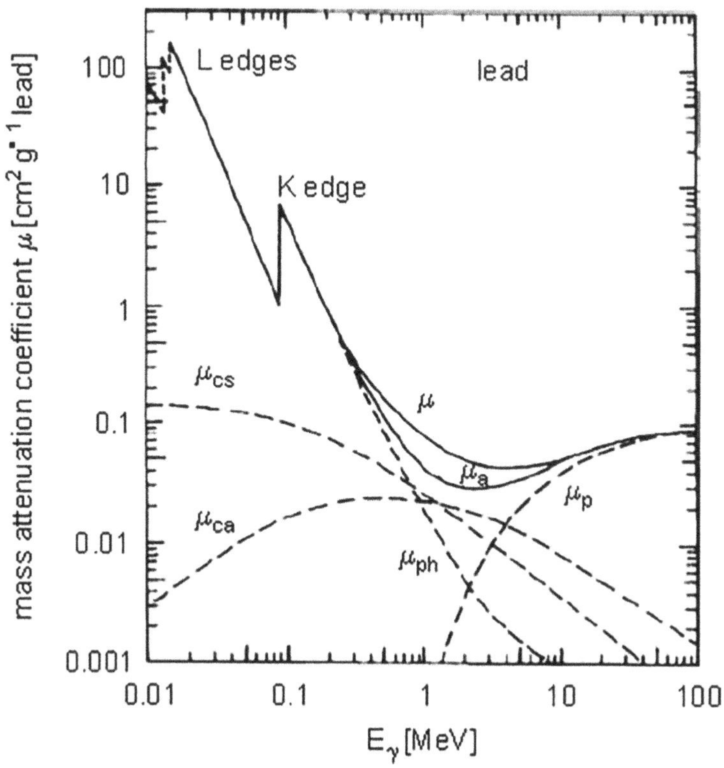

Figure 3.2. Mass attenuation coefficient for photon in lead. Reproduced from Gazis (2019). CC BY 3.0.

in lead is much higher than the density of similar electrons in air. The relevance of this is that placing a typical radiation protection ionisation chamber against the lead gives a very much higher indication with the slide open or cap off compared with the same instrument with the slide closed or cap on. This is because the tissue equivalent cap material absorbs the electrons from the lead and replaces them with a smaller population of electrons, as the dose in tissue is much lower. Hence, the indication drops. Any thin-windowed electron detector such as an end window GM tube will show the same effect. This can lead to the apparently ridiculous situation where a piece of paper is a more effective attenuator of medium energy x-rays than a thin piece of lead.

Compton interactions dominate for low *Z* materials from low energies up to several MeV and for high *Z* materials from hundreds of keV to several MeV. Data for water are shown in figure 3.3.

Note how Compton scatter (μ_{cs}) is more important than the photoelectric effect (μ_{ph}) for energies above 25 keV and more important than pair production up to 9 MeV.

The Compton interaction is where a photon interacts with an electron in an atom, donating a significant fraction of its energy to the electron as it is scattered, resulting in the atom being ionised and the electron generating a track of secondary ions and

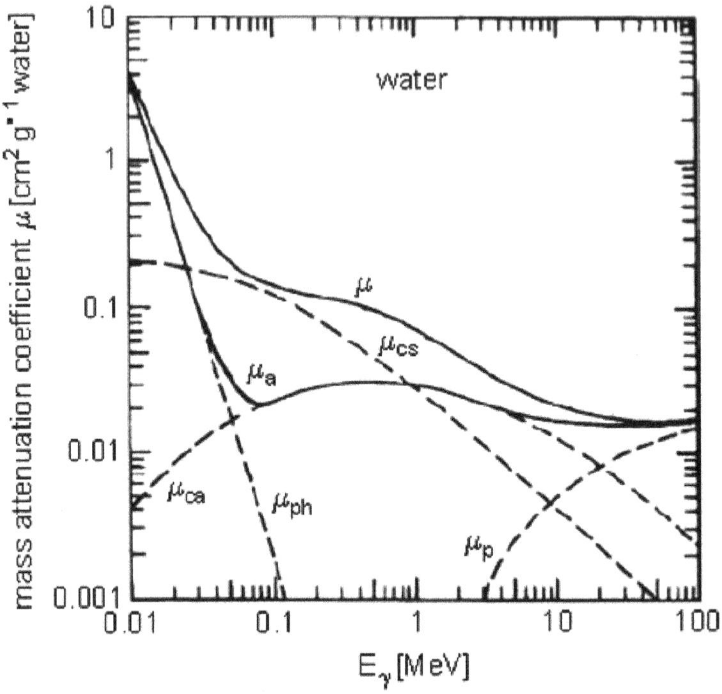

Figure 3.3. Mass attenuation coefficients for photons in water. Reproduced from Gazis (2019). CC BY 3.0.

electrons. Note that whereas the incident photon is absorbed in the photoelectric effect donating all its energy to the electron, in Compton scattering it donates only a fraction of its energy. A lower energy photon is also emitted on a different trajectory to the original photon. As always, energy and momentum are conserved. The energy of the incident photon is equal to the energy of the ejected photon plus the excitation energy of the atom plus the energy of the subsequent, lower energy, photon. The excited atom will de-excite, sometimes emitting x-rays. Momentum along the trajectory of the incident photon equals the momentum of the original photon and is composed of the momentum vector along the original trajectory of the resulting ejected electron and the lower energy photon. The momentum of the ejected electron and that of the lower energy photon normal to the original trajectory must sum to zero. This leads to the following equation for the energy of the scattered photon.

$$Es = \frac{Ep}{\left(1 + \frac{Ep}{511}(1 - \cos\Theta)\right)}$$

where *Es* is the energy of the scattered photon in keV, *Ep* the energy of the incident photon in keV, 511 keV is the rest mass energy of an electron and Θ the angle between the path of the incident photon and the path of the scattered photon.

One important circumstance is when photons are backscattered through 180 degrees, i.e. back along the line of the incident photon. In that case, $\cos\Theta = -1$ and the equation simplifies to

$$Es = \frac{Ep}{\left(1 + \frac{2 \times Ep}{511}\right)}$$

For the Cs-137/Ba-137m emission at 662 keV, this gives Es = 184 keV. In a typical instrument calibration room, the end wall acts as a phantom source of photons with energies close to 184 keV. For a medical exposure where the incident photon has an energy of 70 keV, making the same calculation gives a backscatter energy of 55 keV.

Compton interactions are particularly important for low *Z* materials, as indicated above for water.

In such Compton interactions, the total number of photons is not reduced, only the individual photon energy. The result is that a primary photon may undergo several interactions before the scattered photon, now much reduced in energy, vanishes as a consequence of a photoelectric interaction occurring after it has travelled a significant distance in the medium. Note, too, that for materials such as air and water the probability of interaction also increases only relatively slowly with decrease in energy from 1 MeV down to approximately 40 keV. The result of this is that the ratio of scattered photons to non-interacted photons builds up with distance from the source, reaching an equilibrium as the range of the electrons is less than that of the photons. Dosimetrists use the term 'secondary electron equilibrium' for this condition.

Figure 3.4 (upper panel) shows the result predicted by a statistical modelling process (MCNP) of the relative photon abundance at 40 cm from a Cs-137/Ba-137m (662 keV) gamma source in an infinite volume of water. Note that the ratio of scattered photons to unscattered (662 keV) photons at 40 cm from the source is 22 and that the peak energy is 50–60 keV. A useful way to visualise this is to scale the wavelength of the radiations to the visible range, making the unscattered photons blue and the scattered photons red. The source would look like a blue light in a pink matrix. Note that this is not the dosimetric spectrum. Weighting the spectrum to convert to dose in water produces the result shown in figure 3.4 (lower panel).

Even in dosimetric terms, the scatter is more important than the attenuated 662 keV radiation.

The same process takes place in air, but the lower density of air, 1.2 mg cm^{-3}, means that the distance is expanded by a factor of 800. Hence, the data presented for water at 40 cm would be approximately appropriate for a distance in air of 320 m. For on-plant measurements, where distances between source and instrument are generally measured in metres, the effect is negligible. However, for site perimeter and greater distances the effect is often significant.

The same phenomenon of scattering can be responsible for some counter-intuitive observations. Well-shielded cells are used for processing very high-activity gamma sources such as Co-60 sterilisation sources and the dismantling of nuclear fuel elements (figure 3.5).

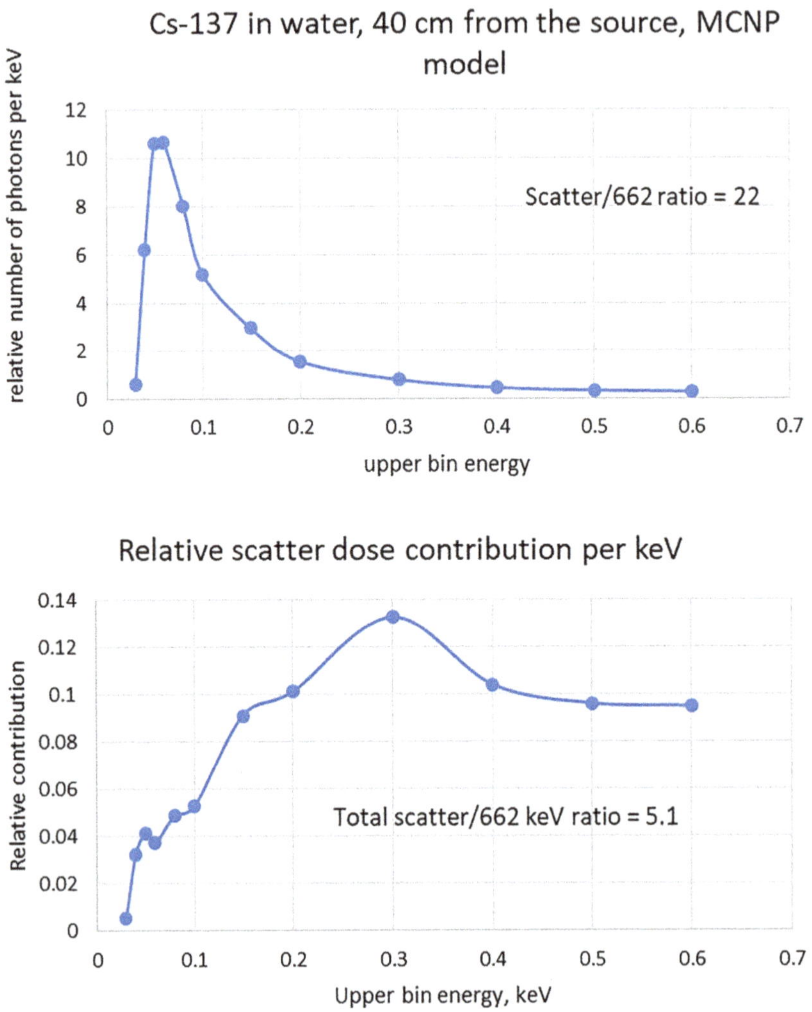

Figure 3.4. Flux and dose rate contributions at 40 cm from a Cs-137/Ba-137m source in water.

Clearly, some radiation will penetrate the lead–glass cell window. However, frequently, particularly in older plant, the operator's dose rate is dominated by radiation that has escaped from the cell through the roof and through the penetrations, particularly for the master–slave manipulators. This does not tend to irradiate the operator directly but can be rescattered from the building roof or the wall behind the operator, leading to the counter-intuitive result that the operator can see the source directly through the window, but is being irradiated mainly from behind.

The same phenomenon is important for any radiation area with thick walls and a thin, or non-existent, roof. The dose rate will be found, initially, to increase as the distance from the wall increases, basically because the wall's shielding of 'sky shine', i.e. the radiation produced by scatter in the air above the walls, decreases.

Figure 3.5. A cell for processing high-activity gamma sources. Image credit: Argonne National Laboratory.

Historically, this has been difficult to calculate, but now programmes such as MCNP can produce credible values of dose rate from sky shine that can then be used to optimise building costs such as the thickness of the roof or the height of the walls.

Compton scattering is an important consideration for instrument calibration facilities. The classic radiation monitor calibration facility uses a heavily shielded container, a shutter, a collimator and a long room ending in a beam stop wall (figure 3.6). It is useful to consider where Compton interactions take place and their consequences. The obvious start is to look at what happens behind the source.

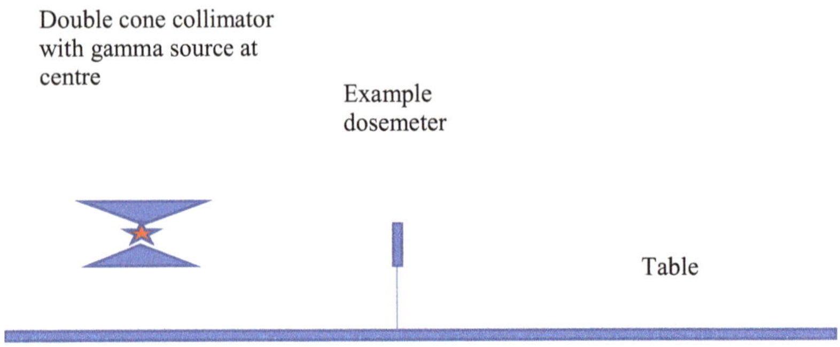

Figure 3.6. Schematic layout of an instrument calibration facility.

The backing is generally lead or tungsten. At an energy of 662 keV (Cs-137) the atomic cross-sections for lead are approximately 22 barns/atom for Compton scatttering and 19 barns/atom for the photoelectric effect. Hence only 54 % of scattered electrons arise from Compton scattering. At that energy, the Compton scatter is forward peaked, i.e. continues within 90 degrees of the original photon. If these Compton-scattered photons interact further into the shielding then their reduced energy encourages photoelectric, rather than Compton, interactions. Primary photons that are backscattered through 180 degrees have an energy of 184 keV. If the backscattering takes place within the shielding, then the back-scattered photon stands a good chance of undergoing photoelectric absorption within the shielding. Finally, backscattered photons are only of significance, in terms of beam quality, if they escape the collimator. Their dosimetric influence is also reduced by a factor of four compared with the 662 keV photons because of the lower energy. The result is that the ratio of dose rate generated by photons backscattered from the shielding to the dose rate generated by the unattenuated beam escaping from the collimator is less than 1%.

The other source of potential contamination of the beam is shallow angle scatter from the collimator. The ISO recommended collimator is a long and complicated construction which is not necessary compared with a much simpler and lower cost simple cone. If we consider a cone with a half-angle of 7 degrees, the resultant energy of an incident 662 keV photon that scatters through 7 degrees to be parallel with the axis of the collimator is 0.656 MeV. The difference in energy is trivial in dosimetric terms. In summary, using a collimated beam results in a trivial degradation of beam quality.

However, there are circumstances where it might be useful to irradiate a ring of dosemeters. Using a totally uncollimated source has several disadvantages. The dosemeters are irradiated by the Compton scatter produced by the whole room surface. This tends to be dosimetrically unimportant where the room is large compared with the source–dosemeter distance. The argument is simple. Assume that all the radiation from the source reaches the walls. Hence, a fixed fraction of the radiation is backscattered, independent of the room size. However, for a spherical room with the source at the centre, the fraction of the radiation returning to the centre follows the inverse square law as we can consider the spherical room to be equivalent to a point source—all the radiating surfaces are at the same distance. The dose rate from such a source has an inverse square law dependence. If the room is small, only a small number of times bigger than the source–dosemeter distance, then scatter can become significant. The other major problem is that all the walls need to offer sufficient shielding to unattenuated radiation from the source. This is potentially more expensive, especially for the ceiling, than producing a beam stop. The way to get over this is to use a double conical absorber, one element below the source and one above, to produce a ring beam. This exploits the property of the minimal reduction in energy generated by shallow angle scatter.

Pair production dominates photon scattering for high Z materials from a few MeV and low Z materials from about 10 MeV. In pair production virtually all the energy of a photon is converted to an electron–positron pair with identical energies. To conserve momentum, a very small amount of energy is transferred to the nucleus of the atom where the process takes place. Given that a single nucleon has a mass 1822 times that of an electron means that momentum can be conserved with a small velocity, which corresponds to a very small energy. This process requires that the incident photon has at least twice the rest mass energy of an electron, i.e. $E > 1.022$ MeV

Pair production is not very important for many nuclear radiation protection workers as very high-energy gamma emitters are unusual and have very short half-lives. The one that is most important is N-16, an activation product of O-16, which is a very short half-life beta emitter and generates 6–7 MeV gamma radiation. Pair production is more important in the medical and industrial fields, where very high-energy sources, such as betatrons and linacs, are used for industrial radiography and linacs are used for radiotherapy.

The products are an electron and a positron. These lose energy in very similar ways because their masses are the same, and the magnitude of their charges is the same, only the sign is different. The difference between the two is that when the positron 'stops', it will annihilate by combining with an electron, generating two 0.511 MeV photons in opposing directions. These are often obvious when using gamma spectrometers in the presence of very high photon energies. The spectrometers detect 0.511 MeV photons generated in the environment but can also detect pair production in the detector itself. This classically shows up as three peaks—the highest energy one having the full energy of the incident photon, with one 0.511 MeV below it, where one photon has escaped (single escape peak), and another a further 0.511 MeV below that, where both photons escaped (double escape peak).

When the photon energy is very high. some much more unusual interactions can take place. One circumstance where they can be important is the target area and collimator of high-energy linacs. Photoneutron production is only important in high Z materials such as lead and tungsten, which are popular as parts of shielding and collimation because of their high density and, less importantly, their higher mass energy absorption coefficient. The energy thresholds for photoneutron production in tungsten and lead are 6.74 and 6.19 MeV. The neutrons act as a contaminant in the therapy beam, generated with intensities of the order of 10^{11} neutrons per gray of x-ray dose.

Table 3.1. Mass energy absorption coefficients for lead and water.

	Mass energy absorption coefficients ($cm^2\ g^{-1}$)	
Material	100 keV	2 MeV
Lead	1.976	0.0236
Water	0.0255	0.0261
Ratio	7.7	0.9

The result of this process is that components irradiated by intense very high-energy x-ray beams can become activated, often incorporating electron capture nuclides because of their neutron deficit. If the component has been out of use for a long time, then activation products can be dominated by longer lived radionuclides other than the ones generated in the high Z components.

There are some popular misconceptions in shielding design and use. For example, lead is often seen as 'better' than water and concrete for shielding. Indeed, at lower energies, i.e. typical x-radiography energies, high Z materials are 'better' all round. They have a higher mass energy absorption coefficient and a high density, which means a few mm of lead, for example, can be a very effective shield. However, at higher energies, for example at 2 MeV, the advantages are less clear-cut. Density is useful, as it allows a smaller thickness for a given attenuation. However, the mass energy absorption coefficients differ less (table 3.1).

Note that, at 2 MeV, the dose rate in water, a measure of its effectiveness as shielding, is higher than in lead. This is because the electron density in electrons per gram is higher at 0.55 electrons per nucleon in water compared with 0.4 in lead—which favours Compton interactions. Hence, at 2 MeV, the radiological advantage of lead over water, and similarly concrete, is its density, rather than a higher level of radiation interactions.

3.2 Electron interactions

The major mechanisms of photon interactions result in energetic electrons. These behave in an identical way to beta radiations of the same energy. The main mechanism for energy loss is interaction with atomic electrons. This leads to an initially surprising result that the energy loss in hydrogen, in MeV $g^{-1}\ cm^{-2}$, is much higher than that in lead. At 1 MeV, the corresponding values are 3.815 and 0.994, a ratio of 3.8, driven mainly by the ratio of 1 electron per nucleon in hydrogen and 0.4 in lead. The second important interaction is the generation of bremsstrahlung, which translates as 'braking radiation' and is x-radiation generated by acceleration of electrons in the field of a nucleus. In contrast to the above, this is far more likely with lead, rather than hydrogen, giving effective energy losses of 0.005 MeV $g^{-1}\ cm^{-2}$ for hydrogen, and 0.129 for lead, a ratio of 0.04. A consequence of the interaction with atomic electrons can be the generation of characteristic x-rays. These three types of interactions are discussed further below.

Collision is in some ways a poor word to describe electron scattering. Electrons are not 'billiard balls'. If they were, the primary electron range would be small as the electron would lose a significant fraction of its energy on each collision. Instead, a typical interaction transfers only a small fraction of the primary electron energy.

Furthermore, interactions are also quite unlikely. In hydrogen, the energy loss rate at 1 MeV by collision is 3.815 MeV g^{-1} cm^{-2}. If we make a reasonable approximation and assume each interaction transfers, on average, 1 keV, then 0.1 g cm^{-2} of hydrogen will see approximately 400 energetic secondary electrons generated. If we imagine a 1 cm^3 volume of compressed hydrogen containing 100 mg then the number of atoms will be 6×10^{22}. This is equivalent to 4×10^7 layers of atoms. Hence the chance of a 1 keV transfer interaction is once in 10^5 layers.

If an interaction takes place, an orbital electron is removed from its atom and sets off losing energy in exactly the same way as the primary, only, because of its much lower energy, many of the interactions either excite an atomic electron, i.e. raise it to a higher energy without removing it from an atom, or achieve the ionisation but leave the electron with a small residual energy that ends up as heat.

At the end of the process, the electron reaches its thermal velocity and may become attached, for example, to an ionised atom. On the other hand, if the target is an insulated block and the electron is part of an intense electron beam, from an electron linac for example, the block can become highly charged, unusually with significant charge deep within the block, rather than on the surface. If the block is, for example, hit with an earthed pointed conductor, the charge will track out from deep within the block, generating an 'electron tree' (figure 3.7).

With bremsstrahlung, energetic electrons interact with the nuclei of atoms, resulting in the generation of x-radiation. This can result in a 'cascade'. An energetic

Figure 3.7. Electron tree.

electron, hitting a linac target, generates x-rays. These can then go on to generate Compton electrons which may, in their turn, generate bremsstrahlung, which can then generate another electron of significant energy.

Bremsstrahlung is most likely with energetic electrons and high Z material, hence the use of tungsten ($Z = 74$) as an anode in conventional x-ray equipment and tantalum ($Z = 73$) as a linac target. It is a very inefficient process, which means that most of the electron beam energy ends up heating the target. Hence, in x-ray tubes, the tungsten target (anode), a relatively small piece of tungsten, is embedded in a copper block which is drilled to allow water (for anodes at ground) or oil (for anodes at high positive potential) to circulate to remove the energy being deposited therein. The heat requiring removal may be as much as several kilowatts. Cooling is vital—a failure in the cooling circuit will result in the anode heating up and expanding. This can destroy whatever seal there is to the tube housing, resulting in loss of vacuum. The efficiency, i.e. the fraction of the total energy emitted as bremsstrahlung, is approximately $EZ/700$, with E in MeV. For a tungsten anode, $Z = 74$, and a typical industrial x-ray set tube potential of 300 keV, the efficiency is only 3%. The Z dependence mean that bremsstrahlung in tissue, for example, is negligible.

A consequence of bremsstrahlung is that high-activity beta sources, which use metal backing and housings, often stainless steel, always generate significant bremsstrahlung within the source itself. When a beta source is mounted in a machine, such as a thickness gauge, the bulk of the radiation exposure outside the machine is generally bremsstrahlung, some from the source itself but also some generated by the beta radiation interacting with the machine components. Beta radiation is easily attenuated and should only escape in areas where shielding is limited or non-existent, such as gaps over conveyor belts. Hence the machine is often far more an x-ray rather than a beta source when it comes to radiation protection.

Taking this further, storage and transport containers for beta sources should have a low Z material close to the source, such as plastic or aluminium, with a high Z layer, such as lead or tungsten, outside. The low Z material minimises the bremsstrahlung generation, takes up a small volume and has a minimal weight. The high Z external material must mainly deal with generation inside the source housing and can be relatively thin, compared with the thickness required if the low Z absorber was left out.

The US National Institute of Standards and Technology (NIST) web site (https://www.nist.gov/) gives details of the energy loss rates for different elements. These are expressed in terms of MeV cm^{-2} g. The energy and Z dependences are worth appreciating (figure 3.8).

At low energies (10 keV), the energy loss rate in carbon is more than twice that in lead. This is mainly because the electron density, in electrons per nucleon, is 0.5 for C-12 and 0.4 for Pb-82. This effect persists up to about 5 MeV. However, at very high energies, beyond those of beta emitters and conventional x-ray equipment, but employed in linacs, the stopping power in lead is higher because of the greater efficiency of bremsstrahlung generation.

Note, too, that the range of a beta particle is very energy dependent. Betas lose energy continuously. Higher energy betas (1 MeV or so) have more energy to lose

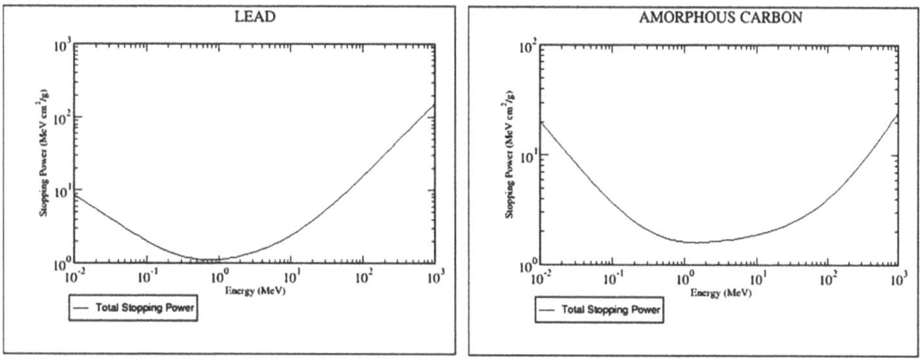

Figure 3.8. Energy loss rates by bremsstrahlung production in lead and amorphous carbon. Reprinted with permission from NIST.

and have (initially) a lower energy loss rate. For aluminium, an accurate fit to published range is $R = 412\ E_0^{(1.265-0.0954\ln Eo)}$ with the range in mg cm^{-2}.

It is this attenuation curve shape that leads to a rather unexpected observation. The basic physics of gamma interaction leads very clearly to an exponential transmission equation—$dN/dd = kN$, where k is a constant, N is the number of photons present in the beam and d is the mass per unit area of the absorber. This leads to the well-known equation $N = N_0 \times e^{-kd}$. However, with beta radiation in contrast to photons, there is not a choice of either interacting or not interacting when passing through a small thickness of material. Beta particles do not lose energy continuously, but on any sort of macroscopic basis the energy loss appears to be continuous. Hence, the maximum energy of a beta spectrum will inevitably reduce with penetration into an absorber. Despite this, in an approximate way, their number, as they penetrate a material, follows an exponential transmission curve. The example below (figure 3.9) uses a Cl-36 (E_{max} = 710 keV) source which is 10 cm × 15 cm in area, in contact with a series of aluminium absorbers and a BP19 detector, which is a large area, thin-windowed, wide-energy-range beta scintillation detector).

The results do not form a perfect exponential but do demonstrate the broad exponential form. If the experiment had been repeated with a small area collimated source, a small area detector with a significant space between them and the absorbers in the centre of the gap then a closer to exponential result would be generated.

Common beta emitters range in energy from tritium (H-3) which has a maximum beta energy of 18.6 keV to Bi-214, which has a maximum beta energy of 3.26 MeV. The corresponding ranges for a maximum energy beta in air are 7 mm and 15 m. In water, these are approximately 9 μm and 19 mm. There is a very important caveat concerning this process. Energetic electrons scatter and pursue a complicated path. They frequently end up heading towards where they came from. The range of a very energetic beta particle in air may be 15 m, but the chance of one of them getting to 15 m is utterly negligible. Hence the practical range, i.e. the distance from the source that they might be detected at, is far less. This applies to any material. The huge

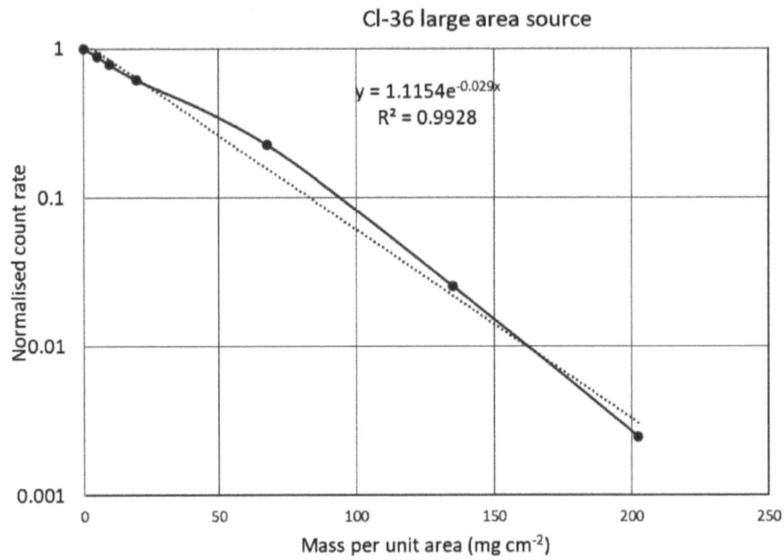

Figure 3.9. Attenuation of the flux from a large-area Cl-36 source through an aluminium absorber.

variation in ranges has implications for monitoring. Detection of tritium demands a windowless counter, i.e. the beta particle has to be able to reach the counting medium with nothing in the way. Liquid scintillation counting is the popular method. On the other hand, for higher energies (>0.5 MeV) then the window of a detector can be quite robust, thicker than, for example, a typical document wallet, with only a small reduction in sensitivity compared with the typical 'thin window' beta detector. The other implication is the sensitivity of low energy beta radiation emitted from a surface to any coating on the surface, such as grease. The lowest energy beta emitter which can be monitored directly, with confidence, is C-14 (E_{max} = 156 keV) but that is only possible on surfaces which look clean, as in fresh from a dishwasher clean. Higher energy betas, such as Sr-90/Y-90, are much more tolerant and can be monitored through a coat of varnish, for example. The distance between a detector and a surface is also critical for C-14, as an example. With a detector which is much smaller than the contaminated area then moving back from the normal recommended separation of 3–10 mm will halve the response, whereas, for Sr-90/Y-90 and a very large active area, then moving back 100 mm will produce a change of not more than 10 %.

3.3 Alpha particles

An alpha particle is a helium nucleus comprising two protons and two neutrons, which forms a very stable entity. Alpha particles are emitted from the nucleus of some heavy atoms. This emission, on average, leads to greater stability in the progeny than the parent. However, there are long chain decays in Nature, a good example being that headed by U-238. This finally reaches stability at Pb-206, a total of 8 alpha and 6 beta decays later. All the intermediate nuclides are less stable than the head of the chain, some dramatically so. The least stable is Po-214, which has a half-life of 1.6×10^{-4} s.

The fundamental thing to understand about alpha particles is how heavy they are compared with beta particles. The mass of a helium nucleus is 7200 times that of an electron. That is the approximate ratio of the masses of a Ford Transit van and a mobile phone. This much greater mass greatly influences the process of energy loss and range. Alpha particles, unlike electrons, rarely scatter. Such scattering, when it does occur, requires a very short-range interaction with a nucleus. It was this property that led to the first credible estimate of the dimensions of a nucleus as less than 1/10 000th of the size of an atom.

The image in figure 3.10 shows alpha particles in a cloud chamber. Note there appear to be two distinct path lengths which means two distinct energies, and that there is one significant deflection.

Consider a 2 MeV alpha particle and a 2 MeV beta particle in water. The range of the alpha particle will be 1.1 mg cm^{-2}, whereas that of the beta is 980 mg cm^{-2}, a difference of a factor of 900. This means that the ionisation density along the path of the alpha particle will be vastly higher. This density is also increased because the average energy of the electrons from an ionisation event is also much less, leading to a 'tight' column of ions. It is the relatively slow movement of the alpha particle that causes this. The velocity of a 2 MeV alpha particle is approximately 3×10^7 m s^{-1}. This limits the energy of an electron with which it collides to about 1 keV. This electron will lose energy very quickly, generating approximately 30 ion pairs. It is these properties which are responsible for the high radiobiological effectiveness of alpha particles. The energy loss rate for alpha particles in water is illustrated in figure 3.11.

Figure 3.10. Alpha particle tracks observed in a cloud chamber. Rutherford *et al* (2010) © Cambridge University Press. Reprinted with permission.

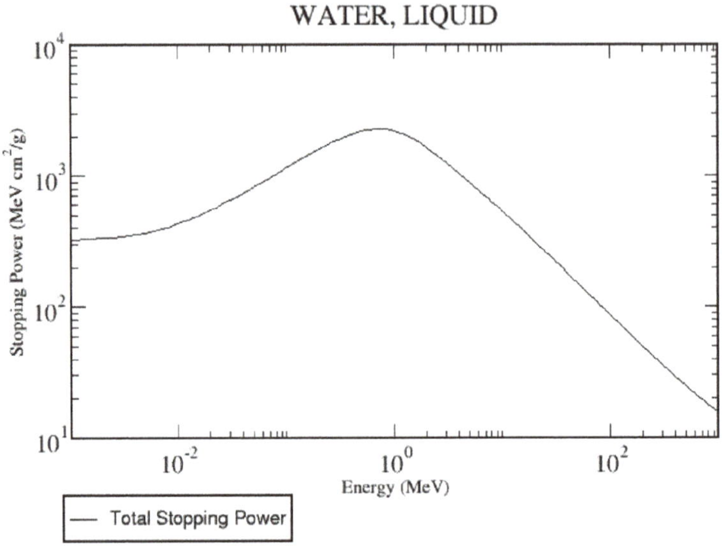

Figure 3.11. Energy loss rate for alpha particles in water. Reprinted with permission from NIST.

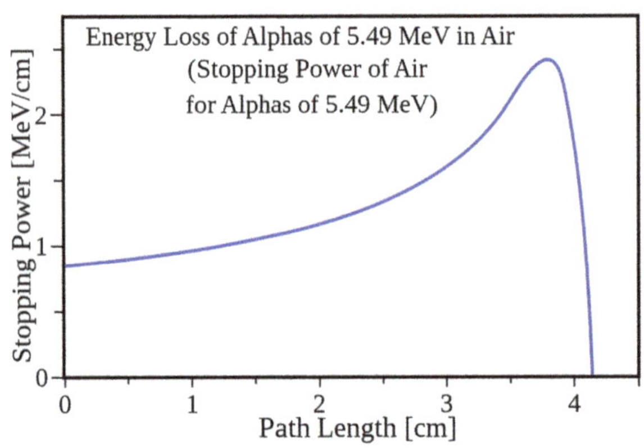

Figure 3.12. Energy loss of 5.49 MeV alpha particles in air. This Bragg Curve for Alphas in Air.png image has been obtained by the author(s) from the Wikimedia website, where it is stated to have been released into the public domain. It is included within this book on that basis.

Following from this is the form of the energy loss curve with depth of penetration into a material that exhibits a peak (the Bragg peak) toward the end of the range of the alpha particle (figure 3.12).

This rapid energy loss has two effects when it comes to detection of alpha contamination. It demands that any detector has a thin widow to allow particles to pass through. This must be a small fraction of the range of an alpha particle because the alpha particles are emitted at random angles to the surface. Figure 3.13 shows

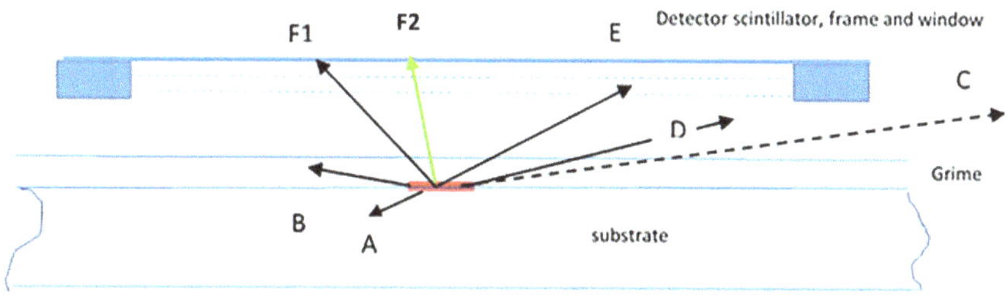

Letter	Description
A	Goes into the substrate
B	Stopped in grime
C	Missing the detector, but would stop in the air gap anyway
D	Stopped in the air gap
E	Stopped in the window
F1	Deposits <0.5 MeV in the scintillator. Not counted
F2	Deposits >0.5 MeV in the scintillator. Counted

Figure 3.13. Potential fate of alpha particles emitted from a greasy surface below a scintillation detector.

the potential fate of alpha particles emitted from a greasy surface below an alpha scintillation detector.

However, for the alpha particles that do reach the scintillator, the rapid energy loss means that only a thin scintillator is required to stop any alpha particle. This greatly helps discrimination against direct interactions with any beta particles also present, interactions with gamma photons and interactions with electrons generated elsewhere by gamma photons as these will deposit far less energy. Hence, large pulses are readily identified as alpha detection events.

3.4 Neutron radiation

Energetic neutrons are emitted by two main mechanisms. One is in the process of fission. For example, when U-235 undergoes fission, on average 2.5 energetic neutrons are also emitted. The other common source is nuclear reactions, particularly α–n reactions. An example of a manufactured neutron source comprises an Am-241 source combined with beryllium, in which a very small fraction (less than 1 in 10^4) of alpha particles emitted leads to the generation of energetic neutrons by the α + Be-9 → C-12 + n reaction. Neutrons are very different from the other radiations considered in one specific way. This is the huge range of energies that can be found in a single workplace. If we start with a typical Am-241/Be source, this emits a spectrum of neutrons as shown in figure 3.14.

This is dominated by neutrons in the MeV region. However, as these neutrons interact, they progressively lose energy. Often, in a workplace, there is a large thermal neutron component, i.e. neutrons that have been scattered so often that their energy is that of a nucleon in thermal equilibrium with its surroundings. These are known as thermal neutrons. At 20 °C, the corresponding 'average kinetic energy' is

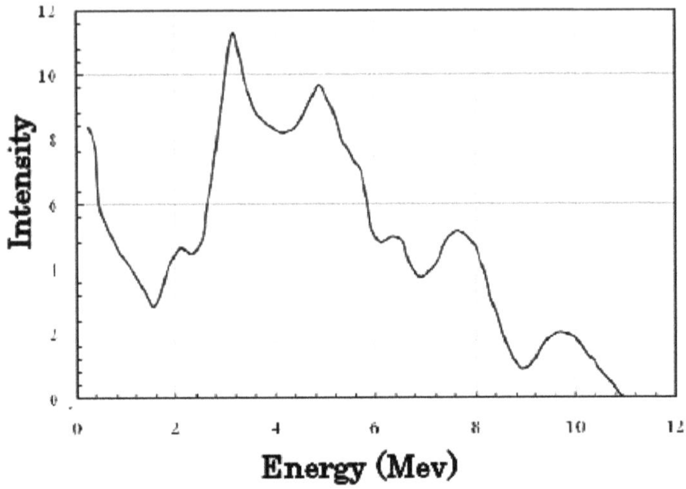

Figure 3.14. The neutron spectrum of an Am-241/Be source. Permission to reproduce extracts from British Standards is granted by BSI.[1]

quoted (usually) as 0.025 eV, a factor of 10^8 lower than the original neutron if generated by an Am-241/Be source. Note that the usual terminology is not correct, because this is not the average kinetic energy but rather the modal energy of the distribution.

Neutrons interact in several ways. The simplest is elastic collision, i.e. all the energy remains kinetic. A neutron interacts with a nucleus and, effectively, 'bounces off'. If the nucleus is heavy, the energy transfer is small. If we consider a U-238 nucleus and a 1 MeV neutron in a 'head on' collision, then the neutron is reflected back along its path with 99% of its original speed and 98% of its original energy. On the other hand, if we consider the same collision with a hydrogen nucleus, i.e. a proton, then the incident neutron stops and the proton continues on the same path with the same energy, i.e. in this case 1 MeV. This process effectively leaves the orbital electron behind, turning the nucleus into a heavily ionising positively charged particle. This is easiest to visualise by looking at the collision from the proton's point of view. At the instant of collision, the electron appears to suddenly move away. Making the calculation shows that a proton gaining an energy of 1 MeV appears to see its electron suddenly gain an energy of 1 keV, far above the 14 eV ionisation potential. Hence the hydrogen atom turns into an energetic proton that will generate ions by electrostatic interaction. This process is the dominant one in tissue for high neutron energies. Note that 'head on' collisions are rare—neutrons are generally deflected from their path, but the same conservation of momentum and conservation of energy equations still apply. The practical effect of this, in tissue, is that most of

[1] British Standards can be obtained in PDF or hard copy formats from BSI Knowledge: https://knowledge.bsigroup.com or by contacting BSI Customer Services for hardcopies only: Tel: +44 (0)20 8996 9001, Email: cservices@bsigroup.com.

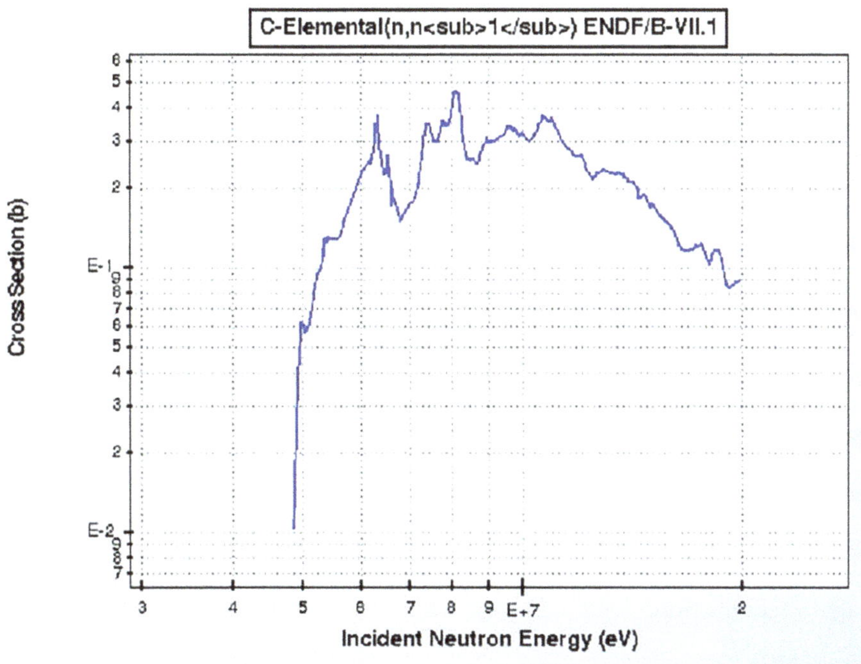

Figure 3.15. Cross-section for inelastic neutron scattering in calcium. Reproduced from Kavetskiy et al [2017]. CC BY 3.0.

the energetic protons generated have energies of a reasonable fraction of the incident neutron energy. Hence, tissue exposed to energetic neutrons will be bombarded by a range of proton energies from close to zero up to the energy of the incident neutron minus the binding energy of the hydrogen electron, 14 eV.

Inelastic collisions can also occur. This where the energy of the scattered neutron is less than would be predicted from an elastic collision. What it has done is to raise the energy of the nucleus with which it has collided. This energy is generally emitted as a gamma ray. This process is most likely for high neutron energies and high Z target atoms and is thus of limited importance in radiation protection. Note the high threshold energy for this interaction for calcium in figure 3.15.

Neutron-capture reactions can also occur that involve the generation of energetic charged particles. The detection of neutrons often employs this process. For example, He-3 proportional counters make use of the reaction:

$$1_0\text{n} + {}^3_2\text{He} \rightarrow {}^3_1\text{H} + {}^1_1\text{p} + 764\ \text{keV}.$$

This reaction is exothermic, i.e. some of the combined rest masses are converted into kinetic energy. For thermal energies, this has a very large cross-section of 5330 barns. The cross-section for He-3 production decreases as the energy increases at about a factor of 2.8 decrease for a factor of 10 energy increase. In higher Z materials, such as silicon, the reaction is endothermic, i.e. to proceed, the neutron has to have a significant kinetic energy and, hence, there is a threshold. For silicon-32, as an example, the threshold for the reaction that generates a proton is approximately 1.8 MeV and the change in cross-section with energy is more complicated.

Neutron-capture reactions involving the generation of energetic charged particles are not an important consideration in interactions in tissues. However, neutron capture without generating an energetic particle is of major importance in radiation protection.

Specifically, hydrogen has a thermal neutron cross-section of 0.33 barns for the reaction ${}^{1}_{0}n + {}^{1}_{1}H \rightarrow {}^{2}_{1}H + 2.22$ MeV gamma. This is not a particularly large cross-section but is important for thermal neutron exposure as the most likely interaction to absorb a neutron in the body. The result is that a person irradiated by thermal neutrons acts as a 2.22 MeV gamma source. Where irradiation is localised, this results in radiation being 'spread around' the body.

Finally, it is worth mentioning that fission reactions are not likely in the human body, but fission is a major cause of neutron exposure from the operation of nuclear reactors. For U-235, capture of a neutron can lead to fission of the nucleus into two major parts, one with a mean atomic mass of around 140 and the other with a mean atomic mass of about 90, together with the several energetic neutrons that can propagate the reaction.

References

Gazis E 2019 in *Medical Imaging - Principles and Applications* (InTech Open) https://doi.org/10.5772/intechopen.84356

Kavetskiy A, Yakubova G, Prior S A and Torbert H A 2017 *New Insights on Gamma Rays* (Intech Open) https://doi.org/10.5772/68014

Rutherford E, Chadwick J and Drummond Ellis C 2010 *Radiations from Radioactive Substances* (Cambridge Univ. Press) https://doi.org/10.1017/CBO9780511707179

Chapter 4

Radiation monitoring and instruments

P Burgess

4.1 Introduction

There is a huge range of situations where a measurement of the radiation level is a legal requirement, a societal obligation or a means to further research. Here, the coverage is restricted to the more popular aspects only. The first important requirement is to determine exactly what measurements need performing. This step is often not taken seriously enough, and the result is an unsatisfactory programme. Another important aspect is to remember that the world moves on, and what made a sensible monitoring programme thirty years ago may now be totally inadequate. Radiation protection in some organisations is seen as a regrettable expense, and virtually all organisations have the need to spend money carefully. This tends to lead to people struggling on with equipment that no longer satisfies (or perhaps never really satisfied) what is wanted from it. So, every now and again, it is worth starting from the beginning and reviewing the programme and the equipment required.

One specific context that is gaining in importance is decommissioning and site clearance. It is vital to recognise that the equipment used for routine operations when the plant was operating will generally either be useless or will have to be significantly reconfigured for use as part of decommissioning.

Instrument choice is basically a process of choosing an instrument that is appropriate to the situation in the workplace.

An instrument must:

(a) be appropriate for the type(s) and energy or energies of the radiations that it is intended to measure,

(b) be reasonably unresponsive to other radiations that may be present, such as cosmic and natural gamma background radiation and that might conceal significant levels of the radiations of interest,

(c) measure the right quantity, such as ambient dose equivalent rate for a gamma monitor,

doi:10.1088/978-0-7503-2424-3ch4

(d) have sufficient sensitivity to enable meaningful measurements at the lowest level of interest,
(e) be acceptable to the user, in terms of size, weight, shape, control functions and display,
(f) be sufficiently robust for day-to-day use,
(g) be maintainable, repairable, inexpensive to live with and have an acceptable useful life.

4.2 Fundamental considerations in instrument choice

The bullet points below identify aspects of instrument choice that are universally relevant and that demand careful thought.

- Compared with most engineering and scientific measurements, the uncertainties in radiation measurement are very large. Even the best measurements are accurate to little better than 1% and a typical 'health physics' measurement uncertainty is probably no better than +/−20%, even in quite simple circumstances. Hence there is little justification for quoting results with a better precision than 3% when reporting a quantity, e.g. ambient dose equivalent rate, but this does not preclude record-keeping having a precision of 1%, which can be useful when one instrument is used by one user in making repeated measurements at a clearly identified point. In that case, the uncertainties, other (generally) than those associated with the ability to estimate the mean indication, will not influence the change in indication.
- Following from this, some instruments deliberately reduce the precision of what is shown. Others, however, can display four-digit precision, the last two digits of which are clearly close to being random numbers when the instrument is in use at low levels.
- The dynamic range of measurement. Generally, the top end is easy to identify. For gamma dose rate, for example, there will be a range that will be encountered in normal operation and in events that the facility operator intends to deal with without changing equipment. This tends to drive choice. However, it is vital to appreciate that almost all measurements are typically made at very low levels. For gamma dose rate measurement and modern plant, dose rates in the working area and outside the facility are normally not more than a very few μSv h^{-1}. It is important to make sure that a survey instrument has a good sensitivity that will allow the user to make measurements at each point in a few seconds. In the past, a desire to have a high useful dose rate has resulted in users being given instruments that had a poor sensitivity, obliging them to spend tens of seconds at each point and still ending up with a statistically weak result that made the identification of trends, such as a gently rising dose rate, difficult.
- Instruments that are pulse counting and can generate a click for each pulse generated are far easier to use when searching for shielding leaks and hot spots than those that do not count pulses or generate an audible signal.

- There is a trade-off between the response time, i.e. effectively the averaging time of an instrument, and the level of statistical fluctuation, i.e. the variability of the indication at any reading. When a manufacturer tries to extend the dose rate or count rate downwards, there is a tendency to make the response time longer. This makes the instrument slow to respond and obliges the user to observe the display over tens of seconds to establish confidence that some sort of mean reading has been reached. In practice, for routine measurement, the maximum response time that is usable is about 4 s.
- The above consideration means that the minimum useful count rate for a pulse counting instrument is three or four counts per second. Below that, either the display will fluctuate too much, or the response time will have to be too long.
- Most modern instruments are digital. For pulse counting instruments this, surprisingly, brings problems that are almost automatically dealt with by conventional analogue instruments. The conventional averaging process is no longer enough. Nearly all modern instruments use a statistical process—the instrument 'looks at' the count rate every counting period (usually about 1 s) and decides if it is statistically different from the previous value. If it is not, it stores the count in a register with the previous one. This process generally continues until at least eight readings are stored. At that point, when a new one is recorded, the oldest one is deleted. This process will accommodate fixed or slowly changing count rates very well. If there is a statistically significant change in count rate, the instrument recognises this and, effectively, starts again. Given that this is a statistical process, even in a fixed field, the instrument will occasionally see what looks like a significant change and will generate a short term high or low indication, particularly at low count rates.
- A worrying problem that has occurred often in the past and can still be seen in some modern instruments is that such instruments fail to danger under overload (i.e. very high) dose rate and count rate conditions. Beyond some point, as the dose rate increases, the indication reduces, usually quite fast. This can result in someone walking towards instead of away from a source. It is vital that, for any facility, the maximum dose rate or contamination level to which an instrument could be exposed is calculated and any instrument has its performance checked to at least that level. There is no technical reason why any instrument, under overload conditions, cannot be held at overload indication. A failure is indicative of poor design.
- It is not necessary always to use an instrument with a good energy response when commissioning a facility or investigating a previously inaccessible area. It is often more effective to survey the facility with an instrument with a very high sensitivity, even though the response is very energy dependent. For example, a 51 mm diameter, 51 mm thick sodium iodide scintillator, a relatively common size, has a response of approximately 900 cps per μSv h^{-1} for Co-60 gamma radiation (mean energy 1.25 MeV) which rises to a maximum of approximately 10 000 cps per μSv h^{-1} at 80 keV. This can be used to identify shielding weakness or hot spots on the outside of a shielded

area and then a more metrologically credible, but less sensitive, instrument can be used to identify whether the level is, in fact, unacceptably high or not.
- There are many situations where it is possible to use a simple instrument with a marked energy dependence to demonstrate compliance with a limit. For example, high voltage valves can generate significant x-radiation, particularly when not conducting, as that is when the anode voltage is at its highest. The radiation energy is typically low, less than 30 keV, and is thus easy to shield. However, there is an overwhelming need for electrical isolation at such high voltages which means that the degree of shielding must be a compromise. The aim, usually, is to confirm that the dose rate is very low. The equipment can be quickly and easily inspected using a thin sodium iodide scintillator, which can have a response that varies between a few hundred and a few thousand cps per $\mu Sv\ h^{-1}$ over the possible energy range. If the maximum count rate is less than 100 cps then it is possible to make the robust statement that the dose rate is less than 1 $\mu Sv\ h^{-1}$.
- Instrument choice is often driven by metrological considerations. This is understandable—the reading has to be accurate enough to serve its purpose—but it can often be important to compromise by balancing metrological performance against ease of use, robustness and, in all honesty, cost.

4.3 External x- and gamma-dose-rate monitoring

The quantity of measurement will be ambient dose equivalent rate, the radionuclides or equipment generating the radiation will generally be known, as will the positions of the sources, which will mostly be shielded and/or some distance from the monitoring point. This may give the impression that the radiation energies and directions at the monitoring point are easy to determine. This can be true, in very simple circumstances, but as soon as a source is shielded the energy spectrum will change. Transmission through intact high-*Z* shielding will 'harden' the spectrum, i.e. if the radiation emitted covers a wide spectrum, such as a typical x-ray beam, then the lower energy components will be attenuated more effectively by lead shielding than the higher energy components. Hardening the spectrum generally makes life easier as it reduces the need for any radiation monitor to have a wide energy response. However, almost all shielding requires penetrations—ducts, cable ways, the door in—and radiation that escapes from these will generally be 'softened', i.e. will have had its energy reduced, as to escape the radiation will have undergone Compton scattering. The result of this is that the external radiation may effectively be downshifted in energy. So, in combination, a survey around a single facility may encounter radiation that has a higher mean energy than the generated energy in one position and one where its mean energy is much lower in another position. The problem with this lack of predictability means that an operator must either assess the spectrum or be cautious and buy an instrument with a wide energy range. The problem with the latter used to be that wide energy range instruments tended to be more vulnerable, less easy to use or more expensive. A way round this was to perform a detailed survey when the facility was commissioned using a wide energy range instrument and to make a point-by-point comparison with one that 'might' be good enough. For example, consider a facility that uses a 150 kV x-ray set. The mean energy of the radiation will be approximately 50 keV, but, if there are no in-beam filters, the spectrum will extend down to 10 keV. Lead shielding will preferentially attenuate the low energies and the radiation escaping on the far side will have a mean energy of more than 100 keV, which falls within the range of a simple energy compensated Geiger–Müller (GM) detector. These instruments are not expensive, they can be useful down to a few μSv h^{-1}, they are robust, and they have the huge advantage of an audio output which makes them far more effective in finding leaks than a silent instrument. The typical approach is to use an ionisation chamber which has a very wide energy response and a good polar response as the instrument that gives the 'right answer'. Comparing the results from the two instruments demonstrates whether the robust energy compensated GM instrument is adequate. This approach still has some merit, but life has been simplified by the development of energy compensated end-window GM tube instruments that have an energy range that extends down to 10 keV and that are only slightly more expensive and slightly less robust. Such instruments are discussed further below.

There are two approaches to measurement. One is based on the principle that the instrument can effectively reproduce the quantity of interest. This includes ionisation chambers—they have defined volume and, hence, a defined mass of air, and the output is a current. The energy required to generate an ion pair for air is well established at 34 eV per ion pair which means the current can be converted to an energy deposited in the detector in joules. Hence, we have joules per kg per second, the dimensions of dose rate. The other approach is to take another type of detector that is attractive for some other reason, such as GM detector, a PiN silicon diode or a sodium iodide scintillator, and modify or adapt it to produce an output that is proportional to dose rate over a useful energy range. A short summary of the design and characteristics of each type is given below.

4.4 External beta dose rate monitoring

In this context, the quantity to be measured is H′(0.07), i.e. 'skin dose' in most scenarios. Again, an attempt can be made to reproduce the quantity of interest, or another detector type can be adapted to give a good enough answer.

There are two obvious ways of reproducing the quantity of interest. One is to use an ionisation chamber, following the same arguments as above. For energetic betas and a thin window, the chamber will have much the same current per unit dose rate, H′(0.07), beta, as current per unit dose rate, H*(10), gamma. However, the ion chamber must be fitted with a window to ensure that the output has the correct beta-energy dependence at lower energies. The apparent solution would be to use a 7 mg cm^{-2} entrance window, the same thickness as the assumed skin depth. This is erroneous, because normal chambers, where the diameter and depth are very similar, have volumes of around 300 cm^3, simply to generate enough current at low dose rates, which results in a depth of 8 cm. As air has a mass per unit volume of 1.23 mg cm^{-3} at normal temperature and pressure, the mass per unit area of air alone to the back of the chamber is typically 9 mg cm^{-2}. In practice, the window thickness is set using a Pm-147 beta dose rate source, the lowest energy beta dose rate source generally available, and selecting the thickness of aluminised plastic that gives close to the right answer. The weakness of the ion-chamber-based approach is the very large volume of the typical detector. The basic rule is that the source must be at least three detector dimensions away from the detector centre before the indication can be trusted as a good measure of the dose rate at the position of the detector centre. That will be approximately 25 cm. This is driven by the inverse square law. If the detector is held closer than that, the indication will increasingly deviate from the true rate at the centre of the detector. For a long, thin, detector, end on, the indication will rise above the true value at the position of the detector centre. For a 'pancake' style detector, i.e. a relatively shallow disc, then the indication will fall below the true value at the detector centre.

It is emphasised that it is impossible to measure 'contact dose rate' without a specific correction factor. For a small area (point or a few mm in diameter) and a typical ionisation chamber, the ratio between the indicated H′(0.07) dose rate and the true value averaged over 1 cm^2, the appropriate measurement quantity for skin dose, is generally taken as 100. This can be very important when measuring the dose rate caused by localised skin contamination. Generally, the best option to determine the average dose rate over 1 cm^2 is to use either an extremity thermoluminescent dosimeter or to estimate the source activity from a more distant measurement and then use tabulated values.

The other obvious way would be to use a 1 cm^2 extremely thin plastic scintillator, which is, for example, 0.06 mm thick, with a thin plastic light tight window and a photomultiplier and to integrate pulse areas above thermal noise. This is not a design that is currently available.

4.5 Neutron dose equivalent rate monitoring

The quantity to be measured is again ambient dose equivalent rate. There are several aspects that make neutron measurements intrinsically more difficult than photon measurements.The consequence of this is that the user is frequently confronted by a very heavy, low sensitivity, instrument with a very large, energy-related uncertainty.

- The effective dose per neutron incident on the human body is, at 1 MeV, approximately 50 times higher than that for 1 MeV gamma radiation. Hence, the number available for detection is approximately 50 times less for the same detector area at a particular dose equivalent rate.
- The energy of neutrons can range from thermal (i.e. 0.025 eV at 20 centigrade) to tens of MeV with machine-generated beams and 3 MeV with radioactive sources. This energy range of at least 10^8 that arises when, for example, monitoring neutron dose equivalent rates generated by an Am-241 + Be neutron source, is clearly more difficult than addressing the 10^3 range encountered in gamma monitoring.

- In many ways, the 'easiest' way to produce an instrument with an operationally acceptable (note, not 'good') energy response is to moderate (i.e. reduce the energy of) incident neutrons. This makes their detection more likely. Unfortunately, this requires a large mass of moderating but not capturing material, typically polyethylene, which makes the instruments very heavy, generally at least 6 kg.
- Even the best instruments have a very poor (compared with photon measurement) energy response. Instruments are generally designed to give good results for high energies and for thermal neutrons, the reason being that in many circumstances the occupational field is dominated by these two regions, with relatively few neutrons present in the intermediate energy region.
- For circumstances where the neutron field is more complicated, workplace neutron spectrum measurements may be required to generate 'calibration factors' that can be applied to routine radiation protection measurements.

Recently, other approaches to neutron measurement have become viable, which are of better sensitivity and greatly reduced weight. However, these start to respond to gamma radiation at a lower (but still well above normal occupational level) dose rate.

4.6 Detector characteristics

4.6.1 Ionisation chambers

An intrinsically very simple detector, comprising a conducting cylinder of low atomic number material, usually a plastic composite with a hard carbon coating, polarised at between 18 and 300 volts depending on type, and a collector composed of a thin disc of carbon-coated plastic or of aluminised plastic close to the centre of the detector volume or, for some types, a collector formed on the back wall of the chamber. Often the end window is a thin layer of aluminised plastic which is to allow the measurement of less-penetrating radiations such as beta and low energy x-rays. The chamber has internal energy correcting components, generally aluminium, to raise the response at lower photon energies to give the correct $H'(0.07)$ energy dependence. There is also an additional absorber, such as slide or cap, that generates the attenuation corresponding to 10 mm of tissue. This slide is often made of the appropriate thickness of plastic and aluminium, which, in combination, behave like 10 mm of tissue in terms of transmission. The aluminium is added mostly because of its density, (2.7) which makes the slide thinner and less clumsy than a 9 mm thick all-plastic slide. However, all-plastic slides are sometimes used. When the slide is in place, the energy dependence should be a good match to that needed for the measurement of ambient dose equivalent rate. It is also relatively simple to produce a good polar response by adding additional aluminium on the outside of the chamber. For the purposes of radiation protection and x-, gamma- and beta-radiation, a good ion-chamber-based instrument essentially generates the 'right answer' for a distant source.

However, it has a huge limitation. The W value for air is 33 eV per ion pair, which means that a secondary electron that deposits 100 keV in the gas generates only about 3000 ion pairs, which in turn leads to a sensitivity of only 10^{-17} A cm^{-3} per $\mu Sv\ h^{-1}$. The maximum volume that can be employed in a portable instrument is of the order of 400 cm^3. Beyond that volume, the instrument becomes cumbersome, and the polarising voltage required for operation at high (100 mSv h^{-1}) dose rates becomes impossible to generate directly by a compact battery. A high voltage generating power supply cannot be switched on and the instrument function correctly within a few seconds as ion chambers take time to polarise and significant currents can flow for minutes, if not hours, after the polarising is applied. One way round this is to leave the polarising supply running even when the rest of the instrument is switched off, but this reduces battery life. Using a 400 cm^3 ionisation chamber, the current at 5 $\mu Sv\ h^{-1}$ is 2×10^{-14} A, which requires not only an unusual amplifier but also special attention to insulation. Higher doses and dose rates are clearly easier.

Figures 4.1 and 4.2 show a schematic of a typical ionisation chamber together with a photograph of the ion-chamber internals of an Eberline RO-2 ionisation chamber survey instrument with the thin aluminised Melinex window removed. This is a very old, but very simple and popular, design.

The electronics in this example is a single operational amplifier in the base of the ion-chamber assembly (figure 4.3). The larger glass components are reed switches that are operated by a magnet on an arm. One is used to zero the indication and the other to change the gain. The smaller are the feedback resistors (2×10^{12} ohms and 2×10^{10} ohms) that set the gain. Note the use of high-quality insulators.

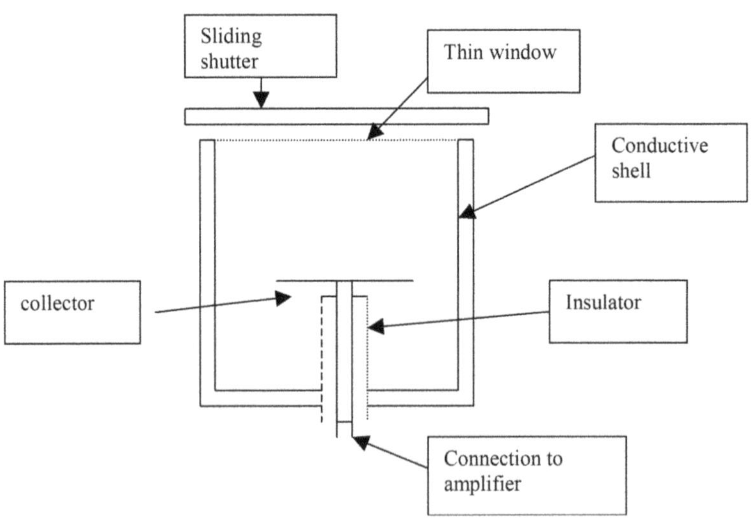

Figure 4.1. Schematic of a typical ionisation chamber.

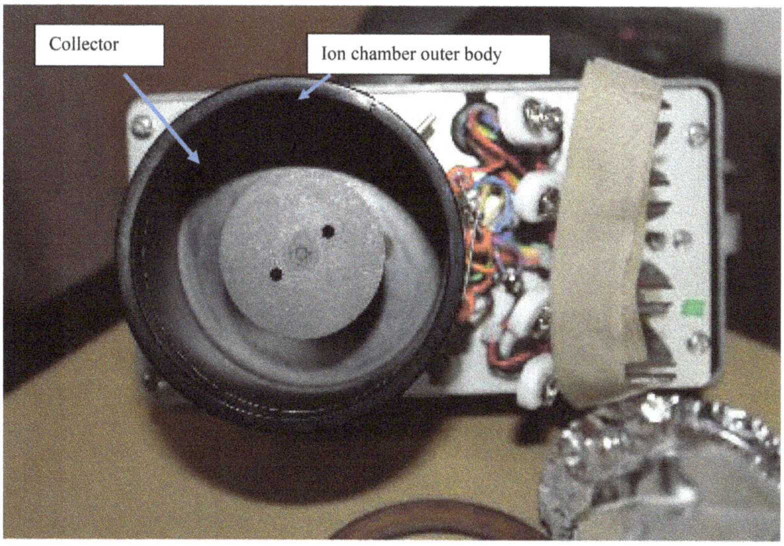

Figure 4.2. Ion-chamber internals of an Eberline RO-2 ionisation chamber survey instrument with the thin aluminised Melinex window removed.

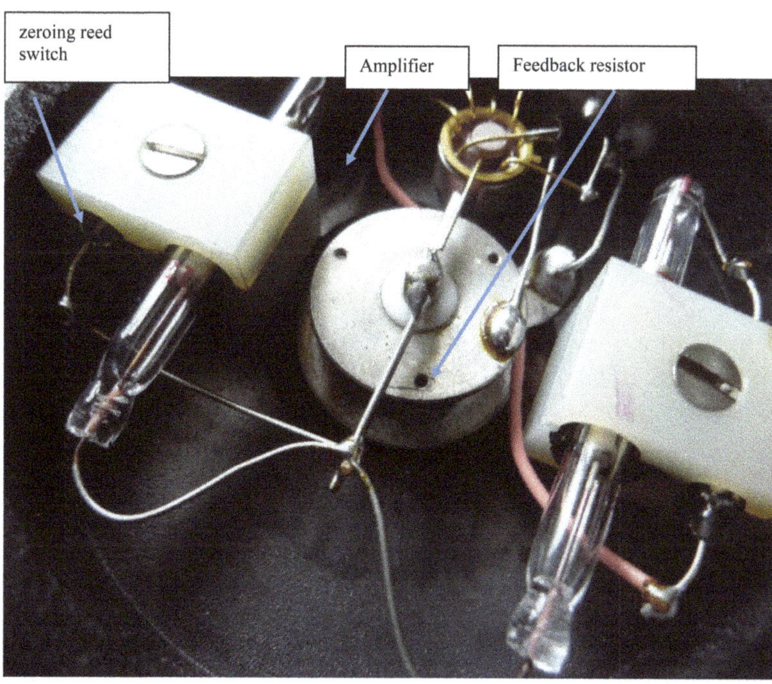

Figure 4.3. Electronic components of an Eberline RO-2 ionisation chamber survey instrument.

With the introduction of end window energy compensated GMs (see below), the need for ion chambers has been reduced and, in radiation protection, they are now mostly used only where the radiation fields are a complex mix of beta and gamma radiation and, particularly, for use with pulsed sources such as Linacs.

4.6.2 The proportional counter

This detector is not often used for gamma dose rate monitoring, but it does have advantages. The important thing to realise is that the output is proportional to the energy **deposited** in the detector, not necessarily the energy of the radiation generating the pulse.

The fundamentals of the proportional counter are as follows.Examples of the different types of instruments are given below.

- A cathode, which is frequently at ground, as it is the outside of the detector. For dose rate measurement, the cathode is normally cylindrical, but for contamination the detector is usually rectangular with dimensions of the order of 400 mm × 300 mm · 30 mm. One of the large sides is relatively transparent to the radiation of interest. For alpha + beta detection this is aluminised plastic with a mass per unit area of the order of 1.2 mg cm^{-2}, corresponding to a thickness of 0.01 mm. For a higher energy beta + x-ray contamination monitor, the window is generally titanium of 5 mg cm^{-2} mass per unit area, i.e. a thickness of 0.01 mm.
- An anode wire or wires. These are extremely thin, often 0.025 mm in diameter, and generally made of tungsten with a good surface finish. For cylindrical detectors, a single wire is used, whereas for contamination an array of parallel wires is employed.
- A counting gas, composed of a gas or gas mixture, where any electrons generated by radiation interactions remain unattached. For gamma dose rate, neutron dose rate and high-energy beta and x-ray contamination measurement, the detector is generally sealed. The fill gas can be methane for the gamma dose rate versions, He-3 or BF_3 for neutron detectors, and xenon for contamination monitors.
- For installed alpha + beta contamination detectors, as used in exit monitors, the gas normally flows at the rate of a few cm^3 per minute. This removes any oxygen which diffuses in through the thin window. The gas is typically argon with between 10% and 2.5% methane to act as a quencher. Alternatively, for portable equipment, propane lighter fuel gas can be used, stored in a reservoir and then occasionally released into the detector. This was once popular but is now rare.
- A positive high voltage (HV) supply, commonly 1.8 kV, connected to the anode via a high resistance. This voltage must be stable, as the gain is very voltage dependent. A good rule of thumb is that 100 V increase in HV gives a factor of 4 change in gain, which corresponds to a 1. 4% increase per volt increase.

- An amplifier with reasonably good noise performance
- A threshold detector that can be set to count pulses over a defined amplitude.

There is a small number of instruments that use a cylindrical sealed proportional counter for the measurement of X and gamma dose rates. They look and behave in a very similar way to the standard energy compensated GM tube. Their merits are a very slightly better low-energy performance and a higher maximum useful dose rate, The low energy response is still inferior to that of an end window energy compensated GM detector.

For measuring neutron dose equivalent rate, a proportional counter like that shown in figure 4.4 may be used.

A sealed proportional counter for contamination monitoring is illustrated in figure 4.5 and a flow-through alpha + beta contamination detector and corresponding monitor are shown in figure 4.6.

The basic physics of the detection process in proportional counters is described below.

For x- and gamma-radiation, a small proportion of the incident photons will generate energetic secondary electrons that will enter the active volume and generate positive ions and free electrons. The free electrons will drift towards the anode wire, picking up energy in the electric field and losing it by collision. The electric field (E) from a long wire in a cylindrical detector (in volts m^{-1}) at radius R is $V/\{R \ln (b/a)\}$ where V is the applied potential, a is the wire radius and b is the cathode radius. Assuming 1000 V, $a = 10^{-5}$ m and $b = 10^{-2}$ m gives $E = 145/R$. Close to the wire, i.e. where $R = 2 \times 10^{-5}$ m, the field strength is 7×10^{6} V m^{-1}. This electric field is sufficient to give an electron a kinetic energy of 10 eV or more between collisions

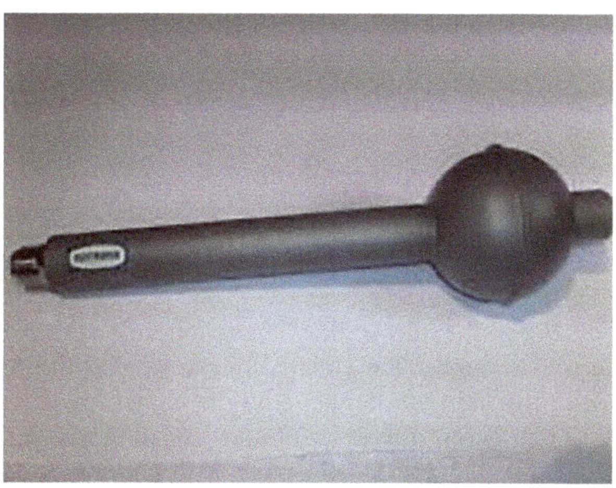

Figure 4.4. A He-3 filled, 33 mm diameter, spherical proportional counter. The fill pressure is two atmospheres, and the recommended operating potential is 800–900 V.

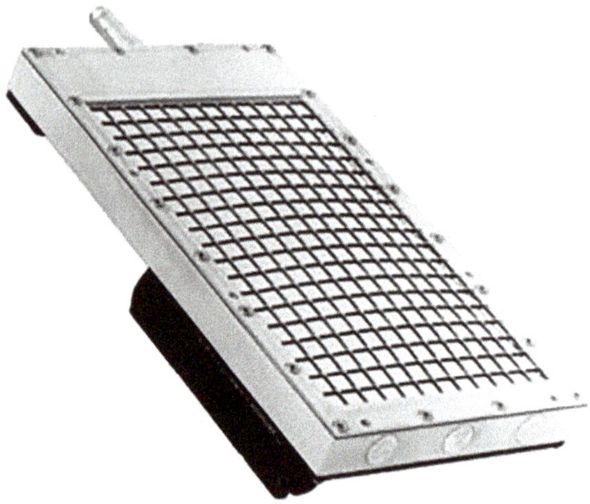

Figure 4.5. Sealed proportional counter for contamination monitoring.

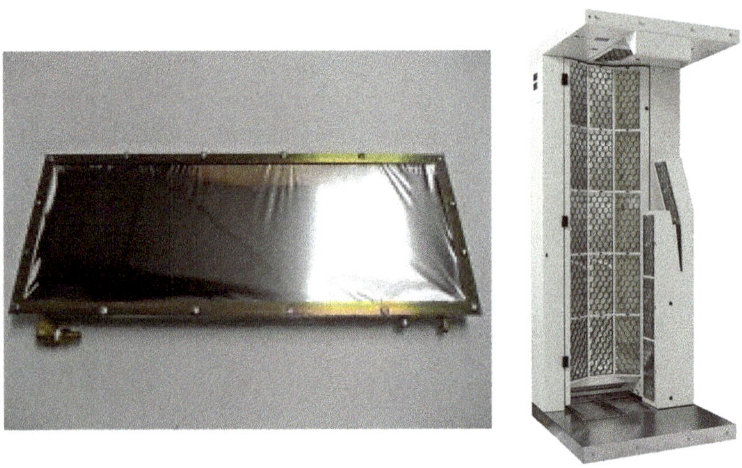

Figure 4.6. Flow-through alpha + beta contamination detector and corresponding monitor.

with gas molecules. When they do collide, ionisation may take place, doubling the number of electrons. The higher the voltage, or the smaller the wire diameter, the further from the wire this starts to happen and the higher the gain. It is possible to produce gains of 10 000 before the detector starts turning into a GM tube and all energy information is lost. The result is a pulse with an amplitude proportional to the energy deposited in the detector.

For energetic beta and gamma radiations, the energy of the primary electrons is so high that only part of the energy is deposited. Consider a 2 MeV beta particle.

This has a range in air of several metres. For a detector thickness of 30 mm, only about 1% of the energy of the beta will be deposited, or about 20 keV. For alpha particles, however, all the energy of the particle will be deposited. This is never the full energy, as the particle will lose energy crossing the air gap and penetrating the window. The shallower the angle, the longer the path length and the lower the energy of the alpha when it gets into the detector. However, even at as low an energy as 1 MeV, that is 50 times the energy of the energetic beta previously considered. Hence, a threshold of a few 100 keV can be set, where all the pulses above that threshold are taken to be generated by alpha particles. Pulses below that threshold will be generated by beta and gamma radiation, but also by alpha particles at the extreme end of their range, where they have only perhaps 100 keV left. If xenon, which is a high atomic number gas, is used as the counter gas then there is a significant chance of photoelectric detection within the gas for low photon energies, e.g. Fe-55 at 5.9 keV. Such detectors can be used as spectrometers at very low (<20 keV) photon energies.

Failure modes of proportional counters are dominated, in the exit contamination monitoring types by window damage, frequently of the 'screwdriver in the back pocket' variety. The window should be removed. Note that the wires are generally close to invisible. Do not run the risk of wire damage as restringing the wire demands patience and significant manual dexterity. The window should have the aluminised plastic replaced, making sure that the inside, at minimum, is conducting and the window replaced on the detector. This requires care, as it is easy for a screwdriver to slip and go through the new window. The detector will then need to be refilled with gas by turning up the flow rate for a short period, following the manufacturer's instructions.

4.6.3 Geiger-Müller (GM) detectors

As discussed above, if you continue to increase the high voltage on a proportional counter, it will eventually turn into a GM detector. However, this is not what happens in practice. GM tubes almost always operate at a lower HV than proportional counters. There are two reasons—the gas pressure is low, which gives a higher mean free path between collisions allowing the electron to pick up more energy and increasing the probability of a productive interaction, and because virtually all GM tubes use an additional amplification mechanism involving dual gases.

Cylindrical GM detectors have a low gas pressure, typically about 0.1 atmospheres, a relatively thick anode, typically 0.5–4 mm diameter stainless steel or tungsten, a steel cathode with a diameter from 6 mm up to about 25 mm and a gas filling which is either a helium/neon (small tubes) or neon/argon mix (the majority), in the ratio of 99–1 or less, and a few % of a halogen quencher, typically chlorine or bromine.

The detection process basically starts like a proportional counter but, in this case, collisions that generate excitation in the filling gas atoms are used. This is where one of the electrons is raised out of its ground state energy to a higher energy bound state. With the right gas, these atoms de-excite quickly, within a microsecond or so, sometimes generating a UV photon. This photon can then be picked up elsewhere in the tube and can ionise the gas, starting the proportional process again. The mechanism operates because the ionisation energy of the gas reduces as the atomic number increases. For example, the lowest energy for helium is 24.6 eV, for neon is 21.6 eV and for argon is 15.8 eV. Hence an excitation energy in the neon atom in the 15.8–21.6 eV region can result in the ionisation of an argon atom.

This process can clearly 'run away', resulting in the ionisation of many atoms and the generation of a large current pulse. If we take an end window GM detector with a mica window which is at least partially transparent, a low-light camera will see the visible light component of the flash very clearly. As an extreme measure, it is possible to operate a glass-walled GM (now quite rare) at a huge over-voltage and see the light pulse by eye. As gases produce different colours, it is possible to work out what is the major component of the gas fill. Neon is a bright red/orange, for example.

If this process was not stopped, the tube would continue to glow. There are two mechanisms that stop the discharge:

- as soon as current starts to flow through the GM tube anode resistor, the voltage across the tube drops, reducing the electric field;
- as soon as the process starts, a sheath of positive ions builds up close to the anode. This increases the electric field between the sheath and the cathode, but this has little effect as no amplification takes place in this region. It also reduces the field close to the anode, which reduces the amplification in the important zone

Both mechanisms can be seen in standard GM tubes. As the anode resistance is reduced, the 'switch off' effect of the voltage reduction is suppressed, but the time taken to recharge the tube capacitance also reduces, leading to a shorter dead time, which is desirable in many ways. (It also has other, less desirable, effects, see below.) If the anode resistance is decreased beyond a certain point the dead time stops falling. This is because the dead time is now being driven by the positive ion sheath. As this is composed of 'heavy' ions, as compared with electrons, it takes much longer for the electric field to clear them back to the cathode where they can be neutralised by gaining an electron.

Most GM tubes found in the UK market are based on Philips designs from the 1950s and operate at 450–575 V, depending on type. Older detectors from the USA run at a standardised 900 V, which is probably the result of a decision during the original weapons' project in 1943. For the Philips designs, anode resistors run from 2.2 Mohms up to 10 Mohms. Some of the USA designs operate with resistances as low as 1 Mohm.

GM detectors should have a very good plateau, i.e. a voltage region where the count rate varies only slightly with HV (figure 4.7).

An important point is that we (almost) never measure the plateau and set it in an operating instrument. What we set is (almost always) the manufacturer's recommended voltage. If the instrument does not work correctly, either the GM

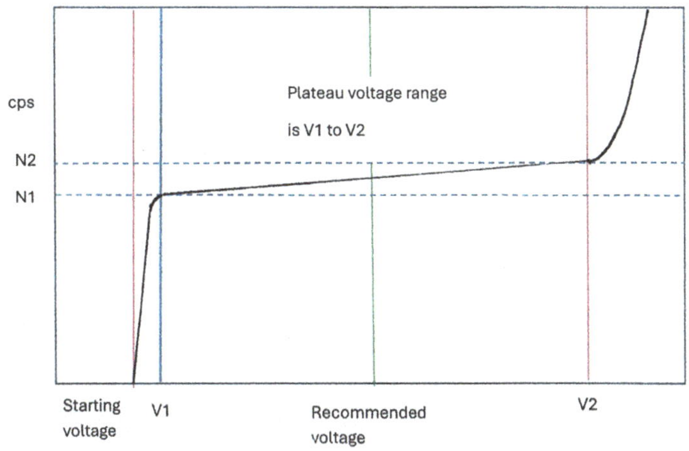

Figure 4.7. Plateau voltage range for a GM tube.

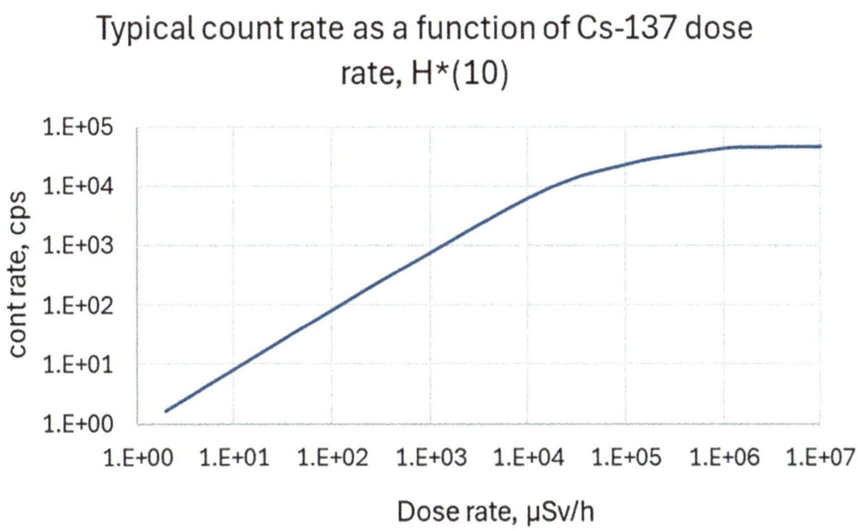

Figure 4.8. Count rate as a function of dose rate for a GM tube.

tube or the electronics is faulty. We do not 'chase' a reduction in performance by increasing the HV.

Figure 4.8 shows a typical count rate/dose rate curve for a good design.

Note that the count rate falls off at high dose rates because of the phenomenon of dead time, the time taken for the detector to deal with any pulse. Note, also, that the count rate effectively reaches a constant value independent of dose rate.

Dead times are size dependent, mainly because the bigger the detector, the larger the capacitance and the longer it takes to recharge. For the smallest, the reference dead times are as low as 5 μs, and for the largest as high as 210 μs. The dead time is more complicated than usually thought. The classic correction equation is: $N = N_o/(1-N_oT)$, where N_o is the observed count rate, T is the dead time and N is the corrected value. In this expression, $(1-N_oT)$ is basically the live time, i.e. the fraction of time over which the detector could actually generate another count. In practice, the dead time tends to reduce as the count rate increases, which means the classic equation does not give accurate results. At the extreme, if we use the dead time we measure at relatively low count rates and operate the detector at high count rates, $(1-N_oT)$ goes negative, which is obviously a nonsense. Hence the mechanism used to correct is generally more complicated. But it is not difficult to work out a good equation using a form such as $N = N_o/(1-aN_oT + bN_o^2T)$ where a and b are optimised by iteration using recorded count rate/dose rate values. This process is circuit- and geometry-dependent, i.e. the same GM detector in a different instrument design will require slightly different values of a and b.

An alternative approach is to deliberately switch off the detector by lowering the anode–cathode voltage difference as soon as a pulse starts. After a few μs, the voltage is raised again. What the instrument actually measures is the time of counting, rather than the count rate. This is slightly electronically tricky but can extend the useful dose rate range to a degree beyond the method above.

Another important mechanism is the action of the quencher gas. The bond energy for the Br_2 molecule is 1.99 eV and for the Cl_2 molecule is 2.51 eV, both of which are below the energy of a photon which could restart the GM pulse. Where excited atoms collide with the quencher, they tend to de-excite and break up the quencher molecule, thus preventing the excited atom from restarting the pulse. The result is two Br or Cl free radicals that are chemically extremely reactive. Hence the limitation on the materials that halogen-quenched GM tubes can be made from to certain grades of steel, some types of glass and mica, which is the only material that has been found to work well as a thin window. The free radicals generally quite quickly pair up to give molecules again, with only a small, and harmless, energy release.

The final mechanism is that the process of neutralising a positive ion by the addition of an electron from the cathode generates significant energy. This can release an 'extra' electron from the cathode, restarting the GM process. This is minimised by oxidising the surface to give a high work function, i.e. energy required to remove an electron.

GM tubes fail at extremely high total counts. This failure is either by adsorption of the quencher into the anode and cathode, leading to a reduction in concentration, or a roughening of the anode, leading to high localised electric fields. Both of these effects tend to be 'fail safe', i.e. the detectors start double counting, increasing the apparent radiation intensity or background.

However, 'good' GM tubes have two very useful characteristics:

- If the high voltage supply can provide enough current, they appear to continue to count, no matter what the dose rate. A very common detector, the Centronic ZP1320, which was used to generate the count rate/dose rate relationship shown in figure 4.8, has been tested at kGy h^{-1}, continued to count and appeared to have suffered no damage.

- They do not produce very small pulses. For the Centronic and some LND types, for example, the minimum pulse size in a well-designed circuit is never less than 20% of the low count rate amplitude. Hence all that is required is a pulse detector that is triggered at about 10% of the low count rate pulse size.

To exploit these characteristics needs carefully designed electronics. There have been plenty of examples of prototypes that failed to danger (fixed before production) and even production units that failed to danger because the high voltage supply was under-powered, or the threshold was set too high.

There are also plenty of 'bad' GM tubes on the market, where the count rate, even in a well-designed circuit, falls off above a given point as the dose rate increases (figure 4.9).

Consideration needs also to be given to the energy dependence and hence compensation required for GM tubes. For metal-walled detectors, a typical 'as made' GM tube energy response is shown in figure 4.10 for a Centronic ZP1320 and the quantity air kerma.

The fall-off at the low energy is due to photon attenuation by the 32–40 mg cm^{-2} steel tube wall, a reduction in the photo-electron range and a reduction in the photon fluence per unit area per gray.

The peak is more subtle. It is a characteristic of x- and gamma-radiation that the total energy that hits a target is more important than the individual photon energy when it comes to dose. For example, for air kerma the factor for 600 keV is 2.84 pGy $photon^{-1}$ cm^{-2} and for 60 keV it is 0.289 pGy $photon^{-1}$ cm^{-2}, i.e. nearly 10 times as many photons per cm^2 are required at 60 keV compared with 600 keV per unit air kerma. Hence a photon counter would give 10 times the count rate at 60 keV than at 600, very like the graph above. Therefore, it seems that a GM detector is close to a photon counter. It is, but at a very low efficiency of about 0.5%, i.e. about 1 photon in 200 hitting the detector will generate a pulse. However, we know that steel interacts more with low energy photons than high, so we might expect the low energy count rate to be higher. This effect does not occur because the range of the secondary electrons is much shorter at 60 keV than at 600 keV. Hence the detector 'sees' further into the tube

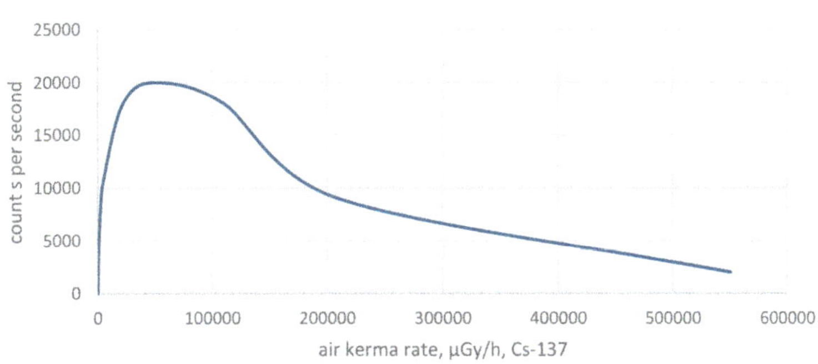

Figure 4.9. Count rate against dose rate for a poorly performing GM tube.

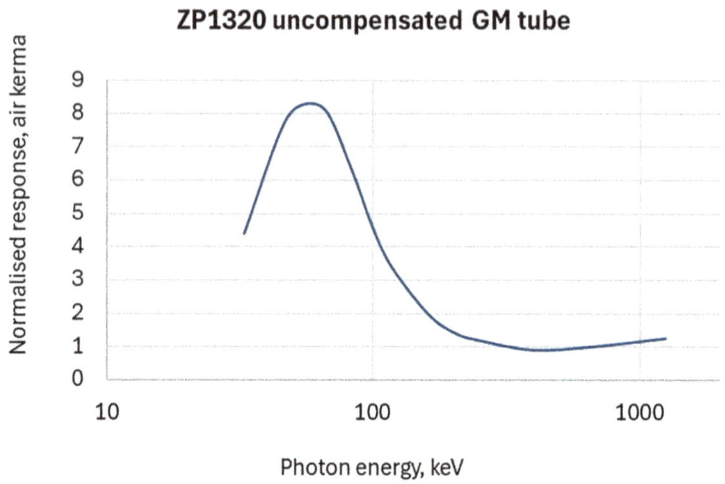

Figure 4.10. Energy response for a Centronic ZP1320.

wall at the higher energy. The two effects broadly balance out above 60 keV to give a detection efficiency with only a relatively small energy dependence.

The response above is of little use for dose rate measurement. Even if you know the energy of the original source, by the time it reaches the monitoring point, for all barring the simplest situations, the energy and direction will have changed in a way that is difficult to predict. Hence, we cannot 'calibrate out' the problem.

The solution is to fit the detector with an energy-compensation filter (figure 4.11).

The steps are as follows. This may sound hard work, but, for the larger detectors, 15 or 20 h will normally suffice to get a good result. Very small detectors are difficult, as the centre gap may become smaller than the filter thickness, thus effectively collimating the detector. It may be necessary to compromise the low energy performance by fitting a copper sleeve, chosen so that the peak response is moved from approximately 60–80 keV. This will reduce the height of the peak, allowing for a wider centre gap. Results of this process are shown in figure 4.12.

- Measure the bare energy response.
- Calculate the ratio, R, of the peak response (generally at 60 keV) to that for Cs-137 at 662 keV. For the ZP1320 it is 8.2.
- Set the final response at that energy to 1.1 rather than 1, to help hold up the low energy response.
- Divide that value by R. The result is 0.13 in this case. Use that value to calculate the width of the gap in the filter based on the sensitive length of the GM.
- Calculate the response that would be predicted for just the radiation that gets through the gap. That will be about 1.1 at the peak, falling to 0.13 for 662 keV.
- Subtract that from 1 to get the 'deficit' and divide that by the 662 keV value, in this case 0.87, to get the desired transmission for the metal filter elements. This will range from, in theory, less than zero at the peak to 1 at 662 keV.

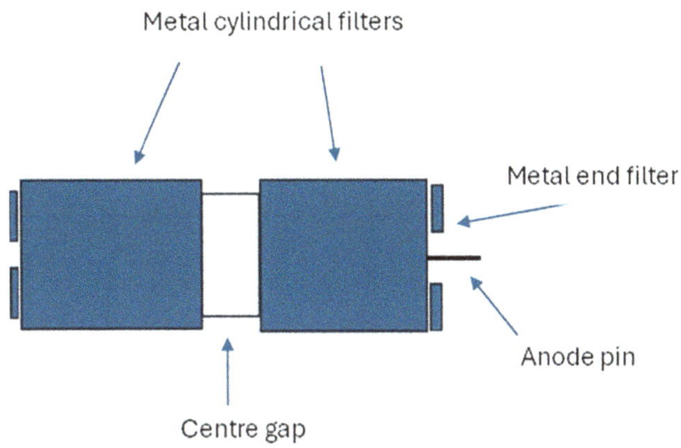

Figure 4.11. Energy compensation filter.

- Clearly transmission of 1.0 at high energies and 0.0 at the peak energy are not possible, but this can be ignored in selecting a material for the filter. Basically, there are only tin and lead and alloys thereof. These are the only practical materials that have a high enough atomic number, can be machined easily and are not expensive.
- Calculate the thickness for each energy, based on normal incidence and the square root of the mass attenuation coefficient and the mass-energy absorption coefficient. This is inevitably an approximation, as the filter is cylindrical. The use of the square root of the mass attenuation coefficient and the mass-energy absorption coefficient is based on the fact that, at higher energies, the mass attenuation coefficient over-estimates the fraction of photons removed as many of the photons will undergo Compton interaction, which results in the emission of a lower energy photon that may then be detected by the GM detector, whereas the mass-energy absorption coefficient is based on the energy deposited locally and under-estimates the number of photons removed from the incident beam as some of the Compton-scattered photons will be scattered at angles which do not intersect the detector.
- Then make a prototype where the gap is smaller than the calculated value and the wall slightly too thick.
- Remeasure the energy response. Given the above approach, the response, when normalised to 1 for the 662 keV gamma radiation from Cs-137, is likely to be less than unity.
- To enhance the low energy region (<118 keV), widen the gap.
- To enhance the energies from 118 keV up to the highest energy below 662 keV, reduce the wall thickness.
- This is an iterative, but not difficult, process.
- Apply the same approach to the ends of the filter.
- Measure the polar response at a range of energies.
- For smaller detectors, achieving a good polar response may require the use of tapered edges to the centre gap.

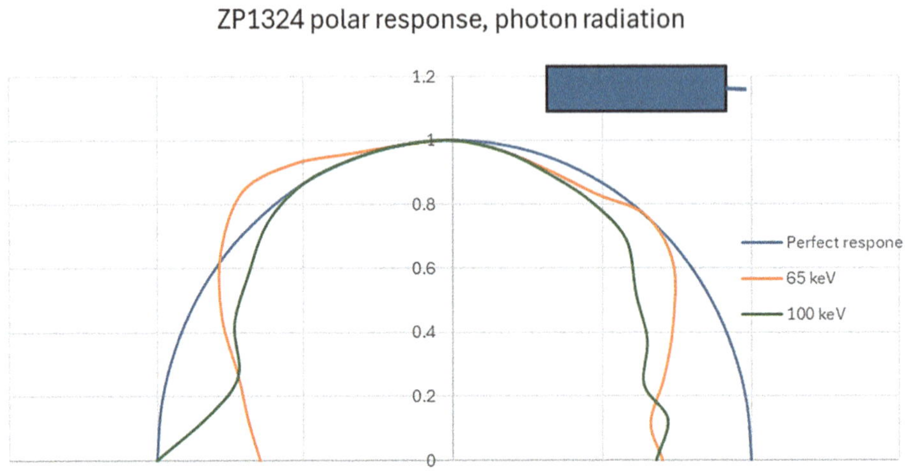

Figure 4.12. Polar response of the ZP1321 at two different energies.

End-window compensated detectors differ from those described above. The difference with these is that the rapid fall-off in response at low energies seen in the steel-walled detectors is much reduced as the mica end window has a lower mass per unit area and the atomic number is less. Effectively, the energy response for H* (10) changes very little from 60 keV down to 10 keV, the lower limit at which calculation of H*(10) is relevant. This means it is possible to energy compensate these using a simple tin washer, sometimes with a thin aluminium disc placed over it to reduce the very low energy response slightly. The method used is identical to that for the metal-walled detectors and is based on the response at 60 keV. The polar response is not easy to modify and there is always a large fall-off at angles over about 60 degrees to the normal at low energies caused by the steel wall. This can be slightly compensated for by allowing radiation to 'shine under' the filter. Figure 4.13 shows the results achieved in a common type.

The filter is often easy to remove. This allows the detector to be used as a beta contamination monitor or a detector with an enhanced low energy detection efficiency for searching for shielding leaks around security x-ray equipment, for example (figure 4.14). The same approach can be used for the quantity air kerma but is not so effective and gives a lower useful energy limit of approximately 30 keV.

Previous to development of these detectors, if there was a significant component present, or even potentially present, below 50 keV then an ion chamber was needed. The introduction of these new instruments has greatly reduced the need for ion-chamber-based instruments, with the advantages of better sensitivity—the sensitivity of the type above is 5 cps $\mu^{-1}Sv^{-1}$ h^{-1}—reduced bulk, an audio output and much reduced maintenance needs.

For end-window GMs, the usual failure is window damage, which lets oxygen in and stops the process. The only solution is replacement. This failure mode is particularly common in pancake GMs used for contamination monitoring (figure 4.15). Note the circular ring anode.

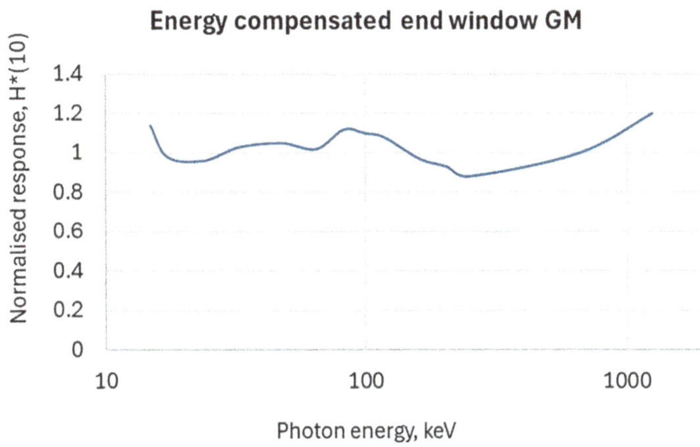

Figure 4.13. Energy response of an end-window compensated detector.

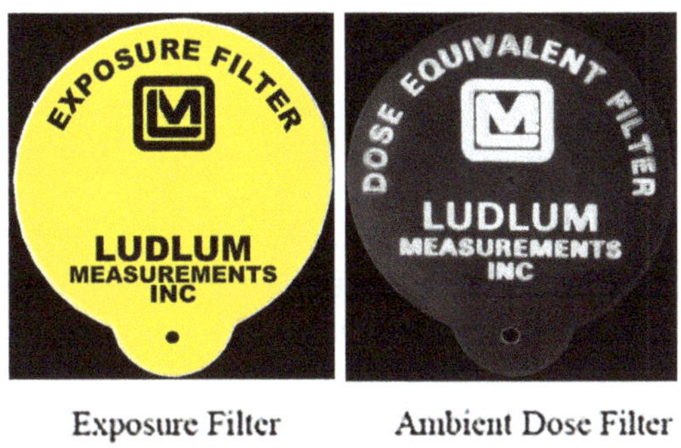

Figure 4.14. Illustrative types of filters.

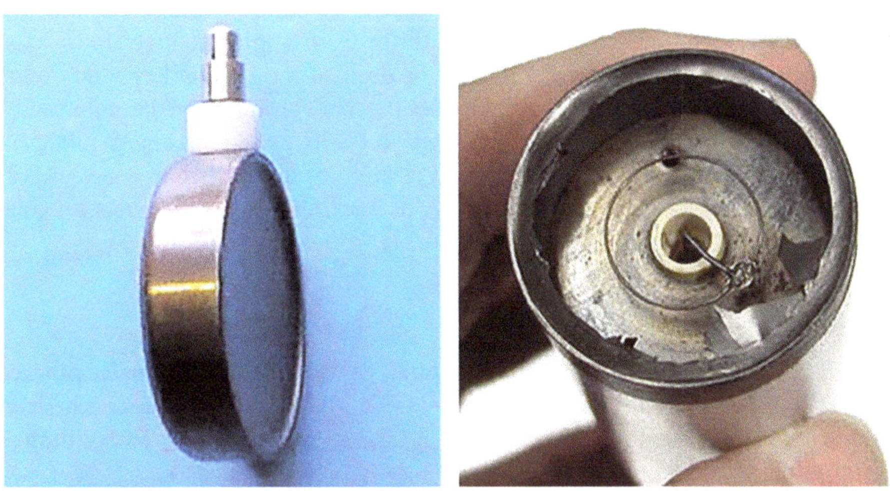

Figure 4.15. Damaged end-window GM.

Figure 4.16. Cylindrical metal-walled detector.

For cylindrical metal-walled detectors, the most common failure mode is cracks in the glass to metal seals. This can be caused by being dropped, but it can also be caused by having too stiff an anode clip connection or too much weight hanging on the connector. Both black components in figure 4.16 are glass. Note the anode pin and clip at the far end.

To check that a GM tube is still working, the obvious first step is to confirm that the background count rate is credible. The background count rate has two components—the contribution from the outside world—gamma radiation from building materials and the ground together with cosmic rays—and internal activity. The latter is dominated by potassium-40 (K-40), a naturally radioactive nuclide found as a fixed fraction of all potassium. This emits beta and gamma radiation. Any beta emitted into the detector volume will be counted. Most glass contains some level of potassium, and most GM detectors contain glass, either obviously, as ends, sometimes as a low melting point solder to assemble ceramic and glass components and sometimes in the ceramic components. Hence, most GMs have an in-built ‘check source’. This works well for larger detectors, where counts per minute can be expected, even in low background environments like submarines at depth. The way it is employed is to reset a clock after every pulse. If the clock counts for the set period without a count, it will generate a FAIL signal. The set period can range from a minute or so up to tens of minutes for the smallest detectors. This may be deemed too long in key safety related equipment to detect a failure, in which case the background can be enhanced by wrapping a piece of ion-exchange cloth soaked in Cl-36, a medium energy, long half-life, beta emitter, round the cathode. This typically has an activity of a few Bq. The most common example is the PDRM82 civil defence monitor, which uses an energy compensated version of the ZP1300 (figure 4.17).

4.6.4 Scintillation detectors

A scintillator is a material that, when irradiated, generates light. The basic physics is complicated but can be summarised as follows. This collection process can be seen if you remove the light detector and place the eye at the same position. Everything you see should be transparent (most scintillators), white or aluminised.

Figure 4.17. The PDRM82 civil defence monitor.

- Radiation enters and generates ions and free electrons.
- Some of the free electrons are captured by atoms of the scintillator and lose their excess energy by emitting visible and near-UV light.
- This must happen quickly, within a μs or so, so that the light can be tied to a particular radiation event, like the passage of an alpha particle or a gamma photon.
- The detector is surrounded either by a white material which is a very efficient scatterer of the light or a reflective aluminised material.
- As far as possible, the only part of the detector which is not white or aluminised is the light detector.
- A good proportion of the light emitted ends up entering the light detector. This may be directly or may be after several reflections and scatters.
- This detector turns the light into an electrical signal that can then be amplified and processed.

There is a huge range of scintillators. To select an appropriate one, the following steps should be followed. Are there any special operating conditions, such as high temperatures or strong magnetic fields? You will also have to choose a window material. Things to consider in making this choice include the following. The need for a thin window for alpha and low energy beta counting results in lots of window damage. The light entering from even a minor defect will cause a high and erratic background or total count rate failure as the current through the photomultiplier (PM) tube increases. It is possible to make spot repairs in moments of need using paint. But you can take this far too far (figure 4.18).

- Decide which radiations you wish to detect.
- Decide which radiations you do not want to detect. For example, in many situations, there will be a gamma background. If you wish to detect alpha contamination at low levels then you do not want the detector also to detect the gammas.

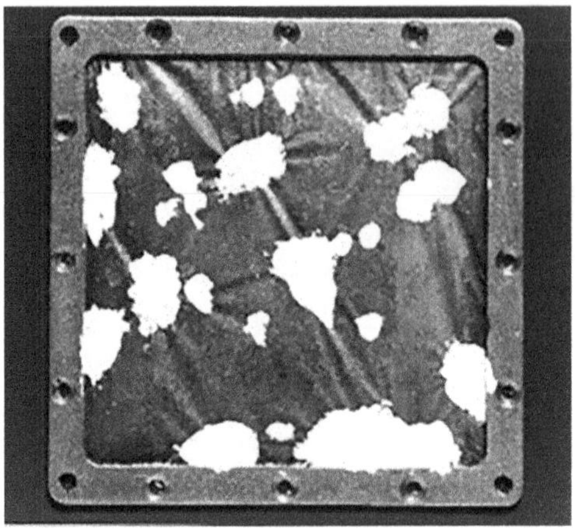

Figure 4.18. Excessive repair of a thin window.

- Decide if you want energy information from the detection process, or if you simply want to record a count.
- If you do want energy information, how detailed do you wish it to be? This can range from the 'big pulses are alphas, small pulses are beta and gamma' to gamma spectrometry, where you split the signal into 512 or more channels.
- How big would you like the detector to be? For example, you may wish to monitor the insides of respirators. This clearly demands a small and handy detector. On the other hand, you may wish to check that a truck is not unexpectedly radioactive, in which case you need detectors with dimensions of the order of metres.
- The radiations of interest and their penetrating power. For example, alpha particles have a very short range so the only practical material at the present is two layers of melinex aluminised on both sides. This gives four layers of vacuum deposited aluminium as a light screen with a total mass per unit area of 1.2 mg cm^{-2}.
- For gamma radiation, except at very low energy, aluminium is a good material because it has a low atomic number and a low density, and it is a good engineering material for making thin cans.
- Some scintillators are hygroscopic and must be kept sealed from moisture. The most common example is sodium iodide. A melinex window will not be good enough. Aluminium will be good enough but is too absorbing for photon radiation below 20 keV. Beryllium is a suitable replacement. It is more expensive than aluminium, not so easy to work and can be toxic as a dust, hence it is only used when needed.

Available scintillators include zinc sulphide, various plastics, sodium iodide and other inorganic solids. These are briefly described below.

For alpha particles, zinc sulphide with a silver doping is, and has been for 100 years, the scintillator of choice. It is very efficient, which is unusual for alpha detection as the very high energy deposition seems to overload many scintillators resulting in a lower light output per MeV than for betas. It is tough, inexpensive and easy to deposit on a plastic backing plate (for an alpha-only detector) or onto a plastic scintillator (for an alpha + beta detector). It is normally not very transparent to its own light, but as it only needs to be in the form of tiny grains, for alpha detection this is not a problem. The thin scintillator means that the energy deposited by electrons (either betas or from gamma interactions) is small and it is easy to set a counting threshold that ignores these.

The plastics used are long chain molecules dissolved in a monomer such as styrene which is then polymerised to make a solid material. They can be cast in thin sheets (down to 0.25 mm) for beta detection or thick blocks (50 mm upwards) for gamma detection. They do not produce as much light as zinc sulphide but have very good energy linearity and tend to be fast. It is easy to detect pulses down to a few 10 s of keV without noise problems. They do eventually degrade if used in very high radiation intensities, but this will never happen for hand-held equipment. Note that there are no optical components at all in some types of beta contamination detector although other designs have mechanisms to make the light detection more uniform over the area of the detector, such as a mirror at the far end and a grey filter over the area directly beneath the PM tube. In the future, there is likely to be a switch from PMs to SiPMs (silicon photomultipliers) which are much more compact and robust. This is being driven by a reduction in cost per unit sensitive area and in electronic noise. An illustrative plastic-based detector is illustrated in figure 4.19.

The most common gamma scintillators are those that use doped sodium iodide crystals. The one shown in figure 4.20 is 51 mm in diameter and 51 mm deep (2 inches · 2 inches). It is encased in aluminium, except for the glass top which couples to the light detector and is coated in magnesium oxide, the whitest material that can be employed).

Such a detector has all the characteristics that are required for gamma detection. It is, however, extremely brittle and very hygroscopic, making it easily damaged (figure 4.21).

- A high density, which gives a high value mass per unit area from front to back.
- A high atomic number, which gives good gamma detection generally and a high photopeak efficiency. This is a measure of the fraction of incident photons that deposit all their energy in a crystal.

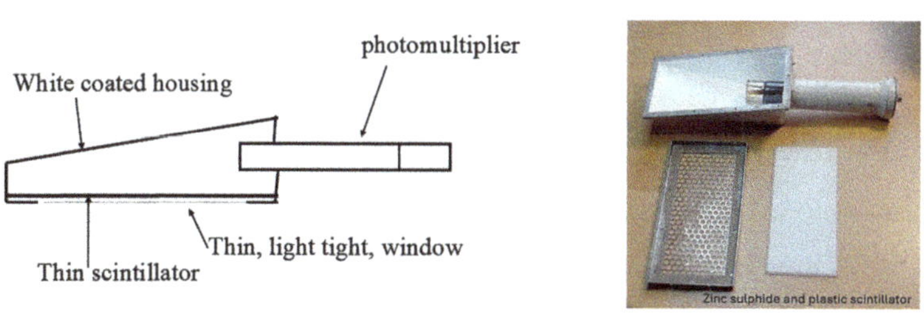

Figure 4.19. Plastic-based scintillation detector.

Figure 4.20. A cased sodium iodide crystal detector.

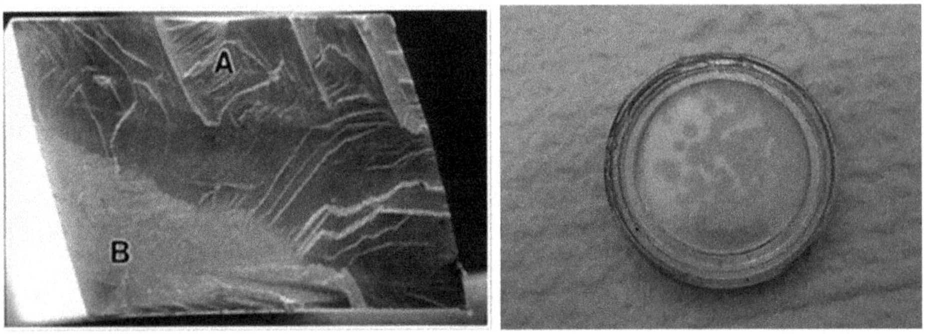

Figure 4.21. Examples of damaged sodium iodide crystals. note: light from beyond the crack cannot reach the light detector. The yellow areas do not scintillate and soak up light from other parts of the crystal.

- Good energy linearity, which indicates that the pulse produced by a photon that undergoes a single photoelectric interaction is close in size to the pulse that arises from a photon of the same energy that first experiences a Compton interaction and then the resulting lower energy photon undergoes a photoelectric interaction.
- As a consequence of the above, it has a reasonably good energy resolution (see below).
- It is extremely transparent to its own light.
- It is possible to make very large (5 l) detectors.
- The light is a good match to a PM tube.

Sodium iodide crystals are frequently used as gamma spectrometers. Figure 4.22 shows the spectrum from a Co-60 source, with its two gamma lines at 1.17 and 1.33 MeV. The energy resolution is the full width at half maximum (FWHM) expressed as a percentage. For these lines, it is about 5%, which is good performance for this material.

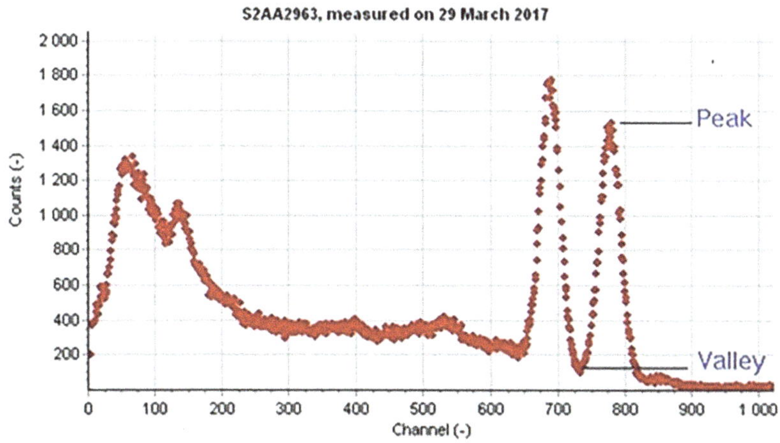

Figure 4.22. Spectrum from a sodium iodide detector exposed to a Co-60 source. This [File:Co60 Spectrum. JPG] image has been obtained by the author from the Wikimedia website where it was made available by [User:HPaul] under a CC BY-SA 3.0 licence. It is included within this article on that basis. It is attributed to [User:HPaul].

4.6.5 Other inorganic scintillators

There are many materials that make good scintillators, and more are being developed. One focus is an improvement in energy resolution—the narrower the peak, the fewer the number of counts directly under the peak from background and Compton scatter from higher energy gammas, and the lower the maximum missable activity (MMA). Another focus is on getting a better light match to SiPMs, as discussed below.

4.6.6 The photomultiplier tube

A typical PM is shown in figure 4.23 (upper). It is glass-walled and is a vacuum device. The dynodes are made from materials such as oxidised beryllium copper. Its electrical characteristics are illustrated in figure 4.23 (middle and lower).

Normally in radiation protection and in distinction from figure 4.23 (lower), the cathode is at ground and the anode is at a positive high voltage.

The photocathode is made from a material with good electron emission when illuminated. The most common type is the bi-alkali type—either Sb–Rb–Cs or Sb–K–Cs. These are efficient and have a spectral response that matches the photons produced by sodium iodide (figure 4.24).

Electrons emitted as a consequence of a light pulse are focused and accelerated towards the first dynode through about +100 V. The collision results in a typical amplification of 3–4. This process then repeats stage-by-stage over 7–11 stages and results in a gain of around 10 000.

PMs are available in a huge range of sizes. This is particularly useful when coupling to large scintillators.

However, there are some limitations and weaknesses of PMs, as listed below.

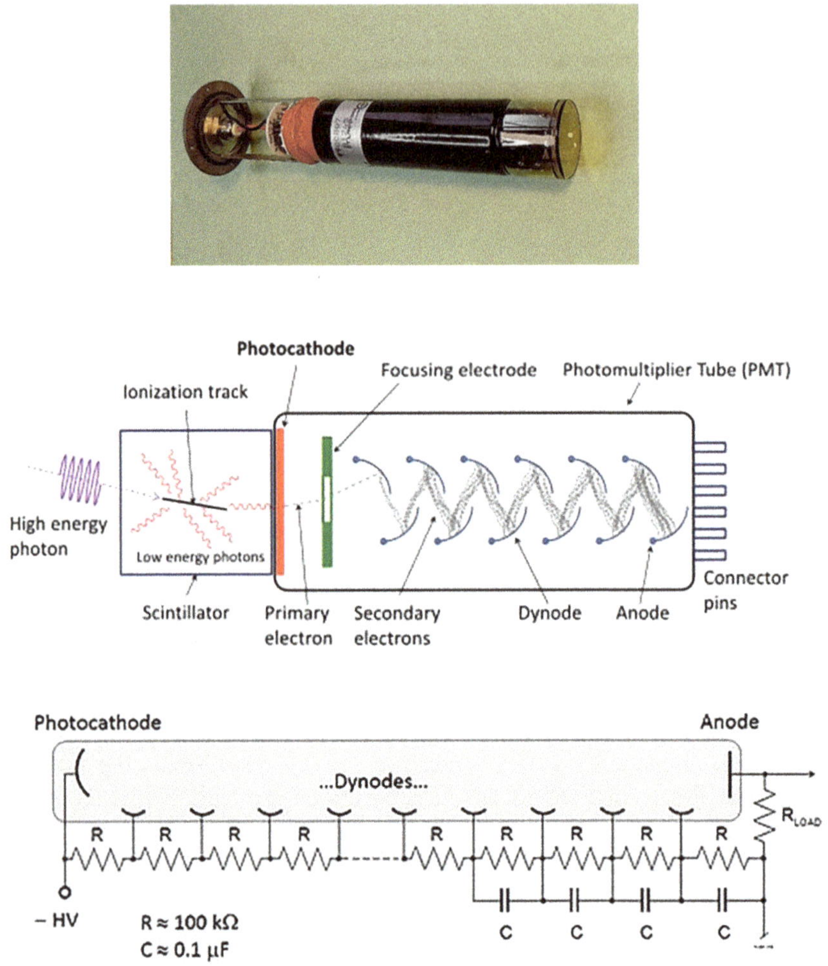

Figure 4.23. Upper panel—a typical PM. Middle panel—schematic of its mode of operation. Bottom panel—electrical arrangements. This [PhotoMultiplierTubeAndScintillator.svg] image has been obtained by the author(s) from the Wikimedia website where it was made available by [User:Qwerty123uiop] under a CC BY 3.0 licence. It is included within this article on that basis. It is attributed to [User:Qwerty123uiop].

- The gain varies widely between examples of the same type. Hence a change of PM tube requires resetting the HV.
- The gain is very voltage dependent, in a very similar way to proportional counters. Again, the rule of thumb is 100 V gives a factor of 4 increase.
- They operate at high voltages. Conventional radiation protection instruments generally operate in the range 600–1300 V.
- They are vacuum glass and metal devices and are thus slightly fragile.
- They are difficult to make, requiring good manual skills and unusual materials.
- They are susceptible to magnetic fields which defocus the electron beam and reduce the gain. Accordingly, they should be magnetically shielded using a mu-metal shield. This is the component marked ‘foil shield’ in figure 4.25.

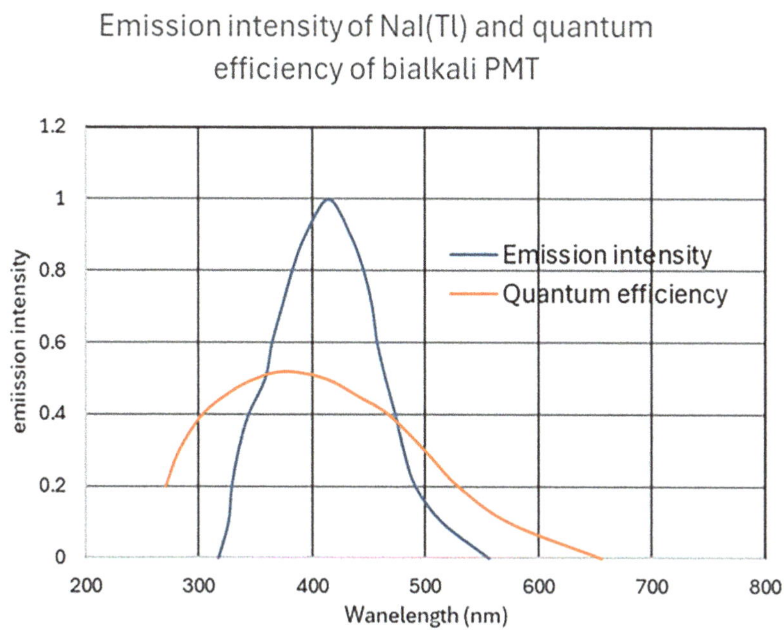

Figure 4.24. Match between emissions from a sodium iodide crystal and sensitivity of a Bi-alkali photomultiplier.

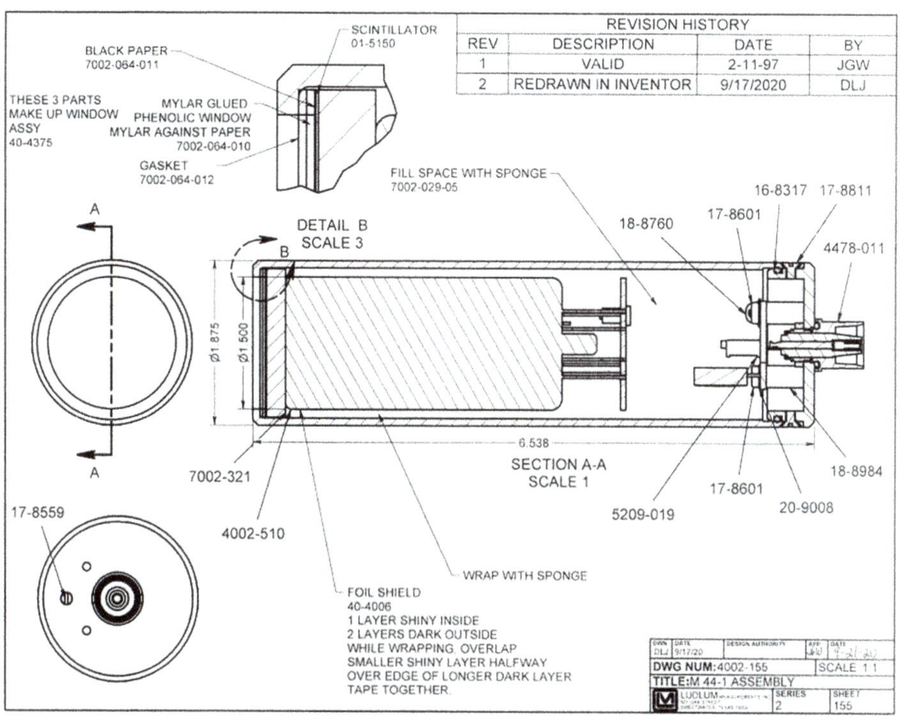

Figure 4.25. Magnetically shielded PM assembly.

- However, some probes do not have magnetic shields and are susceptible to even quite small magnetic fields. They are thus unreliable when checking structural steel, tools etc.
- The interior is a complex spot-welded metal structure. Sometimes the spot welds fail, leading to erratic operation.
- The gain tends to drop gently with time as the photocathode erodes. When new, they are a tea-brown colour, but they gradually fade with use. This means that the HV must be raised by a few volts every year in an instrument which is in frequent use.

4.6.7 The silicon photomultiplier

There are currently only a few radiation protection instruments that use these, but, in many other areas, they are taking over from PMs. Each one is an array of silicon avalanche diodes, which are reverse-biased diodes where conduction can be triggered by light photons (figure 4.26). They are either On or Off, i.e. there is no energy information. Once triggered, they will continue to count until stopped. This is achieved using a series resistor to drop the polarising voltage in the same way as in a GM detector. They are combined into arrays and it is the number or fraction of each array that is triggered per event that generates the energy information.

They have many advantages:

- consistent from example to example;
- compact;
- silicon technology with no need for unusual manual skills or unusual materials;
- low voltage operation, typically about 32 V;
- they have a significant temperature coefficient, but this is controlled from within the module.

Figure 4.26. A silicon PM array. Image provided by Scionix, NL.

The disadvantages are:

- each element is quite small and they have to be tiled to produce a large area;
- the fraction of the front face that is light sensitive is only about 60%;
- the effective noise level is significantly higher than for a good PM tube (but is improving all the time).

4.6.8 Solid state—semiconductor—detectors

These comprise diodes of one form or another. Two main materials have been in use for many years:

- silicon, in alpha and beta in air monitors, in one popular gamma alarm (Canberra G64) and many types of personal gamma dosemeters;
- germanium, used in gamma spectrometers. This uses very much larger hyper-pure germanium diodes (hpGe).

There is a high level of current development, particularly of materials that offer the prospect of gamma spectrometry at room temperature—silicon has too low a Z and germanium needs cooling. The leading contender is CZT, cadmium zinc telluride.

The principle of operation is that of an ionisation chamber. Unlike normal diodes, all the detectors have a relatively thick layer in the middle which is radiation sensitive. When they are reverse biased, electrons and holes caused by radiation can be collected and the resulting charge pulse amplified. Their huge advantage over air is that the W value, the energy to produce a charge pair, is only 3.6 eV in silicon compared with 34 eV in air. This means the charge pulse is about 9 times larger.

4.6.9 Alpha and beta in air monitors

These use passivated implanted planar silicon (PIPS). These are mounted within a few mm of a filter card, through which air is drawn. Particulate activity deposits on the filter paper. The detector is attached to a multi-channel analyser (figure 4.27). A typical spectrum is given in the lower panel of figure 4.27.

Each of the peaks corresponds to an alpha energy. The natural radon progeny produce the pink spectrum, which is effectively 'background'. The black spectrum is Am-241, representative of artificial radionuclides, the target of the monitoring programme generally. The overlap means that the 'alpha channel', i.e. the energy region where artificial radionuclides are detected, requires compensation. Given the high radiotoxicity of most alpha emitters, there is a desire to set very low alarm levels at less than 1 Bq m^{-3}. The radon level in many buildings can be tens of Bq m^{-3} depending on the local geology making compensation demanding. The main problem with these in operation is keeping the window clean. Any build-up will lead to poor spectra and, if radioactive, a high background.

4.6.10 Installed gamma monitors and personal dosemeters

These use PIN diodes. These have an intrinsic layer (the I) which is extremely pure silicon. Generally, they are used as light detectors, where the layer thickness is set to absorb virtually all the light. But this also equips them to absorb useful amounts of

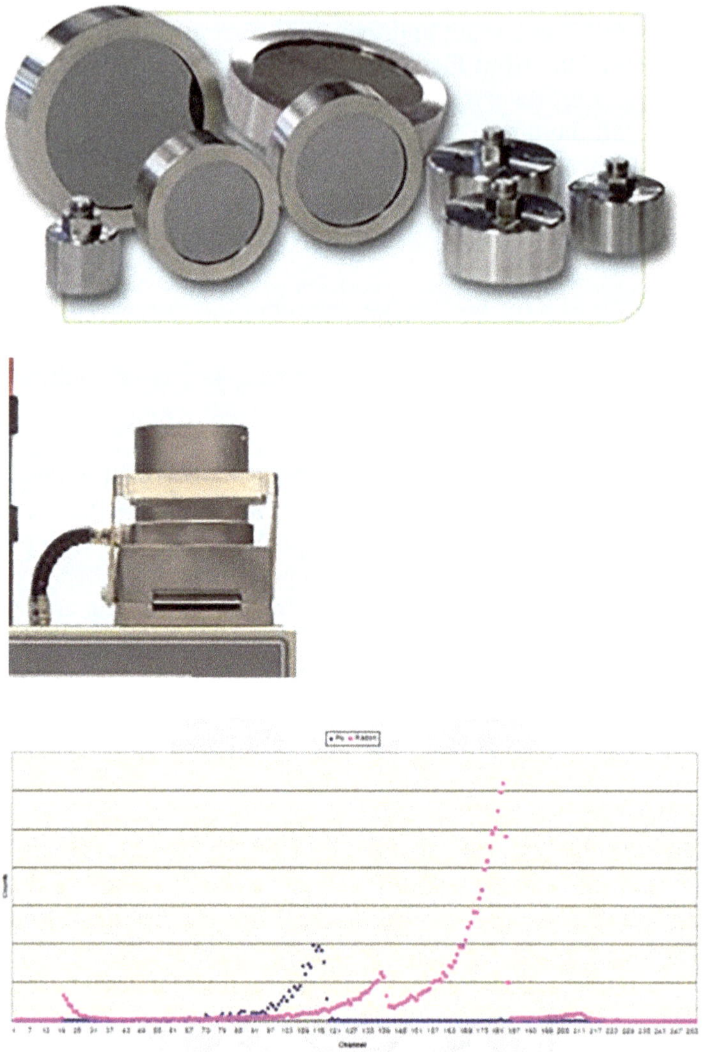

Figure 4.27. PIPS detectors (upper panel), counting arrangement (middle panel) and illustrative spectrum (lower panel). Image credit: Eliot Williams.

energy from Compton and, at lower energies, photoelectric electrons. The charge pulse is collected, amplified and counted. In this context, they can be used as a replacement for GM detectors. Illustrative examples of their use are shown in figure 4.28.

The main problem with these is the very small pulse—the lowest energy for the EPD is 15 keV, which generates only 5000 charge pairs. This means they are extraordinarily sensitive to RF interference—phones, radar, electrical welding etc. Hence the very high level of screening seen on the right. There are two detectors under the large enclosure and one under the righthand screen. The righthand screen is also an energy-compensation filter made of tin.

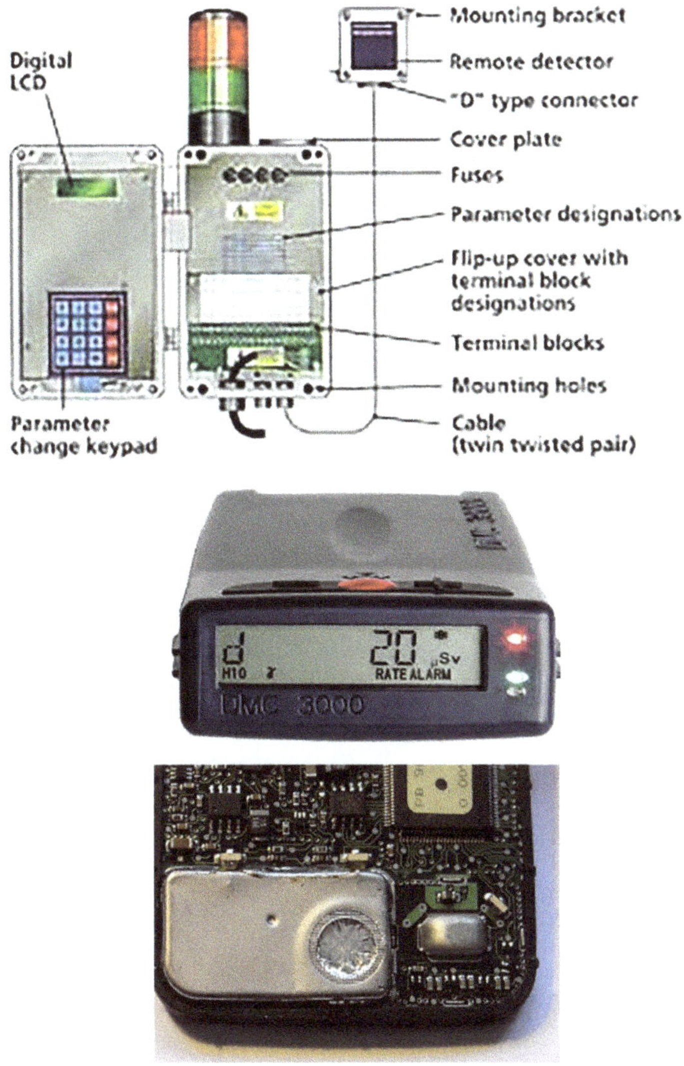

Figure 4.28. Devices that use PIN diodes.

4.6.11 Gamma spectrometry detectors

These are large volumes of hyper-pure germanium. This has a reasonably high *Z* (32) although it would be good to find a suitable higher *Z* material. It also has a reasonable density 5.3). In operation, it must be kept cold to stop diffusion in from outside the hyper-pure layer and also to reduce electronic noise. Cooling always used to be by liquid nitrogen, simply because it was easy to use and not expensive, but now, increasingly, it is by electrical refrigeration. The unit on the left in figure 4.29 is cooled by liquid nitrogen in the large Dewar. The detector is in the cylinder on the right. The one on the right is portable and electrically cooled.

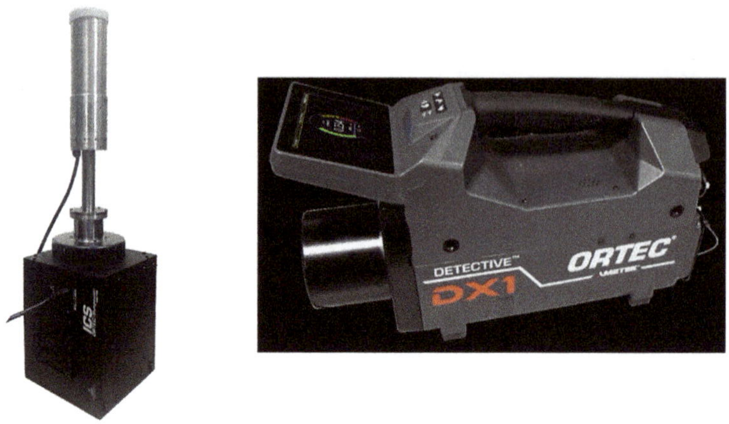

Figure 4.29. Examples of detectors used for gamma spectrometry. Image credit: ORTEC, AMETEK-GB.

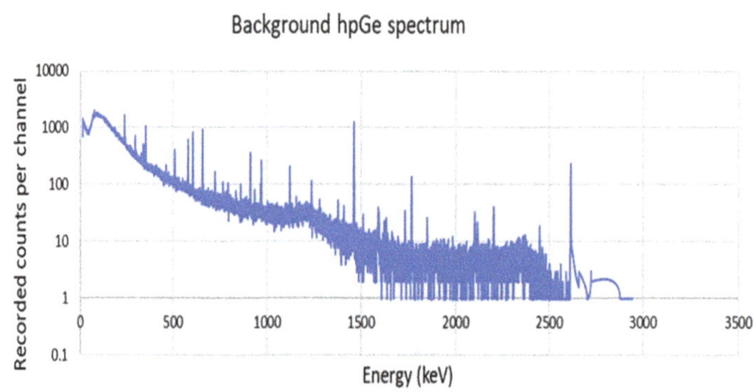

Figure 4.30. Natural background as measured using high resolution gamma spectrometry.

The merit of these is their very good energy resolution, typically about 0.5 keV at 662 keV, which separates gamma energy peaks well and means there is a lower background under each photopeak. This is shown in figure 4.30 for natural background.

4.7 Management of monitoring equipment

Management of monitoring equipment in practice used to range from the totally shambolic to the very well controlled, but increasing recognition in the UK that the Ionising Radiation Regulations 2017 (IRRs) require effective control of instruments has generally improved the way instruments are managed. The employer's duty is defined in regulation 20:

20.—(1) Every employer who designates an area as a controlled or supervised area must take such steps as are necessary (otherwise than by use of assessed doses of individuals), having regard to the nature and extent of the risks resulting from exposure to ionising radiation, to ensure that levels of ionising radiation are adequately monitored for each such area and that working conditions in those areas are kept under review.

Specifically, we have:

(3) The employer upon whom a duty is imposed by paragraph (1) must provide suitable and sufficient equipment for carrying out the monitoring required by that paragraph, which equipment must

(a) be properly maintained so that it remains fit for the purpose for which it was intended; and

(b) be adequately tested and examined at appropriate intervals.

- This means that there must be a process to allow the employer to have confidence that the organisation is meeting these requirements. What aspects are relevant?
- Each instrument must 'belong' to someone, i.e. within the organisation, someone has to ensure that the instrument is basically in sufficiently good condition to be able to make the required measurements effectively. For small organisations, with only one or two instruments, the radiation protection supervisor or equivalent can fulfil that function. In a large organisation, there may be a team assigned to the task, with their own accommodation and, perhaps, their own test facilities.
- Howsoever this is accomplished, the person dealing with an instrument has to have sufficient knowledge to be able to identify any significant degradation in performance.
- This often requires a check source. Some may be opportunistic—there may be an area of enhanced background that can be used for very sensitive equipment—but often a formal check source is best. For many test sources, the nuclides and activities are such that they do not require registration, but they will still have to be looked after and kept secure.
- Someone must make sure that relevant periodic (usually annual) testing is performed to a satisfactory level. There are two aspects —making it happen and making sure that the testing is technically sufficient and credible. Generally, it is the radiation protection advisor or the equivalent who has the technical responsibility but making it happen is often best done by the office manager.
- The organisation may wish to perform its own front-line maintenance, such as the repair of cables and the re-foiling of contamination probes. The staff performing this will need the appropriate facilities and training. Re-foiling does not require a full periodic test. Generally. the questions to be asked are: is the background as expected, does it pass the light leakage test and is the response to a suitable check source as expected? If yes, then the instrument can be returned to service.

- In a larger organisation, the employer may appoint someone as a 'qualified person'. This demands a deeper and more comprehensive knowledge of instrumentation than is common in radiation protection advisors. For smaller organisations, frequently the external test house will perform that service.
- However, ultimately, the instrument users have to have a level of respect for the equipment. Personal issue can be hugely successful, in that people generally will look after personal issue equipment better than common user equipment. It also will identify Mr Clumsy and Ms Don't Really Care quickly, who otherwise can wreak havoc.
- This level of respect can only be engendered if an instrument is delivered back from testing and maintenance in good order: clean, no out of date labels, no loose screws, no loose connectors etc. If you give someone something that looks like junk, they will treat it like junk.
- But in a busy organisation, with several shifts, personal issue can simply be too expensive. This pushes the onus back on the instrument maintenance staff.
- An often-ignored opportunity is to allow staff to see 'the other side of the fence', in the sense of making sure that instrument users have experience of instrument maintenance and repairs, particularly foil replacement, and the maintenance staff appreciate the user's world, which may involve difficult of access areas, confined spaces, poor light, high temperature etc. Experience indicates that this interaction will result in less damage to equipment and a reduced tendency to blame the users for breakages. Ultimately, the only people who don't ever break equipment may not be using it properly or often enough.
- There is a balance between first cost and lifetime cost. The classic example is surface contamination monitoring. In reasonably civilised conditions, such as a university bench, a 2′ pancake GM detector may be the best choice—limited area to monitor, good conditions, only occasional use, low first cost. But the same instrument can have a huge annual maintenance cost, because of window puncturing, in more industrial surroundings and more intensive use. In that case, a more expensive thin-windowed scintillation detector may be the better choice—more expensive to buy but replacing a window is a much cheaper option than replacing a GM detector.

4.8 Design of monitoring programmes

Monitoring is expensive, so the programme should be designed or updated so that the cost is as low as possible, provided it meets the desired aims. The first thing that helps with this is a very good understanding of why the facility is there, what the sources of radioactivity are and what goes on in the facility. Once these have been determined, you can move on to what the radiological ambitions are currently and what is expected to happen in the future. Then, finally, it is likely that there is a risk assessment and/or safety case for the facility that should identify potential failures and their consequences. For existing facilities, there is likely to be an established monitoring regime together with possibly very comprehensive monitoring data, but it is important not to feel obliged simply to 'tweek' the current programme as this may have been developed to satisfy previous dose limits etc or may be a hangover from when the facility was used more intensively. At this point, you have sufficient information to at least begin to plan a monitoring programme, but it is vital to realise that you might have to scrap the first attempt and recommence.

4.8.1 Use of historical data

A building plan is a good start. How is it split up? What are the classifications for each area in terms of access to dose rates and the contamination level? How is the ventilation arranged? Are there discharge points to atmosphere or to drains? Is there the potential for a criticality event? Are there any potential sources of fire? Have there been any serious accidents or incidents, such as a fire, that may have led to contamination in relatively unexpected places or to the replacement of sections of ventilation duct? Are there changing rooms and staff facilities such as tea rooms? What moves in and out and at what frequency? Are there physical risks such as ponds? Have there been major changes of use of the building and has the building been modified significantly since it was first commissioned? What PPE is worn routinely? For a significant programme, use of a Global Information System (GIS) is often very helpful. This allows everything you know about a building to be plotted onto the building drawings, making retrieval of any information about any location very easy.

Once a good understanding of the building and its purpose is clear, the next stage is to look at the sources of radiation. This could really be very simple, such as a laboratory with a single modern electron microscope. The only radiations expected are low energy x-rays that can be reliably removed by isolating the power. Alternatively, it could be a fuel reprocessing plant holding very large activities of material that is associated with alpha, beta, gamma, x-ray and neutron emitting radionuclides. This will offer the potential for life-threatening gamma dose rates, very high levels of airborne activity, leaks of highly radioactive liquor, significant neutron dose equivalent rates and criticality events. If it is an old building, it should also have detailed monitoring results over a long period. These, again, are best added onto the GIS record.

An understanding of occupancy is vital. Some buildings may have large numbers of people working on each of five shifts, whereas others might be visited once a week to check on the function of a pump or to add another drum to a store of low-level waste. One classic thing to be sensitive to is where the only person who normally enters an area is the radiation protection surveyor. What purpose is there to this visit? Frequently there will be good reason, normally that the surveyor is not there simply to measure dose rates but that the survey programme is designed to make sure that, at least occasionally, someone walks around the building who should notice, for example, broken windows, roof damage and dripping pipes. But often it has been possible to reduce the frequency of such visits. The occupancy review will determine, for example, the facilities needed for exit contamination monitoring. Taken with the other information on potential contaminants, monitoring could be by using a ratemeter and probe chosen to match the fingerprint (i.e. the expected nuclide mix) or it might require a row of whole-body surface contamination exit monitors.

4.8.2 Planning

A useful approach is to separate the problem into static and dynamic situations. Static situations are where the radiation field in the occupied area can be expected to remain constant. A classic example is the industrial radiography x-ray room. Once it is built, commissioned and the external walls have been checked for sufficient bulk shielding thickness and the absence of defects, there is little reason to repeat the measurements. The one part that is dynamic, at least potentially, is the door. Shielding can be damaged by collisions with test objects being moved into and out of the room, and the door runners can wear, leading to the door hanging away from the wall. This would indicate more regular checks. But there is no need, generally, for a radiologically high-quality measurement. The edge leakage dose rates should be confirmed at the commissioning stage with an appropriate instrument, i.e. one with a small volume that can get into the edge, and with the most demanding possible set-up. The next step is to use an instrument with a much higher, but energy dependent, sensitivity. The count rates around the door can be recorded along with position. Checking, then, is easy. The most demanding situation is set-up and the sensitive instrument measurement repeated. No or little change means no need for a generally slower radiologically accurate survey = lots of time saved.

Dynamic situations are where the situation can be expected to change. Two alternatives are identified. The easy one is where equipment covers and shielding are removed during maintenance, such as in a baggage x-ray machine or an electron microscope. After maintenance is complete, it is clearly necessary to check that any shielding has been replaced properly. Again, a very sensitive instrument measurement is useful. Sometimes an energy compensated GM detector will do, particularly where dose rates in contact are expected to be a few μSv h^{-1}, but again a confirmation of the expected count rate using a very sensitive instrument can be quicker.

The more demanding context is where the situation is less controlled. The common one is where unsealed radioactive material is dealt with, either as a product (hospital radiopharmacy) or a consequence (decommissioning of the fuel routes on a power station). The staff will generally be very aware of the need to minimise the spread of contamination, but failures can occur—air flows may fail, seals may leak, contamination may exist where none was expected etc. Poor attitude can rapidly lead to contamination spread—sitting on barriers wearing dirty coveralls, poor PPE wear, poor undressing technique etc. In such situations, regular monitoring at the exit point is essential, as is a regular survey of the working area. What can be extremely useful, with high-quality staff, is craftsman-based monitoring. People working in the facility are given basic equipment and training so that they can identify contamination at source and deal with it appropriately. This can be as simple as stopping the job and asking for assistance. This clearly needs its own pool of instruments.

There are situations where dose rates are variable. Industrial radiochemistry is a classic example. Liquids flowing along pipes may start to deposit activity on bends, leading to an increase in dose rate in the working area. Decommissioning is also vulnerable to this problem. When shielding which may have been in place for 30 years is removed, then the dose rate may be higher than expected.

The most common dynamic situation which has led to problems is open area industrial radiography, i.e. where, for example, a structural weld on a new bridge needs to be radiographed. The area should be barriered off, but that is usually at a standard distance. The danger moment is when the source is wound back into the transport container. The big question is whether it is back in the container or whether it has become detached from the drive cable and is somewhere in the exposure tube, unshielded and waiting to fall out and be picked up as a curiosity.

4.8.3 Predicting the radiation fields

The consequences of hardening by shielding and softening by scatter of photon beams have been discussed elsewhere, but an understanding of what is there to be measured is essential. It is very easy to get a low result by using the wrong instrument. The presence of a pulsed source, such as a linac, also imposes specific requirements on the equipment to be used. Hence the planning of any monitoring programme must include a review of the characteristics of the situation. Sometimes, it is not possible to make a robust decision without looking. Often, an onsite comparison of an instrument which definitely 'works', for example an ion chamber, but which is unattractive for routine use—poor sensitivity, no ability to search using audio, expensive etc—with one which is smaller, cheaper, more sensitive, produces clicks but which has a more restricted energy range, will demonstrate the way forward.

Measurement of contamination ranges from very easy to very difficult. In the 'very easy' class are radionuclides such as P-32, an energetic beta emitter with only a moderate radiotoxicity. If a process only uses P-32 then there is a huge range of suitable instruments, there are no worries about absorption in thin layers of dust or wax on a bench and spills on coveralls will be easy to detect. Contrast this with an old nuclear research facility, where difficult mixtures can be expected—long half-life alpha emitters, betas with an energy range from undetectable directly (H-3) to high energy (Rh-106), a mix of fuel, fission products and activation products etc. A good understanding of this—the fingerprint or radionuclide vector—is vital to identify suitable instruments and derive response factors and acceptable upper count rates. The limitations the fingerprint imposes, e.g. determining how sensitive our measurements will be to surface conditions and whether we need to have an easy method of checking those conditions, will need to be evaluated very carefully.

4.8.4 Lowest levels of interest

This is an aspect of monitoring that it often not given enough thought. It is vital that the equipment is sensitive enough to allow rapid measurement at the levels of interest. Clearance monitoring was once memorably described as 'Dull!' in front of a large audience with such a depth of feeling that everyone who was there retains the memory. It is always worthwhile looking for ways of making faster measurements to the required quality and assessing their feasibility. This can release the surveyors, who often have far more skills than simply radiation measurement, to do other tasks. The apparent block to this is sometimes surveyors resist the introduction of instruments that have a higher background because of their higher sensitivity. This is correct to a degree when it comes to the identification of localised activity, in that the peak count rate is generally not that much higher than that of the less sensitive instrument. However, the increase in count rate will generally last longer as the increased sensitivity is often produced by increased area, which means the detector will be directly above the hot spot for longer. This increases the chance of detection. For distributed activity, then that qualification vanishes. The phrase 'It's easier to see 50 above 50 than 5 above 5' will often strike a chord but the best approach is undoubtedly a demonstration.

4.8.5 Feedback

The best way to review a draft monitoring programme is to perform as much of it yourself as possible, probably under the supervision of the responsible surveyor. You will have clear view of what is involved, you may see better ways of achieving the end in view and it gives the opportunity for the surveyor to comment as you work through it. It is extremely good for morale for the boss to be seen on plant and, perhaps, be laughed at a little. If you take the surveyors along with you so that they see their views are taken seriously then they will operate the programme better and be much more inclined to identify weaknesses and suggest improvements.

IOP Publishing

Principles and Techniques of Radiological Protection

Michael Charles Thorne

Chapter 5

The natural radiation environment and human modifications to that environment

N Mitchell and M C Thorne

5.1 Introduction

The natural ionising radiation environment is comprised of cosmic rays, radionuclides produced by interactions of those cosmic rays with the atmosphere, terrestrial and aquatic environments, and radiations from naturally occurring radionuclides present since the formation of the Earth ('primordial radionuclides') plus the progeny of those radionuclides. Whilst these natural radiations are ubiquitous, the amount of radiation to which an organism is exposed largely depends on, and varies with, location and human activity. For most populations on Earth, natural background radiation represents the main source of exposure to ionising radiation.

All organisms on Earth have evolved in this radioactive environment, bathed in a continuous stream of ionising radiations, subject to internal exposure from radionuclides absorbed from their diet or following gas and/or liquid exchanges with substances present in their habitat.

This chapter provides an account of exposure to the natural radiation environment, of factors affecting the variability of that exposure and considers the perceptible consequences of human actions. The following sections loosely follow the evolutionary and technological advances of humanity exploring how and to what extent exposure to natural radiation has changed. Our focus is on the source of radiation, starting with exposure to natural background radiation, leading to exposure from sources of radiation that we use or generate and in certain situations have released into the natural environment and ending with the radiation environment that is encountered away from the Earth.

5.2 Natural radiation

Our hominin ancestors evolved in Africa and began using stone tools about 3 million years ago. Fire was used on a daily basis about 300 000 years ago and the first

doi:10.1088/978-0-7503-2424-3ch5

remains of *Homo sapiens* appear about 200 000 years ago. Other than shaping tools from stone, the actions of humankind at that time had no impact on the distribution of radiation in the environment. The sources of natural radiation in the environment then, and now, comprise primordial radionuclides, cosmic rays and cosmogenic radionuclides.

5.2.1 Primordial radionuclides

Some naturally occurring radionuclides have very long half-lives and are found throughout the Earth's crust. Radioisotopes of naturally occurring radioactive elements (so-called primordial radionuclides) include ^{40}K and uranium (^{235}U and ^{238}U) and thorium (^{232}Th) along with their shorter-lived decay products. Primordial radionuclides were produced during stellar nucleosynthesis in processes summarised by Fowler, in his Nobel prize acceptance lecture describing the development of nuclear astrophysics (Fowler 1984). A complete list of primordial radionuclides is provided by Health Protection Agency, now called the UK Health Security Agency (Watson *et al* 2005).

For human exposure, ^{40}K and the ^{238}U, ^{235}U and ^{232}Th decay series are the most significant radionuclides. In a closed system, radionuclides in a decay series tend to approach an equilibrium at which the rate of supply of a radionuclide from decay of its parent is matched by the rate of its decay to subsequent radionuclides in the decay chain.

The ^{235}U decay chain comprises five radionuclides with half-lives longer than one day, a succession of shorter-lived radionuclides and ends with the stable isotope ^{207}Pb. The chain, neglecting minor branching, is represented in linear form as:

$$^{235}\text{U} \rightarrow {}^{231}\text{Th} \rightarrow {}^{231}\text{Pa} \rightarrow {}^{227}\text{Ac} \rightarrow {}^{227}\text{Th} \rightarrow {}^{223}\text{Ra} \rightarrow$$
$$^{219}\text{Rn} \rightarrow {}^{215}\text{Po} \rightarrow {}^{211}\text{Pb} \rightarrow {}^{211}\text{Bi} \rightarrow {}^{207}\text{Tl} \rightarrow {}^{207}\text{Pb}.$$

The ^{238}U decay chain comprises nine radionuclides with half-lives longer than one day, a succession of shorter-lived radionuclides and ends with the stable isotope ^{206}Pb. The chain, neglecting minor branching is represented in linear form as:

$$^{238}\text{U} \rightarrow {}^{234}\text{Th} \rightarrow {}^{234\text{m}}\text{Pa} \rightarrow {}^{234}\text{U} \rightarrow {}^{230}\text{Th} \rightarrow {}^{226}\text{Ra} \rightarrow {}^{222}\text{Rn} \rightarrow$$
$$^{218}\text{Po} \rightarrow {}^{214}\text{Pb} \rightarrow {}^{214}\text{Bi} \rightarrow {}^{214}\text{Po} \rightarrow {}^{210}\text{Pb} \rightarrow {}^{210}\text{Bi} \rightarrow {}^{210}\text{Po} \rightarrow {}^{206}\text{Pb}.$$

The ^{232}Th decay chain comprises three radionuclides with half-lives longer than one day, a succession of shorter-lived radionuclides and ends with the stable isotope ^{208}Pb. The chain is represented in linear form as:

$$^{232}\text{Th} \rightarrow {}^{228}\text{Ra} \rightarrow {}^{278}\text{Ac} \rightarrow {}^{228}\text{Th} \rightarrow {}^{224}\text{Ra} \rightarrow {}^{220}\text{Rn} \rightarrow {}^{216}\text{Po} \rightarrow$$
$$^{212}\text{Pb} \rightarrow {}^{212}\text{Bi} \rightarrow (0.64{}^{212}\text{Po} \,\&\, 0.36{}^{208}\text{Tl}) \rightarrow {}^{208}\text{Pb}.$$

The ^{238}U and ^{232}Th decay series include ^{222}Rn (radon) and ^{220}Rn (thoron), respectively. Exposure to the noble gases and their short-lived progeny may be important in determining doses from inhalation in some circumstances and are discussed separately below. The relative abundance of the various radionuclides in a decay series can also be variable due to the depletion of one radionuclide relative to another by natural processes. An example of this is the loss of radon gas due to diffusion and release to atmosphere. Fractionation can also occur due to the disruptive nature of the α-decay process leading to recoil of the residual nucleus on α-emission.

Other primordial radionuclides are found in the environment, but these exist at very low concentrations and make little contribution to exposure (Committee on Evaluation of EPA Guidelines 1999). The primordial beta-emitting radionuclide ^{87}Rb with a half-life of 4.75 10^{10} y does contribute and the average annual exposure is estimated to be about 6 μSv due to ingestion (UNSCEAR 1988).

5.2.1.1 External exposure

There is considerable variation in external dose rates from terrestrial gamma radiation and this is due to the mineral characteristics, origin, and development of any given landform. Global measurements of absorbed dose rates in air from terrestrial gamma radiation are compiled and considered by the United Nations Scientific Committee on the Effects of Atomic Radiation (UNSCEAR) (UNSCEAR 2000, 2008) and more recently detailed maps have been generated showing the distribution of natural radiation exposure within Europe (Cinelli *et al* 2019). Information on exposures comes from direct measurement of outdoor external dose rate or is evaluated using measurements of radionuclide concentrations in soil. Additional information on population distribution was used to calculate an average outdoor dose rate of 50–59 nGy h^{-1} (0.44–0.52 mGy y^{-1}). The assessment of external dose rates in Europe showed a population-weighted annual effective dose ranging between 0.31 and 0.85 mSv (Cinelli *et al* 2019).

There are locations associated with high outdoor external dose rates from primordial radionuclides; this is due largely to Monazite sands, uranium-rich mineral deposits or volcanic features on the Earth's surface. For example:

- the black sands on coastal beaches near Guarapari (Brazil) contain high levels of ^{232}Th and ^{226}Ra producing an absorbed dose rate of 329 mGy y^{-1} at Areia Preta (Vasconcelos *et al* 2011);
- the volcanic intrusions at Poços de Caldas (Brazil), within a caldera forming a raised roughly circular plateau about 30 km across, include notable mineral deposits (Bossew *et al* 2015), including the open cast Osamu Utsumi uranium mine (closed in 1995) and at Morro de Ferro there is a thorium-rich rare earths deposit at the surface producing an absorbed dose rate in excess of 100 mGy y^{-1};
- the water from springs near Ramsar (Iran) contains ^{226}Ra (dissolved from an underlying uranium-rich geology) giving a mean dose rate of 10.2 mGy y^{-1} with a maximum of 260 mGy y^{-1} (Karam and Mortazavi 2001);
- the coastal belt of Karunagappally (India) has elevated external dose rates due to thorium-containing Monazite sand with an average of 4.5 mGy y^{-1} and as high as 70 mGy y^{-1} at certain coastal locations (Nair *et al* 2009);
- the European atlas of natural radiation shows a distribution of absorbed dose rates associated with surface uranium and thorium mineral geology (Cinelli *et al* 2019) with peak values located in northwest Portugal, the caldera to the north of Rome and on volcanic islands in the Aegean Sea (up to 2 mGy y^{-1}).

Radionuclides present in the construction material used for dwellings also contribute to absorbed dose rates indoors. Overall, UNSCEAR estimates a world-wide average effective dose rate of 0.48 mSv y^{-1} (0.41 mSv y^{-1} indoors and 0.07 mSv y^{-1} outdoors). In regions of the world where wood-frame construction predominates (e.g. United States, Iceland) the indoor and outdoor rates are similar. When aggregates and rocks are used in construction, external exposure becomes dominated by the indoor environment (UNSCEAR suggest a value of 1.4, ranging from 0.6 to 2.3, for the ratio of indoor to outdoor exposure rates). External exposures for individual countries are generally in the range 0.3 to 1.0 mSv y^{-1} (UNSCEAR 2008).

5.2.1.2 Internal exposure

Internal exposures from the intake of primordial radionuclides occur following inhalation or ingestion of radionuclides. Doses from ingestion are mainly due to ^{40}K and to the ^{238}U and ^{232}Th decay series present in foods and drinking water. Doses from inhalation result from dust particles present in air but also from the short-lived decay products of radon (discussed separately below).

5.2.1.3 Ingestion

The ingestion of radionuclides depends on the consumption rates of food and water and on the concentrations of radionuclides present when consumed. The mobility and transfer of a radionuclide within an ecosystem is dependent on many factors such as chemical form and speciation, the characteristics of the ecosystem producing the food, and on physical, chemical and biological processes associated with a particular source-pathway-receptor relationship. The activity concentration of a radionuclide in a specific food is therefore highly variable.

Global estimates of ingestion dose (UNSCEAR 2000) using reference food consumption and drinking water profiles and average radionuclide concentrations gave an effective dose ranging from 0.11 mSv y^{-1} for an adult to 0.26 mSv y^{-1} for an infant from uranium- and thorium-series radionuclides. For the ^{238}U series, infant dose is greatest from ^{210}Po (0.18 mSv y^{-1}) and ^{210}Pb (0.04 mSv y^{-1}) with a smaller contribution from ^{226}Ra (0.0075 mSv y^{-1}). For the ^{232}Th series, ^{228}Ra is the main contributor to infant dose (0.031 mSv y^{-1}). UNSCEAR did not differentiate the contribution from ^{235}U series radionuclides and presented a total infant dose of 1.1 10^{-5} mSv y^{-1}. Annual intakes estimated in different surveys typically vary by up to a factor of three above or below the reference values. Reported concentrations in food (Watson *et al* 2005) suggests that ingestion doses would be lower for a largely plant-based diet.

Uranium and thorium can occur in Nature in the tetravalent oxidation state and their ions have similar radii (Mitchell *et al* 2013). Consequently, the two elements can substitute extensively for one another, explaining their geochemical coherence. Most environmental transport of thorium is due to physical processes where it is associated with particulate matter (Hinck *et al* 2010). Thorium-containing minerals are generally resistant to weathering and only under very acidic conditions does thorium show any mobility. Thorium is also readily adsorbed by Fe-hydroxides, clay particles and organic matter making the solubilisation of thorium unlikely. Transport in surface waters can occur as ThO_2 in suspended particles or in a colloidal state. This means that thorium intakes are associated with soil and sediment externally adhering to plants or otherwise ingested inadvertently. Rates of inadvertent ingestion of soil have been comprehensively evaluated (Simon 1998). The environmental behaviour of uranium is more complex (IAEA 2023) being sensitive to oxidation state. Under oxidising conditions, uranium forms the uranyl ion (UO_2^{2+}) in which the uranium is present as U(VI). The uranyl ion tends to form compounds that are soluble in water. This means that uranium is often mobile in oxidising conditions and separates from thorium, which exists only in the tetravalent state and whose compounds are generally insoluble in water (Faure and Mensing 2005).

The diffusion of radon gas to atmosphere and subsequent decay can lead to the deposition of ^{210}Pb and ^{210}Po (both directly and through ingrowth) to the Earth's surface. Polonium-210 is one of the most radiotoxic natural radionuclides and readily accumulates within food chains (IAEA 2017). Accumulation of ^{210}Po in seafoods such as shellfish can be an important pathway (Pinero–García *et al* 2023) reaching a dose rate of 0.48 mSv y^{-1} in some population groups with high seafood consumption rates. A study in Japan estimated a dose rate of 0.64 mSv y^{-1} for adult consumption of seafood with ^{210}Po being the main contributor (Ota *et al* 2009). A study of the consumption of marine organisms in Greenland indicated dose rates to high consumption groups could reach 6.3 mSv y^{-1} in children (Hansen *et al* 2022). Elevated exposures can also occur to indigenous Sami and Inuit populations who eat reindeer and caribou that feed on lichens (Salminen–Paatero and Paatero 2021). Lichen is very efficient at collecting deposited radionuclides on its external surfaces and is the main fodder for reindeer during the winter season. A study of caribou consumption estimated dose rates from ^{210}Po at 1.2 mSv y^{-1} from consumption of meat, liver and kidney (Thomas and Gates 1999). The doses presented here relate to intakes of radionuclides based on observed concentrations in diets. Effective dose rates can also be estimated from measurements of radionuclides in human tissues based on autopsy studies (UNSCEAR 2000) that show the main contributor to dose is ^{210}Po, which is mainly distributed in soft tissues.

Potassium-40 (^{40}K), with a half-life of 1.25×10^9 years, occurs everywhere on the Earth. It represents a very small fraction (0.012%) of naturally occurring potassium which exists in large amounts throughout the environment. The amount of total potassium is enriched in acid igneous and metamorphic rocks and can reach 6% in some potassium-rich minerals. In contrast, the content of sedimentary rocks is very variable being dependent on detrital mineralogy, primary chemically precipitated sedimentary minerals and chemical modifications during sediment diagenesis. Similarly, the potassium content of soil is highly variable being related to the bedrock from which the soil derives (Cinelli *et al* 2019).

Potassium is an essential nutrient for most plants and animals. It is a key element for the functioning of all living cells. In humans, potassium is involved in nerve cell transmission and regulating body functions such as digestion, heart rate and the water content of cells. The potassium content of the body is therefore regulated and held constant by homeostatic control mechanisms, such that concentrations in tissues are independent of the rate of intake of the element. However, there are some differences between men and women (Stadt *et al* 2022), and variations as a function of age (UNSCEAR 2008). There are also differences due to variations in the proportion of muscle to total body mass, because potassium is more highly concentrated in muscle than in adipose tissue. The annual effective dose due to ^{40}K in the body is about 0.165 mSv for adults and 0.185 mSv for children (UNSCEAR 2000).

5.2.1.4 Inhalation

Internal exposures from the inhalation of primordial radionuclides make a relatively minor contribution to total exposure. The primordial radionuclides present in air are associated with resuspended soil particles, with any combustion products of fuels (coal or plant material) and, the decay products of radon are present because of radon gas in air. Excluding the contribution from radon (discussed below) the dose rate is dominated by ^{210}Pb (0.0042 mSv y^{-1}) and ^{210}Po (0.0013 mSv y^{-1}), with a small contribution from ^{228}Th (0.0003 mSv y^{-1}) in children (UNSCEAR 2008). These typical dose rates are substantially lower than the contribution by ingestion.

Inhalation of smoke will increase doses due to the inhalation of primordial radionuclides. Studies have estimated the doses from inhalation of ^{210}Pb and ^{210}Po in tobacco smoke (Christopher *et al* 2019). External deposition of ^{210}Pb and ^{210}Po (and subsequent ingrowth) on tobacco leaves from the atmosphere leads to a combined activity concentration of about 30 Bq kg^{-1} in tobacco products resulting in dose rates of about 0.2 mSv y^{-1} of which about 79% is due to ^{210}Po.

5.2.1.5 Exposure to radon

UNSCEAR has published documents on the sources and effects of exposure to radon and thoron (and their progeny) covering exposure of workers and the public in comprehensive reports published in 2000, 2006 and 2019. The International Commission on Radiological Protection (ICRP) has also published regular updates (ICRP 1993, 2010, 2014) and has concluded that the population averaged effective dose from radon and its progeny represent about half the total dose received by members of the public from all natural sources of radiation. Thoron has a very short half-life and does not therefore contribute significantly to dose except in areas dominated by thorium-bearing rocks and sediments.

Radon is present in all buildings and underground locations, but levels can vary considerably depending on factors such as local geology and type of building. People are exposed to radon largely in buildings and the large variability of indoor radon concentrations results in a very heterogeneous distribution of exposures (ICRP 2014). Outdoor radon exposure is generally not a major issue. Radon released from soils, sediments and building materials leads to accumulation of the gas in indoor air. Outdoors, atmospheric transport tends to dilute and disperse the gas. This dispersal is particularly effective in small-island locations, but less effective in continental interiors where geological contexts with high radon emission rates can be spatially very extensive, meaning that downwind losses of radon by dispersion are balanced by gains from upwind.

Radon is inert and nearly all inhaled gas is exhaled. The main contribution to exposure comes from any aerosols containing two of the short-lived radon progeny ^{218}Po and ^{214}Po that emit alpha particles. Radon and its progeny are considered an important risk factor associated with lung cancer, second only to tobacco smoking.

Radon concentrations in air, both indoors and outside, have been mapped throughout Europe and the complexity of the factors that impact estimated doses are described in detail (Cinelli *et al* 2019). Typical outdoor air concentrations of radon range from 1 to over 100 Bq m^{-3} (UNSCEAR 2000). Indoor air concentrations are also highly variable and peak concentrations exceeding 10 000 Bq m^{-3} have been measured in some parts of Europe (Cinelli *et al* 2019).

The global population-weighted average dose rate from radon estimated by UNSCEAR (1.15 mSv y^{-1}) is lower than that estimated for Europe (1.96 mSv y^{-1}). The peak weighted value for a European country is 4.94 mSv y^{-1} in Finland.

For certain occupations, radon and its progeny also dominate occupational radiation exposure. For example, in the nuclear fuel cycle, radon inhaled following its release in uranium mines makes a substantial contribution to doses to the workers.

5.2.2 Cosmic rays

The discovery of cosmic radiation is attributed to Victor Hess who used balloon flights to show changes in the level of radiation with increasing altitude (Hess 1912). During his seventh flight, that took place in August 1912, he measured radiation up to an altitude of 4.7 km using three ionisation chambers. Hess observed an increase in radiation above that measured at ground level by a factor of 2–2.7. Supported by his earlier observation that measured radiation levels were unaffected by a total eclipse of the Sun, Hess concluded that radiation was penetrating the atmosphere from outer space.

Cosmic rays are particles consisting of ionised nuclei with high energies that interact with our environment arriving at Earth from space (Gaisser *et al* 2016). These particles hit the Earth's atmosphere at a rate of about 1000 m^{-2} s^{-1} from all directions at speeds approaching the speed of light.

Cosmic rays include protons, helium nuclei, electrons and the nuclei of heavier atoms (Zyla *et al* 2020). These particles are moving extremely fast, close to the speed of light (3×10^5 km s^{-1}) and possess relatively large amounts of energy. Particles that have not interacted with matter are termed 'primary cosmic rays'. Secondary cosmic rays comprise particles that are produced by the interaction of primary rays with matter and largely consist of muons (mu mesons), electrons and pions (pi mesons). Also produced are various radionuclides termed cosmogenic radionuclides. The sources of cosmic radiation important to Earth's surface are galactic cosmic radiation arriving from outside the Solar System and to a lesser extent solar cosmic radiation (UNSCEAR 2008).

The Sun's corona produces a particle flux that travels away from the Sun and is referred to as the solar wind. It is not constant and as the Sun rotates, the solar wind varies in speed (0.3 to 0.8 km s^{-1}), composition and magnetic field strength. Strong fluctuations in solar cosmic radiation occur as a result of solar flares and coronal mass ejections that vary in frequency with the Sun's 11-year solar cycle (ICRP 2016). The solar particles that reach Earth are mainly protons (~99% of the flux) with energies mainly below 1×10^8 eV and rarely above 1×10^{10} eV (UNSCEAR 2008).

Galactic cosmic radiation consists of protons (85.5%), alpha particles (12%), heavier nuclei (1%) and electrons (1.5%). These particles are largely produced within our own Galaxy, and they range in energy from about 1 10^8 eV to above 1 10^{20} eV. The higher-energy particles (those above about 1 10^{15} eV) are thought to originate in other galaxies. In contrast to solar radiation, galactic cosmic radiation is largely isotropic as it arrives at Earth. The solar wind deflects galactic cosmic radiation from the Earth resulting in a reduction in cosmic radiation corresponding to heightened solar activity.

The Earth's atmosphere and magnetic field also provide shielding against cosmic radiation. The volume of space surrounding Earth, dominated by the Earth's magnetic field, is called the magnetosphere. Solar wind compresses the magnetosphere on the sunward side (to about 6×10^4 km from the centre of the Earth) and there is an elongated tail on the dark side of Earth (to about 4×10^5 km). The boundary between the solar wind and the Earth's magnetic field is called the magnetopause and this zone constantly moves due to buffeting by the ever-changing solar wind. Components of cosmic radiation are diverted by these magnetic fields.

As a consequence, the intensity of cosmic radiation that can reach our atmosphere is at a minimum at the equator and at a maximum at the geomagnetic poles.

Cosmic rays with energy above the magnetic cut-off energy at a given latitude pass through the magnetosphere and reach the upper atmosphere where they interact with atoms and molecules in the air. There follows a cascade of interactions that produce large showers of newly created secondary particles (including pions, muons, muon neutrinos, high-energy photons, electrons and positrons). The flux of particles at ground level is mainly muons and electrons/positrons in the ratio of about 0.75–0.25. With increasing altitude neutrons, electrons/positrons, photons and protons are the most significant contributors. Cosmic-ray intensity increases with altitude to a maximum at about 12 km (towards the top of the troposphere) and then falls to the limit of the atmosphere. Most of the primary cosmic ray particles have therefore interacted with the increasing density of the atmosphere at 12 km altitude and absorption of the secondaries occurs below 12 km. When a solar flare or coronal mass ejection occurs, particles can produce significant dose rates at higher altitudes.

Exposures at ground level will therefore vary with latitude and altitude. A worldwide average annual effective dose from the photon and ionising radiation component of cosmic rays is about 0.27 mSv (about 10% lower at the equator relative to higher latitude), and the neutron component is about 0.1 mSv (but varying by a factor of 5 from the equator to higher latitudes, and by a factor of 32 from sea level to an altitude of 4000 m). The ionising component (0.22 mSv) is also sensitive to altitude, as illustrated by measurements in China showing a factor of 4 variation between sea level and an altitude of 4000 m. Overall the range of the average annual effective dose to the world population is 0.3–2 mSv.

A map showing the distribution of cosmic ray exposures in Europe (Cinelli *et al* 2019) clearly illustrates the impact of altitude, with highest doses occurring in the mountainous regions of central Europe where dose rates can exceed 1.1 mSv y^{-1} with a maximum of 4 mSv y^{-1}.

5.2.3 Cosmogenic radionuclides

When cosmic radiation interacts with nuclei in the atmosphere the cosmogenic radionuclides produced include ^{3}H, ^{7}Be, ^{10}Be, ^{14}C, ^{22}Na, ^{26}Al, ^{32}Si, ^{32}P, ^{33}P, ^{35}S, ^{36}Cl, ^{37}Ar, ^{39}Ar and ^{81}Kr. Annex B of the UNSCEAR 2000 Report (UNSCEAR 2000) details production rates and the global inventory.

The radionuclides of interest in biological systems are ^{3}H, ^{7}Be, ^{10}Be, ^{14}C and ^{22}Na. The global average effective dose contribution from ^{14}C is 12 μSv y^{-1}, from ^{7}Be is 0.03 μSv y^{-1}, from ^{22}Na is 0.15 μSv y^{-1} and from ^{3}H is 0.01 μSv y^{-1}. The others contribute very little to radiation exposure, but they are of interest as tracers in the atmosphere and hydrological systems (e.g., ^{36}Cl is used for determining the ages of groundwaters in subsurface aquifers).

The largest dose contribution is from ^{14}C formed from the interaction of slow cosmic neutrons with ^{14}N producing ^{14}C at a rate of about 1.4 PBq y^{-1}. A global inventory of about 12 750 PBq was reported in UNSCEAR (2000) and this inventory circulates with other carbon atoms exchanged within the biosphere, pedosphere, geosphere, hydrosphere and atmosphere of the Earth. A recent review details the behaviour of ^{14}C in the biosphere (Thorne *et al* 2023). ^{14}C is rapidly distributed throughout the atmosphere and the ratio of ^{14}C/C in atmospheric CO_2 is therefore independent of location and the same ratio is reflected in most living organisms. The two stable isotopes of carbon are ^{12}C and ^{13}C and these have natural abundances of about 98.9% and 1.1%, respectively, in our biosphere. The ratio of ^{14}C/C is of the order of 10^{-12} representing an annual relatively constant production rate to give a global atmospheric inventory of about 140 PBq.

The isotopic composition of carbon is impacted by physical, chemical and biological processes and the ratios are not therefore constant. The impact of humankind on production of ^{14}C is considered later in this chapter. Isotopic fractionation occurs in biological processes leading to the use of carbon isotopic composition to follow the impact of ^{12}C releases from the burning of fossil fuels. Note that fossil fuels are depleted in ^{14}C due to radioactive decay and accumulate ^{13}C with a slightly lower efficiency than ^{12}C due to preferential assimilation of ^{12}C during photosynthesis.

5.3 Human modifications of the natural radiation environment

The growth of knowledge, technology and population has led to the *(currently unsustainable)* exploitation of natural resources which in turn has resulted in the redistribution of natural radionuclides in the biosphere. With the onset of controlled fission, a large variety of artificially created (anthropogenic) radionuclides have been introduced into our environment. There are therefore circumstances that will add to natural radiation exposure, but increases are often limited to a relatively small geographical area (local as opposed to regional or global) or limited to groups of individuals (e.g. radiation workers, patients).

The broad groupings used by PHE (Oatway *et al.* 2016) to summarise average UK exposures (2.7 mSv y^{-1}) are:

- Natural radiation in the environment (2.25 mSv y^{-1}):
    - o Radon and thoron (1.3 mSv y^{-1})
    - o Intake of natural radionuclides, excluding radon (0.27 mSv y^{-1})
    - o Terrestrial gamma radiation (0.35 mSv y^{-1})
    - o Cosmic radiation (0.33 mSv y^{-1});
- Weapons fallout (0.005 mSv y^{-1});
- Other anthropogenic radioactivity released into the environment (0.0008 mSv y^{-1});
- Patient exposure from medical procedures (0.44 mSv y^{-1});
- Occupational exposure from the use of radioactive material (0.0004 mSv y^{-1}).

Similar groupings but identifying the contribution of a specific accident (Chornobyl) and medical procedures (diagnostic and therapeutic), were reported by UNEP (2016) (an average of 3 mSv y^{-1}). The estimates for average exposure in the USA (NCRP 2009) are substantially higher (6.2 mSv y^{-1}) with different types of medical exposure reported in greater detail (computed tomography, nuclear medicine, interventional fluoroscopy, conventional radiography/ fluoroscopy). The higher exposure estimates in the USA are from to radon and thoron doses (2.28 mSv y^{-1}) and patient exposure from medical procedures (3 mSv y^{-1}).

Later chapters consider radiation exposure from the nuclear industry (chapter 14), radiology (chapter 15), nuclear medicine (chapter 16), industries utilising naturally occurring radioactive materials (chapter 17), major accidents that have released radionuclides into the environment (chapter 19) and localised incidents (chapter 20). The rest of this chapter considers how the natural radiation exposure landscape has been modified by nuclear weapons testing before looking at how human activities can change the amount and type of radiation to which an individual may be exposed, specifically as a result of air travel and space exploration.

5.3.1 Nuclear weapons testing

Modification of the natural radiation environment has occurred on a global scale due to the atmospheric testing of nuclear and thermonuclear weapons. Annex C of UNSCEAR (2000) shows the number of tests carried out above and below ground to 2000 CE and the total yield (Mt) associated with these weapons (see figure 5.1).

Over this period, nuclear devices were detonated near the ground surface (120), underground (1877), at different elevations in the atmosphere (374), under water (8) and in space (11). When uranium nuclei undergo fission radioactive fragments are formed and these are released to the surroundings in a nuclear explosion. The fate and potential radiological impact of the radioactive fragments (Whicker and Schulz 1982) depends on the type and composition of the nuclear device (affecting the radionuclides produced) and the location and size of the explosion (affecting the

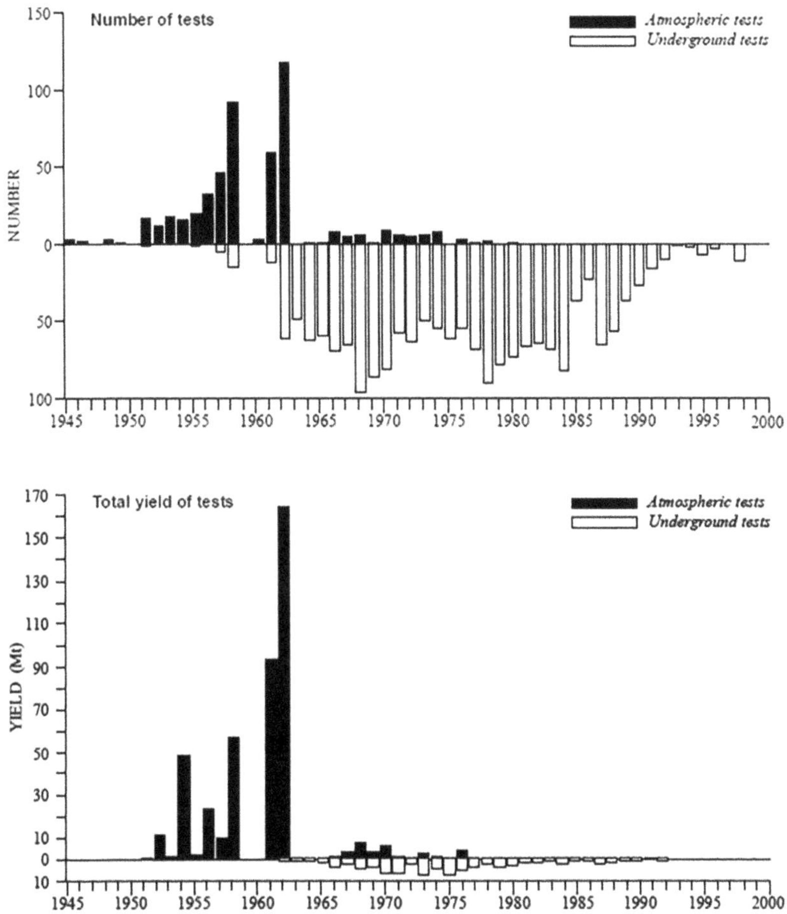

Figure 5.1. History of testing of nuclear and thermonuclear weapons. Reprinted from UNSCEAR (2000), with permission of UNSCEAR.

quantity of radioactivity released). Nuclear devices using only fission produce large amounts of fission products, whereas thermonuclear devices use the fission process to drive a fusion reaction. The large number of neutrons produced by both types of device interact with elements in soil, air, water and other materials surrounding the device to produce radionuclides by neutron activation.

An aboveground nuclear explosion results in a rapidly expanding fireball comprising an extremely hot and initially, visibly bright spherical mass of air and volatised material. The fireball quickly rises in the atmosphere and when the explosion occurs close to the Earth's surface the updraft will draw surface material into the developing mushroom cloud. As the temperature drops, vaporised materials condense, and the resulting particles are dissipated in the atmosphere and eventually fall back to Earth (radioactive fallout). Most of the fallout from lower yielding explosions (kilo tonnes to <1 Mega tonne, Mt) will be deposited locally with

Table 5.1. Population-weighted cumulative deposition densities for fallout radionuclides over the period 1945–99 (from table 11 of Annex C, UNSCEAR, 2000 report).

	Cumulative deposition density (Bq m^{-2})		
Radionuclide	Northern hemisphere	Southern hemisphere	World
^{131}I	510	273	482
^{140}Ba	1520	808	1440
^{141}Ce	3080	1380	2900
^{103}Ru	4660	2100	4380
^{89}Sr	3440	1470	3220
^{91}Y	5560	2490	5220
^{95}Zr	7590	7130	7130
^{144}Ce	50 000	8120	45 400
^{54}Mn	6560	714	5920
^{106}Ru	33 300	5470	30 300
^{125}Sb	8160	1380	7420
^{55}Fe	14 600	1630	13 200
^{90}Sr	52 900	12 600	48 440
^{137}Cs	81 000	19 200	74 100

dispersion occurring largely in the troposphere, i.e. little is dispersed globally. With increasing explosive yields (in particular, above 1 Mt yield) greater dispersion occurs in the stratosphere and this leads to the global distribution of fallout. Atmospheric tests have been conducted mainly in the northern hemisphere leading to greater fallout there relative to the southern hemisphere, as illustrated for 14 radionuclides in table 5.1. There is also enhanced deposition in the temperate regions relative to the polar and equatorial regions, by about a factor of two, due to the air circulation patterns in the troposphere and the dynamics of stratosphere-troposphere exchange.

The detonation of the first experimental nuclear weapon, named Gadget (25 kt; Selby *et al* 2021), at the White Sands Missile Range, Alamogordo, New Mexico, on 16th July 1945, resulted in the dispersion of artificial radionuclides in the atmosphere and transportation of radioactive material over substantial distances. The potential for fallout to be transported large distances was first identified in strawboard produced at a mill on the Wabash River at Vincennes, Indiana, that was found to result in fogged spots on unused x-ray film. Webb (1949) found that strawboard packaging was contaminated with ^{141}Ce thought to have been present in river water used in the manufacturing process. The strawboard was manufactured on 6 August 1945, at a location about 1800 km from the Project Trinity test site where Gadget was detonated. Coincidentally, the strawboard was manufactured on the same day 'Little Boy' was exploded over Hiroshima, Japan.

The mushroom cloud from the Trinity test rose to about 11 000 m and spread to the northeast over the hours that followed (Hoffman 1947). Within the crater formed by the blast, external dose rates 12 h after the explosion exceeded 900 Gy h^{-1}, decreasing rapidly over time ($t^{-1.2}$) to about 60 mGy h^{-1} at the epicentre after

4 weeks (King 1947). Dose rates also decreased with distance from the crater centre and after 4 weeks the dose rate at 300 m had declined to about 1 mGy h^{-1}. Current background radiation in New Mexico is about 0.08 μGy h^{-1} (United States Environmental Protection Agency 2024) and dose rates at the site are now about 10 times greater than local background (US Military Garrisons 2023). Peak measurements in the crater were associated with a green glass-like material, subsequently named trinitite (Eby *et al.* 2015). Peak dose rates radiating from the crater were also associated with trinitite formed as molten sand dropped from the ascending fireball. Measurements show that fallout deposition started to occur at a distance of a few kilometres from ground zero (Quinn 1987). The greatest fallout levels outside the White Sands Missile Range were at Chupadera Mesa, 48 km from the test site, attributed to a localised rain-shower. At 4 weeks the dose rate in localised areas around Chupadera Mesa was greater than 130 mGy h^{-1}, spread over an area 20 km by 1.6 km wide. Maps showing the fallout track show the dose rates (after 4 weeks) at a distance of 150 km from ground zero, on this arc dose rates are x10 above background covering a zone about 340 km wide, in the central area, about 26 km wide, dose rates were greater than 9 mGy h^{-1}. There have subsequently been many studies of the fallout from both the Trinity and the Nevada test sites, with recent re-evaluations suggesting radioactivity spread over much of the US (Bouville *et al.* 2020; Selby *et al.* 2021; Philippe *et al.* 2023).

Very few atmospheric tests have occurred since the Partial Test Ban Treaty of 1963 (signed by Governments of the United States, the United Kingdom of Great Britain and Northern Ireland, and the Union of Soviet Socialist Republics), which prohibits nuclear weapons tests (or any nuclear explosions) in the atmosphere, in outer space, under water, or in any other environment where radioactive debris could spread beyond the territory of the State conducting the explosion. A total of 502 nuclear weapons tests were conducted in the atmosphere between 1945 and 1973 (UNSCEAR 2008). These tests were mainly by the United States (195) and the USSR (219) before 1963, the UK undertook 21 tests over the same period. France (45) and China (22) have conducted atmospheric tests since 1963 although none have occurred since 1980.

The Comprehensive Nuclear-Test-Ban Treaty (CTBT) was established in 1996 and has 187 signatories of which 176 states have ratified the treaty (as of 2023). In November 2023, Russia revoked its ratification of the CTBT. The countries that have signed but not ratified the treaty now include China, Egypt, Israel, Iran, Russia and the United States, countries that have not signed the treaty include Democratic People's Republic of Korea (DPRK, also known as North Korea), India and Pakistan. This group of nine countries need to ratify the Treaty in order for it to come into force in international law. The CTBT has established a regime for detecting nuclear explosions through an International Monitoring System (IMS) and once ratified allows on-site inspections by a Commission established by the Treaty. Since 1996, underground tests have been carried out by India and Pakistan and most recently by the DPRK in 2017.

A summary of local and regional exposures following atmospheric tests is reported by UNSCEAR. Exposure estimates are provided for those sites where

Table 5.2. Doses from local fallout around three test sites.

Test site	Exposed population	Doses
Nevada	180 000 in 300 local communities	Effective doses from external exposure: >3 mSv in 20% of the population; highest values 60–90 mSv; population-weighted average 2.8 mSv
	3545 locally exposed individuals	Absorbed doses to the thyroid 0–4.6 Gy, average 0.098 Gy
Bikini and Enewetak Atolls, Marshall Islands	Populations of Rongalap, Ailinginae and Utirik Atolls from the Castle Bravo test	Effective doses from external exposure 0.1–1.9 Sv; Equivalent doses to the thyroid from external and internal exposure were 12, 22 and 52 Sv on average to adults, 9-year-old and 1-year old children, respectively
	Servicemen on Rongelap from the Castle Bravo test	Effective dose from external exposure: 0.8 Sv
	Japanese fishermen on the Lucky Dragon from the Castle Bravo test	Fallout deposition on deck: 1.7–6 Sv. Absorbed doses to the thyroid: 0.8–4.5 Gy
Semipalatinsk, Kazakstan	Seven villages closest to the test site	0.2–900 mGy external, whole-body dose; 0.3–3.8 Gy for thyroid dose, but possibly up to 8 Gy for children from one settlement

Note: For most other test sites, local doses would have been lower than those listed here.

atmospheric testing was performed and includes both mainland continental and oceanic sites. The reported doses are very variable and range from a few microSievert to a few Sievert, illustrated in table 5.2 by a summary of radiation exposures at three test sites.

The global distribution of nuclear weapons fallout has increased radiation exposure on Earth. UNSCEAR have published estimates of the world-wide doses based on a documented dose assessment methodology using measured or calculated deposition densities. A recent series of six papers (published in 2022) has attempted to provide a basis for the conceptual models, methodology and data needed to undertake a dose assessment for exposure to radioactive fallout from nuclear detonations (Simon *et al.* 2022; Beck *et al.* 2022; Bouville *et al.* 2022; Anspaugh *et al.* 2022; Melo *et al.* 2022; Thiessen *et al.* 2022).

Exposure peaked in 1963 with an estimated world average dose of 113 $\mu Sv\ y^{-1}$ (with approximately equal contributions from inhalation, ingestion and external exposure), declining to less than 7 $\mu Sv\ y^{-1}$ in the 1990s at which time 53% was due to external irradiation and 47% to ingestion. The contributions from different radionuclides are illustrated in the timeline for world-wide average effective dose rates from external irradiation, ingestion and inhalation produced for the UNSCEAR 2000 report. In total, average effective dose rates from fallout peaked in 1963 in the Northern Hemisphere at 125 μSv y^{-1}. In the Southern Hemisphere, the peak was in 1962 at 59 μSv y^{-1}. The world-wide average (a peak of 113 μSv y^{-1} in 1963) is dominated by the Northern Hemisphere contribution and is also weighted heavily by the much larger population of the Northern Hemisphere. By 1999, the Northern Hemisphere, Southern Hemisphere and world-wide averages were 5.87, 2.68 and 5.51 μSv y^{-1}, respectively.

- For external irradiation, average effective dose rates peaked in 1962 and 1963 at around 38 μSv y^{-1}, with Zr-95 contributing 50% of this total. By 1999, the world-wide average was 2.90 μSv y^{-1}, entirely from Cs-137.
- For ingestion, dose rates from radionuclides other than H-3 and C-14 peaked at 35 μSv y^{-1} in 1962, mainly from I-131 (20.4 μSv y^{-1}) and Cs-137 (10.3 μSv y^{-1}) with a contribution from Sr-90 (3.1 μSv y^{-1}). By 1999, the effective dose rate was 0.90 μSv y^{-1}, almost entirely from Cs-137 (0.35 μSv y^{-1}) and Sr-90 (0.56 μSv y^{-1}). The effective dose rate from H-3 and C-14 peaked at 12.7 μSv y^{-1} in 1962, with 7.2 μSv y^{-1} and 5.5 μSv y^{-1} from H-3 and C-14, respectively. By 1999, the total effective dose rate had fallen to 1.7 μSv y^{-1}, almost all from ingestion of C-14.
- For inhalation, dose rates peaked at 36 μSv y^{-1} in 1963, with the major contributions being from Ce-144 (15.3 μSv y^{-1}), Ru-106 (9.3 μSv y^{-1}), Sr-90 (1.9 μSv y^{-1}) and radioisotopes of plutonium and americium (7.7 μSv y^{-1}). By 1985, inhalation doses made only a negligible contribution to exposure from fallout.

5.3.2 Radiation exposure away from Earth's surface

Aviation and space exploration are two activities that lead to an increase in radiation exposure from natural radiation. This occurs when astronauts and aircraft passengers or crew leave the surface of the Earth and the natural radiation environment changes as atmospheric and geomagnetic sheltering effects decline.

5.3.2.1 Aircraft flight

When cosmic radiation interacts with the Earth's atmosphere and an aircraft's structure, it produces secondary cosmic rays that can penetrate the walls of aircraft and expose passengers and crew to radiation. The variability of the dose due to cosmic radiation at ground level was discussed earlier in this chapter. This section considers how the natural radiation environment for the crew and passengers is modified by commercial flights. Consideration is given by UNSCEAR 2022 to the potential doses, and ICRP Publication 132 provides guidance on radiological protection, from cosmic radiation in aviation. The dose from cosmic radiation in aviation is routinely estimated using computer codes that have been validated by monitoring results (see ICRP Publication 132 (ICRP 2016)).

Radiation exposure during a flight is largely dependent on five factors:

1) flight duration;
2) altitude—dose rate increases with the altitude above sea level;

3) geomagnetic effects:
    a. dose rates are reduced by geomagnetic shielding that occurs in the region ± 30° around the equator;
    b. dose rates rise between 60° latitude and the geomagnetic poles where more charged particles follow the geomagnetic field lines into the atmosphere;
    c. dose rates increase in the South Atlantic Anomaly, a region extending from about −50° to 0° latitude and from −90° to 40° longitude, due to the asymmetry between the magnetic and rotation axes of the Earth;
4) solar magnetic activity cycle—the flux of galactic particles arriving at the Earth is disrupted during high solar activity, causing a reduction in dose rate;
5) solar particle events—these can occur throughout the solar magnetic activity cycle but are most frequent in solar maximum years, they result from fast coronal mass ejections.

Close to the equator the geomagnetic field is nearly parallel to the ground and the magnetic shielding effect is greatest resulting in low dose rates. Approaching the magnetic poles, the geomagnetic field becomes almost vertical to the ground and a higher proportion of primary cosmic radiation particles reach the atmosphere and generate secondary radiation that penetrates to aviation altitudes. Rates of exposure to cosmic radiation are therefore higher in polar regions, and lower near the equator. The impact of geomagnetic shielding on effective dose rate is shown in figure 5.2 and clearly illustrates the effects described above.

The dose during a flight therefore depends on the altitude profile for the route, the latitude, the date and the duration of the flight. Dose rates are therefore very variable and are typically in the range of 2–10 μSv/h at typical commercial flight

Figure 5.2. Effects of geomagnetic shielding on effective dose rate at an altitude of 11 km in December 2002. ICRP (2016). Copyright © 2016 by Sage Publications. Reprinted by Permission of Sage Publications.

altitudes. At these dose rates several hundred hours flying time are needed for this dose to approach that from other sources of natural radiation.

UNSCEAR (2022) estimates an annual average effective dose of 2.8 mSv for aircrew when averaged over both long and short haul flights between 1990 and 2014 and also reports estimates of peak annual doses for Finnish aircrew of 9.5 mSv. ICRP Publication 132 reports a study of effective dose for aircraft crew in European countries showing the average annual effective dose varied from 1 mSv for the airline of the Czech Republic to 2.5 mSv for airlines from Finland and Sweden (ICRP 2016). The peak annual effective dose is approximately 6-7 mSv for airlines operating in Denmark, Germany, and Finland.

Radiation exposure at high altitudes during solar storm periods and solar particle events is a particular concern. Estimates indicate that dose rates from these events may reach 1.7 mSv h^{-1}.

The ICRP concludes that most passengers in aircraft are occasional flyers, and their exposure to cosmic radiation is considered to be negligible in the context of their total radiation exposure. For the most frequent flyers and those approaching the flying hours of crew, the ICRP encourages a graded approach of self-assessment and self-adjustment moving to case-by-case agreements where exposure is employment related. For aircraft crew, recommendations to the operating management include:

1) informing the aircraft crew through training programmes;
2) disseminate information to raise awareness about cosmic radiation and support informed decisions;
3) assess the dose of aircraft crew (using a validated model);
4) keep a record of the individual and cumulative dose of aircraft crew; and,
5) adjust flight rosters in relation to a selected dose reference level.

The ICRP also recommends frequently flying pregnant women consider adjusting their flight frequency to reduce the exposure of their embryo/foetus and airline management should have provisions in place to adjust the duties of pregnant aircraft crew.

5.3.2.2 Space travel

The naturally occurring radiation environment in space is very different from that at the Earth's surface. The main sources that have greater intensity in space are:

1) galactic cosmic radiation;
2) solar particle events; and,
3) radiation from particles trapped in the Earth's magnetic field.

Variations in the space environment occur throughout the Solar System primarily driven by solar storm phenomena such as coronal mass ejections, solar flares, solar particle events and solar wind. These phenomena occur on the Sun's surface and solar storms directed towards Earth or its orbit are potential drivers of space weather events on Earth (National Oceanic and Atmospheric Administration 2024). The most visual impacts are the Aurora (Borealis and Australis) formed when electrons collide with the Earth's atmosphere. These electrons follow the magnetic field lines of Earth towards the polar regions and when entering the tail of the magnetosphere the accelerated electrons collide with oxygen and nitrogen atoms and molecules in the Earth's upper atmosphere. The impacted atoms and molecules release energy acquired from these collisions in the form of light. The aurora typically forms at 80–500 km above Earth's surface.

The radiation that astronauts are exposed to whilst they are in space includes electrons, protons, alpha particles and heavier ions. Some of these may have very high energies. Secondary radiations may occur, created by interactions between radiation and materials of the spacecraft. The ICRP considers that exposure assessments for astronauts in space are a special situation relating to a small number of individuals and the guidelines developed by ICRP are specific to a space environment (ICRP 2013).

The US Apollo programme placed 24 people in space beyond low Earth orbit. Other crewed spaceflights have remained within low Earth orbit, typically at an altitude between 160 km and 1000 km. By late 2023, a total of 676 people had travelled into space (50 km or more above the Earth's surface) (Hobbs 2023). Dose records reported by UNSCEAR (2022) for 336 NASA astronauts (based on thermoluminescent dosimeters), estimates cumulative mean total badge dose as 16.8 mGy with a standard deviation of 23.6 mGy. Records for astronauts spending long periods in space on the Mir space station and the International Space Station (ISS), indicate dose rates of 0.4 to 0.8 mSv d^{-1}, leading to effective doses for crew in the range of 100–300 mSv for missions lasting a year. NASA's Spaceflight Human-System Standard (Rev C, 2023) document limits career effective radiation dose due to spaceflight radiation exposure to less than 600 mSv (for all ages and sexes), this is equivalent to a 3% risk of exposure induced death (NASA 2023).

Galactic cosmic radiation originates from outside of the Solar System comprising 98.5% nuclei and 1.5% electrons (Dachev *et al.* 2017). Most of the nuclei are protons (~87%), alpha particles (~12%) and heavier nuclei representing ~1%–2% referred to as HZE particles (i.e. *H*igh nuclear charge *Z* and energy *E*). Measurements from instruments on the ISS from 2014 to 2016 indicate an average daily dose rate of 71.6 μGy d^{-1} from galactic cosmic radiation. The inverse relationship of galactic cosmic radiation with the solar cycle, provides dose rates in the range 50–100 mGy y^{-1} at the solar maximum and 150–300 mGy y^{-1} at the solar minimum (Chancellor *et al.* 2018). Dosimetry analyses for the ISS astronauts have shown that 80% or more of organ dose equivalents incurred on the ISS are from galactic cosmic rays with a small contribution from trapped protons (Cucinotta *et al.* 2008).

The more energetic heavy particles are a cause for concern during long periods in interplanetary space. The risks associated with exposure to high linear energy transfer particles in particular are poorly understood and as these particles travel through spacecraft shielding, they may also fragment and produce secondary radiation (Krittanawong *et al.* 2023).

Solar particle event frequency is related to sunspot activity and these unpredictable events can release large quantities of energetic protons (up to 1 GeV per nucleon) with fluences in excess of 10^9 protons cm^{-2}. Whilst event frequency is related to the solar cycle, the intensity of an event is not. The dose rates associated with solar particle events is therefore unpredictable, but rates as high as 1400–2837 mGy h^{-1} are possible (Chancellor *et al.* 2018). Measurements during four solar particle events experienced by the Mars science laboratory spacecraft in 2012 (Zeitlin *et al.* 2013), recorded doses in the range of 1.2–19.5 mSv per event, representing 5% of the total dose recorded during the mission (500 mSv).

The Earth's magnetic field traps and concentrates particles in two, sometimes three, nested donut shaped rings that are most intense over the Equator. These zones, discovered in 1958, are called the Van Allen belts. The flux of charged particles in these zones is continuous but shows two regions of maximum particle density (see https://spacecenter.org/what-are-the-van-allen-radiation-belts).

The two Van Allen Probes launched by NASA in August 2012 studied the region for about seven years, gathering information that has greatly improved understanding of particle dynamics within the belts. A detailed review of electron dynamics and behaviour in the belts is provided by Li and Hudson (2019) and Baker (2021), who consider the impacts of the magnetosphere, interplanetary shocks, wave–particle interactions and the plasmasphere.

The inner region stretches from about 600 to 10 000 km above the Earth's surface, at lowest altitude over the South Atlantic Anomaly and not extending outside an angle of 60° either side of the celestial equator. The inner belt is dominated by highly energetic protons (energies >10 MeV) and electrons (mostly <1 MeV). The source of the protons is the decay of neutrons produced when high-energy cosmic rays from outside the Solar System collide with atoms and molecules of Earth's atmosphere and, solar energetic protons that are associated with flares and coronal mass ejections (these are the primary source of <50 MeV protons). The inner region electrons are periodically refreshed from the outer region by so-called slot-penetrating events. The decay of neutrons also contributes to energetic electrons found in the inner region. The maximum particle density is centred approximately 4000–6000 km above the Earth's surface.

The outer region is less stable, occurring from about 13 000 to 60 000 km. The maximum particle density is centred at approximately 20 000 km and consists mainly of trapped electrons originating from solar activity. The particle flux in the outer belt is dynamic and highly variable with rapid changes in response to particles released from the Sun.

The potential doses to astronauts from the radiation belts are dominated by the high-energy protons of the inner belt. The South Atlantic Anomaly (off the Brazilian coast) is a region where the inner belt is closest to the Earth's surface, at an altitude of about 200 km (ICRP 2013). This is due to the 11° inclination of the Earth's geomagnetic dipole axis from its axis of rotation and a 500 km displacement of the dipole centre reducing field strength values. The region accounts for a significant fraction of total exposure in low Earth orbit even though the exposure period is relatively short compared with the time in orbit.

Most of the Apollo missions took a short trajectory through the Van Allen Belts and astronaut doses ranged from 1.6 to 5.8 mGy. Apollo 14 spent a longer period in the Van Allen Belts (NASA 1971) contributing to a greater mission dose of

11.2 mGy (Johnston *et al.* 1975). The last lunar landings (Apollo 16 and 17) took place in April and December 1972 with two crew members each spending about 3 days on the surface of the Moon. On the 4 August 1972 there was a major series of solar storms that led to extreme solar flare, solar particle event and geomagnetic storm components (Knipp *et al.* 2018). Assuming shielding only from a space suit, it is estimated that the skin dose rates during these events could have exceeded 1.4 Gy h^{-1} (Parsons and Townsend 2000) and a dose of 6000 mSv is estimated for crew on the lunar surface (NASA 2008). The radiation exposure of astronauts in space during planning for missions outside low Earth orbit remains a major concern (NASA 2025). A manned mission to Mars will not be feasible until improved shielding is developed or transit time is reduced (McPhee and Charles 2009).

References

Anspaugh L R, Bouville A, Thiessen K M, Hoffman F O, Beck H L, Gordeev K I and Simon S L 2022 A methodology for calculation of internal dose following exposure to radioactive fallout from the detonation of a nuclear fission device *Health Phys.* **122** 84–124

Baker D N 2021 Wave-particle interaction effects in the Van Allen belts *EPS* **73** 189

Beck H L, Bouville A, Simon S L, Anspaugh L R, Thiessen K M, Shinkarev S and Gordeev K 2022 A method for estimating the deposition density of fallout on the ground and on vegetation from a low-yield, low-altitude nuclear detonation *Health Phys.* **122** 21–53

Bossew P, Da Silva N C and Oliveira R J 2015 Natural radiation in the geological anomaly of Pocos de Caldas, Minas Gerais, Brazil *Radiation Emergency Med.* **4** 7–12

Bouville A, Beck H L, Thiessen K M, Hoffman F O, Potischman N and Simon S L 2020 The methodology used to assess doses from the first nuclearweapons test (trinity) to the populations of New Mexico *Health Phys.* **119** 400–27

Bouville A, Beck H L, Anspaugh L R, Gordeev K, Shinkarev S, k, Thiessen M, Hoffman F O and Simon S L 2022 A methodology for estimating external doses to individuals and populations exposed to radioactive fallout from nuclear detonations *Health Phys.* **122** 54–83

Chancellor J C, Blue R S, Cengel K A, Auñó S M, Rubins K H, Katzgraber H G and Kennedy A R 2018 Limitations in predicting the space radiation health risk for exploration astronauts *npj Microgravity* **4** 8

Christopher S, Periyasamy M, Athif P, Syed Mohamed H E, Sadiq Bukhari A and Shahul Hameed P 2019 Activity concentration of polonium-210 and lead-210 in tobacco products and annual committed effective dose to tobacco users in Tiruchirappalli District (Tamil Nadu, India) *J. Radioanal. Nucl. Chem.* **323** 1425–9

Cinelli G, De Cort M and Tollefsen T 2019 *European Atlas of Natural Radiation* (Luxembourg: Publication Office of the European Union)

Committee on Evaluation of EPA Guidelines 1999 *Evaluation of Guidelines for Exposures to Technologically Enhanced Naturally Occurring Radioactive Materials* (Washington, DC: National Academy Press)

Cucinotta F A, Kim M-H Y, Willingham V and George K A 2008 Physical and biological organ dosimetry analysis for International Space Station astronauts *Radiat. Res.* **170** 127–38

Dachev T P, Bankov N G, Tomov B T, Matviichuk Y N, Dimitrov P G, Häder D -P and Horneck G 2017 Overview of the ISS radiation environment observed during the ESA EXPOSE-R2 mission in 2014–2016 *Space Weather* **15** 1475–89

Eby G N, Charnley N, Pirrie D, Hermes R, Smoliga J and Rollinson G 2015 Trinitite redux: mineralogy and petrology *Am. Mineral.* **100** 427–41

Faure G and Mensing T M 2005 *Isotopes: Principles and Applications* 3rd edn (New Jersey: Wiley)

Fowler W A 1984 Experimental and theoretical nuclear astrophysics; the quest for the origin of the elements (Nobel lecture) *Rev. Mod. Phys.* **56** 49

Gaisser T K, Engel R and Resconi E 2016 *Cosmic Rays and Particle Physics* (Cambridge: Cambridge University Press)

Hansen V, Mosbech A, Rigét F F, Søgaard-Hansen J, Bjerregaard P, Dietz R, Sonne C *et al* 2022 Background ^{210}Po activity concentrations in Greenland marine biota and dose assessment *Sci. Total Environ.* **806** 150508

Hess V F 1912 Über Beobachtungen der durchdringenden Strahlung bei sieben Freiballonfahrten *Physik. Z.* **13** 1084–91

Hinck J E, Linder G, Finger S, Little E, Tillitt D and Kuhne W 2010 Biological pathways of exposure and ecotoxicity values for uranium and associated radionuclides *Hydrological, Geological, and Biological Site Characterization of Breccia Pipe Uranium Deposits in Northern Arizona* ed E A Alpine (Reston, VA: US Geological Survey) pp 287–384

Hobbs Z 2023 Astronomy: How Many People Have Gone to Space? 8 November https://astronomy.com/space-exploration/how-many-people-have-gone-to-space/ (Accessed 28 August 2024)

Hoffman J G 1947 *Nuclear Explosion 16 July 1945: Health Physics Report on Radioactive Contamination Throughout New Mexico Following the Nuclear Explosion* (Los Alamos, NM: Los Alamos Scientific Laboratory)

IAEA 2017 *The Environmental Behaviour of Polonium* (Vienna: International Atomic Energy Agency, Technical Reports Series No 484)

IAEA 2023 *The Environmental Behavior of Uranium* (Vienna: International Atomic Energy Agency Technical Reports Series No. 488) .

ICRP 1993 *Protection Against Rn-222 at Home and at Work* ICRP Publication 65 (Oxford: International Commission on Radiological Protection)

ICRP 2010 *Lung Cancer Risk from Radon and Progeny and Statement on Radon* ICRP Publication 115 (Oxford: International Commission on Radiological Protection)

ICRP 2013 *Assessment of Radiation Exposure of Astronauts in Space* ICRP Publication 123 (Oxford: The International Commission on Radiological Protection)

ICRP 2014 *Radiological Protection Against Radon Exposure* ICRP Publication 126 (London: International Commission on Radiological Protection)

ICRP 2016 *Radiological Protection from Cosmic Radiation in Aviation* ICRP Publication 132 (London: The International Commission on Radiological Protection)

Johnston R S, Dietlein L F and Berry C A 1975 *Biomedical Results of APOLLO* NASA SP-368 (Washington, DC: NASA)

Karam P A and Mortazavi S M J 2001 The very high background radiation area in Ramsar, Iran: public health risk or signal for a regulatory paradigm shift ? *Valdor 2001. Values in Decisions on Risk. 2. Valdor Symposium Addressing Transparency in Risk Assessment and Decision Making* NEI-SE-436 (Stockholm: Swedish Nuclear Power Inspectorate, Swedish Radiation Protection Inst., UK Environment Agency, UK Nirex Ltd., Commission of the European Communities) pp 495–502

King L 1947 *Ground Activity After the Trinity Shot* LAMS-658 (Los Alamos, NM: Los Alamos Scientific Laboratory)

Knipp D J, Fraser B J, Shea M A and Smart D F 2018 On the little-known consequences of the 4 August 1972 ultra-fast coronal mass ejecta: facts, commentary, and call to action *Space Weather* **16** 1635–43

Krittanawong C, Singh N K, Scheuring R A, Urquieta E, Macaulay T R, Kaplin S *et al* 2023 Human health during space travel: state of the art review *Cells* **12** 40

Li W and Hudson M K 2019 Earth's Van Allen Radiation Belts: From Discovery to the Van Allen Probes *Era. J. Geophys. Res.: Space Physics* **124** 8319–51

McPhee J C and Charles J B 2009 *Human Health and Performance Risks of Space Exploration Missions* (Houston, TX: NASA)

Melo D R, Bertelli L, Ibrahim S A, Anspaugh L R, Bouville A and Simon S L 2022 Dose coefficients for internal dose assessments for exposure to radioactive fallout *Health Phys.* **122** 125–235

Mitchell N G, Pérez-Sánchez D and Thorne M C 2013 A review of the behaviour of U-238 series radionuclides in soils and plants *J. Radiol. Prot.* **33** R17

Nair R R K, Rajan B, Akiba S, Jayalekshmi P, Nair M K, Gangadharan P, Koga T, Morishima H, Nakamura S and Sugahara T 2009 Background radiation and cancer incidence in Kerala, India—Karanagappally cohort study *Health Phys.* **96** 55–66

NASA 1971 *Apollo 14 Mission Report: Biomedical Evaluation (on-line pdf)* https://nasa.gov/history/alsj/a14/a14mr10.htm (Accessed 3 June 2025)

NASA 2008 *Working on the Moon: Lessons from Apollo – Radiation Environment* 8 February https://nasa.gov/history/alsj/WOTM/WOTM-Radiation.html Accessed 3 June 2025.

NASA 2023 *NASA Spaceflight Human-system Standard—Volume 1 Crew Health* Technical Standard NASA-STD-3001 (Washington, DC: National Aeronautics and Space Administration.)

NASA. 2025 *Human Research Roadmap: Risk of Radiation Carcinogenesis.* 13 May, https://humanresearchroadmap.nasa.gov/Risks/risk.aspx?i=96 (Accessed 31 May 2025).

National Oceanic and Atmospheric Administration. 2024 *Space Weather Prediction Center* https://swpc.noaa.gov/phenomena (Accessed 28 August 2025)

NCRP 2009 *Ionizing Radiation Exposure of the Population of the United States* NCRP 160 (Bethesda, MD: National Council on Radiation Protection and Measurement)

Oatway W B, Jones A L, Holmes S, Watson S and Cabianca T 2016 *Ionising Radiation Exposure of the UK Population: 2010 Review* PHE-CRCE-026 (Oxfordshire: Public Health England)

Ota T, Sanada T, Kashiwara Y, Morimoto T and Sato K 2009 Evaluation for committed effective dose due to dietary foods by the intake for Japanese adults *Jpn. J. Health Phys.* **44** 80–8

Parsons J L and Townsend L W 2000 Interplanetary crew dose rates for the August 1972 solar particle event *Radiat. Res.* **153** 729–33

Philippe S, Alzner S, Compo G P, Grimshaw M and Smith M 2023 Fallout from U.S. atmospheric nuclear tests in New Mexico and Nevada (1945–1962) *Physics arXiv preprint* arXiv:2307.11040

Pinero-García F, Thomas R, Mantero J, Forssell-Aronsson E and Isaksson M 2023 Biodistribution of ^{210}Po in seafood and risk assessment for consumers in Sweden *Food Control* **151** 109789

Quinn V 1987 *Analysis of Nuclear Test Trinity Radiological and Meteorological Data* Contract DE-A108-76DP00351 (Las Vegas, NV: US Department of Energy)

Salminen-Paatero S and Paatero J 2021 Transfer of natural radionuclides in terrestrial food chains—a review of investigations in Finland *Int. J. Environ. Res. Public Health* **18** 10577

Selby H D, Hanson S K, Meininger D, Oldham W J, Kinman W S, Miller J L, Reilly S D *et al* 2021 A new yield assessment for the Trinity nuclear test, 75 years later *Nucl. Technol.* **207** S321–5 Supplement 1

Simon S L 1998 Soil ingestion by humans: a review of history, data, and etiology with application to risk assessment of radioactively contaminated soil *Health Phys.* **74** 647–72

Simon S L, Bouville A, Beck H L, Anspaugh L R, Thiessen K M, Hoffman F O and Shinkarev S 2022 Dose estimation for exposure to radioactive fallout from nuclear detonations *Health Phys.* **122** 1–20

Stadt M M, Leete J, Devinyak S and Layton A T 2022 A mathematical model of potassium homeostasis: effect of feedforward and feedback controls *PLoS Comput. Biol.* **18** 1–28

Thiessen K M, Hoffman F O, Bouville A, Anspaugh L R, Beck H L and Simon S L 2022 Parameter values for estimation of internal doses from ingestion of radioactive fallout from nuclear detonations *Health Phys.* **122** 236–68

Thomas P A and Gates T E 1999 Radionuclides in the Lichen-Caribou-human food chain near uranium mining operations in Northern Saskatchewan, Canada *Envron. Health Perspect.* **107** 527–37

Thorne M C, Ikonen A and Smith K 2023 *Transport of C-14 in Terrestrial and Freshwater Environments: Final Report* (Solna: Svensk Kärnbränslehantering AB)

UNEP 2016 *Radiation Effects and Sources* (Vienna: United Nations Environment Programme)

United States Environmental Protection Agency 2024 *RadNet Dashboard.* 27 August. https://epa.gov/radnet/radnet-near-real-time-air-data#radnet-dashboard (Accessed 28 August 2024)

UNSCEAR 1988 *Sources, Effects and Risks of Ionizing Radiation* (New York: United Nations)

UNSCEAR 2000 *Sources and Effects of Ionizing Radiation. Report to the General Assembly with Scientific Annexes. Annex B, Report to the General Assembly with Scientific Annexes* New York: United Nations Scientific Committee on the Effects of Atomic Radiation

UNSCEAR 2008 *Sources and Effects of Ionizing Radiation: 2008 Report - Annex B. Report to the General Assembly with Scientific Annexes* (New York: United Nations Scientific Committee on the Effects of Atomic Radiations)

UNSCEAR 2022 *Sources, effects and risks of ionizing radiation: Evaluation of occupational exposure to ionizing radiation. Volume IV, Annex D* (New York: United Nations)

US Military Garrisons 2023 *Trinity Site. Brochure, White Sands Missile Range* (New Mexico: Public Affairs Office)

Vasconcelos, D C, Pereira C, Oliveira A H, Santos T O, Reis P L and Rocha Z 2011 Natural radioactivity in sand beaches of Guarapari, Espirito Santo State, Brazil—a comparative study *INAC 2011* (Belo Horizonte: ABEN) pp 1–7

Watson S J, Jones A L, Oatway W B and Hughes J S 2005 *Ionising Radiation Exposure of the UK Population: 2005 Review* HPA-RPD-001 (Didcot: Health Protection Agency)

Webb J H 1949 The fogging of photographic film by radioactive contaminants in cardboard packaging materials *Phys. Rev* **76** 375–80

Whicker F W and Schultz V 1982 *Radioecology: Nuclear Energy and the Environment* Vol II (Florida: CRC Press, Boca Raton)

Zeitlin C, Hassler D M, Cucinotta F A, Ehresmann B, Wimmer-Schweingruber R F, Brinza D E, Kang S *et al* 2013 Measurements of energetic particle radiation in transit to Mars on the Mars Science Laboratory *Science* **340** 1080–3

Zyla P A, Particle Data Group *et al* 2020 2020 review of particle physics *Prog. Theor. Exp. Phys.* **2020** 083C01

IOP Publishing

Principles and Techniques of Radiological Protection

Michael Charles Thorne

Chapter 6

Principles of internal dosimetry

M C Thorne

6.1 Introduction

In occupational, medical and environmental contexts, radionuclides can be taken into the human body by various routes. The most usual are ingestion and inhalation. However, uptake across the intact skin can be significant for a few radionuclides, notably H-3 as tritiated water, and uptake from wounds can also occur. In a medical context, radionuclides may be administered for purposes of diagnosis or treatment by ingestion or inhalation, but more usually by injection. In broad terms, issues in internal dosimetry can be distinguished into two categories:

- Given an intake of a specific physicochemical form of a radionuclide, what radiation doses are delivered to the various tissues and organs of the body;
- From a specific pattern of excretion of a radionuclide in urine, faeces and exhaled breath, what can be inferred about the route, magnitude and physicochemical form of the intake of that radionuclide, or its ancestors, and/or what can be said about the radiation doses delivered to the various tissues and organs of the body.

These two issues can be addressed by the application of mathematical models representing the distribution, retention and excretion of radionuclides following their entry into the human body, complemented by models of radiation transport in body tissues that permit dose rates in target tissues and organs to be calculated for unit concentration of a radionuclide in various source tissues and organs, where the source and target tissues and organs may be the same or differ from each other.

The mathematical models used to represent distribution, retention and excretion, often termed biokinetic models, are applied both in estimating doses from a specified intake and in estimating intake from a measured pattern of excretion. It is important that such models are realistic when applied in both contexts. Also, a distinction may need to be made between models applied generically to set

doi:10.1088/978-0-7503-2424-3ch6

occupational or environmental standards (e.g. in defining reference values of committed effective dose per unit intake, or limiting concentrations in environmental media) and adaptations of those models to be applied in specific circumstances (e.g. where the exposed individual has characteristics that differ substantially from those of the reference person used in standards setting, or when the mode of intake and/or physicochemical form differ significantly from the reference assumptions). Model adaptations may relate solely to the parameter values used or may extend to changes in the overall mathematical structure of the model. In modifying the parameter values used in a model, various techniques may be used varying from ad hoc adjustments to specific parameters through to the use of computer-based methods to determine the combinations of parameter values that give optimal fits to the observations. In recent years, there has been increased interest in the use of Bayesian techniques for updating inputs to biokinetic models or their parameterisation, in the light of observations. This can be useful when a reference model exists. Bayesian updating can then be used to adjust the time series of estimated inputs to the model and/or to modify the parameter values used in the model. This provides a clear audit trail and justification for changes to the reference model.

6.2 Dosimetric quantities

An important consideration in internal dosimetry is that the radiation doses received by tissues and organs are protracted in time, even following an essentially instantaneous pulse intake. The timescale over which the dose is protracted depends both on the radioactive half-life of the radionuclide and its biokinetics. Thus, for example, although iodine is retained in the thyroid with a biological half-life of about 80 days, the effective half-life of I-131 in the thyroid is only about 7.3 days, determined primarily by its radioactive half-life of 8.04 days. Conversely, although the radioactive half-life of H-3 is 12.35 years, its effective half-life in the body after intake of tritiated water is only about 10 days because of the rapid turnover of water in the body and the limited incorporation of H-3 present in tritiated water into organic forms of hydrogen in the body. In practice, it is usual to time-integrate the dose rates to tissues and organs over some period following intake and refer to the integrated quantities as committed doses, i.e. the doses to which the individual is committed over the specified period following intake if no action is taken to modify the uptake and retention of the radionuclide. From committed absorbed doses, it is usual to calculate committed equivalent doses and committed effective doses.

As to the period of integration, following ICRP recommendations, it is usually taken as 50 years for adults and to 70 years of age for infants and children.

The committed absorbed dose to target tissue or organ T from source tissue or organ S is conveniently written:

$$D_{\mathrm{T,S}} = 1.602\ 10^{-13} A_{\mathrm{S}} S_{\mathrm{T}\leftarrow\mathrm{S}} \tag{6.1}$$

Where $D_{T,S}$ (Gy) is the committed absorbed dose over the specified time interval in target tissue or organ T from a radionuclide in source tissue or organ S;

A_S (transformations) is the time-integrated activity of the radionuclide over the specified interval in source tissue or organ S;

$S_{T \leftarrow S}$ (MeV kg^{-1} transformation^{-1}) is the energy deposited per unit mass of T for a single transformation (or decay) of the radionuclide in S;

$1.602\ 10^{-13}$ is the conversion factor from MeV to J and the principle adopted is that 1 Gy = 1 J kg^{-1}.

Where the transformation of a radionuclide results in several emissions with different probabilities per transformation, equation (6.1) can be rewritten to apply to each emission i, i.e.:

$$D_{T,S,i} = 1.602\ 10^{-13} A_S S_{T \leftarrow S,i} \tag{6.2}$$

Considering the contributions from all relevant source tissues and organs, the total committed absorbed dose to target tissue or organ T, from emission type i, $D_{T,i}$ is given by:

$$D_{T,i} = \Sigma_S 1.602\ 10^{-13} A_S S_{T \leftarrow S,i} \tag{6.3}$$

Assuming that each emission type i can be associated with a well-defined radiation weighting factor w_i, the total committed equivalent dose to tissue or organ T, H_T (Sv) is given by:

$$H_T = \Sigma_i\, w_i\, D_{T,i} = \Sigma_i\, w_i\, \Sigma_S\ 1.602\ 10^{-13} A_S S_{T \leftarrow S,i} \tag{6.4}$$

The total committed effective dose, H, is given by:

$$H = \Sigma_T\, w_T\, H_T \tag{6.5}$$

where w_T are the tissue and organ weighting factors.

Thus, mathematical modelling for internal dosimetry can be distinguished into two components:

- Biokinetic modelling to estimate $A_S(t, \tau, \varphi)$, where t is the time after a pulse intake, τ is the age at intake and φ is a vector of all other factors influencing retention, e.g. body mass and dietary composition;
- Radiation transport modelling to estimate $S_{T \leftarrow S,i}(M, \theta)$, where M is body mass and θ is a vector of all other factors influencing radiation transport (e.g. the geometry and composition of tissues and organs).

These two aspects are discussed in sections 6.3 and 6.4. Applications of biokinetic models in the context of monitoring for internal contamination and for retrospectively estimating intakes and doses are discussed separately in section 6.5. More detailed accounts of radiation transport modelling and biokinetic modelling, as applied to humans, are provided in chapters 7 and 9, respectively. Radiation transport modelling for other biota is discussed in chapter 12, and modelling the transport of radionuclides in the environment, for various biota, is discussed in chapter 10.

6.3 Biokinetic modelling

The majority of biokinetic models used in radiological protection have been developed for, and published by, the International Commission on Radiological Protection. These models are generally element specific, though a common structure is often used, and they are typically based on a mixture of human and animal data relating mainly to observations on the behaviour of radioisotopes of the element, but also taking into account information on the behaviour of stable isotopes, as appropriate. Although the models are often structured to take account of physiological processes, they are primarily empirical in nature, i.e. they are conceptualised and parameterised with a view to conforming to the observational data. This means that the parameters and parameter values used in the models do not, in general, have any simple correspondence to physiological processes and the rates at which those processes operate.

The models are almost exclusively compartmental in nature. Thus, the body is treated as a set of well-mixed 'pools' with transfer rates between the 'pools' dependent upon age in some cases. It is also assumed that transfer rates between the compartments are independent of the amounts of the radionuclide that are present in the various compartments. This assumption can often be justified because the radionuclides are only present at trace levels. Thus, any time-dependent output from the model can be calculated by convoluting the time-dependent input to the model with response functions that describes the time-varying content of the compartments in response to an instantaneous unit pulse input into each of the compartments to which input can occur, e.g. the stomach for ingestion and various regions of the respiratory tract for inhalation. These response functions do not depend on the size of the input, but they may depend on the age of the individual and any other factors influencing retention. The implications of adopting a compartmental modelling approach and of assuming that system behaviour is independent of the amount of radionuclide present are discussed in detail in chapter 9.

Special considerations arise when radionuclide decay chains must be addressed. Both parent and progeny radionuclides may be taken into the body initially. However, subsequently, there will be ingrowth of the progeny in the body competing with decay of the progeny originally taken in. The biokinetic behaviour of progeny radionuclides originally taken in may differ substantially from that of those produced by ingrowth. For example, if the parent radionuclide is efficiently incorporated into mineral bone, the progeny radionuclide produced by ingrowth may be effectively retained within the bone matrix, even though it has no specific affinity with mineralised tissues. Issues also arise when radionuclides enter the body incorporated in particulate material. In this case, their biokinetic behaviour in the organs of intake (i.e. the respiratory tract for inhalation and the gastrointestinal tract for ingestion) may be determined primarily by the physicochemical properties of the carrier particle, rather than the physicochemical properties of the radionuclide incorporated in that particle.

In general, biokinetic models can be distinguished into those appropriate to the organs of intake (the respiratory system and gastrointestinal tract) and those used to represent the behaviour of radionuclides after they have entered the blood (often referred to as their entry into the systemic circulation). Radionuclides entering the body by intravenous injection can be taken to enter directly into the blood. For entry through wounds and by intramuscular injection, it will generally be appropriate to adopt a radionuclide-specific and case-specific model for transfer from the wound or injection site to extracellular body fluids (blood and lymph).

6.3.1 The respiratory tract

The biokinetic model of the respiratory tract currently in use is closely based on that developed by the ICRP (1994). The focus is largely on the deposition, retention and clearance of inhaled aerosols, but the inhalation of radioactive gases is also addressed.

The pattern of deposition of aerosols in the respiratory system is determined by the size of the particles. Large particles, i.e. those with diameters of more than 10 μm, tend to deposit mainly in the large airways, including the trachea and bronchi, whereas submicron particles penetrate deep into the lungs and deposit mainly in the smaller bronchioles and alveolar sacs. Deposition mechanisms include impaction and gravitational settling, both of which depend upon the density of the particles as well as their size, and diffusion, which does not exhibit the same dependence on density.

In practice, aerosols tend to exhibit lognormal size distributions, with a substantial proportion of the particles sufficiently large for impaction and gravitational settling to dominate. Therefore, the pattern of deposition can generally be estimated based on the Activity Median Aerodynamic Diameter (AMAD) of the aerosol particles. The aerodynamic diameter of a particle is defined as the diameter of a particle of unit density that has the same terminal settling velocity in air as the particle of interest. The ratio of the aerodynamic diameter to the equivalent volume diameter (the diameter of a sphere with the same volume as the particle of interest) is proportional to the square root of the ratio of their densities. Thus, high density materials, e.g. UO_2, of a specific particle size, deposit more centrally in the respiratory system than do low-density materials of the same particle size. These low-density materials deposit more peripherally.

The respiratory compartments in which inhaled particles may be deposited are illustrated in figure 6.1. Compartments into which deposition of inhaled particles occurs are indicated by yellow arrows and clearance routes are indicated by red arrows. Particles deposited in the anterior nasal region are subject to removal only by extrinsic means, e.g. nose blowing. In contrast, most material deposited in the naso-oropharynx and larynx is cleared to the throat in the film of fluid that covers the surfaces of these airways. In this region and in the deeper (more peripheral) parts of the respiratory system, particles can penetrate epithelial tissues, enter the lymphatic system and be transferred to regional lymph nodes. These are well adapted to trapping such particles indefinitely, leaving solubilisation as the main process by which radionuclides in these particles can be released to the systemic circulation.

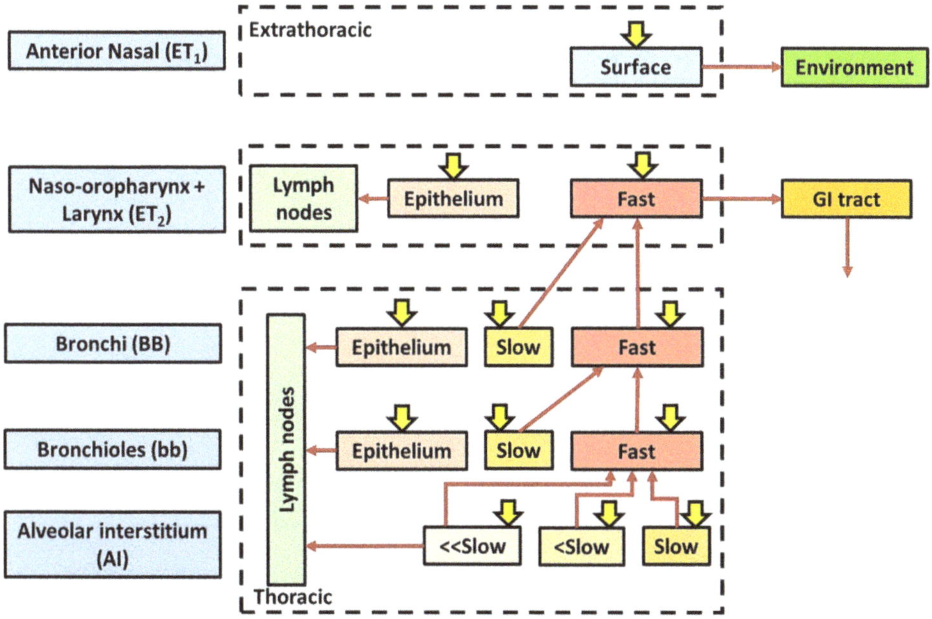

Figure 6.1. Respiratory tract compartments in which inhaled particles may be deposited (based on figure 7 of ICRP (1994)).

As with the naso-oropharynx and larynx, clearance of particles from the deep lung is in the continual stream of mucus that moves upward through the bronchioles and bronchi to the throat where it is swallowed. Particle transport by this process is much less effective within the alveolar region. Thus, inhaled, insoluble particles may remain in the alveolar region for an extended period until they dissolve or are transported to the regional lymph nodes.

Solubilisation in the model competes with the transport processes shown in figure 6.1. Each compartment in that figure, except for 'surface' in ET1, can be thought of as being operated on by solubilisation processes, as illustrated in figure 6.2.

The scheme shown in figure 6.2 allows for physicochemical transformation of the deposited particles. It also allows radionuclides released by dissolution of the particles to be transferred directly to the systemic circulation or to bind to tissue components. Thus, it gives high flexibility in representing clearance.

In terms of radioactive gases, the ICRP (1994) distinguishes three broad classes based on their solubility and reactivity. The most soluble and highly reactive gases, e.g. SO_2, are taken to be completely absorbed in the extra-thoracic tissues, whereas the least soluble and least reactive, e.g. SF_6, are treated as totally unabsorbed and dose rates are calculated simply by assuming that the air spaces of the respiratory tract are filled with the gas at its concentration in the inhaled air. Special considerations apply in respect of Rn-220 (thoron), Rn-222 (radon) and their short-lived progeny, because it is the short-lived alpha-emitting progeny that deliver

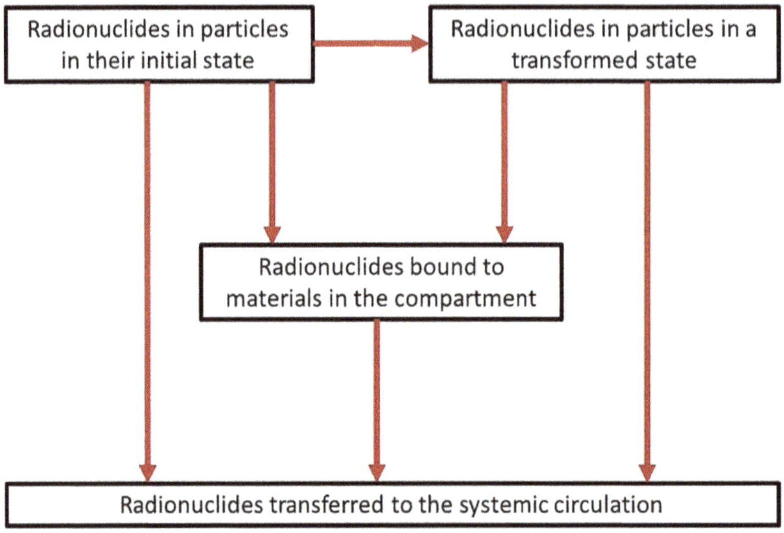

Figure 6.2. Solubilisation of radionuclides from particles deposited in the respiratory system.

most of the effective dose following inhalation of thoron and radon (see Annex B of ICRP (2010)).

The particle-clearance aspects of the ICRP respiratory tract model are independent of the physicochemical form of the inhaled aerosol, except to the extent that the particle-size distribution governs the initial pattern of deposition in the various compartments. However, the physicochemical form governs the solubilisation of the deposited material. In principle, the model parameter values could be specific to each physicochemical form of interest. However, in practice, the data available are insufficient to make such distinctions and most materials are assigned to one of three classes. These are Types F, M and S. Type F is characterised by rapid absorption of almost all material deposited in BB, bb and AI (see figure 6.1) and 50% of material deposited in ET_2. For Type M, about 70% of the deposit in AI eventually reaches the blood, whereas about 10% of the deposit in BB and bb and 5% of the deposit in ET_2 is rapidly solubilised and transported to the blood. For Type S, there is little transfer to blood from any of the compartments, except in the case of AI for which a slow transfer of about 10% of the deposited material occurs.

More details of the ICRP model and its parameterisation are provided in chapter 10.

6.3.2 The gastrointestinal tract

As with the respiratory tract model, the ICRP model of the human alimentary tract (ICRP 2006) is compartmental in nature. It is illustrated in figure 6.3.

In this model, a radionuclide is taken to be transported through the sequential components of the gastrointestinal tract in the contents of the tract. In each component, the radionuclide can be taken up into the wall of the tract and then either returned from the wall to the contents or transferred onward to the systemic circulation. In practice, for most radionuclides, many of the transfer rates between

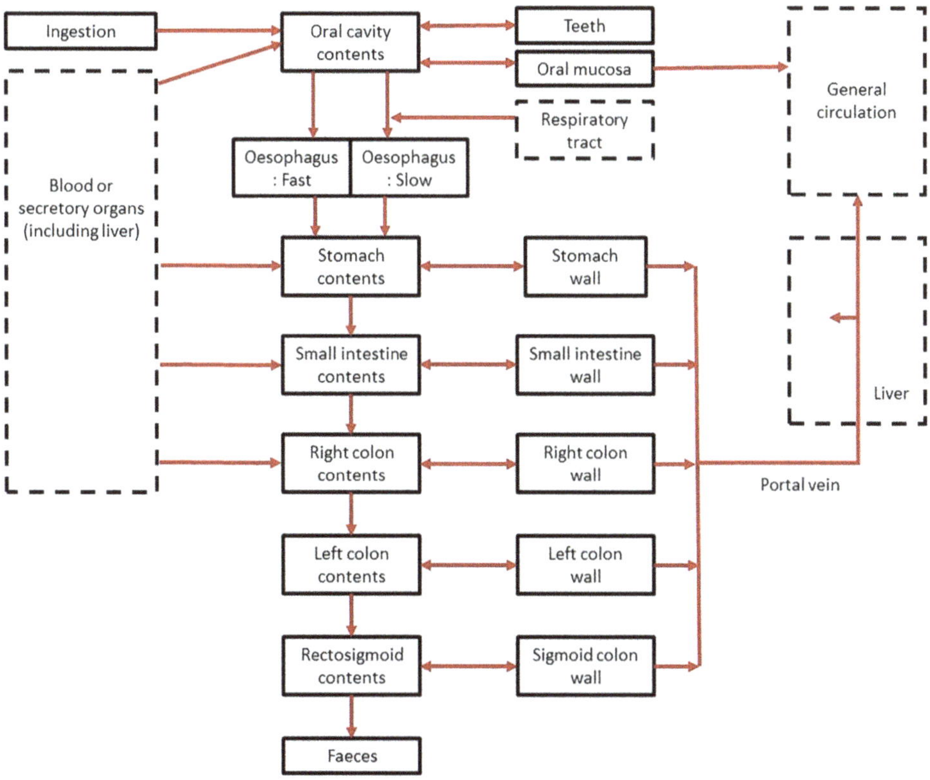

Figure 6.3. Structure of the human alimentary tract model. The dashed boxes are included to show the connections between this model and both the human respiratory tract model and systemic biokinetic models (based on figure 5.1 of ICRP (2006)).

compartments are set to zero. Thus, for example, in general uptake from the gastrointestinal tract is taken to occur only from the small intestine.

A key parameter of the model is the fractional absorption of a radionuclide from the gastrointestinal tract, f_A, which is the sum of the fractional absorption values for the individual components of the tract. Thus, if an element is absorbed from the stomach as well as the small intestine, $f_A = f_{ST} + f_{SI}$. The fractional absorption values are used to compute the rate coefficients, $k_{i,B}$, for uptake to the systemic circulation from each compartment i of the gastrointestinal tract:

$$k_{i,B} = f_i \, k_{i,i+1} / \left\{ 1 - \sum_{1}^{i} f_k \right\} \tag{6.6}$$

In the typical case in which absorption is assumed to occur only from the small intestine:

$$k_{SI,B} = f_{SI} \, k_{SI,RC} / \{ 1 - f_{SI} \} \tag{6.7}$$

In an earlier model of the gastrointestinal tract adopted by the ICRP (1979), absorption was taken to occur only from the small intestine and $f_A = f_{SI}$ was denoted f_1. This terminology is still often used.

Uptake into the walls of the gastrointestinal tract can be of specific relevance in pre-weaning infants, in whom such uptake is enhanced. This is of importance for alpha-emitting actinides because these are only poorly absorbed to the systemic circulation and are of little radiological significance if present in the contents of the gastrointestinal tract because the short range of alpha particles in tissue prevents them from irradiating the cells at risk in the wall of the tract. However, uptake into the wall in pre-weaning infants facilitates such irradiation.

More details of the ICRP model of the gastrointestinal tract are provided in chapter 9.

6.3.3 Element-specific models for radionuclides entering the systemic circulation

As with the models for organs of intake, models for radionuclide retention in the body following their entry into the systemic circulation are compartmental in nature. The detailed configuration of the model is element specific. Here, it is illustrated for three elements with contrasting behaviour in the body, strontium, caesium and chromium.

Strontium is an alkaline earth and behaves biochemically very similarly to calcium, as do the heavier alkaline earths (barium and radium). This means that strontium is a bone-seeking radionuclide. Although initially deposited mainly on bone surfaces, it relatively rapidly exchanges with calcium throughout the mineral matrix of bone. In addition, it is widely distributed throughout the soft tissues of the body, where it substitutes for calcium. Thus, the biokinetic model for strontium is as illustrated in figure 6.4.

Thus, strontium entering the blood following ingestion, inhalation, through a wound or after injection exchanges with other tissues to give a relatively uniform distribution throughout all soft tissues. However, even within soft tissues several

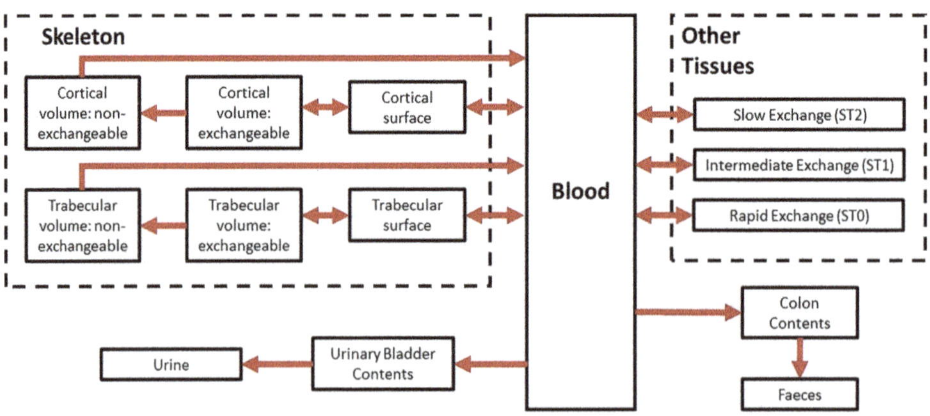

Figure 6.4. The ICRP biokinetic model for strontium (based on figure 10.1 of ICRP (2016)).

'pools' of strontium are recognised with different rates of turnover. Skeletal uptake from blood competes with soft tissue uptake both for the strontium that initially enters the blood and for strontium that is recycled from soft tissues or from the skeleton. Within the skeleton, a distinction is made between cortical and trabecular bone. Cortical bone is the massive component, e.g. the shafts of the long bones of the skeleton, whereas trabecular bone is the more delicate, reticulated bone that occupies the ends of the long bones and much of the interior of other bones, such as the vertebrae. These two types of bone have substantially different turnover rates, but in each case deposition is initially on bone surfaces, but diffusion plus the deposition of new bone mineral then redistributes the strontium throughout the bone volume, with resorption of bone mineral facilitating its recycling back to blood.

For caesium, which is a biochemical analogue of potassium, retention occurs mainly in soft tissues and specifically muscle, with uptake in the skeleton having only a minor role. However, the element is readily exchangeable across the walls of all parts of the gastrointestinal tract. This leads to the model structure shown in figure 6.5.

More recently, Hiller and Leggett (2020) have developed a biokinetic model for chromium. This distinguishes the two main valence states of chromium, Cr(III) and Cr(VI), because these are metabolised differently. Although the model is directed primarily to representing the biokinetics of radioisotopes of chromium, it is also of relevance in studies of the chemical toxicity of the element, bearing in mind that Cr (VI) is much more toxic than Cr(III) (ATSDR 2012). A schematic drawing of the proposed model is illustrated in figure 6.6.

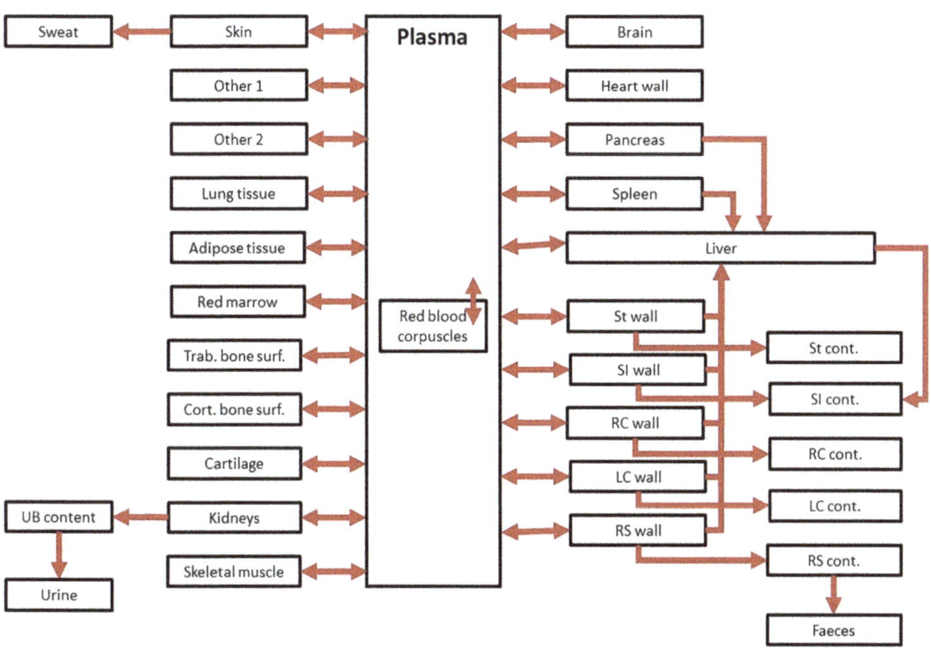

Figure 6.5. The ICRP biokinetic model for caesium (based on figure 6.1 of ICRP (2017)).

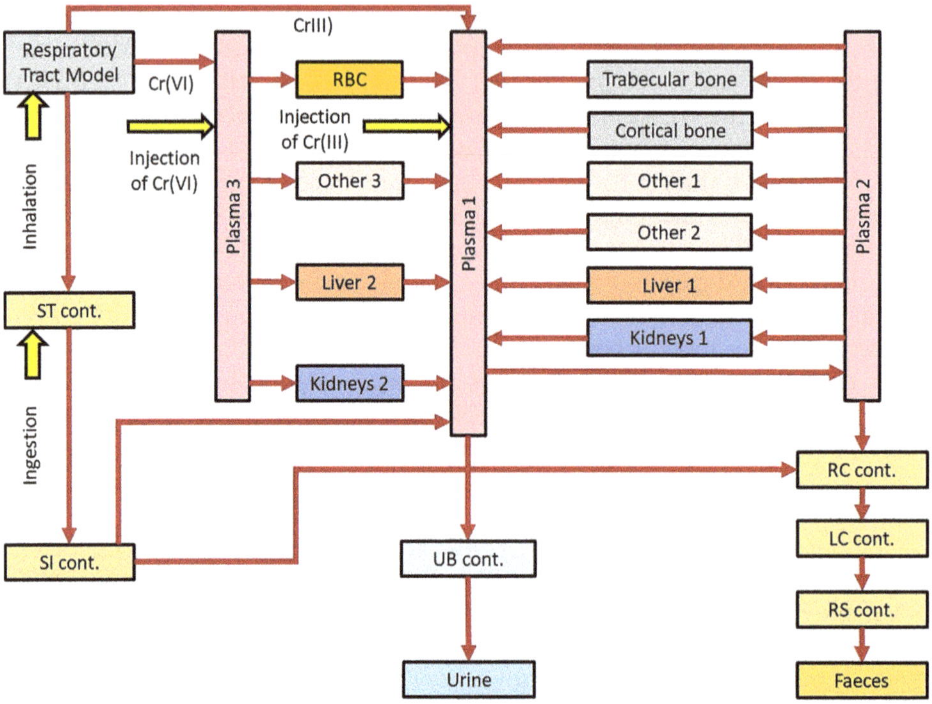

Figure 6.6. Biokinetic model for chromium (based on figure 1 of Hiller and Leggett (2020)). LC = left colon, RBC = red blood corpuscles, RC = right colon, RS = rectosigmoid, SI = small intestine, ST = stomach, UB = urinary bladder.

These three models illustrate clearly how the structures of biokinetic models are determined by a combination of physiological and biochemical considerations.

6.4 Radiation transport modelling

As discussed in section 6.1, radiation transport modelling is directed to estimating $S_{T\leftarrow S,i}$ (MeV kg^{-1} transformation^{-1}), which is the energy deposited per unit mass of target tissue or organ T for a single transformation (or decay) of the radionuclide in source tissue or organ S. Note that this is the energy associated with emission type i and, therefore, must take account of the probability of that emission in each transformation. Formally:

$$S_{T\leftarrow S,i} = \varphi_{T\leftarrow S,i} y_i E_i / M_T \tag{6.8}$$

where $\varphi_{T\leftarrow S,i}$ is the fraction of energy originating as emission i from a transformation in S that is absorbed in T;

y_i is the fractional yield of emission i per transformation;

E_i (MeV) is the average or unique value of the energy of emission i;

M_T (kg) is the mass of the target tissue or organ.

Because alpha particles have a range in tissues of only a few tens of micrometres and beta particles have ranges of, at most, about one centimetre, it is generally a good approximation for these types of emission to assume that all the energy associated with emission i is deposited in the source tissue or organ and that none is deposited in any other tissue or organ. Thus, $\varphi_{\mathrm{T}\leftarrow\mathrm{S,i}} = 1$ for T = S and $\varphi_{\mathrm{T}\leftarrow\mathrm{S,i}} = 0$ for T $\neq$ S. However, there are exceptions to this rule. These arise where the source and target tissues lie close to each other. The primary example is the skeleton. Radioisotopes of many different elements deposit in the skeleton and exhibit long-term retention on the surfaces of mineral bone or distributed throughout the bone matrix. However, the cells that are the targets for cancer induction lie in the bone marrow rather than being distributed within the matrix of mineral bone. Cells within a few tens of micrometres of bone surfaces are the target identified for induction of osteosarcomas, whereas cells distributed throughout the red bone marrow are the target identified for the induction of leukaemia. Therefore, it is necessary to calculate values of $\varphi_{\mathrm{T}\leftarrow\mathrm{S,i}}$ where the source is either the surface or the volume of mineral bone and the target is either a thin layer of tissue on bone surfaces for osteosarcoma induction or the whole of the red bone marrow for leukaemia induction. For example, if a 5 MeV alpha emitter is located on the bone surface, 50% of the emissions will travel into the underlying mineral and 50% will enter the thin layer of tissue overlying the bone surface. Those emissions that enter this thin layer will either stop within it or pass through it and deposit deeper in the bone marrow. The geometry of the situation is shown in figure 6.7, where the range of the alpha particles is taken to be 40 micrometres (μm) in soft tissue and the thickness of the sensitive layer for osteosarcoma induction is taken as 10 μm.

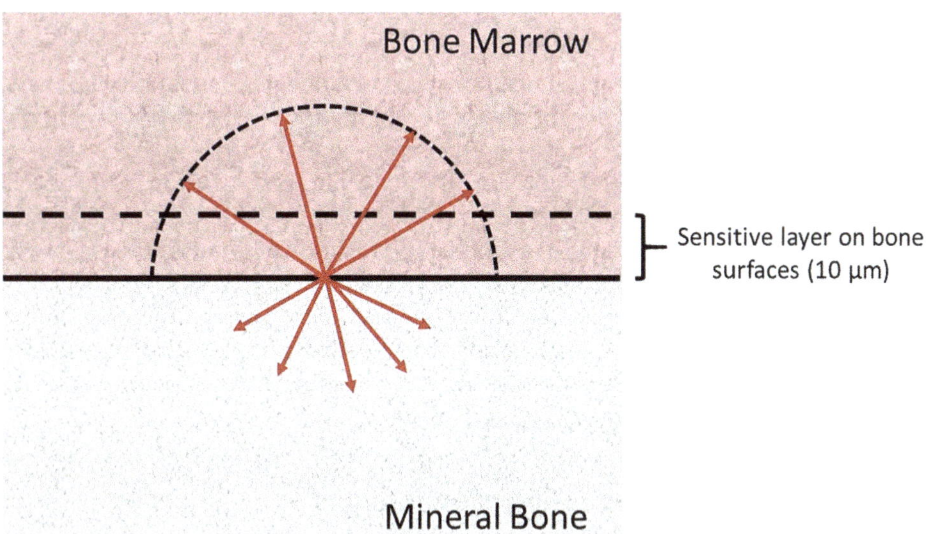

Figure 6.7. Geometry of alpha emissions from the surface of mineral bone (arrows indicate alpha emissions with a range of 40 μm in soft tissue and rather less in mineral bone).

Neglecting variations in the rate of energy deposition along an alpha track, the fraction of energy from emissions into the bone marrow that is absorbed in the 10 μm thick layer is the ratio of the volume of a that layer within 40 μm of the source to the volume of a hemisphere with a radius of 40 μm. Approximating the 10 μm thick layer as a cylinder, this ratio is $\pi R^2 h/2/3R^3 = 3\ h/2\ R$, where h is the thickness of the layer (10 μm) and R is the radius of the hemisphere (i.e. the range of alpha particles in soft tissue, here taken as 40 μm). Thus, this fraction is $3/8 = 0.375$. Therefore, taking account of the consideration that only 50% of the alpha emissions from a bone surface enter the overlying soft tissue, $\varphi_{\mathrm{BM}\leftarrow\mathrm{BS}} = 0.5$ and, $\varphi_{\mathrm{EL}\leftarrow\mathrm{BS}} = 0.375 \times 0.5 = 0.187$, where BM denotes bone marrow, BS denotes bone surface and EL denotes the endosteal layer. Because the sensitive cells for osteosarcoma induction comprise only a small fraction of the endosteal layer, the remainder is taken to comprise part of the bone marrow.

In the gastrointestinal tract, similar considerations arise. However, in this case, alpha emissions from the contents are strongly absorbed at the boundary between the contents and the wall, such that only a small fraction of their energy is transferred to the sensitive cells located in crypts at the base of the villi in the small intestine.

With respect to photon emitters, the situation is different, because photons of energies of up to about 2 MeV are emitted by various radionuclides that can be incorporated into the human body (e.g. Co-60 emits two photons at each transformation with energies of 1.17 and 1.33 MeV, respectively). Such photons are scattered on length scales of centimetres or more in human tissues. This means that much of the energy originating from a source tissue or organ S is either deposited in target tissues or organs other than S or escapes from the body. To calculate values of $\varphi_{\mathrm{T}\leftarrow\mathrm{S,i}}$ for photons, it is usual to define mathematical phantoms to represent the geometry and composition of the various tissues and organs of the body. Each of these in turn is selected as the source tissue and organ. The source tissue or organ is then treated as being randomly but uniformly populated with sites of radionuclide transformation and the transport of photons resulting from those transformations is then simulated using standard radiation transport codes to determine, on average, what fraction of the energy carried away by those photons is deposited in specific target tissues and organs. These calculations are generally made using Monte Carlo radiation transport codes that may either perform full simulations of photon transport or adopt simplified, point-Kernel methods. In the first generation of such models, tissues and organs were represented as simple geometrical shapes. However, it is now possible to use results from scanning studies to generate high-resolution, voxel-based representations of the complex internal geometry of the human body (see, e.g. Xu and Eckerman 2010).

6.5 Monitoring for internal contamination

Whereas routine monitoring for internal contamination may primarily be for reassurance purposes, i.e. to establish that exposures to radioactive materials are being suitably controlled, results from monitoring may also be used directly to estimate internal doses and intakes of radionuclides. The principal methods of monitoring comprise external counting of photons emitted by internally incorporated radionuclides, and measurements of radionuclide concentrations in urine, blood, faeces and exhaled air. Measurements of other materials, such as hair, may be undertaken in a few specific instances.

In general, monitoring for internal contamination is focused on using data collected on a limited number of occasions to reconstruct the intake and dose history of an individual to estimate both the effective dose already incurred and to estimate the committed effective dose that the individual is projected to receive in the future. Such monitoring may also be used to investigate the effectiveness of mitigation measures applied to the individual or to a population group, e.g. excision of material from a wound, use of chelation agents, or blocking of I-131 uptake into the thyroid by the administration of stable iodine.

External counting is generally undertaken in a well-shielded whole-body counter. However, other techniques may be applicable in specific contexts. For example, hand-held instruments may be appropriate for monitoring the uptake and retention of I-131 in the thyroid. Also, in radiation emergency situations, mobile detectors may be used in an unshielded geometry, e.g. in a chair arrangement, to provide initial estimates of internal contamination. Mobile whole-body counting is typically undertaken using low-resolution, NaI(Tl), detectors, but electrically cooled high-purity germanium (HPGe) detectors have the potential to be used in this context and there are on-going developments of crystals that can operate at room temperatures that are likely to have an increasing impact on monitoring practices.

Monitoring of radionuclides in biological samples has generally focused on urine, with blood samples of importance in a medical context. Where exposure to the higher actinides is a potential issue, faecal monitoring may be more useful, because these elements are only very poorly absorbed from the gastrointestinal tract and insoluble forms are only slowly and to a limited degree translocated from the respiratory tract to the systemic circulation following inhalation. Furthermore, after entering the systemic circulation they are efficiently taken up and retained in the liver and skeleton. Therefore, faecal monitoring to detect higher actinides either ingested directly or swallowed after being transported from the respiratory system to the throat may be the preferred method of biological monitoring, because only a very small fraction of the ingested or inhaled radionuclide will be excreted in the urine, with that excretion occurring on long timescales. However, in the long term, urine monitoring may be preferred to estimate the residual body burden of such radionuclides. In practice, the use of personal air samplers to estimate intakes of actinides may be preferred over faecal sampling, as it is less intrusive and gives a relatively direct estimate of intakes. The monitoring of exhaled air for Rn-222 has been used to determine the Ra-226 content of the body. Because Rn-222 is an inert gas, it is not completely retained at its site of production from Ra-226. To some degree, it diffuses to the systemic circulation and is carried in the bloodstream through the lungs, from where it is lost by diffusion into the gas phase and release by exhalation. However, a complication is that the fractional release of Rn-222 by exhalation decreases with time after exposure to Ra-226, because the Ra-226 becomes increasingly incorporated deep in the matrix of mineral bone, making the transfer of Rn-222 to the systemic circulation increasingly difficult.

Having obtained radionuclide concentrations in biological samples, an issue arises as to how those data should best be used in estimating both radiation doses incurred to date and those projected to occur in the future. In general, this requires estimation of both the time-dependent intakes of the monitored individuals and the parameter values of the biokinetic model, as the parameter values appropriate to a specific individual may differ from those adopted in the standard version of the model, as the values used in the standard version will have been selected for a generic, reference individual and considering the specific requirement of setting generally applicable standards for controlling exposure.

In the past, estimation of intakes has often been based on simple rules, e.g. the biokinetic model used was not varied and the intake was assumed to occur at the mid-point of each sampling interval. However, there is now an increasing use of Bayesian techniques in which prior estimates of intakes and of biokinetic model parameters are adjusted in the light of monitoring data. A substantial example of the application of Bayesian methods is provided by Puncher and Riddell (2016).

Bayesian techniques are based on the application of Bayes' theorem. This is based on the reciprocity relationship:

$$P(\mathrm{B}) \times P(\mathrm{A}|\mathrm{B}) = P(\mathrm{A}) \times P(\mathrm{B}|\mathrm{A}) \quad (6.9)$$

where $P(\mathrm{A})$ is the total probability of A and $P(\mathrm{B}|\mathrm{A})$ is the probability of B given that A occurs. i.e. the probability of A given that B occurs (left-hand side) is equal to the probability of B given that A occurs (right-hand side), as both correspond to the probability that A and B occur. Where there is a finite set of outcomes that can occur, this implies:

$$\Sigma_{\mathrm{A}} P(\mathrm{A}|\mathrm{B}) = 1 \quad (6.10)$$

Based on mutually exclusive probabilities, equation (6.10) gives:

$$P(\mathrm{B}) = \Sigma_{\mathrm{A}} P(\mathrm{B}) \times P(\mathrm{A}|\mathrm{B}) \quad (6.11)$$

Thus, from equation (6.9):

$$P(\mathrm{B}) = \Sigma_{\mathrm{A}} P(\mathrm{A}) \times P(\mathrm{B}|\mathrm{A}) \quad (6.12)$$

Hence:

$$P(\mathrm{A}|\mathrm{B}) = P(\mathrm{A}) \times P(\mathrm{B}|\mathrm{A})/\Sigma_{\mathrm{A}} P(\mathrm{A}) \times P(\mathrm{B}|\mathrm{A}) \quad (6.13)$$

This relates the probability that A occurs given B to the probability that A occurs and the probability that B occurs for each possible A.

The form of Bayes' theorem given in equation (6.13) relates to discrete events A and B. However, in the current context it needs to be generalised to continuous distributions of variables. Here, this is illustrated for a single uncertain parameter x and a set of observations of that parameter X_{i} i = 1, N, which is denoted by the vector $\boldsymbol{X}$.

Thus:

$$P(x|\boldsymbol{X}) = P(x) \times P(\boldsymbol{X}|x) / \int P(x) \times P(\boldsymbol{X}|x)dx \tag{6.14}$$

This is more often written:

$$f(x|\boldsymbol{X}) = f(x)L(\boldsymbol{X}|x) / \int f(x)L(\boldsymbol{X}|x)dx \tag{6.15}$$

where $f(x)$ is the prior probability distribution of x;
$L(\boldsymbol{X}|x)$ is the likelihood of observations $\boldsymbol{X}$ occurring given x, known as the likelihood function;
$f(x|\boldsymbol{X})$ is the posterior probability distribution for x given the information $\boldsymbol{X}$.

A simple, but illuminating example arises where x is the expected number of failures of an item and $\boldsymbol{X} = n$ is the observed number of failures. Based on Poisson statistics, this gives a likelihood function:

$$L(n|x) = x^n e^{-x}/n! \tag{6.16}$$

If the item is known not to have failed, $n = 0$, this reduces to:

$$L(0|x) = e^{-x} \tag{6.17}$$

If the prior probability distribution for the expected number of failures is exponentially distributed, i.e. ke^{-kx} then the posterior distribution, $f(x|0)$ is given by:

$$f(x|0) = (k+1)e^{-(k+1)} \tag{6.18}$$

The process is iterative, i.e. if equation (6.18) is taken to be the new prior probability and a second item is found not to fail, the new posterior probability distribution becomes $(k + 2)e^{-(k + 2)}$. This is a general property of Bayesian updating, allowing successive refinement of the prior distribution by using several successive sets of observations. An extensive discussion of Bayesian updating and the choice of prior distributions is given in Williams and Thorne (1997).

References

ATSDR 2012 *Toxicological Profile for Chromium, Agency for Toxic Substances and Disease Registry* (Atlanta, GA: US Department of Health and Human Services, Public Health Service)

Hiller M M and Leggett R W 2020 A biokinetic model for trivalent or hexavalent chromium in adult humans *J. Radiol. Prot.* **40** 19–39

ICRP 1979 Limits for intakes of radionuclides by workers: Part 1, ICRP Publication 30 *Ann. ICRP* **2(3/4)**

ICRP 1994 Human respiratory tract model for radiological protection, ICRP Publication 66 *Ann. ICRP* **24(1–3)**

ICRP 2006 Human alimentary tract model for radiological protection, ICRP Publication 100 *Ann. ICRP* **36(1–2)**

ICRP 2010 Lung cancer risk from radon and progeny and statement on radon, ICRP Publication 115 *Ann. ICRP* **40(1)**

ICRP 2016 Occupational intakes of radionuclides: Part 2, ICRP Publication 134 *Ann. ICRP* **45 (3/4)**

ICRP 2017 Occupational intakes of radionuclides: Part 3, ICRP Publication 137 *Ann. ICRP* **46 (3/4)**

Puncher M and Riddell A E 2016 A Bayesian analysis of plutonium exposures in Sellafield workers *J. Radiol. Prot.* **36** 1–19

Williams M M R and Thorne M C 1997 The estimation of failure rates for low probability events *Prog. Nucl. Energy* **31** 373–476

Xu X G and Eckerman K F (eds) 2010 *Handbook of Anatomical Models for Radiation Dosimetry* (Boca Raton, FL: CRC Press)

Chapter 7

Physical and mathematical phantoms for human dosimetry

M C Thorne

7.1 Introduction

The principal dosimetric quantities used in radiological protection of humans are equivalent doses to tissues and organs, with effective dose being the key quantity derived from these equivalent doses. However, equivalent dose is not a directly measurable quantity but must be inferred by calculation from quantities that can be measured, e.g., particle or photon fluences, or absorbed doses measured in a detector. In the case of intakes of radionuclides, biokinetic models are used to estimate time-integrated amounts or concentrations of radionuclides in individual source tissues and organs, with radiation transport models used to estimate committed equivalent doses in target tissues and organs arising from these time-integrated amounts or concentrations. Thus, dose estimation for both external or internal exposures requires a model of the geometry and composition of the human body, so that relationships can be established between external radiation fields or internal radionuclide distributions and equivalent doses or dose rates.

However, the need for such a model does not imply that it necessarily must represent details of human anatomy. Thus, for example, operational radiological protection quantities (discussed below) are defined at specific locations in a 0.3 m diameter sphere of tissue-equivalent material. Nevertheless, for many applications, there is a need to represent not only the overall geometry of the body, but the geometry and composition of individual tissues and organs.

In this context, the term ‘model’ encompasses two distinct and complementary approaches. A model may be a physical structure into which dosimeters can be inserted. Thus, for example, a cylinder of tissue-equivalent material may have thermoluminescent dosimeters (TLDs) inserted into it at various locations. Insertion of such a cylinder into a CT scanner permits the measurement of dose distributions of relevance to evaluating doses to patients. Alternatively, a model may be a

doi:10.1088/978-0-7503-2424-3ch7

mathematical construct that is used in conjunction with a radiation transport code to estimate doses and dose rates at various locations within it. Those dose rates may arise either from an external radiation field or from a hypothetical radionuclide distribution within the model. Because both the physical and mathematical models are abstractions of the anatomical structure of humans, they are often described as physical or mathematical phantoms.

In the remainder of this chapter, the various types of mathematical and physical models used for radiation transport calculations are described. Section 7.2 covers the use of such models in estimating the operational quantities used in radiological protection. Thus, it addresses the issue as to how an external, inhomogeneous, anisotropic radiation field is related to a limited number of scalar quantities (i.e. doses or dose rates at specific locations). This covers the use of both mathematical and physical models. Section 7.3 then addresses the development and use of mathematical models that represent human anatomy to different levels of detail. Here the focus is on how external radiation fields or internal distributions of radionuclides are used to calculate equivalent doses or dose rates. Section 7.4 then considers the use of physical models that represent human anatomy, with a focus on relating radiation fields generated in various contexts to measured doses and dose rates at various locations in the physical model.

7.2 Models used in estimating operational quantities

Before discussing phantoms appropriate to estimating operational quantities, it is appropriate to describe how those quantities are defined and to make distinctions between different quantities that may be measured. As discussed in Annex B.3 of ICRP (2007), a radiation field of a specific type is fully described by the number of particles, N, their distributions in energy and direction, and the spatial and temporal distributions of these quantities. Radiation field quantities are defined at any point in a radiation field. One such quantity is the fluence, Φ, which is defined by:

$$\Phi = \mathrm{d}N/\mathrm{d}a \tag{7.1}$$

where dN is the number of particles incident upon a small sphere of cross-sectional area da. Equivalently, Φ, can be written:

$$\Phi = \mathrm{d}L/\mathrm{d}V \tag{7.2}$$

where dL is the sum of the lengths of the trajectories passing through a small volume dV.

Where a radiation field is produced by uncharged particles, e.g. photons (x-rays and gamma rays) or neutrons, the transfer of energy to the medium through which they pass is mediated by the generation and slowing down of secondary charged particles, such as electrons. A suitable measure of this transfer of energy is kerma, K, defined by:

$$K = \mathrm{d}E_{tr}/\mathrm{d}m \tag{7.3}$$

where dE_{tr} (J) is the sum of the kinetic energies of all charged particles liberated by uncharged particles in a mass dm (kg). It is emphasised that kerma differs from absorbed dose in that it relates to the energy emitted in a small mass and not the energy absorbed.

In small volumes, fluence, kerma and absorbed dose are subject to statistical fluctuations. However, in practice, these quantities are treated as point values by defining them as expectation values. In micro-dosimetric applications, other quantities are used to express stochastic variations in energy deposition.

As noted above, radiation fields can vary in direction and in space and time. A field is described as homogeneous in space if it is identical (in terms of its distribution in direction and energy) at different locations. It is described as isotropic at a location if it is independent of direction, i.e. if its description is identical irrespective of the rotation of the 3D co-ordinate system used to express that description.

In area monitoring operational dose quantities are used for monitoring external exposures. These are point quantities and are defined such that their value does not depend on the directional distribution of the incident radiation. Furthermore, an operational dose quantity must be additive. This leads to the requirement to define an expanded and aligned radiation field that is related to the actual radiation field at the point of interest. The relevant dose quantities are then calculated for this expanded and aligned field.

An expanded field is a radiation field in which the spectral and angular fluence have the same characteristics at all points of a sufficiently large volume as at the location of interest. This ensures that a detector positioned at different locations within the expanded field will give the same reading. However, it does not ensure that the detector will give the same reading if it points in different directions. To address this, an aligned field is defined. The aligned field can be envisaged as a wide parallel beam with a fluence equal to the integral of the angular distribution of the fluence at the point of interest. With this definition, operational quantities can be defined by specifying the orientation of the detector relative to the axis of the aligned field.

In practice, the monitoring instrument that is used to define standard operational dose quantities is a sphere defined by the International Commission on Radiation Units and Measurement (the ICRU sphere, see Annex B.4 of ICRP (2007)). This is a sphere of tissue-equivalent material 0.3 m in diameter with a density of 1000 kg m^{-3} and mass composition 76.2% oxygen, 11.1% carbon, 10.1% hydrogen and 2.6% nitrogen). The ambient dose equivalent at a depth x mm in the ICRU sphere, $H^*(x)$, is the dose equivalent that would be produced in the sphere by the expanded and aligned field at a depth of x mm on the radius vector opposing the direction of the aligned field. Similarly, the directional dose equivalent, $H'(x, \Omega)$ is the dose equivalent that would be produced by the expanded (but not aligned) field at a depth x mm in the sphere on a radius in a specified direction Ω. For a unidirectional beam, as mainly arises in calibration procedures, the directional dose equivalent can be written as $H'(x, \alpha)$, where α is the angle between the direction Ω and the direction opposite to radiation incidence. The geometry for determination of ambient dose equivalent and directional dose equivalent is illustrated in figure 7.1.

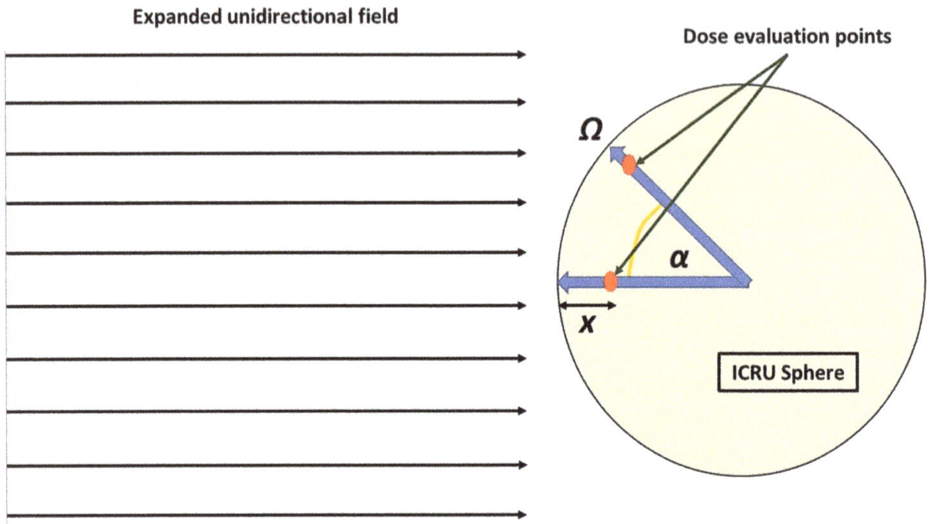

Figure 7.1. Geometry for defining operational quantities.

In selecting reference values of x, consideration must be given to the application of the related operational monitoring quantity. Three values are adopted, $x = 0.07$, 3 and 10 mm.

The value of 0.07 mm is selected as being appropriate to the cells at risk in the skin with respect to cancer induction. These are taken to be in the basal layer of the epidermis (ICRP 1991). As the ICRP (1991) comments at paragraph 297 of its report: 'The choice of skin depth which is appropriate for evaluating skin dose relevant to skin cancer induction is a matter of some controversy, in part because of the difference in thickness of the skin at different body sites and in part because there is not complete agreement on the target cell for the different types of carcinomas. The observation of excess basal cell carcinomas in uranium miners in Czechoslovakia subjected to alpha-particle irradiation of the skin is consistent with a target cell at a depth no greater than the basal layer of the epidermis. Despite some questions about whether the alpha-particle radiation causes the cancers it would be prudent to assume that the basal layer of the epidermis is the site of the target cells for skin cancer induction.'

The value of 3 mm is related to protection of the lens of the eye from the induction of cataracts. In the past, this value was given less consideration than the other two values. However, a re-evaluation of the sensitivity of the lens to cataract induction in recent years and the relatively large doses to the lens of the eye that can be received by staff undertaking various radiology procedures has meant that increased attention is now being given to calculation of this quantity.

The value of 10 mm is a cautious value intended to ensure protection of the deeper tissues and organs of the body. It is located at a sufficient depth to ensure that for most radiation fields, the secondary radiation arising from scattering of the particles impinging on the body has built-up to close to equilibrium but is not so deep as to be associated with substantial attenuation of the incident field. However, as pointed out by the ICRP (paragraph B164 of ICRP (2007)), although in most practical situations of external radiation exposure, the ambient dose equivalent, $H^*(10)$, fulfils the aim of providing a conservative estimate or upper limit for the value of the limiting quantities, this is not always the case for persons in high energy radiation fields such as in the vicinity of high energy accelerators and in cosmic ray fields. For such fields, a depth of 10 mm in ICRU tissue is not sufficient to complete the charged particle build-up in front of that point and hence the operational quantities will under-estimate effective dose.

The effects of attenuation of the primary field and secondary charged particle build-up are considered in the operational dose quantity defined for individual monitoring. This is the personal dose equivalent $H_p(x)$, which is the dose equivalent in ICRU tissue at a distance x below a specified point on the human body (usually the location where the dosimeter is normally worn. As discussed by the ICRP (paragraph B170 of ICRP (2007)), an operational quantity for individual monitoring should allow the effective dose to be assessed or should provide a conservative estimate under nearly all irradiation conditions. This, however, requires that the personal dosimeter be worn at a position on the body which is representative with respect to the exposure. The ICRP points out that for a dosimeter position in front of the trunk, the quantity $H_p(10)$ mostly furnishes a conservative estimate of effective dose, but that for exposure from the back only, a dosimeter worn at the front and correctly measuring $H_p(10)$, will not appropriately assess effective dose. Similar issues arise with partial-body irradiation.

The ICRU sphere can be implemented as a physical system or a mathematical model of it can be used in conjunction with a radiation transport code for dose estimation. It constitutes the simplest physical or mathematical phantom addressed in this chapter. However, it should be noted that other phantoms of simple geometry may also be used. Thus, for example a cuboidal water phantom might be exposed to a monodirectional radiation field oriented perpendicular to one of the sides of the cuboid to give an estimate of the variation of absorbed dose with depth in the phantom. More complex mathematical and physical phantoms are discussed in the following subsections of this chapter.

7.3 Mathematical models of human anatomy

As illustrated in section 7.2, the simplest representation of the human body used in radiological protection is a 0.3 m diameter sphere. This is useful in defining measurable quantities that can be used to monitor exposures to external radiation fields, but it is not adequate for evaluating absorbed or equivalent doses to individual tissues and organs for many of the radiation fields of relevance, e.g. partial-body irradiation from external narrow beams. Furthermore, it is of no relevance to internally incorporated radionuclides, for which it is necessary to estimate dose rates in target tissues and organs from activity distributions in source tissues and organs. For these applications, more realistic representations of human anatomy are required.

In models developed in the 1960s and 1970s, mainly at the Oak Ridge National Laboratory, USA, the structure of the human body was represented using constructive solid geometry (CSG) techniques. CSG modelling allows the user to create a representation of a complex solid object by using Boolean operators or equivalent to combine sets of simple objects, e.g. spheres and ellipsoids, cuboids, cylinders and cones. The first such phantoms were the Fisher–Snyder representations of adults and children (Xu 2010).

In 1969, Snyder and his colleagues reported the first heterogeneous phantom, the MIRD-5 Phantom, named for the Medical Internal Radiation Dose (MIRD) Committee of the Society of Nuclear Medicine that adopted it. This phantom distinguished the skeleton, lungs, and all other tissues. However, by 1978, Snyder and his colleagues had developed an improved version of the adult phantom that contained more than 20 organs (Xu 2010). It was this phantom that was used to calculate the Specific Effective Energy (SEE) values used in setting limits on intakes of radionuclides by workers reported in ICRP Publication 30 (see ICRP (1979) for an account of the approach adopted).

Subsequently, during the 1980s, Cristy and Eckerman at ORNL developed an enhanced set of phantoms comprising a new-born, individuals aged 1, 5, 10 and 15 years, and an adult male (figure 7.2). An adaptation of the 15-year-old was used to represent an adult female. Over the same period, a team from GSF, Germany developed a pair of gender-specific adult phantoms (ADAM and EVA) (Xu 2010).

By the early 1990s, the use of stylised phantoms, such as those described above, was falling out of favour, largely because they compromised anatomical detail leading to inaccurate results. This was of specific concern in radiotherapy and nuclear medicine, where accurate dose estimation was required. Also, with the

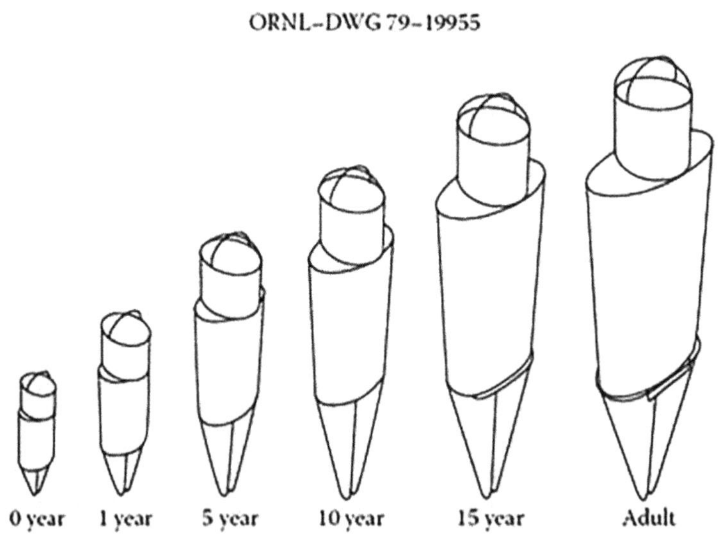

Figure 7.2. External geometry of Cristy–Eckerman phantoms. Xu (2010), Taylor & Francis Ltd. http://tandfonline.com.

advent of computed tomography (CT) and magnetic resonance imaging (MRI) techniques, it became possible to visualise and digitalise the internal structures of the body in detail and in 3D. This led to the development of voxel-based phantoms.

A voxel is the 3D generalisation of a pixel. From imaging using CT and MRI, and from anatomical photography, a tomographic image dataset comprising multiple slices is developed. Each slice comprises a 2D-set of pixels and voxels are formed by attributing to each pixel the thickness of the slice in which it occurs. Thus, a voxel-based phantom comprises many cuboidal elements arranged to form a space-filling structure. Numerous voxel-based models have subsequently been developed, and these are listed in table 1.2 of Xu (2010).

In defining a voxel-based model, each voxel must be assigned to a specific tissue or organ by giving it an appropriate identification number. Each tissue or organ is associated with a specific density and chemical composition, and this defines the radiation transport properties of the voxels associated with it. Thus, a comprehensive definition of the spatially distributed radiation transport properties of the phantom is obtained.

Intrinsically, a voxel-based phantom is grainy, with transport properties exhibiting step changes between one voxel and an adjacent voxel across the boundaries of the tissues and organs that are represented. Another limitation is that voxel-based phantoms are not readily deformable, so there are difficulties in adapting such a phantom to represent different age groups or different stages of pregnancy.

In principle, a voxel-based model is based on CSG techniques, with the idealised, geometrical elements comprising small cuboids. However, modern computer-aided design (CAD) software systems are based on more powerful boundary representation (BREP) techniques. In a BREP approach the surfaces of objects are represented either as polygons or as smooth curves fitted through a set of control points (non-uniform rational B-splines, NURBS, are typically adopted for fitting, because they can be used to achieve very smooth surfaces). BREP-based approaches can readily accommodate complex anatomical features. Also, by adjusting the positions of the control points, BREP-based approaches facilitate surface deformations, which is necessary if adjusting for organ size or for taking account of organ motions (Xu 2010). For N control points, such a transformation is written:

$$C_{\text{new}} = S \times R \times C_{\text{old}} + T \tag{7.4}$$

where C_{new} is a $3 \times N$ matrix of the revised control points; S is a scalar, R is a 3×3 matrix defining rotation, C_{old} is a $3 \times N$ matrix of the original control points, and T is a $3 \times N$ matrix that defines the translation of each control point. The value of 3 arises in equation (7.4) because the transformation occurs in 3D (Xu *et al* 2010a). The flexibility of a BREP-based approach is illustrated using a 3D-polygonal-mesh-based approach to insert a model of the fetus into a model of the pregnant mother (Xu *et al* 2010b).

Although BREP-based approaches provide greater flexibility in model definition than CSG-based approaches, the results obtained have to be used in conjunction with radiation transport codes, such as the MCNP-series of codes (https://mcnp.lanl.gov/).

Thus, BREP-based models must be converted to CSG-based (typically voxel-based) models for application. Therefore, in the following discussion the focus is on radiation transport simulation in voxel-based models.

For radiological protection purposes, the most recent reference adult phantoms for males and females adopted by the ICRP are those described in ICRP (2009). The phantoms were defined to correspond as closely as possible to the characteristics of the ICRP Reference Male and Reference Female (ICRP 2002, 2007). These voxel phantoms are based on computed tomographic data of real people, using datasets obtained for individuals with external dimensions close to the reference data. These phantoms are used for applications relating both to external and internal irradiation.

The male phantom was based on a whole-body clinical CT image dataset of an individual who suffered from leukaemia and had to undergo whole-body irradiation. This individual did not have obvious signs of illness that would appear in the image data. The imaging data comprised 220 slices of 256 × 256 pixels giving an original voxel size of 8 mm in height with an in-plane resolution of 2.08 mm. The female reference computational phantom was based on a scan performed at a relatively high resolution preparatory to whole-body irradiation for the treatment of leukaemia. The dataset consisted of 174 slices of 5 mm width (head and trunk) and 43 slices of 20-mm width (legs), each with 256 × 256 pixels. From the 20-mm-slice images, intermediate slices of 5 mm thickness were obtained by interpolation. The resulting dataset consisted of 346 slices. Minor adjustments were subsequently made to these datasets, e.g. by scaling of voxels, to bring them into closer conformity with the characteristics of the reference adults (ICRP 2002, 2007). Coronal and sagittal images of the reference female phantom are illustrated in figure 7.3.

Radiation transport in phantoms is usually simulated using Monte Carlo approaches. For external exposures, spatially extended fields are generally adopted for reference calculations. Standard irradiation geometries are anterior-posterior (AP), posterior-anterior (PA), left lateral (LLAT), right-lateral (RLAT), rotational (ROT) and isotropic (ISO). In addition, for environmental exposure situations, the geometry that is modelled may be irradiation when standing on an infinite uniformly contaminated plane or on a semi-infinite uniformly contaminated slab. In general, to achieve converged results, the Monte Carlo codes do not use simple, random sampling techniques. Instead, weighted sampling is used to ensure that all aspects of transport are represented that could significantly affect the results obtained. Xu (2010) gives a brief description of some of these codes that are extensively used. These include the Monte Carlo N-Particle (MCNP) and Monte Carlo N-Particle eXtended (MCNPX) codes that cover neutron, photon and electron transport, including coupled transport of all three radiation types. EGS (electron gamma shower) simulates coupled transport of electrons and photons at energies ranging from a few keV up to tens of TeV. GEANT4 is a toolkit for simulation of electromagnetic, hadronic and optical processes at energies up to several TeV. PENELOPE addresses coupled photon–electron transport at energies from a few hundred eV to 1 GeV. Finally, FLUKA models a wide range of particles including photons, electrons, neutrinos, muons, hadrons, neutrons and heavy ions. MCNP, EGS, FLUKA, PHITS (a particle and heavy ion transport code) and GEANT4

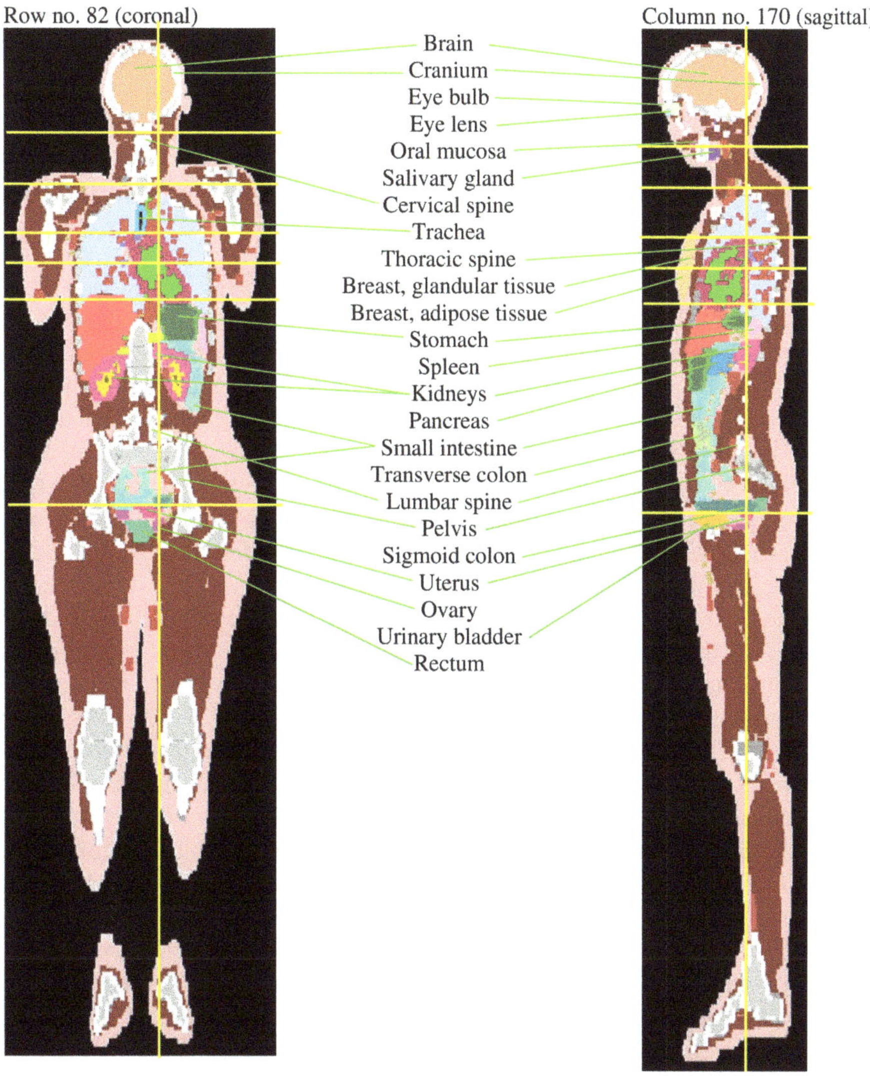

Figure 7.3. Coronal and sagittal images of the ICRP reference female adult phantom. Reproduced with permission from ICRP (2009). Adult Reference Computational Phantoms. ICRP Publication 110. Ann. ICRP 39(2).

were used by the ICRP (2009) in their calculations for external and internal dosimetry.

Confidence in the results from mathematical models of human anatomy and radiation transport can be enhanced by comparing the results obtained from measurements made in the appropriate radiation fields using physical phantoms. These phantoms are typically used for external radiation dosimetry, internal radiation dosimetry and evaluations of image quality in diagnostic imaging procedures. For external radiation dosimetry, a phantom is designed so that small radiation dosimeters can be inserted into different locations within it. In contrast, a phantom used for internal dosimetry is designed to contain either removable organs doped with long-lived radionuclides or includes hollow regions that can be filled with liquids in which short-lived radionuclides are dissolved. Such phantoms are typically used to calibrate external counting devices (e.g. whole-body counters or thyroid monitors) or imaging devices used in nuclear medicine. Physical phantoms are discussed in more detail in section 7.4.

7.4 Physical phantoms

As pointed out by Hintenlang *et al* (2010), a wide variety of physical phantoms has been developed. Many of these are primarily used for the quality assurance of medical devices and these do not require specific anatomical detail. However, for other purposes, e.g. in image quality evaluation or for dosimetry, an accurate representation of human anatomy is required. Many of the general considerations of relevance to the design and construction of a physical phantom can be illustrated by reference to the RANDO phantom, which is one of the oldest and most widely used examples, having been originally designed in 1959. The phantom exists in male and female versions and has traditionally been used to provide detailed dose distributions for use in evaluating radiotherapy treatment plans. These phantoms were originally constructed around a natural human skeleton (but without arms and legs). An external view of the RANDO phantom is shown in figure 7.4. The RANDO soft-tissue-equivalent material was designed to replicate the effective atomic number, electron density and physical density of muscle tissue with randomly distributed fat. Similar considerations applied in specifying the material used to represent the lungs (taking into account its air spaces through use of a lower density than for other soft tissues). However, the RANDO soft-tissue-equivalent materials have a large difference in the weight percentages of carbon and oxygen compared with human tissues. This arises from the low oxygen content of most polymers and resins and is a general issue because most tissue-equivalent materials are formulated from epoxy resins combined with various other molecular compounds. A consequence of this difference in composition is that the RANDO soft-tissue-equivalent material has similar radiation transport properties to human soft tissues above 0.1 MeV but performs less well below 0.1 MeV. Therefore, it is sometimes less satisfactory in diagnostic than in therapeutic applications (Hinterlang *et al* 2010).

In the RANDO phantom, skeletal tissues were traditionally represented using human material. Although this approach gives a good representation of bone structure and of the mineral content of the skeleton, the loss of bone marrow and the drying of the material mean that it is not necessarily as good a representation of living bone as some simulants. Thus, for example the University of Florida phantom series use a rigid epoxy-resin-based material with a density of 1220–1400 $kg\,m^{-3}$, depending on age, to represent the skeleton and pliable, urethane-based materials of density 1040 $kg\,m^{-3}$ to represent soft tissues (Hinterlang *et al* 2010).

As shown in figure 7.4, the RANDO phantom, like most physical phantoms, is segmented into slices. In the case of the RANDO phantom, each slice is 0.025 m

Figure 7.4. View of the exterior of the RANDO male phantom. Reproduced from Ning *et al* (2014). CC0 1.0.

thick and grids of holes may be drilled into each slice for the installation of dosimeters. The dosimeters used with physical phantoms are typically TLDs, but optically stimulated luminescence (OSL) dosimeters may also be used. In either case, the dosimeters must be removed from the phantom for readout. In contrast, a metal-oxide semiconductor field-effect transistor (MOSFET) dosimetry system can be used. This provides readout from *in situ* dosimeters but requires fine wires to be inserted into the phantom and this can lead to imaging artefacts. An alternative approach that has been studied uses small scintillator elements with readout via fibre-optic cables (Hinterlang *et al* 2010).

Various physical phantoms that are available are discussed by Hinterlang *et al* (2010) and are summarised by Xu (2010). They include the RANDO phantoms for Caucasian adults, the ATOM phantoms for Caucasian new-born, one-year-old, five-year-old, ten-year-old, and adult male and female individuals, the PBU-50 3D sectional torso phantom for Caucasian adult males, the CTU-41 torso phantom for Japanese adult males, and the LLNL phantom for Caucasian male adults with chest plates to simulate different thicknesses of the chest wall. Other phantoms from the University of Florida (UF), the Radiation Health Research Institute of Korea (RHRI) and the Radiation Protection Bureau of Canada (RPD) represent individuals of various ages (UF), a typical Korean adult male (RHRI), and individuals of varying age and stature (RPD).

7.5 Considerations in monitoring to determine operational dose quantities

7.5.1 Ambient dose equivalent rate

This is the primary operational quantity for instrument-based monitoring of *x*-, gamma- and neutron radiation in the workplace environment. The definition of the quantity illustrates its essential impracticality—it is impossible to align and expand a radiation field, and it would be clearly impractical to perform monitoring using a 30 cm tissue-equivalent sphere. However, getting out of this corner is, for many circumstances, surprisingly easy. The first step is to accept that any radiation detector averages its signal over the volume of active detector in some way. It is important to recognise this, particularly when it comes to narrow beams and close approach to sources. The onus then falls on the person making the measurement to choose a 'sensible' size for the detector so that the reported measurement, deemed to be at a point at the marked centre of the instrument in the instrument's absence, satisfies the radiation protection needs. Hence, working close to sources or with narrow beams tends to drive towards smaller detectors, whereas measurement at 1 m above relatively uniformly spread contamination allows for detector volumes of litres if required to get the sensitivity.

The second step is that it must be realised that there is no reason to have a sphere. All that is required is to generate the effect of the sphere together with alignment of the radiation field. Dealing with alignment is very easy. If a detector is designed to have an isotropic response then there is no need for alignment. A perfectly isotropic response is not possible, but a spherical ionisation chamber gets close to it—only the stem interferes. However, for hand-held radiation monitors, there is the influence of the person. This is generally ignored and is of little importance where the dominant direction of irradiation is obvious—you may be looking at the source—but, in complicated situations, the dominant direction may be very difficult to identify just by looking. A skilled surveyor, if he or she thinks the direction may be difficult to determine, will often hold the instrument close to the body and gently rotate. The attenuation of the body will be strongest where the radiation has a dominant direction from directly behind. The other influence of the person holding an instrument is backscatter, but this drops off quite fast with separation between body and instrument. Hence, with an instrument, the aim is to have an acceptable polar response over the forward 2π solid angle and to avoid dramatic over-responses anywhere in the rearward 2π.

The other aspect that the instrument should reproduce is the H*(10) to air kerma relationship. In this context, here are several important aspects to recognise.

- The ratio of the mass energy absorption coefficients between air and tissue is close to constant over a wide range of photon energies, so for a small mass of tissue the graph would be flat at about 1.1. The reason it is higher is that the hydrogen content in tissue leads to a higher electron number per unit mass.
- The 10 mm of tissue over the reference point in the sphere will scatter and attenuate the incident radiation. An important aspect of tissue is that it is, over a wide energy range (60 keV–2 MeV) a 'grey absorber', i.e. the chance of an interaction is not particularly dependent on the photon energy. This means that the mean free path of a Compton-scattered photon is little different from that of the original photon. Attenuation, even at 30 keV, is only by a factor of approximately 22%, but rises rapidly below this energy.
- The bulk of the 30 cm sphere will contribute an energy dependent backscatter component generated by Compton interactions. It is this process that produces the rising ratio as the energy reduces from high energies. The directional nature of Compton interactions means that at higher (1 MeV) energy most of the photons scatter forward and will not generally pass back through the reference position in the sphere. As the energy decreases, then the probability of scattering into the backward 2π solid angle increases, leading to a rise in the backscatter contribution.
- Below 50 keV, the mass attenuation coefficient starts rising, which means that Compton-scattered photons have a reduced chance of backscattering through the reference position.
- The shape of this curve is very similar to that of the normal incidence values of Hp(10), the personal dose monitoring quantity for penetrating radiation, which is based on a slab phantom.

This shape of curve is relatively easy to reproduce in an instrument design. It is possible to produce an instrument with a mass of 100 g that will, within the fairly large acceptable uncertainties accepted in radiation protection, reproduce the relevant characteristics of the sphere, which has a mass of 14 kg.

Operationally, H*(10) measurements are used to demonstrate compliance with limits, help drive ALARA and establish trends. However, possibly the most important use is the prediction of dose increment rates on personal dosemeters in high dose tasks. A good H*(10) measurement, with the surveyor standing just behind the approximate position of someone who will be performing the task and holding the detector centre close to the future position of the person's dosemeter, will predict the maximum Hp(10) dose increment rate on the dosemeter. This allows the prediction of 'task doses'. Generally, recorded doses are slightly below those predicted, partly because the time taken is often less than assumed, but also because the personal dosemeter is better shielded for radiation incident from behind a wearer than the survey instrument, which will always be held away from the body.

For neutrons, the same basic ideas apply but in this case the reference energy response is based on the neutron fluence per Sv. The reason neutrons are different is that there is no convenient relatively easily measurable dosimetric quantity like air kerma for photons. Unfortunately, neutron instruments do not have the level of metrological performance that photon and beta monitoring instruments achieve.

7.5.2 Directional dose equivalent rate

This, again, is a sphere-based quantity but, in this case, there is no requirement for alignment as the quantity is intended for the measurement of skin dose, particularly where less-penetrating radiations are limiting, i.e. beta and low energy x-radiation. This means that, in theory, where we can define the position of an ambient dose equivalent measurement in space and time, for directional dose equivalent we should also specify rotation and elevation, i.e. two angles. This is hardly ever done as the purpose of a survey is normally to identify the maximum value of directional dose equivalent at any position. This is often easy—the source of the radiation may be obvious, such as a contaminated surface or a beta thickness gauge—and a good surveyor will always seek for the maximum if the source is not obvious or is widely distributed. The other aspect that is sometimes forgotten is that penetrating radiations will also contribute to a measured directional dose equivalent rate, generally at a level very close to the ambient dose equivalent rate value.

7.5.3 Individual dose equivalents

These are 'the badge quantities', i.e. the quantities that personal dosemeters aim to reproduce. They are not sphere quantities, but are based on a 30 cm × 30 cm × 15 cm tissue-equivalent slab with two reference depths at the centre of one of the large faces, currently 10 and 0.07 mm. These are the analogues of ambient and directional dose equivalent but very clearly without the addition of alignment and expansion. As described above, a careful survey of an operational area before a task should lead to no surprises when the personal dosemeter is read out, i.e. the recorded doses should never exceed dose estimates based on the survey instrument readings and exposure time. This robust interrelation is extremely useful in operational radiation protection.

7.5.4 The influence of dosemeter position

In theory, the dosemeter should be positioned where the individual dose increment rate is likely to be highest. However, this is generally impractical because:

- On a complicated plant, with wearers moving around, it is very difficult to predict that position. In addition, the act of moving around in complicated situations tends to make the person's exposure close to rotationally isotropic, which means that any bias will be small.

- A dosemeter that is not on the wearer's front will stand a good chance of being knocked off without the wearer noticing.
- For an electronic dosemeter, it is important that a wearer can hear and see alarms, and can read the dosemeter, perhaps *in situ*, but generally by removing it and holding it up in front of the eyes.
- A fundamental point is that often in an area everyone should be wearing a dosemeter. Colleagues and supervisors must be able to check easily that a person is wearing one and that demands a standardised position. This is often ensured by having a 'dosemeter pocket' or a dosemeter loop on coveralls.

The commonest situation for unusual wearer position is transport drivers, who spend most of their lives with their backs to the radiation field. Some RPAs have asked drivers to wear their dosemeters in the normal position for loading and unloading, but then fix them to the seat back while driving.

Occasionally, circumstances arise where the field has been dominated by an unexpected direction. For example, sometimes it has been produced by active pipes that were supposed to be empty. One of the clues that an unexpected dominant direction has happened is a low value of Hp(10) compared with prediction. If this effect is observed, then it can be rewarding to identify the unusual direction.

References

Hinterlang D E, Moloney W E and Winslow J 2010 *Handbook of Anatomical Models for Radiation Dosimetry* ed X G Xu and K F Eckerman (Boca Raton, FL: CRC Press) pp 389–409

ICRP 1979 Limits for intakes of radionuclides by workers: Part 1, ICRP Publication 30 *Ann. ICRP* **2(3/4)**

ICRP 1991 The biological basis for dose limitation in the skin, ICRP Publication 59 *Ann. ICRP* **22(2)**

ICRP 2002 Basic anatomical and physiological data for use in radiological protection: reference values, ICRP Publication 89 *Ann. ICRP* **32(3–4)**

ICRP 2007 The 2007 recommendations of the international commission on radiological protection, ICRP Publication 103 *Ann. ICRP* **37(2–4)**

ICRP 2009 Adult reference computational phantoms, ICRP Publication 110 *Ann. ICRP* **39(2)**

Ning P, Zhu S, Shi D, Guo Y and Sun M 2014 X-ray dose reduction in abdominal computed tomography using advanced iterative reconstruction algorithms *PLoS ONE* **9** e92568

Xu X G 2010 Computational phantoms for radiation dosimetry: a 40-year history of evolution *Handbook of Anatomical Models for Radiation Dosimetry* ed X G Xu and K F Eckerman (Boca Raton, FL: CRC Press) pp 3–41

Xu X G, Chao T-C E, Bozkurt A, Shi C and Zhang J 2010a The 3D and 4D VIP-man computational phantoms *Handbook of Anatomical Models for Radiation Dosimetry* ed X G Xu and K F Eckerman (Boca Raton, FL: CRC Press) pp 135–62

Xu X G, Shi C, Stabin M G and Taranenko V 2010b Pregnant female/fetus computational phantoms and the latest RPI-R series representing 3, 6, and 9 month gestational periods *Handbook of Anatomical Models for Radiation Dosimetry* ed X G Xu and K F Eckerman (Boca Raton, FL: CRC Press) pp 305–35

IOP Publishing

Principles and Techniques of Radiological Protection

Michael Charles Thorne

Chapter 8

General principles of mathematical models for the transport of radionuclides

M C Thorne

8.1 Introduction

In assessing the radiological impact of releases of radionuclides to the environment, use is often made of mathematical models of the transport of those radionuclides to determine their time-evolving distribution. Similarly, mathematical models are used to represent the transport, distribution and excretion of radionuclides in and by humans as an integral part of the process of estimating values of committed equivalent and effective doses following the intake of radionuclides, e.g. by ingestion, inhalation or from wounds. In this chapter, some general principles relating to the development, application and testing of such models are described. Specific applications in environmental impact assessment and internal dosimetry are addressed in subsequent chapters. Section 8.2 considers how the context in which the model is to be used affects its structure and describes techniques for constructing a conceptual model appropriate to a specific context. Section 8.3 then describes the types of mathematical models often used to implement those conceptual models, while section 8.4 considers the verification and validation of those mathematical models and section 8.5 addresses the application of those models taking parametric uncertainty into account through the application of sensitivity and uncertainty analyses.

8.2 Developing and justifying a conceptual model

Confronted with the need to develop or apply a mathematical model in a specific context, it is tempting to plunge immediately into the computational component (formulating equations, writing software or parameterising an existing model). However, it is important to establish that the model to be developed or applied is suited to the application. That is, the model must be fit-for-purpose in the context in which it is to be applied. Thus, prior to model development that context must be

doi:10.1088/978-0-7503-2424-3ch8

clearly established. This has long been recognised in the context of post-closure safety assessments for solid radioactive waste disposal, where the various components of the assessment context are specified in detail (IAEA 2003, 2020, SKB 2018, Klos and Thorne 2020). Relevant factors include, for example, whether a deterministic or a stochastic approach is to be adopted, and whether cautious or realistic assumptions are to be made.

Having established the context of model application, that context can be used to create a conceptual model of the system that is to be represented mathematically. The conceptual model defines the chief components of the system (often termed 'features' of the system) and the events or processes that give rise to interactions between those chief components and that influence the transport of radionuclides within and between those components. In addition, it is often found useful to assign properties to the chief components of the system (e.g. temperature, degree of water saturation, radionuclide concentration) and consider how the various events and processes are influenced by the current values of those properties and how, in turn, those events and processes operate to alter those properties. For example, a difference in radionuclide concentration (a property) between two adjacent components will result in diffusion of the radionuclide (a process) from one to the other changing the concentrations in both. The rate of diffusion is controlled by the difference in concentration while reciprocally acting to reduce that difference.

A key consideration in conceptual model development is to demonstrate that the model is comprehensive within the relevant assessment context, while being no more complex than that assessment context requires. This can be addressed, in part, through the development of structured lists of potentially relevant features, events and processes (FEPs) and the screening of existing lists to determine those FEPs that are relevant in a specific assessment context. Such FEP lists are typically developed by reviews of relevant national and international experience, as has been done by the NEA (NEA 1999, 2006).

Although such screened FEP lists are useful in determining those that should be included in a conceptual model, they do not address the issue of how those FEPs should be configured within the model, i.e. they do not make explicit the relationships between the FEPs. However, there are other tools that allow such relationships to be identified. One such approach, which is becoming increasingly adopted, is to use an interaction matrix (IM) (Hudson 1992, IUR 2006, Thorne 2012). Figure 8.1 illustrates the overall approach adopted.

Features of the system are arranged on the lead diagonal of the IM. Each of these features may be associated with a set of properties. The off-diagonal elements are then examined in turn to identify those events and processes that could cause the originating feature to influence the recipient feature. The IM adopts a clockwise convention in distinguishing the originating and recipient features.

It is generally desirable that IMs are no larger than about 15 × 15 or visual scrutiny of their structure becomes difficult. However, this does not limit the method, since a nested approach can be adopted in which both lead-diagonal and off-diagonal elements at the highest level can be expanded into IMs at a more detailed level. As Klos and Thorne (2020) discuss, once an IM has been fully

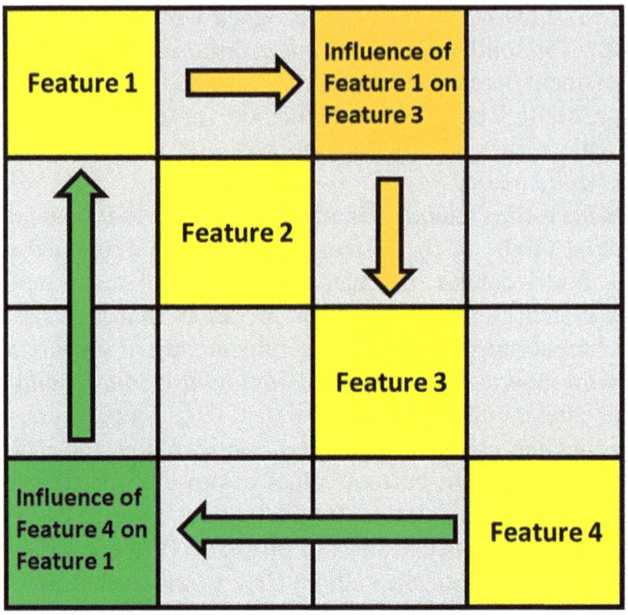

Figure 8.1. Structure of an interaction matrix showing the clockwise representation of influences (events and processes) on features of the system.

populated, it forms a suitable basis for mathematical model development. The steps involved comprise the following (Klos and Thorne 2020).

a) Mapping each element of the lead diagonal onto one or more compartments of the mathematical model, noting that the lead diagonal elements, or the compartments into which they are decomposed, may be spatially distinct or co-extensive. For example, a single 'soils' element could be mapped into two distinct, but spatially co-extensive, elements comprising soil solids and soil solution. The two-element formulation would facilitate the explicit identification and modelling of sorption and desorption processes that would be internal to a single 'soils' element.
b) Events and processes operating between two diagonal elements of the conceptual model should be considered potentially to operate between all combinations of the two sets of compartments into which those diagonal elements are decomposed in the mathematical model.
c) Each compartment in the mathematical model should have one or more properties associated with it. Conserved quantities are preferred, because non-conserved quantities can be computed from conserved quantities, e.g. temperature from heat content and mass using the concept of specific heat or activity from mass using a radioactive decay constant. Other properties of a compartment may also be required, e.g. material densities may be required to relate concentrations expressed on mass and volumetric bases.
d) Equations should be defined to represent the events and processes linking those compartments.

8.3 Common types of mathematical model

Mathematical models for representing the transport of radionuclides and other contaminants through the environment or within the human body can broadly be classified into three types:

- equilibrium;
- compartmental;
- advective-diffusive.

Models of radionuclide transport in the human body are generally multi-compartmental in nature, whereas advective-dispersive models are typically used for dispersal in fluids, e.g. in relation to atmospheric dispersion or the development of plumes of groundwater contamination.

Models of the different types may be used in combination, e.g. dispersal of radionuclides downstream of a release into a river might use an advective-dispersive model, whereas uptake into fish might use either an equilibrium approach or a compartmental approach if the kinetics of uptake were of significance.

8.3.1 Equilibrium models

In an equilibrium model, the concentration of a radionuclide or other contaminant in one component of the system is taken to be directly proportional to the concentration of that contaminant in another component. Thus, for example, if the time-dependent concentration of a radionuclide in soil, C_{soil} (Bq kg^{-1}) is calculated using an advective-dispersive model, the corresponding concentration in plants, C_{plants} (Bq kg^{-1}) may be calculated using:

$$C_{\text{plants}} = F_{\text{v}} C_{\text{soil}} \tag{8.1}$$

The plant:soil concentration ratio, F_{v}, is an equilibrium model of the transfer process. Although technically dimensionless, it is best to be explicit about the units as plant concentrations are sometimes expressed on a dry mass basis and sometimes on a fresh mass basis. Similar types of concentration ratio are used for aquatic organisms, but, in this case, concentrations are usually expressed per unit volume of water and per unit fresh mass of organism. Thus, typical units of the concentration ratio are either L kg^{-1} or m^3 kg^{-1}.

A slightly different type of equilibrium model is used to estimate radionuclide concentrations in animal tissues. In this case, the ratio is between the concentration in the tissue (Bq per kg fresh weight) and the rate of intake of the radionuclide by ingestion (Bq d^{-1}). Thus, the transfer factor has units of d kg^{-1}. A disadvantage of this approach is that intake rates are strongly dependent upon body mass, whereas tissue concentrations are not necessarily so dependent. Therefore, transfer factors for animal tissues typically vary inversely with the mass of the animal. For this reason, more recent compilations of data give both estimates of transfer factors and of ratios of concentrations between animal tissues and the diet of those animals (IAEA 2014).

Concentration ratios are also used in less obvious contexts. For example, in estimating the speed of radionuclide migration in groundwater, it is usual to estimate the velocity of radionuclide transport as the velocity of groundwater divided by a retardation factor (see section 8.3.3). The retardation factor is calculated using a distribution coefficient that is the ratio of the concentration of a radionuclide reversibly adsorbed to the solid phase to the concentration in solution in the groundwater. This model assumes that equilibrium always exists between the liquid and solid phases.

A special type of equilibrium model is termed a specific-activity model. Such models are generally applied to radioisotopes of elements involved in major biogeochemical cycles. Thus, the most familiar examples are for H-3 (hydrogen cycle, both as water and in organic forms) and C-14 (carbon cycle). However, such models have also been used for other radionuclides, e.g. Cl-36 and I-129.

In a specific-activity model, the underlying argument is that the biogeochemical processes involved make little distinction between the radionuclide of interest and the associated stable element. Therefore, if the ratio of radionuclide to stable element is known in the various sources of supply of that element to an organism, the ratio of the radionuclide to the stable element in the organism can be computed as the weighted average of the ratios in the sources of supply, with the weighting factors being the fractions of the element obtained by the organism from each of the sources of supply, assuming that these fractions remain approximately constant. Thus, key considerations in developing a specific-activity model are determining the ratios of radionuclide to stable element in the various sources of supply and in estimating the relative importance of those various sources of supply.

In some cases, stable element data can be used directly in formulating a specific-activity model. For example, carbon has two stable isotopes, C-12 and C-13. Because environmental media vary in their C-13:C-12 ratios, differences in those ratios can sometimes be used to infer the relative contributions from various sources. Similarly, ratios of naturally occurring and fallout-derived radionuclide concentrations to stable element concentrations, i.e. C-14:C-12 ratios, can also be used to infer the sources from which they arose.

Extensive compilations of concentration ratios, animal transfer factors and distribution coefficients are available for use in environmental modelling (e.g. IAEA 2004, 2009, 2010, 2014). A detailed discussion of specific-activity models is given in IAEA (2009).

8.3.2 Compartmental models

The basic principle in compartmental modelling is that the system to be modelled can be represented as a set of interconnected, well-mixed compartments, with the rate of transfer of a radionuclide or other contaminant from one compartment to another, or out of the system, taken to be directly proportional to the amount of contaminant present in the donor compartment. In traditional compartmental modelling the rates of transfer are taken to be time independent. However, time-dependent transfer rates can also be accommodated, although the mathematical approaches to evaluating such systems may differ from those applied to systems involving only time-independent rates.

The mathematical basis of compartmental modelling is discussed in detail by Klos and Thorne (2020). The approach can be concisely represented using matrix notation. Briefly, if the activity content of an n-compartment model is written as a vector $\boldsymbol{a} = (a_1, a_2, \ldots a_n)$, then the general representation of such a model is:

$$\frac{d\boldsymbol{a}(t)}{dt} = \boldsymbol{s}(t) + \boldsymbol{\Lambda}(t)\boldsymbol{a}(t), \tag{8.2}$$

where, the vector $\boldsymbol{s}$ represents the *external* inputs to each of the compartments and the transfers plus loss rates are written as $\boldsymbol{\Lambda}$, a square matrix. Note that all the quantities involved are time dependent. For $\boldsymbol{s}$ and $\boldsymbol{\Lambda}$, this is by specification, but for $\boldsymbol{a}$ it is determined by the form of the set of equations implied by equation (8.2), i.e. $\boldsymbol{a}$ would be time-dependent even if $\boldsymbol{s}$ and $\boldsymbol{\Lambda}$ were time independent over some finite interval. For $j \neq i$, λ_{ij} is positive and is the rate coefficient for transfer from compartment i to compartment j.

However, λ_{ii} is negative and is the sum of all losses from compartment i (both to other compartments, i.e. the negative of the sum of λ_{ij} over all $j \neq i$, and to outside the system, λ_L):

$$\lambda_{ii} = -\left(\lambda_L + \sum_{j\neq i}\lambda_{ij}\right). \tag{8.3}$$

An important case is when $\boldsymbol{s}$ and $\boldsymbol{\Lambda}$ are time-independent and the system exists for sufficiently long that equilibrium is achieved. In this case, the governing equation becomes:

$$\boldsymbol{\Lambda a} = -\boldsymbol{s} \tag{8.4}$$

If $\boldsymbol{\Lambda}$ has an inverse $\boldsymbol{\Lambda}^{-1}$ (which will be the case if the system has either direct or indirect connections between all the compartments and at least one of them exhibits losses to outside the system) then:

$$\boldsymbol{a}\big|_{\text{equilib}} = -\boldsymbol{\Lambda}^{-1}\boldsymbol{s} \tag{8.5}$$

This equilibrium solution does not involve the initial compartmental contents, $\boldsymbol{a}(0)$, i.e. if inputs persist for sufficiently long, the system forgets its initial state.

A further generally useful result relates to the behaviour of a tracer that does not perturb the characteristics of the system as applied to a time-independent system (Klos and Thorne 2020). In this case, the response of the system, $F(t)$, to a short-term pulse input, Q, at time zero is:

$$F(t) = QR(t) \tag{8.6}$$

where the response is independent of the time of input of the pulse and for a continuous input at rate $q(t)$:

$$F(T) = \int_0^T dtq(t)R(T-t) \tag{8.7}$$

This relationship depends only on the tracer and time-independence assumptions. It does not necessarily imply a compartmental system.

For a constant release rate q over the interval (0, U):

$$F(T) = q\int_0^U dtR(T-t) = q\int_{T-U}^T d\tau R(\tau) \tag{8.8}$$

For the special case of evaluation at the end of the period of release, $T = U$ and the integral runs from 0 to T. Also, the equilibrium situation can be determined by letting the time T tend to infinity. Thus:

$$F|_{\text{equilib}} = q\int_0^\infty d\tau R(\tau) \tag{8.9}$$

More generally, equation (8.8) can be rewritten:

$$F(T) = q\int_{T-U}^T d\tau R(\tau) = q\left[\int_0^T d\tau R(\tau) - \int_0^{T-U} d\tau R(\tau)\right] \tag{8.10}$$

Thus, for fixed T, as the duration of the release, U, increases, the second integral decreases and $F(T)$ increases. Thus, the longer the period of release, the greater the impact after the period of release has ceased.

However, for a fixed period of release, U, as T increases both the first and second integrals increase. Thus, $F(T)$ can either decrease or increase as the time of evaluation increases after a fixed period of release (see Klos and Thorne (2020) for examples of the application of equation (8.10)).

For multi-compartmental systems with time-independent transfer rates, specified initial compartmental contents and no subsequent inputs, general analytical solutions are available (e.g. Birchall and James 1989, Klos and Thorne 2020). These can be used together with the convolution approach of equation (8.7) to generate an analytical solution for any time-dependent pattern of input. However, in practice, multi-compartmental models, including those involving time-dependent transfer rates, are now generally simulated using numerical techniques in which the compartmental quantities at $t_i + \Delta t_i$ are calculated from their values at t_i (or, more generally, at t_i and at previous times $t_{i-1}, t_{i-2}\ldots$). Various computational packages are available for undertaking such simulations. These include general-purpose packages that can handle a wide variety of different types of ordinary differential equations (e.g. ModelMaker 4, https://www.apbenson.com/about-modelmaker-4) and packages specifically designed for modelling the transport of radionuclides and other contaminants (these include Ecolego, http://ecolego.facilia.se/ecolego/show/HomePage, GoldSim, https://www.goldsim.com/web/home/ and AMBER, https://www.quintessa.org/software/AMBER).

8.3.3 Advective-diffusive

Advective-diffusive or advective-dispersive modelling is used when a contaminant is being transported in a moving fluid. That fluid may be the sole phase of interest, e.g. in atmospheric dispersion or downstream transport in a river. However, in other contexts, the fluid may move though pores or fractures in a solid medium, as with groundwater transport through rock and soil. The term advective-dispersive modelling is often used, as mechanical processes of dispersion, due to effects such as turbulence or transport through pathways with different flow characteristics, is often treated as if it were a Fickian diffusion process, with a dispersion coefficient replacing or augmenting the diffusion coefficient.

Advective-dispersive modelling is well-illustrated by considering contaminant transport in groundwater in one dimension, taking adsorption to solids into account. Thus, the concentration of radionuclide i in pore waters at location x (m) and time t (s) is defined as $C_L{}^i(x,t)$ (Bq m^{-3}) and the corresponding concentration on solids is defined as $C_S{}^i(x,t)$ (Bq kg^{-1}). This implies that the total concentration of radionuclide i, $C_T{}^i(x,t)$ (Bq m^{-3}) is given by:

$$C_T{}^i(x, t) = \phi C_L{}^i(x, t) + \rho_d C_S{}^i(x, t) \tag{8.11}$$

where ϕ is the water-filled effective porosity of the rock, sediment or soil; and ρ_d (kg m^{-3}) is the dry bulk density of the material.

The rate of change in $C_T{}^i(x,t)$ with time is given by

$$dC_T{}^i(x, t)/dt = -\phi v dC_L{}^i(x, t)/dx + \phi(D_m + av)d^2C_L{}^i(x, t)/dx^2 - \lambda_i C_T{}^i(x, t) \tag{8.12}$$

where v (m s^{-1}) is the groundwater velocity (i.e. the groundwater flux per unit area divided by the water-filled porosity);

D_m (m^2 s^{-1}) is the diffusion coefficient for the radionuclide, which is less than the value in the bulk liquid because of the need for ions to diffuse through the tortuous porosity of the matrix);

α (m) is a scale length for dispersion, which takes account of the consideration that the degree of dispersion with time depends on the rate of groundwater flow; and

λ_i (s^{-1}) is the decay constant for the radionuclide.

Equation (8.12) does not consider *in situ* production from a parent radionuclide. This could be added as a time- and location-dependent source term on the right-hand side. The three terms on the right-hand side represent advective transport in groundwater, diffusion and dispersion in groundwater, and radioactive decay both in groundwater and on the solid phase.

It remains to relate $C_L{}^i(x,t)$ and $C_S{}^i(x,t)$. Generally, on the timescales of interest, kinetic effects can be neglected and a Freundlich isotherm approach can be adopted:

$$C_S{}^i(x, t) = K[C_L{}^i(x, t)]^n \tag{8.13}$$

Typically, n is taken to be unity and K is replaced by the symbol K_d, also known as the distribution coefficient. As defined here, it has units of m^3 kg^{-1}, but dimensionless versions and L kg^{-1} or m^3 te^{-1} are also used.

On this basis and using equation (8.11):

$$C_T{}^i(x, t) = \phi C_L{}^i(x, t) + \rho_d K_d C_L{}^i(x, t) \tag{8.14}$$

It is now convenient to define a retardation factor, R:

$$R = (\phi + \rho_d K_d)/\phi = 1 + \rho_d K_d/\phi \tag{8.15}$$

Thus, by substitution:

$$\begin{aligned} dC_T{}^i(x, t)/dt = -(v/R)dC_T{}^i(x, t)/dx \\ + [(D_m + \alpha v)/R]d^2C_T{}^i(x, t)/dx^2 - \lambda_i C_T{}^i(x, t) \end{aligned} \tag{8.16}$$

In this form, the effects of sorption are made explicit. Radionuclide migration velocities, whether by advection, molecular diffusion or dispersion, are reduced by a factor $1/R$ relative to conservative water tracers such as tritium.

The above description of advection-dispersion is given in one dimension. However, it is readily generalised to three dimensions by replacing the operators d/dx and d^2/dx^2 by their three-dimensional equivalents.

In order to use equation (8.16) or its three-dimensional generalisation, it is also necessary to calculate the groundwater velocity v. This may be time- and space-dependent and is, therefore, more appropriately written as $v(x,t)$. In saturated media, $v(x,t)$, can be calculated using Darcy's Law.

In one dimension, neglecting density and osmotic effects, this can be written as:

$$Q = AK\Delta h/\Delta x \tag{8.17}$$

where Q ($m^3\ s^{-1}$) is the groundwater flow;
K ($m\ s^{-1}$) is the hydraulic conductivity of the medium;
$\Delta h/\Delta x$ is the hydraulic head gradient; and
A (m^2) is the cross-sectional area of the rock under consideration.

The hydraulic head, h, is the height at which a column of water could be maintained by the pressure of the groundwater at a specific point within the formation. An extension to this approach applies in unsaturated media, for which Darcy's Law is replaced by Richards' equation, in which the hydraulic conductivity becomes a function of the degree of saturation and the hydraulic head gradient is replaced by the gradient in matric potential, with the matric potential being a quantity that becomes increasingly negative as the degree of undersaturation of the medium increases, i.e. as it dries. This arises because the residual water in the medium is bound to surfaces and the strength of the binding increases as the residual water content decreases. Richards' equation is typically used in modelling water flow and associated radionuclide transport in soils (Wheater *et al* 2007), although simplified approaches have also been found to be useful (Pérez-Sánchez and Thorne 2014, Thorne 2014).

From the above, the specific groundwater flow, q ($m\ s^{-1}$) is given by $q = Q/A$. Although q has the dimensions of a velocity, it is not the velocity of groundwater in the rock. This is given by $v = q/\phi$, where v (m s^{-1}) is the groundwater velocity and ϕ is the effective porosity.

In practice, ϕ is typically less than the total porosity. This is because, in many rocks and soils, there are preferred routes for groundwater flow, with other components of the water essentially stationary in 'dead-end pores'.

In general, groundwater flow patterns are modelled in two or three dimensions and for heterogeneous anisotropic media using various finite-element or finite-difference codes. In such models, complex geometrical networks of elements or nodes are set up with specific patterns of material properties.

Typically, a fundamental assumption of such modelling is that the rock or soil can be treated as a continuous porous medium. This requires that a representative elementary volume (REV) of the medium can be defined. The need for a well-defined REV can be seen when it is recognised that all geological media are heterogeneous at one or more length scales. Thus, for example, clays contain sub-micron intra-particle pores with dimensions of up to a few microns. At larger length scales, they are relatively uniform, until one reaches scales where inter-bedded sand and gravel layers or lenses become important. In contrast, hard rocks, such as granite, are fractured and faulted on a variety of length scales ranging from less than a metre to more than a kilometre. For such fractured systems, discrete-feature-network (DFN) or hybrid finite-element/finite-volume modelling approaches are now often used, and continuum models are less often adopted than was formerly the case. The most advanced of these models incorporate explicit representations of both rock fractures and the rock matrix and may use both asynchronous time-stepping and parallelisation to increase computational speeds (Shao *et al* 2019).

However, in biosphere applications, water flow and radionuclide transport models are typically continuum models, treating the model space as continuous, albeit with spatially varying properties. The concept of a REV forms a link between the real world and its modelled representation. The REV is chosen such that heterogeneities of smaller spatial scale can be represented in the effective parameter values assigned at the scale of the REV. In contrast, heterogeneities at larger spatial scales must be modelled explicitly.

For more detailed discussion of hydrogeological issues, standard textbooks may be consulted, e.g. Bear (1979) or Hiscock and Bense (2014).

8.4 Model verification and validation

Important considerations in implementing and applying a mathematical model are whether it conforms to specification and whether it comprises an adequate representation of the system under consideration. Verification and validation address these considerations.

Verification seeks to confirm that a model is correctly implemented, i.e. that it solves the equations that it sets out to solve and that it does so sufficiently accurately. At the simplest level, this could be through independent scrutiny of the model coding, recognising that this process is facilitated in modern simulation packages by facilities to visualise the model and to integrate the documentation with the executable software. However, verification may also involve comparing general-purpose numerical solutions with analytical solutions for special cases to which those analytical solutions apply. It may also involve grid-refinement techniques in which the model is run with finer spatial meshes until the results converge. Temporal refinement of model calculations is not generally an issue, as modern techniques of solution typically employ techniques that ensure enough accuracy, e.g. though iterative predictor–corrector techniques.

Validation of a mathematical model involves a comparison of results from the model with observations from the system of interest, or from a component of that system. The validation of radiological assessment models can be problematic, because they are primarily empirical in nature, with all relevant data used in their parameterisation. However, where a model is based on equations purporting to represent physical processes, validation is a possibility.

Validation studies involve using one set of observations to parameterise the model and then using the model to predict a set of results of observations not used in the model parameterisation step. For example, a time-series of observations may be distinguished into two subsets. All the data from the first subset are used to parameterise the model. In contrast, the second subset of data are distinguished into two categories. The first category comprises input data for the model and the second category corresponds to outputs from the model. In validation, the model is run using the first category of data as input and the results obtained are compared with the second category of data. Alternatively, experimental studies may be used to parameterise a model, which is then validated by using the model with a field dataset that includes both input data for the model and observations that can be compared with model outputs.

The time-series example given in the previous paragraph illustrates a potential issue with validation, which is that the dataset used for calibration may be very similar to the dataset used for validation. Thus, if a time-series of climate data is used in the validation of a hydrological model, simply splitting the time series into earlier years and later years may result in two subsets of data with similar characteristics. If this is the case, the validation comparison provides little justification for the general applicability of the model, but only establishes that it performs satisfactorily within the range of the dataset used for calibration. Thus, in developing a validation test, it is appropriate to consider how the available data can best be distinguished to enhance the power of the test. For example, with a time-series of climate data, it may be appropriate to split the time series into wet years and dry years, calibrate the model using data for wet years and then determine whether the calibrated model provides accurate predictions of hydrological responses when used with the climate data for dry years. This version of the validation test determines whether the model is useful for extrapolation, whereas splitting into earlier and later years only shows that the model is useful for interpolation.

In undertaking validation studies, it is preferable if the modelling is performed 'blind', i.e. in ignorance of the observations with which the model outputs are to be compared. Therefore, it is desirable if the relevant datasets are held by an independent referee, who releases the data needed for model calibration and input to the validation calculations, and who compares the results of those calculations with the observations that have been reserved for this purpose. It is also useful if the criteria defining success of the validation test are specified in advance of making the comparison, so that an unequivocal statement of success or failure can be made.

Having performed a validation test, as described above, the data used in the test are available for enhancing the model. However, this implies that they cannot be reused for validation, so new datasets must be identified or acquired if there is a requirement to validate the enhanced model.

As discussed in section 8.5, it is increasingly common for modelling to be undertaken under uncertainty, examining how uncertainties in the input data affect model outputs. This means that validation tests may need to be formulated in probabilistic terms. For example, output results could be specified as a 5th to 95th percentile range and it could then be determined whether about 90% of the model predictions lie within the 5th to 95th percentile range.

Ewen and Parkin (1996) provide a detailed account of an approach to validation under uncertainty of a hydrogeological model used in support of radiological impact assessment studies (Parkin *et al* 1999).

8.5 Modelling under uncertainty

Modelling the transport and distribution of radionuclides in the environment or in the human body is subject to uncertainties of various types. These include uncertainties in the most appropriate conceptual model to adopt, uncertainties in the equations to be used to represent the events and processes included in the conceptual model, limitations to the accuracy with which those equations can be solved, and both uncertainty and variability in the parameter values used with those equations. Issues related to constructing conceptual and mathematical models have been addressed in previous subsections, so here the focus is on addressing parametric variability and uncertainty. Although these are distinct considerations, it is common for both to be described as uncertainty. An alternative terminology distinguishes epistemic uncertainty (that arising from a lack of knowledge) from aleatory uncertainty (arising due to variability in a characteristic of the system being modelled).

Epistemic uncertainty typically arises because relevant data are sparse or because their applicability to the situation under consideration is unclear. For example, distribution coefficients (K_d values) for a radionuclide in soils may vary by several orders of magnitude, as illustrated for uranium in figure 8.2.

For uranium, the available data are sufficiently detailed that distinctions in K_d values can be made between different pH regimes. However, other soil characteristics that determine the range of variation within each pH regime have not been established. Therefore, when considering a soil in a specific assessment context, knowledge of its pH will determine which of the distributions shown in figure 8.2 is appropriate, but even a detailed characterisation of that soil will not determine where it should be located on the associated cumulative distribution. Thus, the K_d value to be adopted for assessment purposes is associated with an epistemic

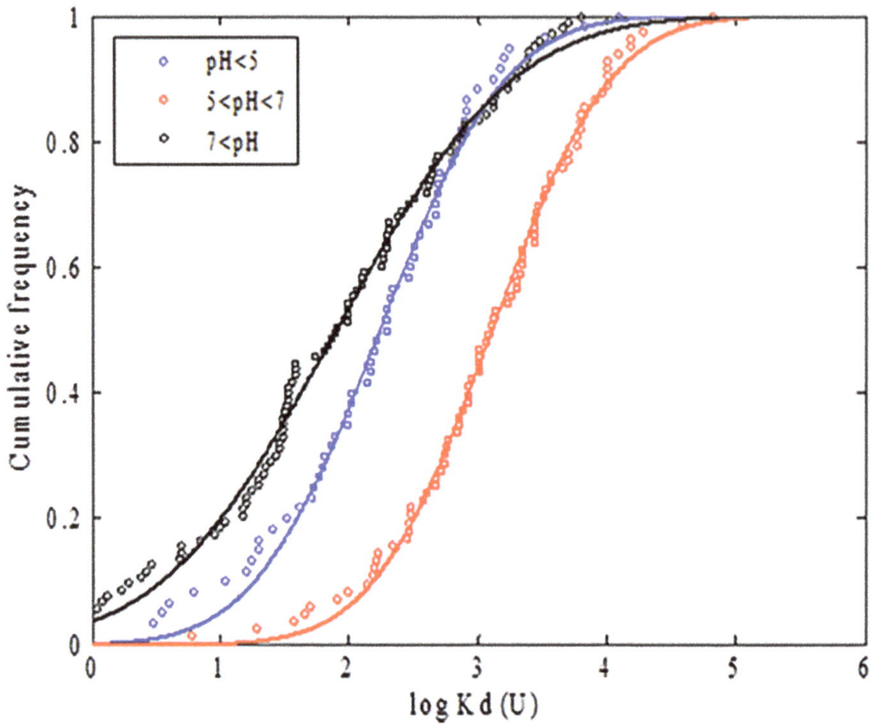

Figure 8.2. Cumulative distribution functions of K_d values for uranium for various soil pH groups. Note that the lognormal distribution at pH > 7 may be an artefact, because the K_d value changes rapidly above pH 7, so the observed lognormal distribution may arise only from the pattern of sampling at different pH values above pH 7. IAEA (2023), © IAEA, 2023, p. 212, ISBN 978-92-0-126722-1

uncertainty relating to a lack of knowledge as to how soil properties influence K_d values for uranium. This uncertainty can be captured by running the model many times (i.e. as multiple realisations) with the K_d value in each realisation sampled from the cumulative probability distribution.

Aleatory uncertainty arises when a characteristic of the assessment model varies between locations, individuals or with time. A typical example is human diet, which varies between individuals. Thus, two individuals located in the same environment will have different intakes of a radionuclide, and hence receive different committed effective doses, because of the differences in their diet. Unlike epistemic uncertainties, aleatory uncertainties cannot be reduced by the acquisition of more knowledge. In some assessment studies, the results are presented as distributions that distinguish between epistemic and aleatory uncertainties.

In deterministic studies, best-estimate, point values for model input parameters are often used and point estimates of results are generated. However, this can give an inappropriate impression of confidence in the result obtained. Also, in non-linear models, selection of best-estimate parameter values for all inputs will not generally result in a best-estimate value for the output. For example, if the inputs are characterised by probability density functions (pdfs), taking arithmetic average values for each pdf and using the averages as inputs will not, in general, give the same result as sampling from the input distributions and computing the arithmetic average of the output pdf.

In general, it is good practice to explore the range of results that arise through sensitivity and uncertainty analyses. In a sensitivity analysis, the input parameter values are varied singly or in combination and the resultant changes in the values of the output parameters are determined. In such a sensitivity analysis, the ranges over which the parameters are varied are typically determined by the ranges that have been observed or estimated to apply, but wider ranges may be used to determine whether there are tipping points where the behaviour of the system changes markedly. In an uncertainty analysis, the input parameter values are sampled from their associated pdfs to generate pdfs of output parameter values. An uncertainty analysis may be used to investigate the sensitivity of the model to changes in specific input parameters in the presence of uncertainties in the others. However, it is often used only to provide statistical measures relating to the output pdfs, e.g. mean, median and various percentile values.

References

Bear J 1979 *Hydraulics of Groundwater* (New York: McGraw-Hill)

Birchall A and James A C 1989 A microcomputer algorithm for solving first-order compartmental models involving recycling *Health Phys.* **56** 857–68

Ewen J and Parkin G 1996 Validation of catchment models for predicting land-use and climate change impacts *J. Hydrol.* **175** 583–94

Hiscock K M and Bense V F 2014 *Hydrogeology: Principles and Practice* 2nd edn (Oxford: Wiley Blackwell)

Hudson J A 1992 *Rock Engineering Systems: Theory and Practice* Ellis Horwood Series in Civil Engineering (Ellis Horwood)

IAEA 2003 *'Reference Biospheres' for Solid Radioactive Waste Disposal, Report of BIOMASS Theme 1 of the IAEA's BIOsphere Modelling and ASSessment Programme* (Vienna: International Atomic Energy Agency) IAEA-BIOMASS-6 (2003)

IAEA 2004 *Sediment Distribution Coefficients and Concentration Factors for Biota in the Marine Environment* (Vienna: International Atomic Energy Agency) IAEA-TRS-422

IAEA 2009 *Quantification of Radionuclide Transfers in Terrestrial and Freshwater Environments for Radiological Assessments* (Vienna: International Atomic Energy Agency) IAEA-TECDOC-1616

IAEA 2010 *Handbook of Parameter Values for the Prediction of Radionuclide Transfer in Terrestrial and Freshwater Environments* (Vienna: International Atomic Energy Agency) IAEA-TRS-472

IAEA 2014 *Handbook of Parameter Values for the Prediction of Radionuclide Transfer to Wildlife* (Vienna: International Atomic Energy Agency) Technical Reports Series No. 479

IAEA 2019 *The Environmental Behaviour of Uranium* Technical Reports Series No. 488 International Atomic Energy Agency, Vienna 200–65

IAEA 2020 BIOMASS 2020, IAEA Report

IAEA 2023 Quantification of environmental transfer process of uranium *The Environmental Behaviour of Uranium* Technical Reports Series No. 488 (IAEA) 200–265

IUR 2006 Recommendations for improving predictions of the long-term environmental behaviour of ^{14}C, ^{36}Cl, ^{99}Tc, ^{237}Np and ^{238}U, Report 06 (International Union of Radioecologists) www.iur-uir.org

Klos R and Thorne M C 2020 Use of interaction matrices to formalise the development of conceptual models of contaminant transport in the biosphere and the translation of those conceptual models into mathematical models *J. Radiol. Prot.* **40** 40–67

NEA 1999 Safety Assessment of Radioactive Waste Repositories—An International Database of Features, Events and Processes. A Report on of the NEA Working Group on Development of a Database of Features, Events and Processes Relevant to the Assessment of Post-closure Safety of Radioactive Waste Repositories. Nuclear Agency of the Organisation for Economic Cooperation and Development (OECD/NEA), Paris. Electronic version 1.2 of the NEA FEP Database developed on behalf of the Nuclear Energy Agency by Safety Assessment Management Ltd. with support of Quintessa Ltd.

NEA 2006 Electronic version 2.1 of the NEA FEP database developed on behalf of the Nuclear Energy Agency by Safety Assessment Management Ltd.

Parkin G, Anderton S P, Ewen J, O'Donnell G M, Thorne M C and Crossland I G 1999 A physically based approach to modelling radionuclide transport in the biosphere *J. Radiol. Prot.* **19** 319–32

Pérez-Sánchez D and Thorne M C 2014 Modelling the behaviour of uranium-series radionuclides in soils and plants taking into account seasonal variations in soil hydrology *J. Environ. Radioact.* **131** 19–30

Shao Q, Matthäi S K and Gross L 2019 Efficient modelling of solute transport in heterogeneous media with discrete event simulation *J. Comput. Phys.* **384** 134–50

SKB 2018 BIOMASS 2020: Interim Report, Report R-18-02, March 2018, Svensk Kärnbränslehantering AB, Solna, Sweden

Thorne M C 2012 Modelling radionuclide transport in the environment and calculating radiation doses *Radionuclide Behaviour in the Natural Environment: Science, Implications and Lessons for the Nuclear Industry* ed C Poinssot and H Geckeis (Cambridge: Woodhead Publishing) ch 14

Thorne M C 2014 Kinetic models for representing the uptake of radionuclides in plants *Radionuclide Contamination and Remediation through Plants* ed D K Gupta and C Walther (Heidelberg: Springer) pp 215–33

Wheater H S, Bell J N B, Butler A P, Jackson B M, Ciciani L, Ashworth D J and Shaw G G 2007 *Biosphere Implications of Deep Disposal of Nuclear Waste: The Upwards Migration of Radionuclides in Vegetated Soils* (London: Imperial College Press)

Chapter 9

Biokinetic models for humans

M C Thorne

9.1 Introduction

Chapter 6 provided a largely qualitative introduction to the biokinetic models used for assessing the distribution and retention of radionuclides entering the body by various routes. In this chapter, a comprehensive quantitative account of these models is provided, including examples of their application in evaluating the radiological significance of intakes of selected radionuclides and in estimating intakes from measurements of radionuclide concentrations in biological samples, with an emphasis on urinalysis. In sections 9.2–9.4, detailed descriptions are given of the behaviour of inhaled radionuclides in the respiratory system (section 9.2), of ingested radionuclides in the alimentary tract (section 9.3) and of retention of radionuclides in wounds and their transfer to the systemic circulation (section 9.4). Section 9.5 then addresses retention of radionuclides in the body after they have entered the systemic circulation and their loss by urinary and faecal excretion, and by exhalation. Section 9.6 then considers the special issues that arise in modelling the behaviour of radionuclides that are isotopes of elements with an important physiological role. Specifically, consideration is given to H-3 and C-14. Section 9.7 then addresses the distribution and retention of Rn-222 produced in the body from its parent Ra-226. Finally, section 9.8 describes how biokinetic models can be used to interpret monitoring data to infer earlier intakes of radionuclides and the radiation doses arising from those intakes.

9.2 Behaviour of inhaled radionuclides in the respiratory system

Radionuclides are usually inhaled either incorporated in, or attached to, aerosols. These can be small solid particles or liquid droplets. Here the term 'particle' is used to refer both to solid particles and droplets. To remain suspended in air for long enough to be available for inhalation, the individual particles or droplets can be no more than a few tens of micrometres in diameter. Thus, the sizes of interest range from substantially less than 1 μm up to about 100 μm. In some contexts,

doi:10.1088/978-0-7503-2424-3ch9

consideration must be given to hygroscopic aerosols. On entering the high humidity environment of the respiratory tract, the individual hygroscopic aerosol particles take up water and increase in size, which affects their pattern of deposition.

The behaviour of aerosol particles both in the atmosphere and after their entry into the respiratory system is affected by both their physical diameter and by their aerodynamic diameter. In addition, the shape of the particles is significant, and this is of special importance for fibrous materials, such as asbestos. The aerodynamic diameter of a particle is defined as the diameter of a particle of unit density that has the same settling velocity in air as the particle of interest. The settling velocity is determined by equating Stokes' drag on the particle to the force exerted by gravity, i.e. for a spherical particle:

$$6\pi\mu Rv = 4\pi R^3\rho g \tag{9.1}$$

where μ (kg m^{-1} s^{-1}) is the dynamic viscosity of the air;
R (m) is the radius of the particle;
v (m s^{-1}) is the terminal settling velocity;
ρ (kg m^{-3}) is the density of the particle;
g (m s^{-2}) is the acceleration due to gravity.

Rearranging equation (9.1) gives:

$$v = 2R^2\rho g/3\mu \tag{9.2}$$

Defining v_u and ρ_u as relating to a unit-density sphere, the aerodynamic diameter of a particle is given by:

$$R_a{}^2\rho_u = R^2\rho \text{ or } R_a = R\{\rho/\rho_u\}^{1/2} \tag{9.3}$$

Thus, for high-density particles, e.g. UO_2, the aerodynamic diameter is larger than the physical diameter.

For sub-micron particles, equation (9.3) is not precise, because it neglects the consideration that air is not a continuous fluid but is composed of individual molecules. To take account of this, equation (9.3) can be modified by the inclusion of a slip-correction factor. However, this is of limited significance, because the deposition of sub-micron particles is dominated by processes that depend on their physical diameter rather than their aerodynamic diameter.

To be inhaled, particles must be present in the breathing zone of the exposed individual. However, the concentration of particles in the breathing zone may differ substantially from the concentration in the general area. This arises both because the activity of the individual can create a local aerosol, e.g. by resuspension from surfaces, and because the individual can affect the pattern of local turbulence, such that particles are preferentially lost from the breathing zone if the presence of the individual reduces the degree of turbulence in that zone. This is a reason why personal air samplers can give substantially different results from those obtained from fixed area samplers.

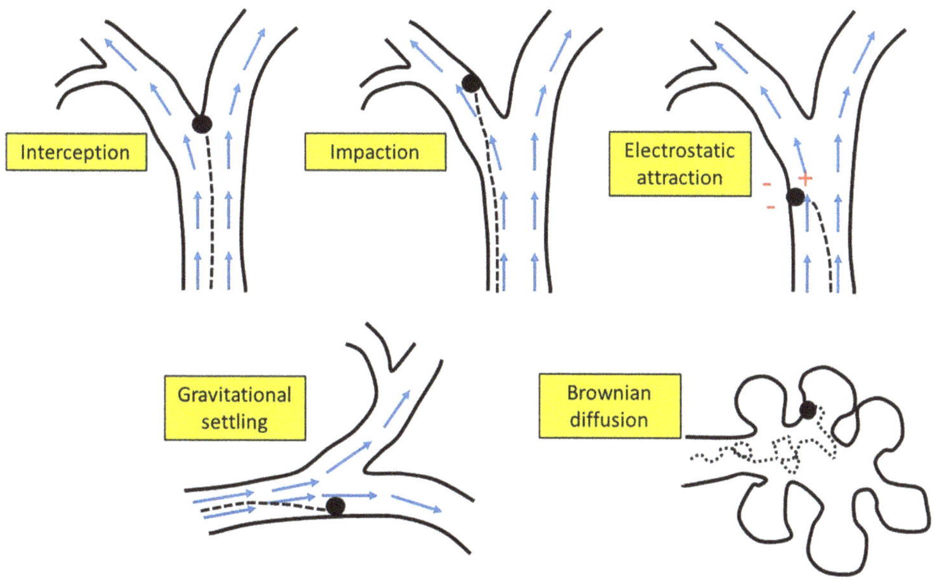

Figure 9.1. Mechanisms for the deposition of aerosols in the respiratory tract (blue arrows illustrate the pattern of airflow and dotted lines indicate the paths of aerosol particles or droplets).

The various mechanisms that influence deposition comprise interception, impaction, gravitational settling and Brownian diffusion. These are illustrated in figure 9.1.

Interception is the incidental meeting of a particle with a surface without the path of the particle deviating from that of the airflow. The larger the particle, the more likely interception is to occur. Impaction reflects the finite inertia of a particle, such that when the airflow changes direction, the particle tends to continue its original course. The susceptibility to impaction increases with particle mass. Gravitational settling also causes particles to deviate from the airflow direction, whereas electrostatic attraction affects charged particles by inducing an opposite charge in the walls of the respiratory tract. Brownian diffusion due to the aerosol particles being buffeted by air molecules is a random walk process. Its significance increases as the size of the particle decreases. In all cases, deposition occurs when a particle encounters the wall of the respiratory tract and remains bound to it.

The ICRP (1994) model of the respiratory tract calculates the deposition of aerosols in different regions of the tract by assuming that it acts as a sequence of filters both on inspiration and expiration. Each filter is characterised by its volume and its overall efficiency in removing airborne particles. On inspiration, the volume of air passing through successive filters decreases due to storage in the preceding filters, i.e. it is the volume of inspired air less the cumulative volume of the preceding filters. On exhalation, the volume passing through each filter is equal to the volume that passed through on inspiration. Once the flows have been calculated, the efficiencies can be used together with the fractions of the inspired air passing through each filter to estimate the fractions of the inhaled aerosol deposited in each

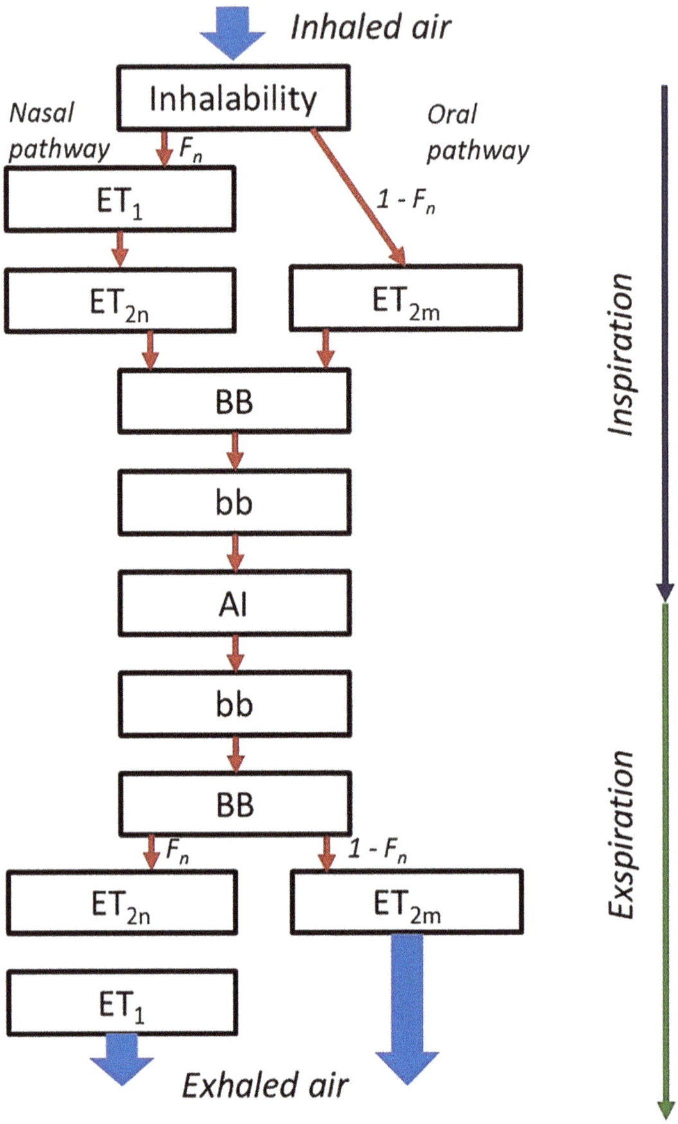

Figure 9.2. Representation of deposition in the respiratory system in terms of a series of filters (ET = extra-thoracic, BB = bronchial, bb = bronchiolar, AI = alveolar-interstitial, F_n = fraction of inhalation partitioned to the nasal pathway).

successive filter, taking the depletion of the aerosol by preceding filters into account. The filtration system, separating the inhalation and exhalation phases, is illustrated in figure 9.2.

The pattern and amount of deposition depends on respiratory parameters and on the size distribution of the inhaled aerosol. With respect to respiratory parameters, there are substantial differences in deposition for nose and mouth breathers,

Table 9.1. Reference pattern of deposition of inhaled aerosols for occupational and environmental exposures (from table 28 of ICRP (1994)).

Region	Deposition for occupational exposure (% of intake)	Deposition for environmental exposure (% of intake)
ET_1	33.9	14.2
ET_2	33.9	17.9
BB	1.8	1.1
Bb	1.1	2.1
AI	5.3	11.9
Total	82.0	47.3

particularly for larger particles, as these penetrate more deeply into the respiratory system in mouth breathers. Other respiratory parameters of relevance include tidal volume, volumetric flow rate of the inspired air and ventilation rate. These vary with age and level of activity. ICRP (1994) provides reference values for 3-month-old and 1-year-old infants, 5-year-old and 10-year-old children, and separately for males and females of age 15 and adult. Distinctions are made between the exercise levels of sleep, sitting, light exercise and heavy exercise. Reference patterns of deposition are given by the ICRP (1994) for occupational and environmental exposures. These are listed in table 9.1. However, it is emphasised that these are for aerosols of specific sizes (AMAD = 5 μm for occupational exposures and 1 μm for environmental exposures) and specific patterns of activity (sitting plus light exercise for occupational exposures and a mix of sitting, sleeping and light exercise for environmental exposures). For other aerosol sizes and patterns of activity, the amount and distribution of deposited activity would be somewhat different.

The smaller aerosol size adopted for environmental exposure results in deeper penetration into the respiratory tract. However, it also reduces the overall filtration efficiency such that a larger proportion of the inhaled aerosol fails to deposit in the respiratory tract and is lost on exhalation.

Once an aerosol has been deposited in the respiratory tract, it is subject to the competing processes of particle transport and solubilisation. To represent clearance by these competing mechanisms, the respiratory tract is represented by a set of interconnected compartments, as described in chapter 6 and illustrated in figures 6.1 and 6.2. In respect of deposition in regions ET_2, BB and bb a distinction is made between material that is available for particle transport to the throat and that which is sequestered in tissue and transported to the extra-thoracic or thoracic lymph nodes. For the bronchial and bronchiolar regions, a further distinction is made between unsequestered material that is transported rapidly or more slowly to the throat. This transport takes place in the mucus layer on bronchial and bronchiolar surfaces, with the mucus being transported upward through the respiratory system by the movement of cilia developed from specialised cells. This mechanism does not exist in the alveolar region. Thus, particle transport from that region is much slower.

Table 9.2. Partitioning of deposition between compartments of the respiratory tract (from table 29 of ICRP (1994)).

Region	Compartment	Fraction of deposit in region assigned to this compartment
ET_2	ET_2	0.9995
	ET_{seq}	0.0005
BB	BB_1	$0.993 - f_s$
	BB_2	f_s
	BB_{seq}	0.007
bb	bb_1	$0.993 - f_s$
	bb_2	f_s
	bb_{seq}	0.007
AI	AI_1	0.3
	AI_2	0.6
	AI_3	0.1

Table 9.3. Clearance rates by particle transport from the compartments of the ICRP model of the respiratory tract (from table 29 of ICRP (1994)).

Source compartment	Receptor compartment	Rate (d^{-1})	Half time (d)
AI_1	bb_1	0.02	35
AI_2	bb_1	0.001	700
AI_3	bb_1	0.0001	7000
AI_3	Thoracic lymph nodes	0.000 02	Not stated
bb_1	BB_1	2	0.35
bb_2	BB_1	0.03	23
bb_{seq}	Thoracic lymph nodes	0.01	70
BB_1	ET_2	10	0.07
BB_2	ET_2	0.03	23
BB_{seq}	Thoracic lymph nodes	0.01	70
ET_2	Gastrointestinal tract	100	0.007
ET_{seq}	Extra-thoracic lymph nodes	0.001	700
ET_1	Environment	1	0.7

The partitioning of deposition between these various compartments is summarised in table 9.2.

The slowly cleared fraction is assumed to be size dependent. It is 0.5 for small particles and decreases exponentially for particles larger than a critical size.

Clearance rates from the various compartments are listed in table 9.3.

Table 9.4. Default values of solubilisation parameters for use with the ICRP respiratory tract model (from table 18 of ICRP (1994)).

Parameter (units)	Type F	Type M	Type S
Fraction dissolving rapidly (–)	1	0.1	0.001
Rapid dissolution rate (d^{-1})	100	100	100
Slow dissolution rate (d^{-1})	Not applicable	0.005	0.0001

Where clearance half times are less than one day, particle clearance is likely to dominate for all but the most soluble materials. In contrast, where clearance times are hundreds or thousands of days, solubilisation is likely to be the dominant removal mechanism for all but the most insoluble materials.

In terms of solubilisation, the model shown in figure 6.2 can be approximated as a two-compartment system, with the two compartments corresponding to fast and slow dissolution, respectively. This system is characterised by three parameters. These are the fraction of the deposited material that dissolves rapidly, the rate of rapid dissolution, and the rate of dissolution of the remaining fraction of the deposit that dissolves more slowly. As discussed in chapter 6, there are three default categories of material defined by their rates of solubilisation. These are F (fast), M (moderate) and S (slow). Default parameter values for these categories of material are listed in table 9.4.

9.3 Behaviour of ingested radionuclides in the gastrointestinal tract

The compartmental structure of the ICRP (2006) model of the gastrointestinal tract is described in chapter 6 and illustrated in figure 6.3. To parameterise this model, consideration must be given to the rates at which material is transported through the various parts of the gastrointestinal tract. This topic is reviewed in chapter 6 of ICRP (2006) in terms of transit times through the various components of the tract. Although clearance of some materials from some parts of the tract, e.g. liquids from the stomach, is well described by a single exponential, in other cases, e.g. transfer of solids from the stomach and transfer of all materials through the small intestine, the process is more linear, with boluses of material being transported rather than the material becoming well mixed. Mean transit times for any component of the gastrointestinal tract typically depend on age, gender and whether the ingested material is a solid or a liquid. Also, particularly for the stomach, there are distinctions between calorific liquids and non-calorific liquids such as water.

Although the kinetics of transfers through some components of the gastrointestinal tract do not conform closely to compartmental, first-order kinetics, the ICRP (2006) model is formulated in compartmental terms. Therefore, transfers through the gastrointestinal tract are expressed in terms of rate coefficients derived from the observed transit times. Reference values are set out in table 9.5.

Table 9.5. Transfer coefficients (d^{-1}) for the movement of alimentary tract contents between regions (from table 6.6 of ICRP (2006)).

Region	Age 3 months	Age 1 year	Age 5–15 years	Adult male	Adult female
Mouth					
Solids		5760	5760	5760	5760
Liquids	43 200	43 200	43 200	43 200	43 200
Total diet	43 200	7200	7200	7200	7200
Oesophagus (fast)					
Solids		10 800	10 800	10 800	10 800
Liquids	21 600	17 280	17 280	17 280	17 280
Total diet	21 600	12 343	12 343	12 343	12 343
Oesophagus (slow)					
Solids		1920	1920	1920	1920
Liquids	2880	2880	2880	2880	2880
Total diet	2880	2160	2160	2160	2160
Stomach					
Solids		19.2	19.2	19.2	13.71
Caloric liquids	19.2	32	32	32	24
Non-caloric liquids	144	48	48	48	48
Total diet	19.2	20.57	20.57	20.57	15.16
Small intestine	6	6	6	6	6
Right colon	3	2.4	2.182	2	1.5
Left colon	3	2.4	2.182	2	1.5
Rectosigmoid	2	2	2	2	1.5

In most radiological protection contexts, the short residence times (seconds to minutes) in the mouth and oesophagus, and differences by age and sex, are of little significance.

Transit through the gastrointestinal tract competes with uptake from the tract in determining whether radionuclides enter the systemic circulation or remain unabsorbed and are lost in faecal excretion. This is discussed in chapter 6, where it is shown that, in the typical case in which absorption is assumed to occur only from the small intestine:

$$k_{\mathrm{SI,B}} = f_{\mathrm{SI}}\, k_{\mathrm{SI,RC}} / \left(1 - f_{\mathrm{SI}}\right) \tag{9.4}$$

where $k_{\mathrm{SI,B}}$ (d^{-1}) is the rate coefficient from the small intestine to blood, f_{SI} (–) is the fractional absorption from the small intestine and $k_{\mathrm{SI,RC}}$ (d^{-1}) is the rate coefficient for transfer of the radionuclide from the small intestine to the right colon. From table 10.5, this has a reference value of 6 d^{-1}, independent of age and sex.

Values of fractional gastrointestinal absorption have been estimated by various methods in both humans and other vertebrates (mainly mammals and birds) over several decades. The values obtained depend on the radionuclide, its physiochemical form, the species under consideration and the method adopted. Values can be obtained after the administration of a radionuclide or from studies based on administration of the corresponding stable element, or from balance studies in which the rate of ingestion of the stable element is compared with rates of loss by various excretory pathways, primarily urine and faeces.

A preferred method of estimating fractional gastrointestinal absorption is to study whole-body or tissue retention after either ingestion or intravenous injection. If the chemical form of the radionuclide on entry into the blood is the same for the two routes of exposure, then the ratio of activity concentrations in organs and tissues for the two modes of administration at times long after exposure will be a measure of the fractional gastrointestinal absorption. However, this is not the situation in the first few hours to days following administration, because uptake from the gastrointestinal tract to blood is delayed and protracted compared with uptake following intravenous injection. Therefore, in using this method, the kinetics of retention over at least the first few days should be followed to ensure that data from the early period, when the kinetics of retention differ between modes of administration, are excluded from consideration.

The issue of whether the chemical form entering the blood following ingestion is the same as that delivered by intravenous injection requires consideration not only of the chemical form administered, but also its behaviour following administration. For example, a radionuclide incorporated in an organic compound or a variety of such compounds found in the diet will be subject to digestive processes and may, therefore, enter the blood in a simple ionic form. Conversely, intravenous injection of low specific activity radionuclides or radionuclides together with a stable carrier, may involve the injection of a significant mass of material and result in processes such as colloid formation that do not occur following ingestion.

A second, popular method of estimating fractional gastrointestinal absorption is to administer a radionuclide or stable element orally and subsequently determine the amounts excreted in the urine and the faeces. However, two issues arise with application of this method. First, if the radionuclide is well-retained in the body, only a small fraction of that taken up into the systemic circulation may be excreted over the duration of the study. Second, some of the material that enters the systemic circulation may be subsequently re-excreted back into the gastrointestinal tract. Both these issues can result in systematic underestimation of the degree of fractional gastrointestinal absorption. In some situations, it may be possible to overcome these issues by taking blood samples and estimating the uptake to blood directly, rather than inferring it indirectly from losses in urinary excretion. Alternatively, whole-body counting may be used to estimate the total body burden of a radionuclide after oral administration.

For some radionuclides, excretion in urine may not be the only, or even the predominant, route by which a radionuclide entering the systemic circulation is lost from the body. Sweat and exhaled air may also be important, e.g. for H-3 and C-14.

Table 9.6. Recommended values of fractional gastrointestinal absorption for adult humans (from table 25 of IAEA (2010)).

Elements	Fractional absorption
H, C, Cs, S, Mo, I	1.0
Se	0.8
Zn, Tc, Po	0.5
Te, Sr, Ca	0.3
Ba, Ra, Pb	0.2
Co, Fe, Sb	0.1
Ru, Ni, Ag	0.05
U	0.02
Zr, Nb	0.01
Ce, Th, Np, Pu, Am, Cm	0.0005

In some contexts, stable element data from dietary balance studies may be used to estimate fractional gastrointestinal absorption. This involves measuring stable element concentrations in diet and excreta. However, such studies have similar issues to the oral administration studies discussed above. Specifically, there can be problems of interpretation where intake and excretion are not in balance and the body is either accumulating or becoming depleted in the element under consideration.

The ICRP (2006) has provided recommended gastrointestinal fractional absorption values for adult humans and these are conveniently listed in IAEA (2010). The IAEA table is reproduced as table 9.6.

Hydrogen is typically present as tritiated water or incorporated in organic forms. Similarly, carbon is typically present in organic forms. Therefore, the recommended fractional gastrointestinal absorption values for these elements reflect the consideration that dietary materials are efficiently digested and the nutrients contained within them are well absorbed. A high degree of absorption also arises for other essential elements, e.g. sulphur and iodine, as well as for elements that are biochemical analogues for essential elements (e.g. caesium is a close biochemical analogue of potassium, whereas technetium is a rather more remote biochemical analogue of iodine). Selenium is both a biochemical analogue of sulphur and an essential trace element. Calcium is an important essential element and is moderately well absorbed, as are the other alkaline earths (strontium, barium and radium). At the other end of the spectrum, the lanthanides (here represented by cerium) and the higher actinides are only very poorly absorbed, even when incorporated in the diet. The transition metals, e.g. cobalt, nickel and iron, exhibit an intermediate degree of adsorption. These elements are essential to health, but not in large amounts.

It is important to recognise that these recommended values cannot be applied to all chemical forms of each element. For example, C-14 incorporated in carbides is only poorly adsorbed from the gastrointestinal tract, as is $SrTiO_3$ (strontium titanate). Similarly, high-fired oxides of uranium and plutonium (UO_2 and PuO_2) exhibit values of fractional gastrointestinal absorption substantially lower than the recommended values listed in table 9.6. Also, absorption can be much increased in the very young.

In respect of uptake, it is important to recognise that this involves two stages. In the first stage, the element is absorbed from the lumen of the gastrointestinal tract into the cells lining its walls. In the second stage, the element is transferred from the interior of these cells to the systemic circulation. For an element that is a distant biochemical analogue of an essential element, the first of these stages may occur efficiently, but not the second. Thus, plutonium can follow the pathways of iron metabolism in being taken up into gastrointestinal cells but cannot follow the second stage of iron metabolism and is, therefore, retained in those cells rather than being transferred to the systemic circulation. It is then released back into the lumen of the tract on a timescale of a few days governed by the lifetime of intestinal epithelial cells. This process of uptake and retention in the walls of the tract can be of importance in individuals of age less than one year.

9.4 Behaviour of radionuclides present in wounds

There are extensive data on the behaviour of radionuclides in wounds that have allowed the NCRP (2006) to develop a comprehensive model for radionuclide retention in and transport from the sites of wounds and to parameterise that model for a wide variety of radionuclides. The model comprises seven compartments, as illustrated in figure 9.3.

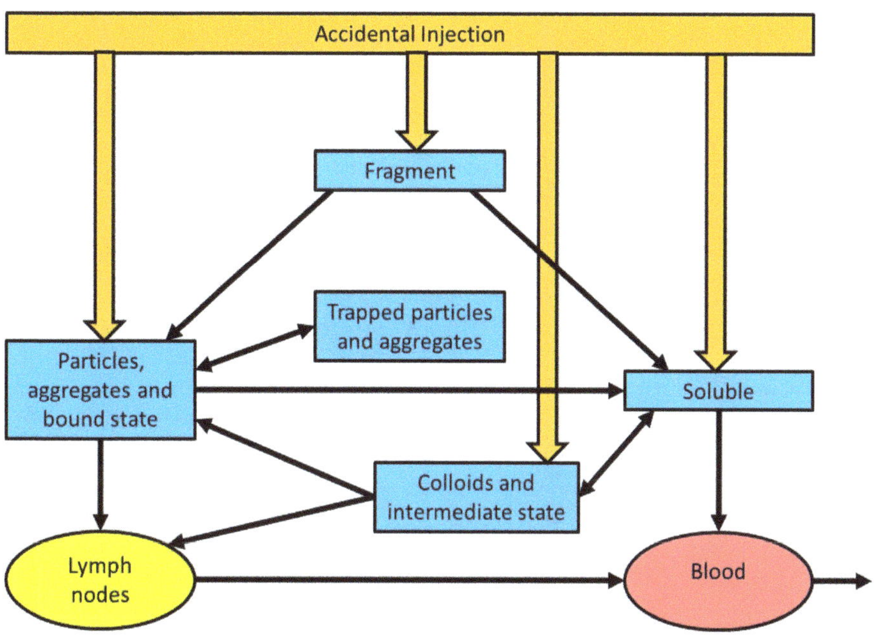

Figure 9.3. Generic model for radionuclide retention in and transport from wounds. NCRP (2006). Reproduced with permission of the National Council on Radiation Protection and Measurements, http://NCRPonline.org.

Because of limitations in the data available the model is independent of the site of the wound. However, because most wounds occur in the skin, subcutaneous tissue or muscle, this is a reasonable approximation. Also because of a lack of relevant data, no distinction is made between punctures, abrasions, lacerations or burns, even though the degree of tissue injury is likely to affect the kinetics of radionuclide transport.

Radionuclides may enter a wound incorporated in small particles or in larger fragments of material. Also, although a radionuclide may enter a wound in soluble form, a tendency to hydrolyse at the neutral pH of tissue fluids may lead to the production of a colloidal form. The compartment for trapped particles and aggregates represents the effects of fibrous encapsulation of particles and aggregates. The compartment described as comprising particles, aggregates and bound state includes radionuclides that have been transformed beyond the colloidal state, e.g. comprising robust hydroxide precipitates and polymers. As this material has recognisable particulate characteristics, it is susceptible to transfer to the regional lymph nodes.

This wound model was parameterised by adjusting the various rate constants to give good fits to a wide range of human and animal data. Based on this analysis, default transfer rates for the wound model were developed for seven categories of radionuclides in wounds. These default rates are listed in table 9.7. Weak, moderate, strong and avid describe the retention at the site of the wound for radionuclides entering the wound in solution. This initial retention in the wound is represented for

Table 9.7. Default transfer rates used in the NCRP biokinetic model for wounds (from table 4.12 of NCRP (2006)).

	Transfer rate (d^{-1})						
	Radionuclides initially in solution						
Transfer	Weak	Moderate	Strong	Avid	Colloids	Particles	Fragments
Soluble to blood	45	45	0.67	7.0	0.5	100	—
Soluble to CIS	20	30	0.6	30	2.5	—	—
CIS to soluble	2.8	0.4	0.024	0.03	0.025	—	—
CIS to PABS	0.25	0.065	0.01	10	0.05	—	—
CIS to lymph nodes	$2\ 10^{-5}$	$2\ 10^{-5}$	$2\ 10^{-5}$	$2\ 10^{-5}$	0.002	—	—
PABS to soluble	0.08	0.02	0.0012	0.005	0.0015	$2\ 10^{-4}$	0.0
PABS to lymph nodes	$2\ 10^{-5}$	$2\ 10^{-5}$	$2\ 10^{-5}$	$2\ 10^{-5}$	$4\ 10^{-4}$	0.0036	0.004
PABS to TPA	—	—	—	—	—	0.04	0.7
TPA to PABS	—	—	—	—	—	0.0036	$5\ 10^{-4}$
Lymph nodes to blood	—	—	—	—	0.03	$6\ 10^{-4}$	0.03
Fragment to soluble	—	—	—	—	—	—	0.0
Fragment to PABS	—	—	—	—	—	—	0.008

Notes: CIS = Colloids and intermediate state; PABS = Particles, aggregates and bound state; TPA = Trapped particles and aggregates; Colloids are taken to be injected into the CIS compartment; Particles are taken to be injected into the PABS compartment; Fragments are taken to be injected into the fragment compartment.

Table 9.8. Parameter values for retention of radionuclides at the site of the contaminated wound.

	Wound retention equation parameters					
Category	A_1 (%)	λ_1 (d^{-1})	A_2 (%)	λ_2 (d^{-1})	A_3 (%)	λ_3 (d^{-1})
Weak	55	55	40	6.0	5.0	0.1
Moderate	55	55	35	0.5	10	0.02
Strong	50	1.0	30	0.03	20	0.001
Avid	19	37	81	0.001	—	—
Colloid	15	3.0	8	0.055	77	$7\ 10^{-4}$
Particle	5	0.05	95	$4\ 10^{-4}$	—	—
Fragment	0.5	0.009	99.5	$6.5\ 10^{-6}$	—	—

Notes: Weak, Moderate, Strong and Avid refer to radionuclides initially in solution. The retention function is $A_1\exp(-\lambda_1 t) + A_2\exp(-\lambda_2 t) + A_3\exp(-\lambda_3 t)$.

both radionuclides in solution and other physicochemical forms using either two or three exponential components with the parameter values listed in table 9.8. Similar multi-exponential retention functions are also provided for the other physicochemical forms of radionuclides listed in table 9.7.

9.5 Retention after entry into the systemic circulation

As illustrated in chapter 6, radionuclide retention following entry into the systemic circulation is typically represented using a multi-compartmental model. Both the structure of the model and its parameterisation are element dependent. Additionally, the parameter values adopted may depend upon the age of the exposed individual, or other factors, such as the rate of intake of the stable element. The latter factor is of relevance when concentrations of the stable element in tissues and organs are homeostatically controlled. In such cases, the biological half-life of a radioisotope of the element in the tissues and organs of the body varies inversely with the rate of intake of the stable element. This is exemplified by the behaviour of Cl-36. The body content of stable chlorine (as chloride) is about 95 g and it is relatively uniformly distributed throughout the tissues and organs of the body, freely exchanging between them in ionic form. Because of this free and rapid exchange, the body is reasonably approximated by a single, well-mixed compartment for chloride retention. In ICRP (1980), the daily intake of chloride in food and fluids was estimated as 5.2 g. To maintain a constant body content of 95 g, this meant that 5.2 g d^{-1} had to be lost from a single well-mixed compartment of 95 g. Therefore, the rate coefficient for loss was estimated as $5.2/95 = 0.0055\ d^{-1}$, corresponding to a biological half-life of 12.7 d. For comparison, the ICRP (1980) commented that studies on two healthy humans had found a biological half-life for Cl-36 retention of six days when they were fed on a diet high in chloride and 30 days when they were fed a diet low in chloride. The ICRP (1980) developed similar arguments for potassium for which a body content of 140 g and an intake rate of 3.3 g d^{-1} were used to derive a loss rate of $3.3/140 = 0.0024\ d^{-1}$, corresponding to a biological half-life of 29.4 d, which the ICRP rounded to 30 d.

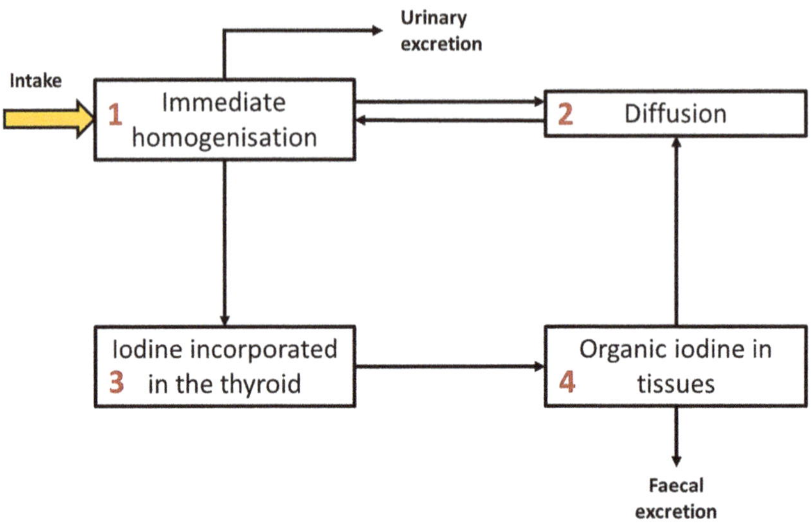

Figure 9.4. Four-compartment model for iodine metabolism.

Homeostatic control may be more complex than in the two cases illustrated above. A good illustration of this arises with iodine. Although more up-to-date models for iodine metabolism are available (e.g. ICRP (2017)), a four-compartment model, as described in Coughtrey *et al* (1983), illustrates the principles that are involved. Such a model is illustrated in figure 9.4.

This model can be characterised in terms of four, first-order ordinary differential equations:

$$dQ_1/dt = I + k_{21}Q_2 - (k_{12} + k_{13} + k_{1U} + k_R)Q_1 \tag{9.5a}$$

$$dQ_2/dt = k_{12}Q_1 + k_{42}Q_4 - (k_{21} + k_R)Q_2 \tag{9.5b}$$

$$dQ_3/dt = k_{13}Q_1 - (k_{34} + k_R)Q_3 \tag{9.5c}$$

$$dQ_4/dt = k_{34}Q_3 - (k_{42} + k_{4F} + k_R)Q_4 \tag{9.5d}$$

where Q_i is the radioisotope or stable iodine content of compartment i, I is the rate of intake of the radionuclide or stable iodine, k_{ij} is the rate of transfer of radioisotope or stable iodine from compartment i to compartment j, to urinary excretion (U) or to faecal excretion (F), and k_R is the radioactive decay constant for the radioisotope or zero for the stable element.

Assuming that the body is in balance for the stable element, the left-hand side of each of these equations can be set to zero, as can k_R. This gives:

$$I = (k_{12} + k_{13} + k_{1U})Q_1 - k_{21}Q_2 \tag{9.6a}$$

$$k_{12}Q_1 - k_{21}Q_2 = -k_{42}Q_4 \tag{9.6b}$$

$$k_{34}Q_3 = k_{13}Q_1 \tag{9.6c}$$

$$k_{34}Q_3 = (k_{42} + k_{4F})Q_4 \tag{9.6d}$$

The concentration of organic iodine in tissues is required to be constant because it has a hormonal function. This constancy is maintained by varying the efficiency with which iodine is taken up into the thyroid. Therefore, Q_4 is maintained constant by varying k_{13}. By solving equation (9.6), it can be shown that:

$$k_{13} = k_{1U}(k_{42} + k_{4F})Q_4/\left(I - k_{4F}Q_4\right) \tag{9.7}$$

Coughtrey *et al* (1983) give typical values of $k_{12} = 48\ \text{d}^{-1}$, $k_{21} = 18.7\ \text{d}^{-1}$, $k_{34} = 0.008\,75\ \text{d}^{-1}$, $k_{42} = 0.0525\ \text{d}^{-1}$, $k_{4\text{F}} = 0.005\,83\ \text{d}^{-1}$, $k_{1\text{U}} = 6.9\ \text{d}^{-1}$ and $Q_4 = 1.2$ mg. With these values:

$$k_{13} = 0.483/(I - 0.007) \tag{9.8}$$

where I has units of mg d^{-1}.

Thus, there is a minimum value for the intake of iodine below which the production of thyroidal hormones cannot be maintained. At iodine intakes well above that threshold, the rate of uptake of iodine from the systemic circulation to the thyroid varies inversely with the rate of intake of the stable element. Thus, administration of stable iodine can be effective in limiting exposures following accidents in which substantial amounts of I-131 are released.

9.6 Special issues with H-3 and C-14

Both H-3 and C-14 may enter the body in either organic or inorganic forms. In the case of H-3 (also given the special name of tritium and symbolised T), the inorganic form of greatest interest is tritiated water (HTO). On either ingestion or inhalation, this is rapidly and completely absorbed and enters the systemic circulation. Tritiated water is also readily taken up across the intact skin and this can be a significant contributor to overall uptake.

Once tritiated water enters the systemic circulation, it mixes with the overall 'pool' of body water and has a turnover time that is determined by the rate of intake of water, both in fluids and as a component of diet. The total water content of an adult male is about 55% of the total body mass. For an adult female, the proportion is rather lower at about 45%. Thus, for an adult male of mass 70 kg, the total body water content is about 38.5 kg or 38.5 l (though 42.0 kg has also been cited (ICRP 1979)). The total water intake for such an individual has been estimated as 3.0 l per day, comprising 1.95 l as fluids, 0.70 l in food and 0.35 l from the oxidation of food (ICRP 1975). Thus, the rate constant for loss of tritium from the body of an adult male is about $3.0/38.5 = 0.078\ \text{d}^{-1}$. This corresponds to a biological half-life of $\ln(2)/0.078 = 8.9$ days. A rounded value of 10 days is often used in assessment modelling. However, a small fraction of tritium that enters the body as tritiated water becomes incorporated in organic compounds that can be well retained in tissues. This means that the whole-body retention of tritium exhibits long-term components that have been estimated to add up to about 10% to the committed effective dose arising from tritiated water (ICRP 1979). The reason for this additional contribution is readily seen using a specific-activity approach. As discussed above, for an adult male of mass 70 kg, the total body water content is 38.5 kg, corresponding to 4.28 kg(H). However, the total body content of hydrogen in the soft tissues of such an individual is 6.30 kg(H). Therefore, if as a result of chronic exposure to tritiated water both the water in the body and hydrogen in soft tissues came to have the same specific activity as the tritiated water to which the individual was exposed, the activity concentration in tissues would be a factor of 6.30/4.28 = 1.47 larger than if only the water in the body came to have that specific activity.

When tritium-labelled organic compounds are ingested, a large fraction may be broken down in the gastrointestinal tract producing tritiated water. Also, organic compounds that are taken up from the gastrointestinal tract may also be catabolised to tritiated water following their uptake (ICRP 1979). Therefore, committed effective dose per unit intake values for organic compounds of tritium will often be of similar magnitude to those for tritiated water. Larger distinctions in values of committed effective dose arise for ingestion of different forms of C-14, as discussed below.

When C-14 is ingested in dietary form, it is used for respiration and as a source of carbon in the maintenance and growth of tissues. The average biological half-life of C-14 in the body can be derived using similar arguments to those developed for tritiated water. In a 70 kg adult, the total content of carbon is 16.0 kg(C) and the daily intake in food and fluids is 0.30 kg(C) (ICRP 1975). Assuming that this dietary carbon is completely absorbed from the gastrointestinal tract, the biological half-life is $\ln(2) \times 16.0/0.30 = 37$ days. However, this average conceals a great deal of variation, with bone collagen and bone mineral retaining C-14 with a biological half-life of more than five years (ICRP 1981).

In contrast, C-14 dissolved in drinking water will be predominantly present as dissolved carbon dioxide or dissolved bicarbonate or carbonate. In these forms, although absorption from the gastrointestinal tract will be close to complete, almost all the C-14 will be rapidly lost by exhalation of carbon dioxide without ever having been metabolised and incorporated in body tissues (Smith and Thorne 2015). In this context, the biological half-life for retention of most of the C-14 in the body is about one hour (ICRP 1981).

9.7 Loss of radon-222 from the body

A topic of particular historical interest, because of the uptake of Ra-226 in the radium-dial painters and the associated induction of cancer in those workers, is the retention of Rn-222 in the body following its production due to the decay of Ra-226 present in both the mineralised tissues of the skeleton and in soft tissues. Because Rn-222 is an inert gas, it will tend to diffuse away from its site of production. The characteristic diffusion length, L (m), is given by $L \sim \sqrt{Dt}$, where D (m^2 s^{-1}) is the diffusion coefficient appropriate to the gas and medium under consideration and t (s) is the time over which diffusion occurs. For Rn-222, the characteristic time is its radioactive half-life of 3.8 days (3.3×10^5 s) and its diffusion coefficient in water is 1×10^{-9} m^2 s^{-1}. Assuming Rn-222 diffusion in soft tissues is the same as in water, its characteristic diffusion length is about 0.02 m. This illustrates that some of the Rn-222 produced in the body is likely to decay close to its site of production, but that much of the Rn-222 that is produced in soft tissues will diffuse into the blood circulation and be exhaled, as is the case for C-14 labelled carbon dioxide discussed in section 9.6. In mineral bone, the situation is rather different, because much of the Rn-222 will remain trapped in the bone matrix, particularly at times long after the original deposition of Ra-226 onto bone surfaces by which time it will have become redistributed throughout the bone volume. Observationally, the fraction of Rn-222 retained in mineral bone increases from 0.06 at one day to 0.40 at 27 years after deposition of the parent Ra-226 (ICRP 1979).

9.8 The interpretation of monitoring data

Various types of monitoring data may be used to infer the timing and amounts of radionuclides taken into the body, and to estimate the committed effective doses arising in consequence. External counting is the most straightforward method. Thus, whole-body counting is often used for estimating the amounts of Cs-134 and Cs-137 present, whereas external counting over the thyroid is an effective procedure for estimating the uptake of I-131. For long-lived isotopes of plutonium and americium, external counting of the chest region can be used to estimate deposition and retention in the respiratory tract, particularly in the context of inhalation of highly insoluble oxides that give rise to only limited uptake of the radionuclides to the systemic circulation and hence little excretion in urine. Monitoring of radionuclide concentrations in faeces can also be useful, both in the case of ingestion of physicochemical forms of some radionuclides that are only poorly absorbed from the gastrointestinal tract and in the case of inhalation of such physicochemical forms, because particle-transport processes can be effective in translocating such particles from the respiratory tract to the throat, where they will be swallowed. Monitoring of exhaled air is useful in specific circumstances, notably in relation to C-14 and Rn-222 derived from Ra-226 deposited in the body. It can also be useful in a clinical context in which short-lived radionuclides are inhaled as a component of some diagnostic procedures. However, overall, urinary monitoring is probably the most widely used procedure.

A distinction needs to be made between routine and special monitoring. Routine monitoring occurs on a regular basis and aims to determine the overall chronic level of exposure to which an individual is subject and to identify anomalies in monitoring results that could imply an incident resulting in internal contamination. Special monitoring could be triggered by such an anomaly, or by other factors, e.g. occurrence of a wound to an individual while present in a contaminated area, or the occurrence of an unusual release of radionuclides to which individuals may have been exposed. Special monitoring is typically characterised by the taking of time-series of samples. This permits estimates of the intakes of radionuclides that may have occurred and estimates of parameter values of the biokinetic models used to estimate retention of those radionuclides in the body. Time-series data arise also from routine monitoring, but the time intervals may be too long to adequately resolve the kinetics of radionuclide behaviour in the exposed individual.

In interpreting monitoring data, it is important to be aware that the kinetics of radionuclide retention and loss introduce substantial uncertainties into the interpretation of these data in respect of estimation of the timing and magnitude of the intake of a radionuclide and in determining the effective dose arising from that intake. This is illustrated here by considering a case of exposure to tritiated water, with retention in the body represented by treating the whole-body as a single, well-mixed compartment, with a biological half-life determined by the rate of water intake by the individual. Thus, the biological half-life, $T_{1/2}$ (d), is given by:

$$T_{1/2} = \ln(2) \times M_{\text{water}}/I_{\text{water}} \tag{9.9}$$

where M_{water} (L) is the body content of water and I_{water} (L d^{-1}) is the rate of intake of water by all routes (ingestion, inhalation and oxidation of water in food).

With this model, an intake of Q (Bq) of H-3 at time zero will give an initial loss rate from the body of $Q \times I_{\text{water}}/M_{\text{water}}$. Thus, the time-dependent loss rate, $R(t)$ (Bq d^{-1}), is given by:

$$R(t) = \{Q \times I_{\text{water}}/M_{\text{water}}\} \times \exp\{-I_{\text{water}}t/M_{\text{water}}\} \tag{9.10}$$

Note that the integral of equation (9.10) over the interval $(0, \infty)$ is Q.

It is assumed that routine sampling in undertaken at an interval of T days. At the beginning of the interval no loss of tritium is detected, but at the end of the interval a rate of loss of U Bq d^{-1} is determined. If it is presumed that this derives from a single intake at some time during the interval, the magnitude of that intake, Q_U (Bq) is given by:

$$Q_U = U\{M_{\text{water}}/I_{\text{water}}\} \times \exp\{I_{\text{water}}x/M_{\text{water}}\} \tag{9.11}$$

where x (d) on the range $(0,T)$ is the time difference between the time of intake and the end of the monitoring interval. Thus, Q_U lies on the range $(U\{M_{\text{water}}/I_{\text{water}}\}, U\{M_{\text{water}}/I_{\text{water}}\} \times \exp\{I_{\text{water}}T/M_{\text{water}}\})$, i.e. the range of uncertainty is $\exp\{I_{\text{water}}T/M_{\text{water}}\}$. For small values of T, this can be approximated as $1 + I_{\text{water}}T/M_{\text{water}}$, i.e. the uncertainty increases in proportion to the length of the sampling interval. The uncertainty in committed effective dose varies in the same way as does to uncertainty in intake, because the kinetics of retention in the body is taken as fixed. However, a different situation arises if the timing of the intake is well-defined, but the kinetics of retention is uncertain.

In this case, U is measured at a well-defined time after intake X, but the rate of intake of water is unknown, lying on the interval $(I_{\text{water,min}}, I_{\text{water,max}})$. This means

Q_U lies on $(U\{M_{water}/I_{water,min}\} \times \exp\{I_{water,min}X/M_{water}\}, U\{M_{water}/I_{water,max}\} \times \exp\{I_{water,max}X/M_{water}\})$.

This gives an uncertainty range for intake, V_{intake}:

$$V_{intake} = \{I_{water,min}/I_{water,max}\} \times \exp\{I_{water,max}X/M_{water}\}/\exp\{I_{water,min}X/M_{water}\} \quad (9.12)$$

However, the committed effective dose, E (Sv), is proportional to the time-integrated retention in the body. The time-integrated retention is QM_{water}/I_{water} (Bq d). Therefore:

$$E = AQ_U M_{water}/I_{water} = AU\{M_{water}/I_{water}\}^2 \times \exp\{I_{water}X/M_{water}\} \quad (9.13)$$

Thus, the uncertainty range, V_E, is:

$$V_E = \{I_{water,min}/I_{water,max}\}^2 \times \exp\{I_{water,max}X/M_{water}\}/\exp\{I_{water,min}X/M_{water}\} \quad (9.14)$$

Comparing expressions (9.12) and (9.14), it is found that an extra factor $I_{water,min}/I_{water,max}$ enters into the expression for uncertainty in committed effective dose compared with intake. This is because increasing the rate of water intake increases the initial rate of loss of the tritium and so reduces the size of intake required to give a specified loss rate. However, it also reduces the residence time in the body and hence decreases the committed effective dose for a specified intake. At times longer after intake, an increased rate of water intake means that greater excretory losses will have occurred prior to monitoring and hence an increased estimate of intake is required to give the observed loss rate.

The principles outlined above can be extended to any biokinetic model. A multi-compartmental model will generally provide multi-exponential tissue and organ retention functions and multi-exponential excretion rates for a pulse input. Measured excretion data will permit evaluation of the parameters of the multi-exponential excretion equation and an estimate of the pattern of intake. The parameter values obtained for the excretion function will, in turn, partly constrain the parameter values of the tissue and organ retention functions. However, in real cases, the number of monitoring data are likely to be limited meaning that ambiguities and uncertainties will arise in estimating the parameter values to adopt. In these circumstances, it may be appropriate to fix some of the parameters at reference values and only vary a limited number of remaining parameters. Typically, the parameter values characterising the biokinetic model might be fixed, as in equation (9.11), with attention focused on determining uncertainties in the intake arising from uncertainties as to the timing of that intake.

Finally, it is noted that ambiguities tend to arise in fitting multiple exponentials to empirical data. These arise because the controlling variables can be adjusted simultaneously to give similar fits to sparse data sets. This is illustrated in figure 9.5, which shows the function:

$$F(x) = A_1 \exp(-k_1 x) + (1 - A_1)\exp(-k_2 x) \quad (9.15)$$

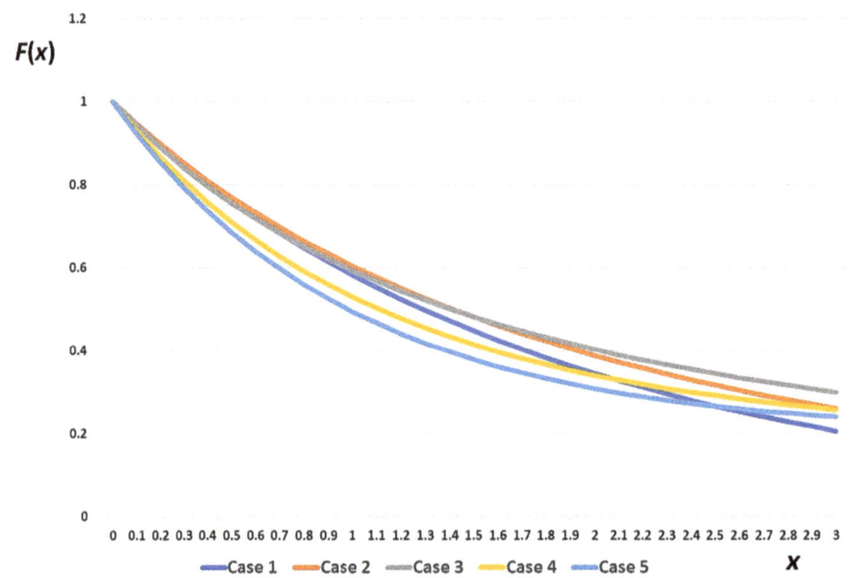

Figure 9.5. Effect of varying parameter values in equation (9.15). The value of k_1 is 1.0 throughout. Case 1 has $A_1 = 0.1$, $k_2 = 0.5$; case 2 has $A_1 = 0.3$, $k_2 = 0.35$; case 3 has $A_1 = 0.5$, $k_2 = 0.20$; case 4 has $A_1 = 0.7$, $k_2 = 0.10$; and case 5 has $A_1 = 0.8$, $k_2 = 0.0$.

The figure shows that substantial changes in the parameter values give similar curves. In this case, increasing the intercept of the rapidly declining component 1 can be compensated by decreasing the rate at which component 2 declines.

References

Coughtrey P J, Jackson D and Thorne M C 1983 *Radionuclide Distribution and Transport in Terrestrial and Aquatic Ecosystems* **vol 3** (Rotterdam: A A Balkema)

IAEA 2010 *Handbook of Parameter Values for the Prediction of Radionuclide Transfer in Terrestrial and Freshwater Environments* IAEA Technical Reports Series No. 472 Vienna: International Atomic Energy Agency

ICRP 1975 *Report of the Task Group on Reference Man, ICRP Publication 23* (Oxford: Pergamon)

ICRP 1979 Limits for intakes of radionuclides by workers: Part 1, ICRP Publication 30 *Ann. ICRP* **2(3/4)**

ICRP 1980 Limits for intakes of radionuclides by workers: Part 2, ICRP Publication 30 *Ann. ICRP* **4(3/4)**

ICRP 1981 Limits for intakes of radionuclides by workers: Part 3, ICRP Publication 30 *Ann. ICRP* **6(2/3)**

ICRP 1994 Human respiratory tract model for radiological protection, ICRP Publication 66 *Ann. ICRP* **24(1–3)**

ICRP 2006 Human alimentary tract model for radiological protection, ICRP Publication 100 *Ann. ICRP* **36(1–2)**

ICRP 2017 Occupational intakes of radionuclides: Part 3, ICRP Publication 137 *Ann. ICRP* **46(3/4)**

NCRP 2006 *Development of a Biokinetic Model for Radionuclide-contaminated Wounds and Procedures for their Assessment* Dosimetry and Treatment, NCRP Report No. 156 Bethesda, MD:National Council on Radiation Protection and Measurements

Smith G M and Thorne M C 2015 Appropriate selection of dose coefficients in radiological assessments: C-14 and Cl-36 *J. Radiol. Prot.* **35** 737–40

Chapter 10

Models for the environmental transport of radionuclides

R Klos and M C Thorne

10.1 Introduction

Radionuclides may be released to the environment on a routine basis, e.g., in liquid and gaseous discharges, or because of accidents at nuclear installations. These discharges may occur at the present day or far into the future if they originate from geological repositories for radioactive wastes. In addition, releases may arise from legacy sites, e.g., facilities formerly associated with the mining and milling of uranium ore. In all these contexts, there is a need to estimate the radiological impacts of existing or projected future discharges on human health and the environment.

In cases where discharges are ongoing and both their characteristics and those of the local environment do not change substantially, radionuclide concentrations in environmental media and the radiological impacts of those concentrations may be estimated by using empirical regression relationships relating past rates of discharge to observed environmental concentrations. However, even in this context complications may arise because concentrations of some radionuclides will be related to the history of discharges over many years rather than to the current rate of discharge. Where either discharge characteristics or the environment change substantially, reliance will necessarily be placed on mathematical models to estimate radionuclide concentrations and hence radiological impacts. This chapter addresses the structure of such models, covering both models of the transport of radionuclides through the environment and the relationships between radionuclide concentrations obtained using such models of transport and radiological impacts on human health and the environment.

Section 10.2 goes into more detail on the role of models in radiological impact assessment and introduces the various types of model that are available, which are described in section 10.3. In many contexts, the first stage of such modelling is the

doi:10.1088/978-0-7503-2424-3ch10

simulation of the dispersion and deposition of radionuclides released to the atmosphere. This is addressed in section 10.4. Dispersion and accumulation in terrestrial and aquatic environments are addressed in sections 10.5–10.7. Section 10.5 addresses terrestrial and freshwater environments, focusing on the natural scale for such modelling, which is the surface-water catchment. Section 10.6 covers the estuarine environment and the special issues that arise in modelling a system that varies rapidly both in space and time. Section 10.7 is focused on the marine environment and considers coastal, offshore and deep ocean environments. Finally, section 10.8 describes how the environmental concentrations estimated using the various dispersion models can be used to calculate radiological impacts on human health and the environment.

10.2 The role of models in radiological impact assessment

In prospective radiological impact assessment, the objective is to assess the adverse effects of releases of radionuclides from a source on both human health and the environment. The source will often be a discharge from an engineered facility, such as a nuclear power station, but it may also be a site contaminated because of earlier activities, e.g. a former uranium mining and milling facility or a site where radium paint was used. Thus, some sources may be closely monitored and controlled, whereas others may be poorly characterised and subject to little or no control. Also, releases may be routine, taking place continuously or periodically, or they may be short-term, arising from incidents and accidents. In terms of evaluating effects on human health, doses and dose rates will generally be low, except in the most severe accidents. Where doses and dose rates are low, effective dose and committed effective dose are the primary indicators of radiological impact on human health. However, in the context of major accidents, absorbed doses to individual tissues and organs may also be of relevance, e.g. absorbed doses to the thyroid from internally incorporated ^{131}I following the Chornobyl accident.

In evaluating radiological impacts on the environment, the fundamental concern is with the higher levels of ecosystem characteristics, i.e. whether populations, communities, habitats or overall biodiversity are affected. However, the available data are not sufficient to evaluate effects at these higher levels across the wide range of ecosystems that may need to be considered, bearing in mind that the rules of engagement that govern the stability of communities vary greatly between communities and have only been poorly characterised. Therefore, criteria for protecting the environment have been defined in terms of dose rates below which it is considered that populations of most species would not be harmed, though this does not preclude harm to individual organisms. In this case, the criteria are expressed in terms of absorbed dose rate. Although high-LET radiations are known to be more effective than low-LET radiations in inducing various effects in a wide range of organisms, a single set of radiation weighting factors appropriate to converting all absorbed doses to an organism to a single measure of effective dose has not, as yet, been adopted.

In undertaking a radiological assessment, a transparent methodology is required that shows clearly that a comprehensive approach has been adopted, i.e. there is no possibility that significant contributions to dose have been overlooked. It may be helpful to adopt a source-pathway-receptor framework. First, all the distinct components of the source are identified (these could comprise the individual heaps of waste rock at a former uranium mine). For each such component, all the potential pathways by which radionuclides may be transported through the environment are then identified. Finally, the receptors are identified, e.g. humans and populations of other biota that could make use of environmental media contaminated because of the operation of the potential pathways. It can also be helpful to apply this process in reverse, i.e. by first identifying all the potential receptors in the local environment then tracing back the transport pathways associated with those receptors to see if they connect with one or more source components. Habits surveys around installations, such as those that are undertaken in the UK (e.g. Clyne *et al* 2013), can be helpful in identifying both potential receptors and the pathways associated with them (see section 10.8.4). Other approaches to demonstrating comprehensiveness include the use of interaction matrices to ensure that all the features, events and processes of relevance are identified and incorporated in an appropriate conceptual framework that can be used as a basis for mathematical modelling (see section 8.2 and Klos and Thorne 2020).

In applying an assessment model, two broad stages can be distinguished. The first stage is to calculate the transport of radionuclides from the source and determine their concentrations in the environmental media to which humans and non-human biota are exposed. Having obtained these concentrations, dose rates to humans and non-human biota can be estimated. For external exposure, the estimation of dose rates can be based on radiation-transport calculations, whereas for internal exposure models for the uptake and retention of radionuclides in the body are also required, as discussed for humans in chapter 9. The various types of model that may be used to represent radionuclide transport through the environment are described in section 10.3.

10.3 Types of model

A first consideration in selecting an assessment model is the dimensionality of the model to be adopted. Atmospheric dispersion is intrinsically a 3D process, with radionuclide concentrations varying with distance downwind, height above the ground and transverse distance from the plume axis (see section 10.4). However, when evaluating the transport of radionuclides deposited from such a plume in an underlying soil–plant system, it may be sufficient to adopt a 1D vertical model of the soil and plants, i.e. neglecting lateral transfers of the deposited activity. Again, where a discharge occurs to a river, it may be adequate to model the advection and dispersion downstream using a 1D model. However, if the activity distribution across the stream channel is also relevant a 2D model may be required.

Irrespective of dimensionality, consideration must also be given to whether a lumped or spatially distributed model is to be adopted. Lumped models typically treat the system under consideration as a set of well-mixed compartments and return only the total activity contents of the compartments, which can then be used to derive average concentrations in the compartments, but no estimate of spatial variability of the concentration within a compartment. In contrast, spatially distributed models represent the system as continuously variable. In practice, the distinction between the two types of model is not absolute. For example, 3D models of surface-water catchments are generally based upon underlying continuous equations, but those equations are solved by finite-element or finite-difference methods that involve discretising them onto finite sets of elements or nodes.

Models can also be primarily based on empirical data or on a representation of the governing processes. A good example of the former is a model of root uptake by plants from contaminated soils that is based on the use of concentration ratios. In this approach, the average concentration of a radionuclide is computed in a topsoil layer of specified thickness (typically 0.1 m for pasture and 0.2 m for soils growing agricultural crops). Plants growing in the contaminated soil are taken to exhibit a specific ratio of their concentration to that in the soil, either because equilibrium has been achieved or because the ratio has been estimated at the time of harvest. Thus, in this simple model C_{plant} is calculated as $F_{\mathrm{plant:soil}}C_{\mathrm{soil}}$, where C_{plant} (Bq kg(fresh weight)$^{-1}$) is the concentration in the plants; C_{soil} (Bq kg(dry weight)$^{-1}$) is the concentration in the soil; and $F_{\mathrm{plant:soil}}$ (kg (dry weight soil) kg (fresh weight plant)$^{-1}$) is the plant-to-soil-concentration ratio.

Values of $F_{\mathrm{plant:soil}}$ are determined by empirical measurements of radionuclide and stable element concentrations in various combinations of soils and plants. Often, simple classification schemes are adopted that show an appreciation of underlying controlling factors, e.g. values may differ according to soil texture, pH or cation exchange capacity (CEC), but those controlling processes are not represented. In contrast, a process-based model would represent explicitly factors such as the changing hydrology of the soil, growth of the root system of a crop and the kinetics of partitioning of a radionuclide between soil solution and the various organic and inorganic components that comprise soil solids (see, e.g. Wheater *et al* 2007, Thorne 2014).

In practice, models used for assessment purposes often combine process-based and empirically based components. This is exemplified in models used to represent radionuclide migration in surface-water catchments. In such models, flows of water on the surface and in the subsurface are explicitly calculated using the underlying hydrological equations, whereas the model for radionuclide transport is typically based on combining these calculated flows with a retardation factor calculated using an empirical, equilibrium distribution factor that governs the partitioning of the radionuclide between soil solution and soil solids (see section 10.5 for details). Furthermore, process-based components necessarily have a basis in empirical studies, e.g. the equations governing flows of water in porous media are observationally based. However, the process-based components are based on laws or hypotheses of generic application, whereas the empirically based components are based on observations either of the system of interest or of a limited range of systems related to it.

An issue with empirically based models is that all the data of relevance are generally used in parameterising the model. This means that validation of the model, in the sense of predicting the outcome of observations not used in its formulation, is often impossible. In contrast, with process-based models, validation, of at least of some aspects of the model, can often be undertaken. Various approaches to validation can be undertaken (see section 8.4), with the most rigorous comprising blind predictions of observational results where the input data used to make the predictions are qualitatively different from the results predicted, e.g. in a surface-water catchment model predicting time-dependent stream flows on the basis of data relating to the topography of the catchment, time-series of temperature and precipitation, and the hydrological characteristics of the soils (see, e.g. Ewen and Parkin 1996). Other approaches to validation may support use of a model in interpolating between observations, but not in extrapolating beyond them. For example, a validation that uses a split-sample technique in which one half of the sample is used for calibration and the other half is used for validation, can confirm that the model is adequately calibrated across the sample range, but does not address whether it is adequately calibrated outside that range (see also section 8.4).

The common types of mathematical models that are used in assessment studies are described in section 8.3. These comprise equilibrium, compartmental and advective-diffusive or advective-dispersive models. As mentioned above, these models may be used in combination. With equilibrium and compartmental models, the dimensionality of the system does not enter explicitly into the model formulation, though it may be implicit, e.g. in the definitions of the compartments. However, with advective-diffusive models, the underlying equations may be solved in 1D, 2D or 3D. The 1D form is often used to model radionuclide transport from geological repositories for solid radioactive waste to the near-surface environment, whereas the 3D form is typically used to model transport once the radionuclides have reached the near-surface environment. This change of dimensionality needs careful handling to ensure consistency across the interface between the two modelling domains (see, e.g., Smith *et al* 2014).

Specific-activity-based models are discussed as a special case in section 8.3.1. These are generally applied to radioisotopes of elements involved in major biogeochemical cycles and are often implemented using an equilibrium approach in which isotopic ratios in two different environmental media are assumed to be the same or to differ only by a small isotopic discrimination factor. However, they can also be applied in kinetic modelling in which the isotopic ratio in the flow of an element from one component of the model to another is taken to be the same or similar to that in the source component. In this approach the isotopic ratio in the receptor component of the model changes with time as the original elemental content of that component is replaced by the incoming flow of the element from the source.

10.4 Atmospheric dispersion models

Radionuclides may be released to the atmosphere from stacks or by other controlled routes, or as fugitive releases from buildings or contaminated areas. Routine releases are often quasi-continuous, but they may be highly variable, depending on the nature of the operations that give rise to them. Where stack releases occur, they can be at a higher temperature than the surrounding air and may be associated with a significant amount of vertical momentum. Thus, thermally induced or momentum-driven plume rise may result in the effective height of the release being significantly greater than the height of the stack. However, if a release occurs from or close to a building, it may be entrained in the turbulent airflow downstream of the building and experience immediate dilution and dispersion. In accident conditions that involve fires, as at Chornobyl, the degree of thermally induced plume rise may be several hundred metres or more.

The physicochemical forms of radionuclides released to the atmosphere include gases and vapours (e.g. $^{14}CO_2$, tritiated water, ^{41}Ar and ^{85}Kr), but also liquid and solid aerosols. The sizes of the aerosol droplets and particles range from less than 1 μm up to 10 μm or more. However, as they are transported downwind, the larger droplets and particles are rapidly lost from the plume by gravitational settling. Thus, at distances of more than a few hundred metres from the source, aerosols are typically associated with an activity median aerodynamic diameter (AMAD) of 1–5 μm. The concept of AMAD is described in detail in section 9.2.

Releases from a stack can typically be treated as point releases, with thermal buoyancy and upward momentum being considered by defining the height of the point release to be greater than the height of the stack. The simplest models of dispersion that are routinely used treat the plume as exhibiting a Gaussian profile in both the horizontal and vertical directions. However, they also represent reflection of the plume at the ground surface and at the top of the mixing layer (see figure 10.1 for the parameters characterising a Gaussian plume, neglecting reflection).

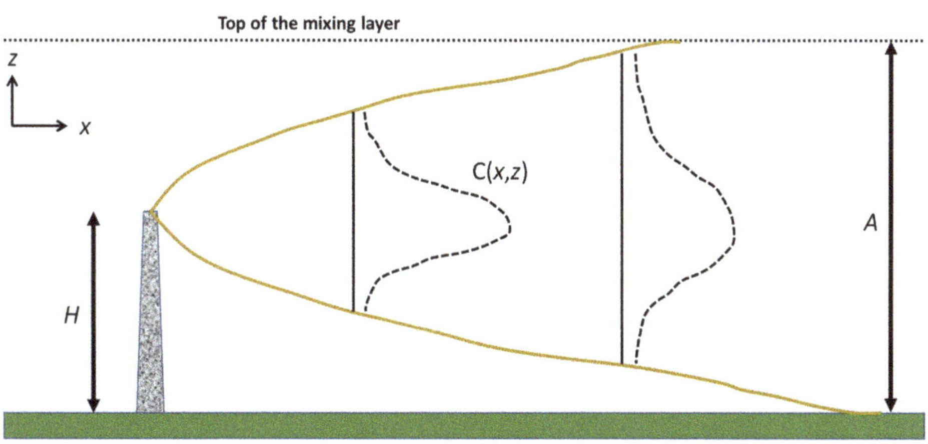

Figure 10.1. Vertical cross-section through a plume.

Thus, the usual form adopted (Clarke 1979) is:

$$C(x, y, z) = \left\{Q/2\pi u_{10}\sigma_z\sigma_y\right\} \times \exp\left[-y^2/2\sigma_y{}^2\right] \times F(H, z, A] \tag{10.1}$$

where $C(x, y, z)$ (Bq m^{-3} or Bq s m^{-3}) is the air-concentration or time-integrated air concentration at location x, y, z depending upon how the source strength Q is defined;

Q (Bq s^{-1} or Bq) is either the release rate or the total release for a pulse release;
u_{10} (m s^{-1}) is the wind speed at a reference height of 10 m above the ground;
σ_z (m) is the degree of dispersion of the plume in the vertical and is a function of the downwind distance x;
σ_y is the degree of dispersion of the plume in the horizontal perpendicular to the wind direction and is a function of the downwind distance x;
$\exp(-y^2/2\sigma_y{}^2)$ is the horizontal dispersion with a Gaussian profile being adopted;
$F(H, z, A)$ is the corresponding vertical profile that takes account of reflection of the plume from the ground and the top of the mixing layer.

A suitable form for $F(H, z, A)$ that takes first- and second-order reflections into account is:

$$\begin{aligned} F(H, z, A) &= \exp[-(z - H)^2/2\sigma_z{}^2] + \exp[-(z + H)^2/2\sigma_z{}^2] \\ &+ \exp[-(2A + z + H)^2/2\sigma_z{}^2] + \exp[-(2A + z - H)^2/2\sigma_z{}^2] \end{aligned} \tag{10.2}$$

However, at greater distances downwind, uniform vertical dispersion within the mixing layer can be assumed and equation (10.1) reduces to:

$$C(x, y, z) = \left\{Q/(2\pi)^{0.5}u_{10}A\sigma_y\right\} \times \exp\left[-y^2/2\sigma_y{}^2\right] \tag{10.3}$$

These equations assume a constant, mean wind direction, although some variability about that mean can be accommodated through the specification of σ_y, as discussed below. However, for chronic releases, it is more usual to evaluate average radionuclide concentrations in 30- or 45-degree sectors. This approach replaces equation (10.1) by:

$$C(x, y, z) = \{Q/(2\pi)^{0.5}x\alpha u_{10}\sigma_z\} \times F(H, z, A) \tag{10.4}$$

where α (radians) is the angular width of the sector.

Equation (10.3) simplifies to:

$$C(x, y, z) = Q/x\alpha u_{10}A \tag{10.5}$$

In equation (10.5), the concentration at downwind distance x (Bq m^{-3}) is simply the release rate divided by the volume of air passing through the outer boundary of the sector at that distance, i.e. the area $x\alpha A$ (m^2) multiplied by wind velocity u_{10} (m s^{-1}).

In simple assessment calculations, typical values of the wind speed and mixing layer depth are used for various categories of atmospheric stability. The Pasquill

Table 10.1. Typical wind speeds and mixing heights for Pasquill atmospheric stability categories.

Stability category	Typical wind speed at 10 m height (m s^{-1})	Typical mixing layer depth (m)
A	1	1300
B	2	900
C	5	850
D	5	800
E	3	400
F	2	100
G	1	100

Table 10.2. Coefficients used to estimate σ_z.

Stability category	*A*	*b*	*c*	*d*
A	0.112	1.060	$5.38\ 10^{-4}$	0.815
B	0.130	0.950	$6.52\ 10^{-4}$	0.750
C	0.112	0.920	$9.05\ 10^{-4}$	0.718
D	0.098	0.889	$1.35\ 10^{-3}$	0.688
E	0.0609	0.895	$1.96\ 10^{-3}$	0.684
F	0.0638	0.783	$1.36\ 10^{-3}$	0.672

stability scheme has often been used and reference values for these parameters for the standard Pasquill stability categories are given in table 10.1.

The least stable of these categories is A and the most stable is G. Categories C and D are the most common, with the most stable categories (F and G) occurring only at night.

Values of σ_z are often estimated using a function of the form:

$$\sigma_z = (ax^b/1 + cx^d) \times F(z_0, x) \tag{10.6}$$

where *a*, *b*, *c* and *d* are coefficients that depend on the atmospheric stability category (see table 10.2) and $F(z_0, x)$ is a correction factor that takes account of the roughness of the ground, which, in turn, influences the turbulence of the near-surface atmosphere.

$F(z_0, x)$ takes two alternative forms depending upon the value of the roughness length (z_0):

$$F(z_0, x) = \ln\{fx^g[1 + (hx^j)^{-1}]\} \text{ for } z_0 \geqslant 0.1 \text{ m} \tag{10.7a}$$

$$F(z_0, x) = \ln\{fx^g[1 + hx^j]^{-1}\} \text{ for } z_0 < 0.1 \text{ m} \tag{10.7b}$$

Values of the coefficients *f*, *g*, *h* and *j* are given in table 10.3.

Table 10.3. Coefficients for the roughness correction factor.

Roughness length (m)	F	g	h	j
0.01	1.56	0.0480	$6.25\ 10^{-4}$	0.45
0.04	2.02	0.0269	$7.76\ 10^{-4}$	0.37
0.1	2.72	0	0	0
0.4	5.16	−0.098	18.6	−0.225
1.0	7.37	−0.0957	$4.29\ 10^{3}$	−0.60
4.0	11.7	−0.128	$4.59\ 10^{4}$	−0.78

Values of z_0 are typically 1×10^{-4} m for the sea, 1×10^{-3} m for sandy desert, 5×10^{-3} m for short grass, 2×10^{-2} m for open grassland, 0.1 to 0.3 m for agricultural areas, 0.5 m for parkland and open suburbia, and 1.0 m for cities and woodlands.

Values of σ_y typically include both a turbulence-related term and a contribution from variations in the wind direction. These are treated as independent. Thus:

$$\sigma_y{}^2 = \sigma_{y-t}{}^2 + \sigma_{y-w}{}^2 \tag{10.8}$$

The contribution due to variations in wind direction, σ_{y-w}, can be estimated from the standard deviation of horizontal wind directions when averaged over consecutive three minute periods for the duration of the release. However, if these data are not available, a useful approximation is:

$$\sigma_{y-w} = 0.065\{7T/u_{10}\}^{0.5}x \tag{10.9}$$

where T is the release duration in hours.

Assuming that turbulent effects are similar in the horizontal and vertical directions, it is reasonable to approximate that $\sigma_{y-t} = \sigma_z$.

The basic principles of plume dispersion are appropriately illustrated by the simple model described above. However, more complex models are available for use in appropriate circumstances (e.g. the ADMS 5 software package; https://www.cerc.co.uk/environmental-software/ADMS-model.html). For example, the above model can be generalised to a 3D puff model, where the release is treated as a series of pulse releases with dispersion represented along the direction of travel as well as transverse to it. The effects of changing wind direction can be represented in such a model by altering the direction of travel of each puff as the wind direction changes. Such models can be driven by real meteorological data, which are typically available at hourly or sub-hourly sampling intervals. For large-scale dispersion modelling, releases can be fed into regional-scale models of the atmospheric circulation. These can be retrospective reconstructions based on data from the global network of weather stations or projections made using the type of models adopted in forecasting future weather patterns. An example of such an approach is the NAME model developed by the UK Meteorological Office (https://www.metoffice.gov.uk/research/approach/modelling-systems/dispersion-model, see also Jones *et al* (2007)).

Where release occurs into the turbulent zone downwind of a building, sometimes known as the building wake, a simple approach can be to treat the release as being at the mid-point of the downstream side of the building and to add in initial contributions to σ_y and σ_z with values of similar magnitude to the width and height of the building, respectively. As in equation (10.8), these initial contributions can be taken to be independent of the increasing downwind dispersion and added in quadrature. Alternatively, a virtual point source can be adopted located upstream of the actual release location.

In complex environments, e.g. in urban areas, it may be necessary to employ techniques based on a computational fluid dynamics approach or to scale estimates of dispersion based on wind-tunnel measurements of air flows and tracer dispersion using scale models of the environment of interest (Gallacher *et al* 2016a, 2016b).

Whatever type of atmospheric dispersion model is adopted, the output comprises a 3D-spatial, time-varying concentration of a radionuclide or other contaminant that can be denoted $C(x, y, z, t)$. This information is directly relevant to assessing the radiological impact of some exposure pathways, e.g. external irradiation from the plume and inhalation. However, other pathways, such as those relating to foodchain transfers, require estimation of the deposition of the contaminant. The main modes are dry and wet deposition. In dry deposition of aerosols, the droplets or particles comprising the aerosol located close to the ground surface interact with the ground and vegetation, e.g. by interception, impaction, gravitational settling and diffusion. In general, it is not possible to model these processes explicitly, so their effect is represented using an empirical deposition velocity, v_d (m s^{-1}). Thus, if the air concentration of a radionuclide at the ground surface is $C(x, y, 0, t)$ Bq m^{-3}, the rate of deposition, $D(x, y, t)$ Bq m^{-2} s^{-1} is given by:

$$D(x, y, t) = v_d C(x, y, 0, t) \tag{10.10}$$

Typical deposition velocities for aerosols are 0.001 to 0.01 m s^{-1}.

In the case of wet deposition, rain or snow is envisaged as falling through the full height of the plume and capturing aerosol droplets or particles as it falls. Therefore, the rate of deposition is given by:

$$D(x, y, t) = \Lambda \int_0^A C(x, y, z, t) dz \tag{10.11}$$

where the washout coefficient, Λ s^{-1} depend on the rate of precipitation. A typical value is 1×10^{-4} s^{-1}.

Both wet and dry deposition deplete the plume of radionuclides and that depletion increases with distance downwind. This can be addressed by computing the degree of depletion as a function of distance and decreasing the source term, Q, in equation (10.1) by a factor that depends on the downwind distance. Under conditions of heavy rain, much of the activity may be deposited within a few kilometres of the release. Conversely, as illustrated by the Chornobyl accident, a plume with a large amount of thermal buoyancy may travel several thousand kilometres at an elevated altitude before a precipitation event results in high local deposition.

Where deposition occurs in the terrestrial environment, it needs to be partitioned between the vegetation and the underlying soil. This is of specific relevance in agricultural systems. For crops, the initial deposition on the foliage determines the degree of external contamination and provides the source for translocation into internal plant parts. Both external and internal contamination can contribute to the ingestion of activity by humans, with the amounts of activity ingested depending upon the amounts of different plant parts that are consumed, modified by the procedures adopted in food preparation. For pasture, the initial contamination of the foliage provides a source of activity to animals, either through grazing on the pasture or through the subsequent consumption of contaminated hay. In the longer-term, soil contamination provides a long-term reservoir for root uptake by plants, with a timescale determined by the degree of sorption of the radionuclide to the soil and its radioactive half-life. The uptake and retention of radionuclides in soils and plants is discussed further in section 10.5.

The partitioning of radionuclide dry deposition between plants and soils is generally estimated using the Chamberlain approach. This was originally developed for pasture but was then extended to a wide variety of crop types. The fraction of deposition associated with plants, f_p, is given by:

$$f_p = 1 - \exp(-\mu B) \qquad (10.12)$$

where B (kg (dry mass) m^{-2}) is the above-ground biomass of the crop at the time of deposition and μ m^2 kg (dry mass)$^{-1}$ is an empirical coefficient that is typically ≈ 3. This form has appropriate behaviour in that, at $\mu B \ll 1, f_p \approx \mu B$, whereas at $\mu B \gg 1, f_p$ approaches an asymptotic value of 1.0. Some compilations of data for both dry and wet deposition, adopt the small standing biomass approximation and report values of f_p/B.

A disadvantage of basing the approach on above-ground biomass is that it takes no account of foliar morphology. This can be addressed by replacing B by the leaf area index (LAI), here denoted L. The LAI can be envisaged by considering vertical lines originating at the soil surface and upwardly directed through the plant canopy (see figure 10.2). The LAI is the average number of times that such a vertical line intersects the canopy. Thus, for a recently planted crop, where there are substantial areas of bare soil between the plants, the LAI $\ll 1$. In contrast, for a well-developed canopy with multiple layers of leaves, the LAI $\gg 1$ (ranging typically up to about 5).

In this approach:

$$f_p = 1 - \exp(-\zeta L) \qquad (10.13)$$

where ζ is a dimensionless coefficient of order one.

Finally, it is relevant to note that after a radionuclide has been deposited from an atmospheric plume, it may be subject to resuspension. Such resuspension can be induced by the wind or it can be due to human actions, such as ploughing. Resuspension is typically represented using one of two approaches. These are a resuspension-factor-based approach and a mass-loading-based approach.

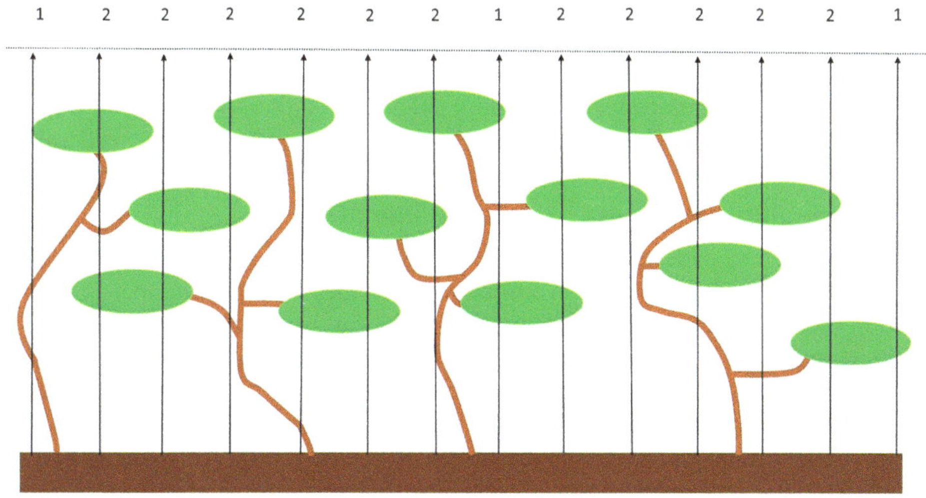

Figure 10.2. Calculation of LAI. In this example, the LAI is 25/14 = 1.79.

Following a deposition event at $t = 0$, the resuspension-factor approach estimates resuspension at time t (d) using:

$$R(t) = A\,\exp(-\lambda t) + B \tag{10.14}$$

where $A + B$ (m^{-1}) gives the initial resuspension factor, B is the resuspension factor measured in the long term and λ (d^{-1}) is the rate of decrease in the short-term component. Under European conditions, the value for $A + B$ measured after the Chornobyl accident ranged from 4×10^{-9} to 5×10^{-8} m^{-1}. The initial rate of decrease in the resuspension factor corresponded to a half time of about 200 to 800 d (Thorne 2012), i.e. the value of λ was in the range 3.5×10^{-3} to 8.7×10^{-4} d.

In the mass-loading-approach, the radionuclide concentration is calculated per unit mass of soil or sediment, e,g. as Bq kg (dry weight)$^{-1}$. This is then combined with an estimate of the dust concentration in air (kg (dry weight) m^{-3}) to give an estimate of the activity concentration in air (Bq m^{-3}) that can be used, for example, to estimate rates of radionuclide intake by inhalation. This approach is of limited usefulness at times soon after a deposition event, because the activity is likely to be present mainly as a thin surface deposit (particularly in urban areas) and it is difficult to define a layer of material in which the activity concentration should be calculated. Hence the resuspension-factor-approach is preferred at early times after deposition and where resuspension from hard surfaces needs to be estimated. However, in rural areas at times long after deposition, or where soil disturbance is involved, the mass-loading approach is preferred.

In applying the mass-loading approach, the main consideration is the concentration of dust in air that should be adopted. This can be estimated by air sampling, keeping in mind that some of the dust will be organic material derived from materials other than soils and sediments, and that some of the mineral dust present will be of distant origin rather than being derived from a local contaminated area.

The US NCRP (1998), who provide a detailed account of alternative resuspension models, cite annual average values of 28 and 75 $\mu g\, m^{-3}$ for rural and urban conditions, respectively, but cite a value of 15 $mg\, m^{-3}$ for measurements made behind a tractor. For assessment screening calculations, the NCRP (1998) adopted values in the range 20–150 $\mu g\, m^{-3}$ depending on the scenario under consideration.

10.5 Modelling of surface-water catchments

Eventually even radionuclides released to the atmosphere are deposited at the Earth's surface, whether on the terrain or water bodies and what happens then is a function of the driving forces, the features, events and processes (FEPs), that act to transport material in the terrestrial and aquatic environments. Radiological assessment models started out as a means of determining the accumulation of radionuclides in soils following deposition from weapons testing, routine releases from nuclear installations or accidents, but disposal of low-level radioactive waste in near-surface facilities added migration of radionuclides through soils to loss of radionuclides from surface soils to the pathways to be considered. The need to model releases from the underlying geology is therefore apparent, in addition to the representation of transport of radionuclides deposited at the soil surface.

The need for long-term isolation of radioactive waste from the biosphere has seen the development of engineered repositories, with a range of types of waste disposed, sophistication of design, host rock characteristics and depth. In all cases, planning and regulation for disposal systems recognises that the containment characteristics of the system will eventually fail, and radionuclides will be able to return to the surface environment. Transport of material can affect radionuclides in solution, on solid material and in gaseous form, but the most significant process is that of solute transport, whether in the deeper geology or in soils and sediments in the surface environment. Water fluxes in the surface hydrological and drainage system can transport radionuclides far downstream from the release location—the geosphere–biosphere interface—where radionuclide-bearing groundwaters first enter surface ecosystems, eventually being transported to estuarine, coastal and marine ecosystems.

Compartmental models have had a long history in radiological modelling. The off-diagonal elements of the transfer matrix (see section 8.3.2) represent the combined transfer of material from the *i*th homogeneous spatial volume to the *j*th, so that the fractional transfer rate, λ_{ij} year^{-1}, of radionuclides from compartment i to compartment j is given by Klos and Thorne (2020) as:

$$\lambda_{ij} = \frac{F_{ij} + k_i^d M_{ij} + D_{ij}}{\varepsilon_i S_i R_i V_i} \tag{10.15}$$

This form might be considered the 'standard model' for radionuclide transport in soils and sediments, since it is applicable to many radionuclides that are likely to be present in small amounts in the biosphere, such that the k_d-concept can be used as an approximation for the distribution of the chemical species between solute and solids in the compartment. The parameterisation is:

F_{ij} m^3 y^{-1}	volumetric water (solute) flux from compartment i to compartment j,
M_{ij}	kg y^{-1} solid material flux, with
k_i^d (m^3 kg^{-1} dw)	the solid–liquid distribution coefficient,
D_{ij}	m^3 y^{-1} the effective exchange of solute by diffusion,
V_i	m^3 physical volume of compartment i,
ε_i unitless	porosity of the medium in the compartment, and
S_i unitless	the degree of saturation.

The retardation factor, R_i, is given by:

$$R_i = 1 + \frac{\rho_i}{\varepsilon_i S_i} k_i^d \tag{10.16}$$

It accounts for the retention of contaminants bound to fixed solid material in the compartment so that their migration is slowed with respect to bulk water fluxes.

The hydrological balance of the landscape determines transport and accumulation of radionuclides in the landscape in many site contexts. The compartmental approximation is flexible, and parameterisation of transfer rates is linked to measurable characteristics of terrestrial and aquatic media, such as precipitation, evapotranspiration and applied irrigation. The simple well model (figure 10.3) is a typical simple model with a broad range of applicability. In situations where the contaminated aquifer may be considered unconnected to the local surface hydrology, for example in the case of deep disposal in clay layers at Meuse/Haute–Marne in France (ANDRA 2015) it is a suitable basis for detailed dose assessment modelling.

Radionuclide concentrations in the aquifer, at depth or nearer to the surface, are determined by the regional groundwater circulation and an understanding of the

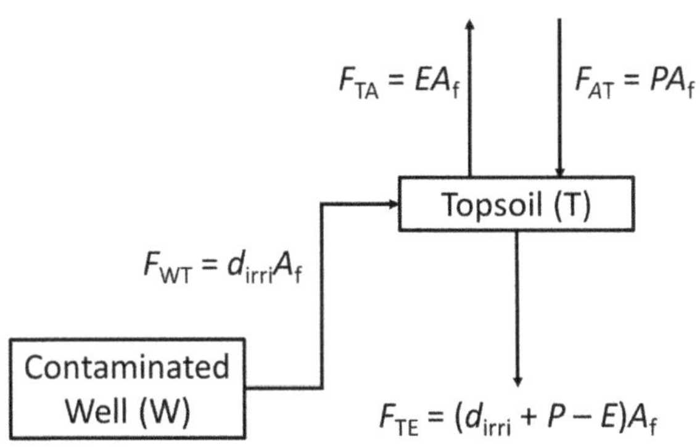

Figure 10.3. Water balance for a simple well model. Area cultivated is A_f m^2, precipitation is P, F_{AT} is the transfer from atmosphere to topsoil, evapotranspiration E governs F_{TA} (topsoil to atmosphere) and annual irrigation d_{irri} governs F_{WT} (well water to topsoil), while the loss term from soil is F_{TE} (topsoil to elsewhere). Water fluxes are characterised by an annual depth, m y^{-1}.

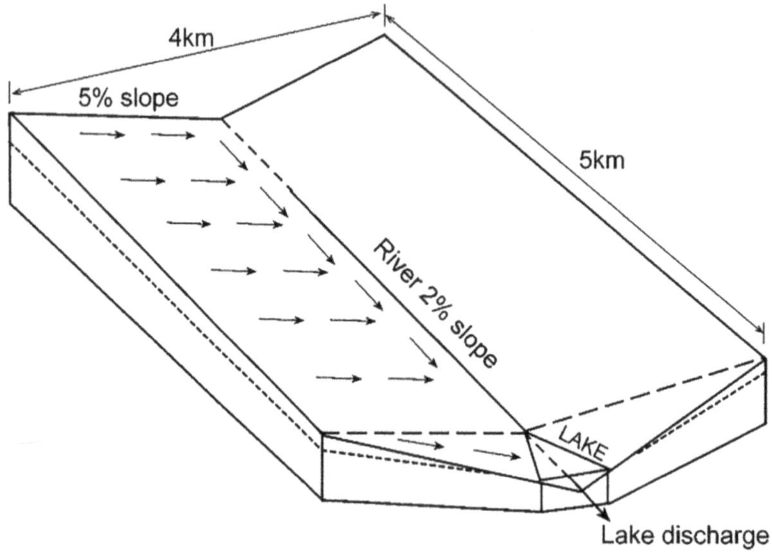

Figure 10.4. Landscape dose objects for a stylised site context. Illustration of the BIO-MASS Example Reference Biosphere 2b catchment. The illustrative representation of the model area shows the spatial extent of the object and the drainage features (river and lake). The overall catchment area is 2×10^7 m^2. The total area is delineated into different ecosystems: woodland, arable land, grassland, as well as shrubland and marshland around the lake. The position of the water table relative to the topographic surface is shown as the dashed line and the arrows represent the implied direction of flow of accumulated net precipitation in the model. IAEA (2003), © IAEA, 2003, p. 388, ISBN 92–0–106303–2

significance of the site context to the near-surface and deeper hydrogeology was developed for dose assessment modelling in the example reference biospheres (ERB) in the IAEA (2003) BIOMASS methodology. The hypothetical ERB2b site context is as shown in figure 10.4. More sophisticated in concept than the simple well model, it illustrates the broadly applicable features—the overall *catchment* (watershed) area, the interaction of the groundwater table with the topography to define the stream bed and lake area, the topographic gradient and the implicit flow direction of the net precipitation (runoff). With the ERB2b concept there are many fluxes, F_{ij}, that can be identified in the system depending on the location at which the contaminated groundwater enters the dose assessment model domain, and these are all governed by the overall water balance within the boundaries of that domain. It may be noted that adequate evaluation of well water concentrations in the simple model shown in figure 10.3 requires a similar understanding of the local hydrology as is needed for the ERB 2b level of detail.

Real landscapes are not so conveniently rectilinear as that depicted in the BIOMASS example. Figure 10.5 shows a portion of a digital elevation model (DEM) rendered as a three-dimesional surface using geographic information systems (GIS) software. In the landscape context from which this DEM comes (POSIVA 2013) the geosphere–biosphere interface is expected to be at local

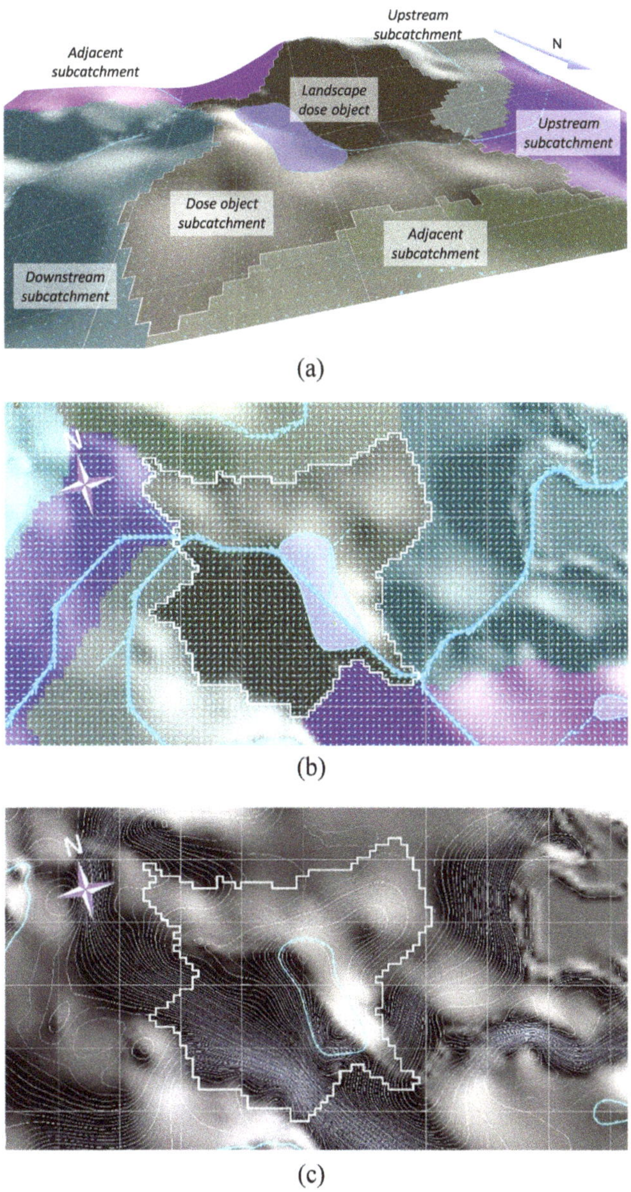

Figure 10.5. Portion of a landscape visualised using a digital terrain model (POSIVA 2013). Example of a low relief landscape. Taken from Guerfi *et al* (2019).

topographic minima, i.e. depressions in the landscape that will form natural lakes, evolving into wetlands as vegetation grows and dies *in situ*.

Figure 10.5 illustrates several important features of the landscape relevant to radiological assessments: (1) a landscape dose object, identified as a lake/wetland area; (2) the sub-catchment area over which net precipitation at the surface drains towards the landscape dose object; and (3) the drainage system that passes through the object, fed by captured net precipitation from the upstream areas. Using GIS methods. it is possible to provide the basis for a water balance scheme for the landscape objects in question.

Figure 10.5(b) shows how the GIS software identifies the potential surface drainage system. An eight-point pour algorithm determines where particles dropped at the topographic surface would move under gravity, and the drainage 'stream' indicates the flow path through the topography for the particles. In practice, this can be seen to be a good approximation for the position of streams in the landscape, whether flowing above the surface or as near-surface groundwater. The watershed boundaries from the analysis delimit the distinct hydrological features of the system allowing focussing on conditions around the landscape dose object.

The landscape context of the example here is northern Europe where wetland areas have, historically, been drained and subsequently cultivated (Saetre *et al* 2013). The landscape dose object can be understood by the *closed-contour method* illustrated in figure 10.5(c). The natural development of ecosystems is such that lakes form and accumulate vegetation, transitioning to wetlands with full terrestrialisation. Formed in this way, these wetlands are naturally level areas in the landscape and, with rich organic soils, are suitable for cultivation.

A reasonable approximation for the area of such an object is the maximum contour of the lake. GIS software can measure the area of the watershed in which the landscape dose object is embedded, as well as the area of the dose object itself. Furthermore, the volume of the depression within the lake/wetland area so defined can be calculated, thereby giving the volume of the lake at the start of the transition to wetland and the amount of soil that will have accumulated as it approaches dynamic equilibrium.

Understanding the local landscape context helps characterise the hydrology of the system. Just as the simple soil model in figure 10.3 uses the balance of meteoric water entering the soil, the same principle can be applied to the overall sub-catchment. Water accumulated at upper elevations will flow as surface runoff or subsurface flow towards the stream and lake or corresponding wetland. The streamflow from the upstream catchment system can also be estimated.

Groundwater circulation is driven by topography (Wörman *et al* 2006, 2007, Neff *et al* 2020). For the basins illustrated here, the water balance is driven mainly by the net precipitation captured in the local watershed, allowing for the upstream inflow in the local and regional drainage system. As a first approximation, it is possible to estimate the circulation of groundwater in the overburden of the watershed and landscape dose object by assuming some functional distribution of the infiltrating meteoric water captured by the overall watershed area, using water balance considerations. This is a practical approach for scoping calculations and is more representative of local conditions. In assessments requiring more detail, there are many groundwater modelling tools available to give more precise expressions of groundwater circulation based on detailed site characterisation methods (POSIVA 2013, Werner *et al* 2014) but in practice, in terms of results for dose assessment estimates, the results are similar to within an order of magnitude or better since it is the accumulation of radionuclides in the accessible biosphere that is of interest and retention in the overburden and water bodies can approach dynamic equilibrium in many cases.

This technique for characterising landscape dose objects can also be applied to the simple well scenarios. The context in this example is of an assessment of radionuclide release into an aquifer that could potentially be used as a source of domestic and agricultural water supplies for a typical exposure group. This kind of assessment is typical of the kind of scoping assessment that has been carried out for decades for generic site contexts or with little site-specific information. With the methods described here it is possible to include details of the local hydrological system in a way that can place practical bounds on the well dilution parameters in dose assessment models. The example is taken from Xu *et al* (2021), with details in figure 10.6.

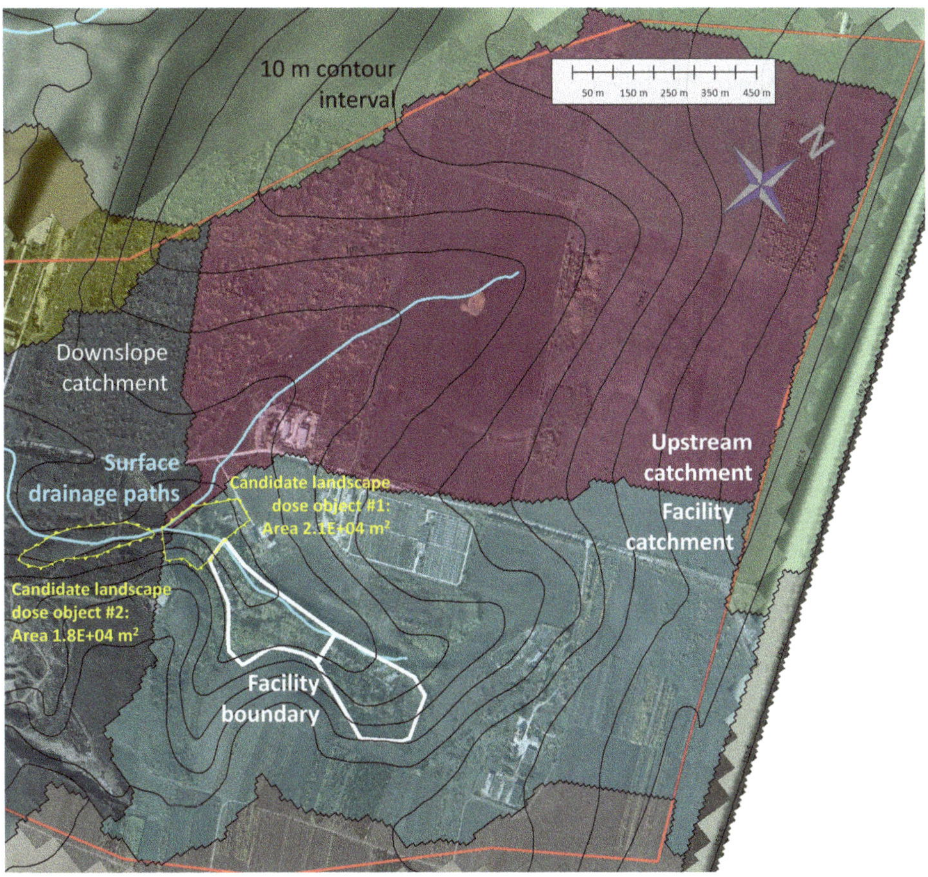

Figure 10.6. Local catchments and biosphere dose objects in a higher relief landscape. Topographic map with iso-photograph and watersheds and preferential flow system. The facility boundary is shown in white, and two candidate landscape dose objects are delineated in yellow. Details taken from Xu *et al* (2021).

The disposal facility is situated in a small valley within a single watershed. Accumulated runoff flows downslope through the valley with an intermittent surface stream during periods of high precipitation with the flow confined to a near-surface aquifer within a few metres of the surface. The preferential flow path defined by the GIS software implies that the stream is closest to the surface near the northern edge of the facility boundary. The total water flow in this catchment can be estimated by the net precipitation over the catchment area because there is no other upstream area; the topography slopes away from the facility to the right (northeast) of the catchment in which the facility is situated. Thereby, a first approximation to the maximum flow in the local aquifer can be defined.

Candidate biosphere dose objects are identified by Xu *et al* (2021) using a combination of the ISAM (IAEA 2004) and BIOMASS (IAEA 2003) methodologies and the techniques developed for the updated BIOMASS methodologies (IAEA in preparation). They are defined as areas close to the facility boundary that is large enough to support a single family of four adults in a single farm. As with the traditional well model, a well is used for domestic and agricultural purposes. These areas are not cultivated in the present-day biosphere but, for the purposes of radiological assessment, they are illustrative of locations that, given their location in relation to the facility boundary and the possibility that they could be developed by a future human population, are likely to give rise to conservative estimates of dose.

Characterisation of the traditional well model has been difficult because the landscape context was not always clear so that the well dilution parameter—the water flux in the well—is ill-defined. In the context illustrated here, it is possible to bound the dilution at least for scoping purposes. The maximum flow is the net precipitation captured by the upstream catchment. A conservative estimate for the dilution in the well can be obtained by the simple assumption that the well water dilution is defined by the total well water usage for domestic and agricultural purposes. In the landscape context illustrated here irrigation abstraction over the area of the biosphere object is used to provide a lower limit to the well dilution parameters for the dose assessment.

In summary, the site context for a radiological protection model is defined by landscape and climate that together define the local hydrological and hydrogeological context. Site descriptive modelling is required to a greater or lesser degree to define a model for a specific application. What has changed in the formulation of models from the earlier examples to the complex and detailed landscape dose models of contemporary license applications is not the details of the features, events and processes in the models but how they are expressed in respect of the landscape context.

GIS offer a method of quantifying details of the local landscape around locations where radionuclides can accumulate in the accessible environment and from where dose can arise. The difference between *simple* and *complex* models lies not in the representation of the FEPs in the models but in the incorporation of site- and region-specific information. A simpler assessment context might mean more straightforward methods of defining local characteristics such as those outlined here. A more detailed assessment content might lead to detailed site investigation studies to better determine local hydrology and hydrogeology and the effects of landscape.

Essentially, a description of the surface-water catchments in which the dose model is situated and those upstream and downstream can provide all relevant information concerning the transport and accumulation of radionuclides in the surface ecosystems.

10.6 Modelling of estuaries

In the context of modelling radionuclide transport in the environment, detailed modelling of estuaries is a somewhat neglected topic. Estuaries are typically developed in soft sediment and the spatial distribution of that sediment changes rapidly with time, affecting the patterns of water flow, which themselves vary diurnally with the tidal cycle and on longer timescales. Therefore, modelling radionuclide transport through an estuary implies the need to simulate sediment dynamics, water flows and contaminant migration in an integrated way.

Estuaries also comprise several distinct components (see figure 10.7).

A convenient simplification for compartmental modelling distinguishes an estuary into upper, middle and lower components. The upper estuary is of low salinity and is strongly conditioned by unidirectional input from the river. In contrast, the lower estuary is of high salinity and is dominated by the influx and efflux of tidal waters. The middle estuary comprises a mixing zone dominated by river flows at low tide and tidal flows at around high tide. Typically, estuaries developed in soft sediments are funnel shaped, implying that the largest exchanges of water are in the lower estuary, where the area of the estuary is largest, and the tidal

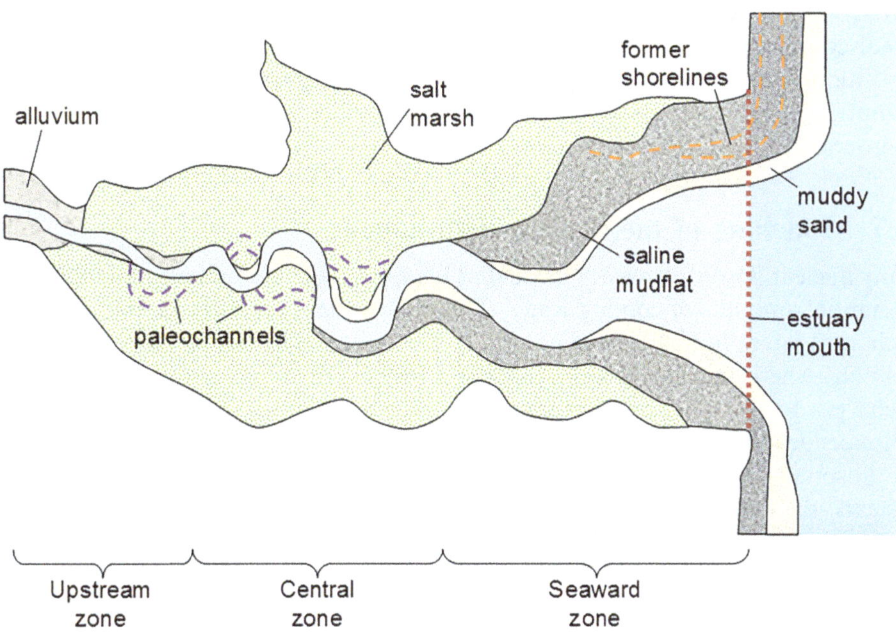

Figure 10.7. Components of the estuarine environment. Reproduced with permission from Walke *et al* (2013). © Nuclear Decommissioning Authority.

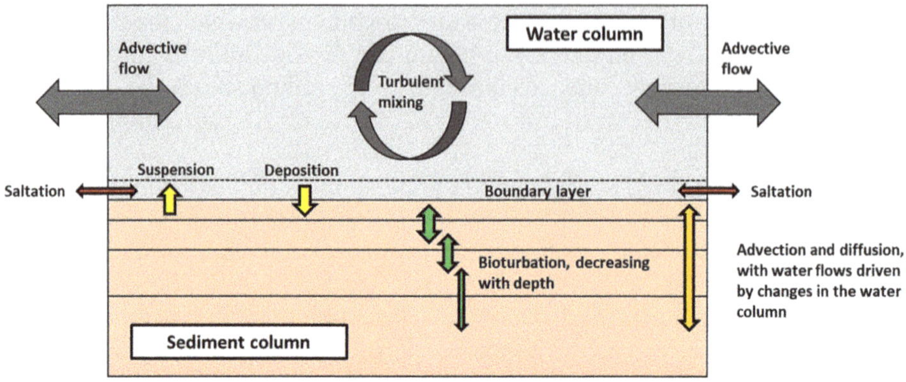

Figure 10.8. Radionuclide transport processes in an estuarine environment.

range is greatest. The sediments underlying the estuary can often be simplified and represented as a stratified column. As illustrated in figure 10.8, radionuclide transport in this column can be due to advection or diffusion in the pore water but can also be due to bioturbation due to sediment-inhabiting organisms. Transfers between the upper layer of sediment and the water column are due to suspension and deposition of particles, but exchanges between radionuclides sorbed to solids and present in solution mean that transfers occur both in the solid and liquid phases. In the water column, radionuclide transport is largely due to advection in the water flows, with this advective transport giving rise to turbulent mixing in the different areas of the estuary. Particles can also be transported in the boundary layer at the bottom of the water column by the process of saltation in which the particle moves in a discontinuous manner with suspension followed by short-distance transport and deposition, with the process repeating as flows in the boundary layer change. A quantitative compartmental model for radionuclide transport in the estuarine environment is described by Walke *et al* (2013).

10.7 Modelling of the marine environment

Many nuclear power plants are located at the coast, because of their requirement for substantial amounts of cooling water. This means that both planned and accidental discharges of radionuclides may occur directly to the marine environment. In addition, where nuclear power plants are located on river systems, radionuclide discharges to the marine environment can arise indirectly. Indirect discharges of radionuclides to the marine environment can also arise from other types of facility, e.g. hospitals, research establishments and sewage-treatment plants. In all these contexts, the area of discharge lies in the coastal or near-shore environment rather than in the offshore environment or the deep ocean.

In general terms, the coastal environment comprises a beach (if present) and the foreshore down to the lowest low water mark. An important feature of this sub-system is that this region is inundated by the sea with a frequency ranging from every tidal cycle down to a few times per year. It is also the component of the overall system that is most significantly influenced by wave action (though waves also influence the transport of bottom sediments in the near-shore coastal region where water depths are less than about 10 m). The storm beach is generally composed of coarse material, though some recent eroded finer material may also be present. The foreshore may be rocky, sandy or composed of abraded shells. In general, it contains little fine-grained material (clay and silt) as this is rapidly transported offshore. Where the foreshore is primarily sandy, it may also include gravel and cobbles. This coarser material may form a lag deposit which tends to protect the beach from erosion.

The marine environment is conveniently distinguished into a near-shore component immediately seaward of the lowest low water mark and extending out to a depth of 10–20 m, where the effects of wave action on bottom sediments become negligible, and an offshore component that extends from this boundary further out to sea (Walke *et al* 2013).

Radionuclide releases to the beach and foreshore can occur in upwelling groundwater. That groundwater may emerge as springs or seeps. These could be on the cliff face above or below the level of the surface of the storm beach or on the foreshore. Springs or seeps at the cliff face may result in streams of fresh water running over the surface of the beach. On rocky foreshores upwelling groundwater may contribute to the water in rock pools, though a major component of such pools will generally be sea water that is retained as the tide falls. Thus, radionuclide concentrations and salinities in rock pools may vary substantially over each tidal cycle. This may be of particular importance in evaluating radiation exposures of non-human biota and their ecological significance.

Radionuclide interactions with the coarse material of storm beaches or surface lags will be limited because of the small surface-to-volume ratio of this material and the ease with which water percolates through it. Although there will be a greater degree of interaction with sand-sized material, sorption will generally be less than it would be on silt or clay materials. Therefore, the degree of hold up of radionuclides in the beach and foreshore is likely to be limited.

Most beaches are comprised of predominantly sand-sized material. However, gravel is the dominant component of a significant number of UK beaches. Gravel on some of these beaches has been derived from coarse glacial deposits that cover the landscape or has been reworked by rivers or weathering from cliffs. Beaches tend to develop a cross-shore equilibrium profile that may be due to a competition between the effects of waves tending to move grains onshore and gravity tending to move them offshore. Beaches tend to develop their longshore morphology in response to the dominant wave energy regime. They can be long and straight where sediment supply has been sufficient to completely veneer the underlying topography or they can comprise a series of pocket beaches separated by headlands on embayed coasts with considerable inheritance from the underlying bedrock. There is a tendency for the beach shoreline to be either swash-aligned or drift-aligned. Swash-aligned shorelines are those on which waves arrive parallel to the shore. They tend to be curved and are likely to have negligible longshore transport. Drift-aligned shorelines are those on which waves break at a sharp angle. Such shorelines are usually longer and straighter than swash-aligned shorelines, and they are characterised by longshore transport. Compartmentalisation of the shore occurs where there are major obstacles, particularly headlands on deeply embayed coasts, to the longshore transport of sediment. The most enclosed beaches are pocket beaches on which the sediment volume remains constant (Woodroffe 2003).

The concept of a sediment budget is often useful in characterising and managing coastal compartments. Input of sediment can occur from a river, from cliff erosion, from offshore, or alongshore from other compartments. Although rivers can bring large quantities of sediment to the coast, this may be deposited in an estuary before reaching the littoral zone. Cliff erosion can be a significant source where the cliffs are eroding rapidly. Further inputs occur because of *in situ* production by calcareous organisms and because of the erosion of foredunes.

In general, the shore face adjusts to a time-averaged profile that is relatively constant over decades to millennia. However, North American and European studies have recognised erosional storm (winter) and constructional swell (summer) profiles. More generally, three different beach types have been recognised. These are dissipative, intermediate and reflective. Dissipative beaches are flat and are typified by northern-hemisphere, mid-latitude winter or storm-profile beaches, whereas summer or berm profile beaches are reflective and are steeper. A dissipative beach is typically composed of fine sand. It has a relatively small volume of sand within the beach itself, compared with the volume of sand stored in the shore-face. It is dominated by large, high-energy waves. The energy of these is dissipated across a broad surf zone and secondary waves can be formed in that surf zone. In contrast, reflective beaches are steep, dominated by an accretionary berm and low breakers (generally < 1 m high) that travel right up to the beach, surge up it, and rapidly return. Reflective beaches are generally composed of coarser material than dissipative beaches. Intermediate beaches include characteristics of both, with a steep reflective profile at the top of the beach and a flatter dissipative profile lower down.

Coastal dunes develop behind beaches on some coasts. Such dunes represent a sand reserve from which material can be borrowed under extreme conditions to reshape the near-shore profile, with return of the sand under more usual conditions. The relationship between dunes and near-shore can become decoupled, e.g. through growth of vegetation on the dunes or armouring of the beach surface.

In contrast to the estuarine and beach environments, the marine environment has been represented in widely used radiological impact assessment models for several decades. In general, a local (and sometimes regional) compartment is defined that is linked to a standard representation of the rest of the world's oceans, expressed in terms of a set of interlinked water column compartments underlain by bottom sediments, see, e.g. Smith and Simmonds (2009).

Dispersion in the marine environment at a local scale of up to a few kilometres from the discharge point is typically modelled by a single, well-mixed water compartment together with associated underlying sediment compartments (see figure 10.9).

As required, this local compartment is interfaced with a regional marine model and exchanges water and suspended sediment with the adjacent regional marine compartment. Smith and Simmonds (2009) emphasise that, if detailed assessments

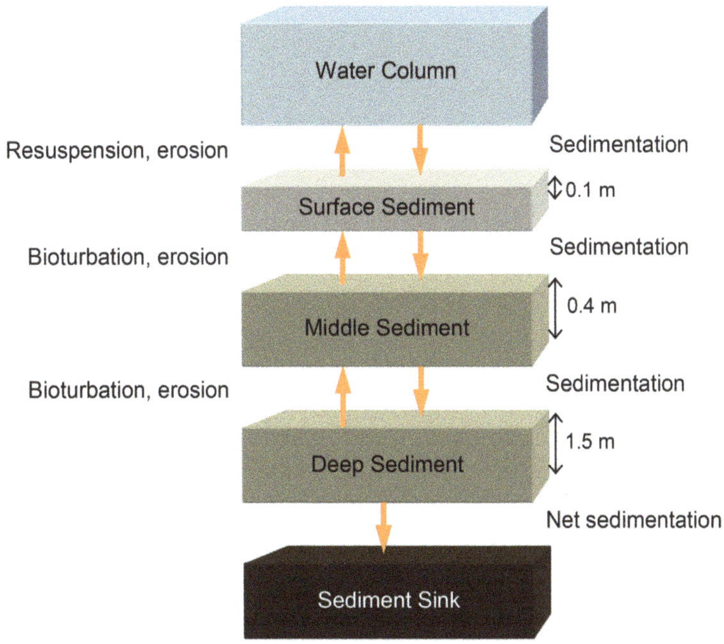

Figure 10.9. The multi-layer marine model structure adopted by Walke *et al* (2013). The upper sediment reflects loosely packed, oxic sediments that interact directly with the water column. The middle sediment reflects anoxic, average-packed sediments. The lower sediment reflects anoxic, densely packed sediment. The depths and properties used for the sediment layers will reflect the type of system being modelled. Reproduced with permission from Walke *et al* (2013). © Nuclear Decommissioning Authority.

of the local radiological impact of discharges from a particular site are to be undertaken, then site-specific hydrographic data will be required. In the UK, the Environment Agency (Dewar *et al* 2011) has provided such local compartment parameters for a large set of locations around the UK coastline. Detailed analysis of these data shows a close correlation between compartmental volume and rate of water exchange, reflecting similar water residence times irrespective of the size of the local mixing compartment (figure 10.10).

As discussed above, modelling of radionuclide transport in the marine environment on a regional or global scale is typically compartmentally based. The global and regional European compartments adopted in PC CREAM 08 (Smith and Simmonds 2009) are illustrated in figures 10.11 and 10.12. Additional compartments are adopted to represent the waters of the Irish Sea and English Channel reflecting a need for greater spatial and temporal resolution in the areas in which discharges occur in typical applications of the model. For parameter values characterising all the compartments included in PC CREAM 08 see tables 4.7 and 4.8 of Smith and Simmonds (2009).

Although compartmental models at the local and regional scales are generally fit-for-purpose in radiological impact assessments of routine releases, they are unlikely to be able to address the full range of issues that arise in relation to accidents

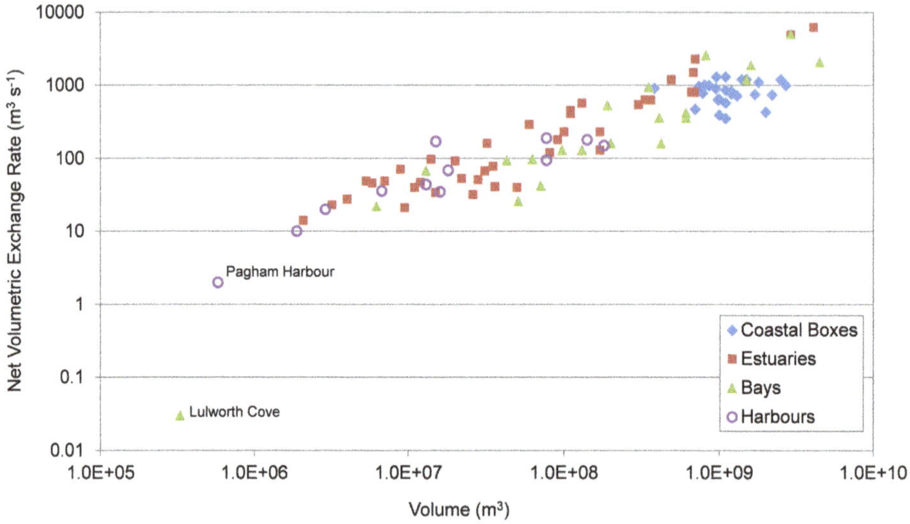

Figure 10.10. Relationship between volume and water exchange rate for coastal environments in the UK. Reproduced with permission from Walke *et al* (2013). © Nuclear Decommissioning Authority.

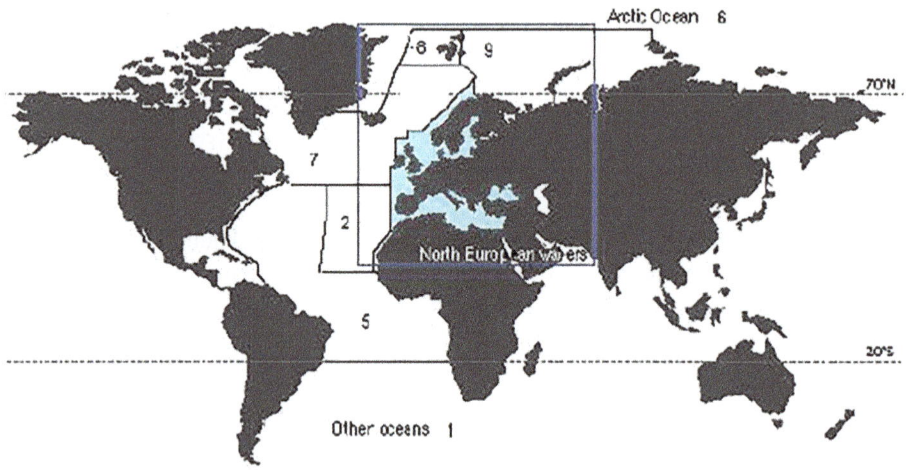

Figure 10.11. Global-scale compartments adopted in PC CREAM 08. Reproduced from Smith and Simmonds (2009). OGLv3.0.

resulting in substantial discharges to the marine environment. In this context, reliance needs to be placed on using underpinning models of water flows and sediment transport, with the dispersing radionuclides treated as tracers that move with the waters and sediments (see Estournel *et al* (2012) and the discussion in UNSCEAR (2014)). Estournel *et al* (2012) provide a good illustration of the approach. They used the SYMPHONIE model with 30 vertical levels of varying thickness, with increased resolution near the sea surface. The horizontal grid was based on an orthogonal curvilinear system with variable horizontal resolution that

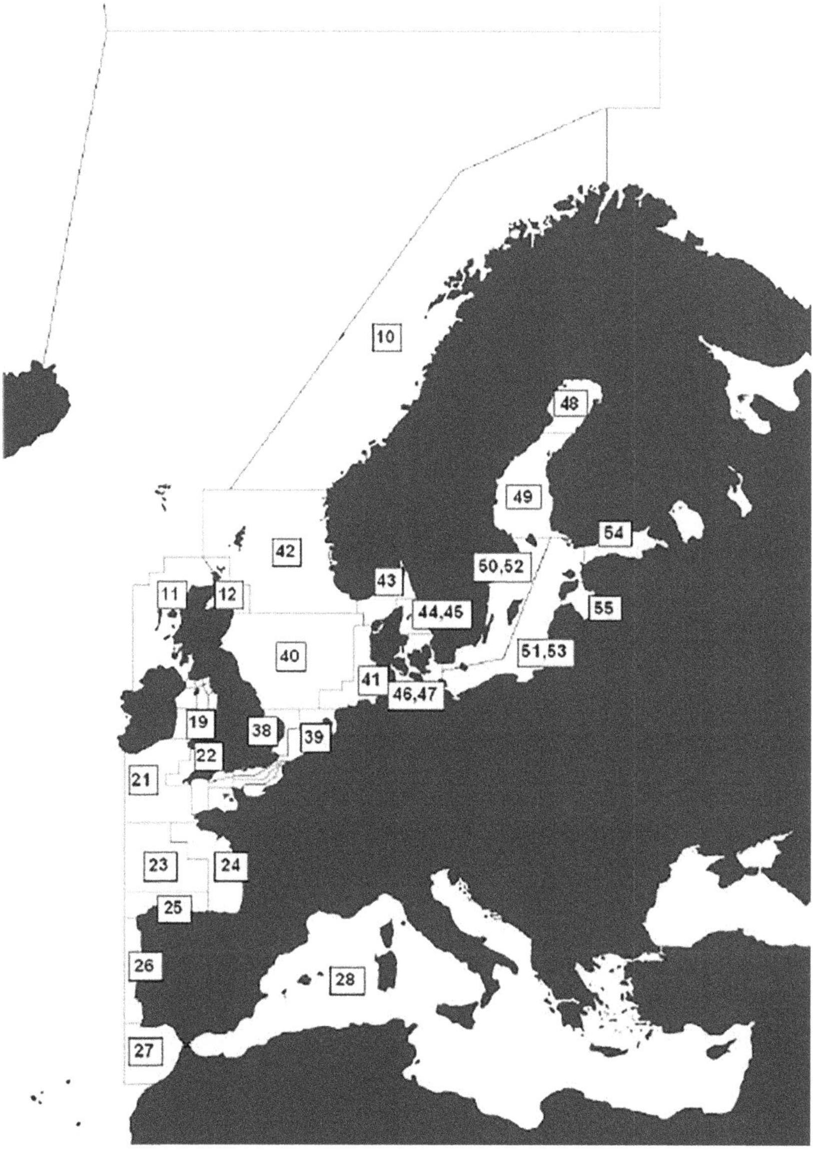

Figure 10.12. Regional-scale compartments adopted in PC CREAM 08. Reproduced from Smith and Simmonds (2009). OGLv3.0.

increased linearly with the distance from the Fukushima Nuclear Power Plant, where the size of the grid mesh was about 600 m. At the open lateral boundaries, the horizontal resolution reached its maximum value of 5.5 km. Illustrative results from the model are shown in figure 10.13.

In both large-scale compartmental marine models and hydrographic dispersion models, there is a requirement to estimate water fluxes either across compartmental boundaries or at the resolution of a single finite element. In this context, it is

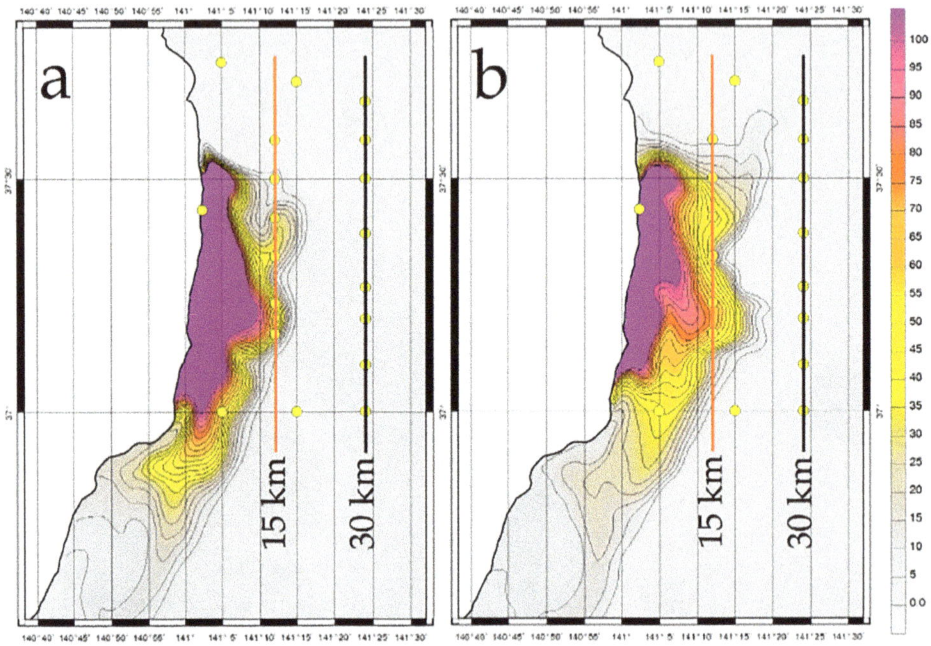

Figure 10.13. Caesium-137 concentration at the ocean surface simulated for 15 April 2011. (a) reference simulation and (b) simulation with increased river discharges. The two lines join the stations located at 15 and 30 km offshore. Estournel *et al* (2012), John Wiley & Sons. ©2012. American Geophysical Union. All Rights Reserved.

important to adopt an appropriate temporal scale such that the periodic effects of diurnal tides are averaged out, because they only move packets of water short distances on a daily timescale and average out to zero movement on a longer timescale. Therefore, the water flows addressed in such models are residual currents averaged over timescales substantially longer than a day. These residual currents result from various causes, including the asymmetric aspects of tidal flows (originating, for example, because tidal flows are governed by the geometry of the coastline) and the Coriolis forces arising from rotation of the Earth.

10.8 Radiological impact assessment

Earlier sections of this chapter describe approaches to estimating radionuclide concentrations in environmental media, including both abiotic and biotic materials. Here, an outline is given as to how those concentrations are used to estimate radiological impacts on both humans and non-human biota. In the case of humans, consideration is first given to the pathways of exposure and approaches for estimating radiation doses from each of those pathways before turning to the complementary requirements of assessing radiological impacts on individuals and larger populations, and for propagating uncertainties through the modelling process. The issue of appropriately representing uncertainties requires consideration as to how reference values and ranges or distributions of parameter values should be selected for radionuclide transport modelling, and of how appropriate patterns of human habits should be defined. Finally, a commentary is provided on how radiological impacts on non-human biota can be assessed, identifying similarities to and differences from the approach adopted for assessing radiological impacts on humans.

10.8.1 Pathways of exposure

Pathways of exposure of humans can be broadly identified as external irradiation, inhalation and ingestion. In addition, in some contexts, uptake through the intact skin or via wounds can be of importance.

External irradiation of members of the public can arise directly from nuclear fuel cycle facilities, e.g. due to authorised discharges to air or water or where radioactive wastes are being disposed in surface trenches. In accident conditions, drainage of a spent-fuel storage pond could lead to high gamma dose rates in the vicinity. In this context, the scattering of gamma photons in the air above the exposed source (skyshine) may be the main route of exposure if the source is shielded along the direct line of sight. However, in most contexts, external irradiation arises either from an atmospheric plume of radioactive material, or from radionuclides deposited on the ground.

Where an atmospheric plume is widely dispersed through the mixing layer, it is often adequate to treat it as a semi-infinite source surrounding the exposed individual. However, in circumstances where an elevated plume has only dispersed to a limited degree, the most highly exposed individuals may be in uncontaminated air below the contaminant plume. This arises because photons with energies around 1 MeV have a mean free path in air of several tens of metres. For such situations, the preferred approach is to use a radiation-transport model to estimate the dose rate at the location of the exposed individual.

Radionuclides deposited on soils and sediments may be present as a thin surface layer or may be distributed over depth. Typically, concentration profiles in undisturbed soils and sediments exhibit an exponential decrease with depth. However, concentration profiles in disturbed soils, such as those subject to ploughing, are more uniform over the depth to which disturbance occurs. Dose rates to an individual standing on contaminated soil or sediment is mainly determined by radionuclide concentrations within a radius of a few metres of the individual.

For assessment purposes, it is generally adequate to assume simplified geometrical configurations when evaluating external dose rates. In particular, the US Environment Protection Agency (EPA) has produced a comprehensive tabulation of dose rates from surface sources and from semi-infinite slab sources (a semi-infinite slab source is infinite in lateral extent and is sufficiently thick that photons emitted at depth from within the source cannot reach its upper surface) (US EPA 2019). However, if there is a need to compute external dose rates from sources of finite extent or for complex geometrical configurations, a commercial radiation-transport code such as MicroShield (https://radiationsoftware.com/microshield) can be used.

Inhalation may be directly from an atmospheric plume or of material resuspended from the ground. In addition, Rn-220 (thoron), Rn-222 (radon) and their short-lived progeny may be inhaled. As exposures to thoron and radon are of relevance in the context of natural background radiation, the special issues relating to their dosimetry are discussed in chapter 5. In general, the committed effective dose from inhalation can be computed by multiplying the concentration of the radionuclide in air (Bq m^{-3}) by the breathing rate (m^3 s^{-1}), the period of exposure(s) and the committed effective dose per unit intake (Sv Bq^{-1}). Breathing rates vary with age and level of activity, as discussed in ICRP (1994). Reference values of committed effective dose per unit intake are provided by the ICRP for various age groups ranging from three-month-old infants to adults (see ICRP 2012). Special considerations apply to inhalation of inert gases. Although these are not retained in the body, they deliver an internal dose from gas present in the lungs and from gas dissolved in tissues, particularly adipose tissues.

For ingestion, the main contributors to radionuclide intakes are generally foods and drinking water. However, ingestion of radionuclides in soil and sediment can also be important for radioisotopes of elements such as the lanthanides and actinides that are strongly bioexcluded both from plants and animals. Annual committed effective doses by ingestion are computed by multiplying the concentration of the radionuclide in the ingested material (Bq kg^{-1} or Bq m^{-3} for fluids such as water and milk) by the consumption rate (kg y^{-1} or m^3 y^{-1}) and the committed effective dose per unit intake (Sv Bq^{-1}). It is important to perform these calculations for different age groups, because values of committed effective dose per unit intake are generally larger for infants and children than for adults (due to their smaller body masses), but intake rates for some foods, e.g. milk, can be as large or larger for infants and children than for adults.

Absorption across the intact skin is potentially of importance for tritiated water, whereas uptake through damaged areas of skin can be important for a wide variety of radionuclides. Such damage may comprise wounds and abrasions but may also be deliberately induced for ritual and/or decorative purposes.

10.8.2 Individual and collective doses

In assessing the radiological impact of distributions of radionuclides present in the environment, consideration needs to be given both to those members of the public who are more highly exposed and to the distribution of doses among the wider population. In respect of limiting the exposure of individuals, the ICRP (2007) makes use of the concept of the Representative Person. The Reference Person 'may be hypothetical, but it is important that the habits (e.g. consumption of foodstuffs, breathing rate, location, usage of local resources) used to characterise the Reference Person are typical habits of a small number of individuals representative of those most highly exposed and not the extreme habits of a single member of the population.' (ICRP (2007), paragraph 194). In practice, as discussed below, this leads to the use of high percentiles of the distributions of occupancies of contaminated areas and of the distributions of rates of consumption of foodstuffs from those areas to define a Representative Person.

Within the wider population, individuals are likely to receive a wide range of annual committed effective doses. Furthermore, where a release leads to persistent contamination of the environment, annual committed effective doses may be delivered over many generations to a changing population. In optimising radiological protection, consideration needs to be given to the overall radiological impact of an option integrated over both space and time relative to the benefits arising from realising that option. To characterise the overall radiological impact, the concept of collective effective dose can be used. The collective effective dose comprises the sum of individual committed effective doses over a defined population, i.e. it has units of person-Sv. When this concept was originally introduced, there was a tendency to aggregate collective effective dose over all space and time. However, this obscures important distinctions between options that are of potential significance in decision making, e.g. between doses delivered in the near- and far-future. Therefore, it is generally good practice to present a matrix of collective doses distinguished by factors such as the exposed population, the timeframe of exposure or the magnitude of annual or lifetime committed effective doses to individuals.

A controversial issue that arises with collective dose is whether it is permissible to multiply a value of collective dose by the risk per unit effective dose used in setting radiological protection standards for individuals. If it is assumed that the linear no-threshold model adopted in radiological protection is realistic and applies down to the lowest doses of interest, then such a multiplication yields an expectation value of the total detriment to health arising in the defined population from the exposure situation. In contrast, if it is assumed that the linear no-threshold model includes a cautious element and that, in practice, a dose threshold for health effects exists, then such a multiplication is not legitimate. However, if this is the case, then collective dose does not have a straightforward relationship to risks to health or to the burden of ill-health on society and it is, therefore, difficult to see how it can be used in optimisation.

10.8.3 Handling uncertainties

In assessing radiological impacts on individuals, consideration must be given to uncertainties in the radionuclide concentrations to which they are exposed and variations in the habits of the exposed individuals. This suggests a clear separation between uncertainty and variability, but this is not the case. Thus, spatial and temporal variability in radionuclide concentrations gives rise to uncertainty as to how this variability should be averaged to define the radionuclide concentrations to which individuals are exposed. However, the habits to be adopted for exposed individuals can, in principle, be constrained by using a well-specified definition for the Reference Individual.

In older assessments, both the parameter values used to describe the transport and distribution of radionuclides in the environment, and the parameter values used to describe human habits, were defined as point estimates. This meant that the results from the assessment were also point estimates that could be compared with regulatory criteria expressed in deterministic terms. However, it is now becoming more common to take uncertainties in parameter values into account. This can be done though different approaches. For example, in single parameter sensitivity studies, one input parameter is varied, while holding all the others fixed, and the effects on model outputs are determined. This approach can be extended to explorations in which several input parameter values are varied. In such sensitivity analyses, the ranges of variation of the input parameters do not necessarily need to be realistic, e.g. extreme ranges may be used to demonstrate limited effects on key outputs. In contrast, in uncertainty analyses, input parameter values are sampled from realistic probability density functions (pdfs) and used to provide distributions of output parameter values. Ideally, the input parameters are independent, meaning that sampling from the pdfs does not need to consider correlations between parameter values. However, if two or more parameter values are correlated this can be addressed by explicitly representing the correlation matrix between them or by defining underlying independent parameters that can be used to compute the higher-level correlated parameters, e.g. if plant uptake is anti-correlated with retention in soil both may be computed after sampling the distribution coefficient that determines both retardation of the radionuclide in soil and the fraction of the activity that is present in soil solution and, therefore, available for plant uptake.

Sensitivity and uncertainty analyses can be combined. For example, an uncertainty analysis can be conducted as described above, but the results can then be shown on a scatter plot in which the abscissa is the value of a selected input parameter and the ordinate is the value of the output parameter. Such a plot shows the effects of varying one input parameter in the presence of uncertainty in all the other input parameters.

In many practical cases, an assessment model may have a substantial number of different input parameters meaning that it is difficult to ensure that key combinations of parameter values are sampled. Simple Monte Carlo sampling techniques tend to be inadequate in such circumstances, with assessment results converging only slowly as the number of samples increases. Therefore, stratified sampling schemes, such as Latin hypercube sampling are typically adopted. In 2D Latin hypercube sampling, the range of each of the two variables is divided into N intervals. A total of N samples is then selected so that no two samples fall within the same interval of either variable. This is illustrated in figure 10.14.

10.8.4 Defining human habits

The human habits required for assessment purposes include occupancies of contaminated areas and consumption rates of water, soil/sediment and various food types.

Information on water intake rates is available from several sources. Values derived from (ICRP 1975) are commonly used (Smith and Jones 2003). Total fluid intakes by ingestion for male adults and children are given by the ICRP (1975) as 3 l day^{-1} and 2 l day^{-1}, respectively. The adult value, although greater than other

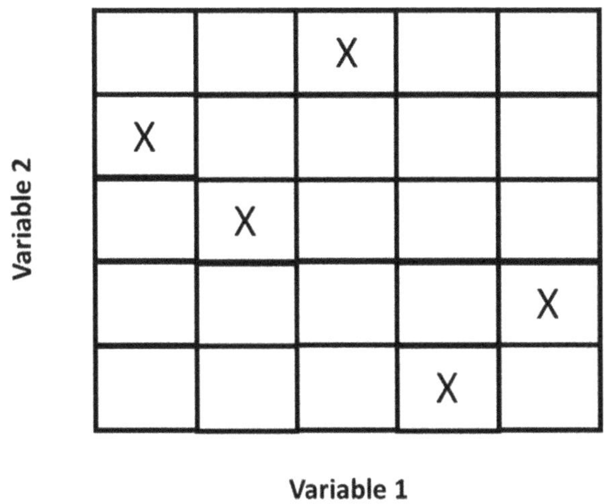

Figure 10.14. Illustrative Latin hypercube sample for two variables and a sample size of five.

published fluid intake rates, is based on consideration of total water balance and the physiological premise that 1 ml of water is required for each kcal of energy expended.

Neglecting any water ingested in food and milk, and produced by the oxidation of food, the ICRP (1975) suggested a water intake rate of 1.65 l day^{-1} for adults and 0.95 l day^{-1} for a ten-year-old child. The ICRP did not give a total fluid intake for infants. However, a value may be derived from an energy expenditure of 1200 kcal day^{-1} for a one-year-old (ICRP 1975). This gives a daily intake of 1.2 l. By assuming that the average milk intake for an infant is 0.5 l day^{-1}, and neglecting intake from food, the daily water intake rate has been estimated to be 0.7 l (Smith and Jones 2003).

Additionally, Hopkin and Ellis (1980) provide some information on tap water consumption. The mean and 90th percentile tap water intake rates for their whole sample, which included children, were approximately 1 l day^{-1} and 1.6 l day^{-1}, respectively.

Information on soil ingestion rates is available from a few small studies carried out in various countries. Summaries of the literature provide data from which indicative rates can be obtained (see Simon 1998, Smith and Jones 2003, Smith and Bedwell 2005 and Davis and Mirick 2006). The results from these studies are sometimes referred to as inadvertent soil ingestion rates. However, this is somewhat misleading. It is well known that relatively short-term deliberate ingestion (exploratory mouthing) is widespread among young children and that such behaviour may be regarded as a 'normal' temporary phenomenon among some young children.

In contrast, the rates from these studies exclude the deliberate consumption of large quantities of soil or sediment (pica) as too rare a condition to warrant inclusion in an environmental exposure assessment.

In terms of food consumption, a typical approach is to adopt 95th or 97.5th percentile values of consumption rates from national dietary surveys. However, this gives rise to two issues. First, the national survey data may not be representative of the locally exposed community, e.g. the inhabitants of a coastal village are likely to have much higher consumption rates of various seafoods than would be estimated from national survey data. Second, no individual is likely to consume more than a few types of food at high rates, i.e. high rates for preferred food types will tend to be compensated by low rates for other food types.

In the UK, the issue of relying on national survey data is avoided by relying on data from targeted local surveys. These surveys provide both occupancy and consumption rate data for almost all individuals living close to nuclear licensed sites (see, e.g. Clyne *et al* 2013). Because such detailed data are available, various approaches can be used in applying these data in defining a representative individual. The UK National Dose Assessment Working Group (NDAWG 2009, 2013) identified two preferred methods that can be used for detailed short-term assessments. These are an unadjusted 'profiling' and a 'top-two' method (described in detail below). However, for longer-term prospective assessments, the NDAWG considered that it might also be appropriate to make upward adjustments to some consumption rates and occupancies for future-proofing of the assessment.

In the profiling method, the habits survey data are first obtained and tabulated by observation number and pathway. For the first pathway, the observations are ordered by descending consumption rate or occupancy for this pathway. Observations for individuals who have consumption or occupancy rates between the maximum and one third of the maximum are identified and retained for the next step of the analysis. For this subset of observations, the mean rates for all pathways are calculated. These mean rates are adopted for the group associated with the first pathway, referred to as the pathway 1 profile, e.g. if pathway 1 was fish consumption, this would be the fish profile.

The process is then repeated for each of the other pathways to give as many profiles as there are pathways. The dose for each profile can then be calculated and the values compared, with the maximum value adopted for compliance purposes.

In the top-two method, the assessment is initially made with all the relevant habits data set to critical levels (95th or 97.5th percentile values considering non-zero values only in defining these percentiles). The assessment results identify the two habits that give rise to the highest doses (the 'top two'). These habits are retained at the critical levels, whereas the other habits are reduced to 50th percentile values and the results of the assessment are recalculated.

10.8.5 Assessing radiological impacts on non-human biota

Whereas detailed information on the uptake and retention of radionuclides in humans is available, for other organisms the data are far less extensive. Therefore, absorbed dose rates to non-human biota are calculated by simple methods. Concentrations in these organisms are calculated as whole-body-average values using empirical concentration ratios (ICRP 2009). This means that they are treated as being in instantaneous equilibrium with the environmental media that they inhabit. Dose rates are also calculated as whole-body average values based on the concentrations in the organisms and in the surrounding media. For these calculations, the organisms are generally treated as being ellipsoidal in shape, with different size ellipsoids corresponding to the various reference animals and plants (RAPs) defined by the ICRP (2008). The dose rates calculated in this way are compared with threshold dose rates below which populations of most types of organisms are unlikely to be significantly adversely affected. Software tools for assessing radiological impacts on non-human biota include RESRAD-BIOTA (https://resrad.evs.anl.gov/codes/resrad-biota/) and ERICA (http://erica-tool.com/).

References

ANDRA 2015 Safety Options Report—Post-Closure Part (DOS-AF), ANDRA Report CG-TE-D-NTE-AMOA-SR2-0000-r 555515-0062, Paris (2015). www.andra.fr

Clarke R H 1979 *A Model for Short and Medium Range Dispersion of Radionuclides Released to the Atmosphere, The First Report of a Working Group on Atmospheric Dispersion* Report NRPB-R91 Harwell, Didcot, Oxfordshire: National Radiological Protection Board

Clyne F J, Garrod C J, Ly V E, Dewar A and Papworth G P 2013 *Radiological Habits Survey: Low Level Waste Repository, 2012* RL 02a/13 Lowestoft, Suffolk: CEFAS

Davis S and Mirick D K 2006 Soil ingestion in children and adults in the same family *J. Expos. Sci. Environ. Epidem.* **16** 63–75

Dewar A, Jenkinson S and Smedley C 2011 *Parameter values used in coastal dispersion modelling for radiological assessments* Report SC060080/R3 Environment Agency

Estournel C, Bosc E, Bocquet M, Ulses C, Marsaleix P *et al* 2012 Assessment of the amount of Cesium-137 released into the Pacific Ocean after the Fukushima accident and analysis of its dispersion in Japanese coastal waters *J. Geophys. Res. Oceans* **117** C11014

Ewen J and Parkin G 1996 Validation of catchment models for predicting land-use and climate change impacts: I. Method *J. Hydrol.* **175** 583–94

Gallacher D J, Robins A G, Burt A, Chadwick S, Hayden P and Williams M 2016a Dispersion of positron emitting radioactive gases in a complex urban building array: a comparison of dose modelling approaches *J. Radiol. Prot.* **36** 746–84

Gallacher D J, Robins A G and Hayden P 2016b Conversion of simulated radioactive pollutant gas concentrations for a complex building array into radiation dose *J. Radiol. Prot.* **36** 785–818

Guerfi R, Dverstorp B, Kłos R A, Nordén M and Xu S 2019 A simple and transparent modelling approach for biosphere assessment in post-closure safety assessment *Proc. of the 14th Int. High-Level Radioactive Waste Management Conf. (IHLRWMC) (Knoxville, TN, April 14–18, 2019)* (American Nuclear Society) pp 546–53

Hopkin S M and Ellis J C 1980 *Drinking water consumption in Great Britain—a survey of drinking habits with special reference to tap-water-based beverages* Technical Report TR137 Water Research Centre

IAEA 2003 *'Reference biospheres' for solid radioactive waste disposal, report of BIOMASS Theme 1 of the BIOsphere Modelling and ASSessment Programme* Report IAEA-BIOMASS-6 Vienna: International Atomic Energy Agency

IAEA 2004 *Safety Assessment Methodologies for Near Surface Disposal Facilities (ISAM)—Results of a Co-ordinated Research Project—Vol.1: Review and Enhancement of Safety Assessment Approaches and Tools* (Vienna: International Atomic Energy Agency)

IAEA in preparation *Biosphere Modelling for Long Term Safety Assessments of Solid Radioactive Waste Disposal Facilities: The Enhanced BIOMASS Methodology* (Vienna: International Atomic Energy Agency)

ICRP 1975 Report of the Task Group on Reference Man *ICRP Publication 23* (Oxford: Pergamon)

ICRP 1994 Human respiratory tract model for radiological protection, ICRP Publication 66 *Ann. ICRP* **24(1–3)**

ICRP 2007 The 2007 recommendations of the international commission on radiological protection: user's edition, ICRP Publication 103 *Ann. ICRP* **37(2–4)**

ICRP 2008 Environmental protection: the concept and use of reference animals and plants, ICRP Publication 108 *Ann. ICRP* **38(4–6)**

ICRP 2009 Environmental protection: transfer parameters for reference animals and plants, ICRP Publication 114 *Ann. ICRP* **39(6)**

ICRP 2012 Compendium of dose coefficients based on ICRP Publication 60, ICRP Publication 119 *Ann. ICRP* **41(Supplement 1)**

Jones A R, Thomson D J, Hort M and Devenish B 2007 The U.K. met office's next-generation atmospheric dispersion model, NAME III *Air Pollution Modeling and its Application XVII (Proc. of the 27th NATO/CCMS Int. Technical Meeting on Air Pollution Modelling and its Application)* ed C Borrego and A-L Norman (Berlin: Springer) pp 580–9

Klos R and Thorne M C 2020 Use of interaction matrices to formalise the development of conceptual models of contaminant transport in the biosphere and the translation of those conceptual models into mathematical models *J. Radiol. Prot.* **40** 40–67

NCRP 1998 *Recommended screening limits for contaminated surface soil and review of factors relevant to site-specific studies* NCRP Report No. 129 Bethesda, MD: National Council on Radiation Protection and Measurements

NDAWG 2009 National Dose Assessment Working Group Paper NDAWG/2/2009: Acquisition and Use of Habits Data for Prospective Assessments, https://srp-uk.org/resources/national-dose-assessment

NDAWG 2013 National Dose Assessment Working Group Guidance Note 7: Use of Habits Data in Prospective Dose Assessments, Version 1.0, Final, May 2013, https://srp-uk.org/resources/national-dose-assessment

Neff B P, Rosenberry D O, Leibowitz S G, Mushet D M, Golden H E, Rains M C, Brooks J R and Lane C R 2020 A hydrologic landscapes perspective on groundwater connectivity of depressional wetlands *Water* **2020** 50

POSIVA 2013 Terrain and Ecosystems Development Modelling in the Biosphere Assessment BSA-2012. Safety Case for the Disposal of Spent Nuclear Fuel at Olkiluoto, POSIVA Report 2012-29, www.posiva.fi

Saetre P, Valentin J, Lagerås P, Avila R and Kautsky U 2013 Land use and food intake of future inhabitants: outlining a representative individual of the most exposed group for dose assessment *Ambio* **42** 488–96

Simon S L 1998 Soil ingestion by humans: a review of history, data and etiology with application to risk assessment of radioactively contaminated soil *Health Phys.* **74** 647–72

Smith G M, Smith K L, Kowe R, Pérez-Sánchez D, Thorne M, Thiery Y, Read D and Molinero J 2014 Recent developments in assessment of long-term radionuclide behavior in the geosphere biosphere subsystem *J. Environ. Radioact.* **131** 89–109

Smith J G and Simmonds J R 2009 The Methodology for Assessing the Radiological Consequences of Routine Releases of Radionuclides to the Environment Used in PC CREAM 08, HPA-RPD-058

Smith K R and Bedwell P 2005 *Public Health Implications of Fragments of Irradiated Fuel—Module 3: The Likelihood of Encountering a Fuel Fragment on Sandside Beach, RPD-EA-9-2005* (Chilton, Didcot, Oxfordshire: HPA)

Smith K R and Jones A L 2003 *Generalised Habit Data for Radiological Assessments, NRPB-W41* (Chilton, Didcot, Oxfordshire: National Radiological Protection Board)

Thorne M C 2012 Modelling radionuclide transport in the environment and calculating radiation doses *Radionuclide Behaviour in the Natural Environment: Science, Implications and Lessons for the Nuclear Industry* ed C Poinssot and H Geckeis (Cambridge: Woodhead Publishing) ch 14

Thorne M C 2014 Kinetic models for representing the uptake of radionuclides in plants *Radionuclide Contamination and Remediation Through Plants* ed D K Gupta and C Walther (Cham: Springer) pp 215–33

UNSCEAR 2014 *UNSCEAR 2013 Report Volume 1, Scientific Annex A, Levels and Effects of Radiation Exposure Due to the Nuclear Accident after the 2011 Great East-Japan Earthquake and Tsunami* (New York: United Nations)

US EPA 2019 Federal Guidance Report 15, External Exposure to Radionuclides in Air, Water and Soil, EPA-402/R19/002, August 2019

Walke R, Thorne M and Smith J 2013 RWMD Biosphere Assessment Model: Marine Component, Quintessa and AMEC Report to NDA RWMD, AMEC Reference 18025/TR/001, Quintessa reference QRS-1628A-1: Issue 2, UK

Werner K, Sassner M and Johansson E 2014 *Hydrology and Near-surface Hydrogeology at Forsmark–synthesis for the SR-PSU Project* (Stockholm: SR-PSU Biosphere. SKB R-13–19. Svensk Kärnbränslehantering AB) www.skb.se

Wheater H S, Bell J N B, Butler A P, Jackson B M, Ciciani L, Ashworth D J and Shaw G G 2007 *Biosphere Implications of Deep Disposal of Nuclear Waste: The Upward Migration of Radionuclides in Vegetated Soils* (London: Imperial College Press)

Woodroffe C D 2003 *Coasts: Form, Process and Evolution* (Cambridge: Cambridge University Press)

Wörman A, Packman A I, Marklund L, Harvey J W and Stone S H 2006 Exact three-dimensional spectral solution to surface-groundwater interactions with arbitrary surface topography *Geophys. Res. Lett.* **33** L07402

Wörman A, Packman A I, Marklund L, Harvey J W and Stone S H 2007 Fractal topography and subsurface water flows from fluvial bedforms to the continental shield *Geophys. Res. Lett.* **34** L07402

Xu S, Kłos R, Dverstorp B, Sandberg V, Stark K and Gisca I 2021 Experience in radiological risk assessment of a surface waste disposal facility in Chişinău, Moldova *J. Radiol. Prot.* **42** 011515

IOP Publishing

Principles and Techniques of Radiological Protection

Michael Charles Thorne

Chapter 11

Radiochemical analysis

D Read and B Russell

11.1 Introduction

There are many reasons for determining the level of radionuclides in the environment, from geochemical prospecting and surveying to geochronology, forensics, discharge monitoring, pollution control and remediation. In all circumstances, some common considerations will apply: the nature and type of environment being considered, the likely range of concentrations encountered, their distribution, and levels of quantification required from the analysis. These factors will determine not only the choice of analytical method(s) but also sampling strategies and sample preparation protocols. Recent advances in the application of *in situ* methods, notably portable gamma spectrometry, may render these a practical choice for reasons of speed and cost over traditional laboratory-based techniques. Nevertheless, laboratory analysis is essential for high quality determination of radionuclides at low concentrations and therefore, is the primary focus of this chapter.

Some basic principles relating to the occurrence of radionuclides in the environment are described in section 11.2 as these have an important bearing on sampling strategies and the preparation of samples for analysis, covered in section 11.3. Section 11.4 briefly outlines the main analytical tools available to a radiochemist and considers whether, and which, of the more common techniques might be favoured in a given application. Section 11.5 addresses aspects of radiation metrology and the crucial issues of quality assurance and uncertainty analysis.

11.2 Radionuclides in the environment

Prior to the advent of nuclear weapons and subsequently, nuclear power generation the requirement for environmental radiochemical analysis was very limited and restricted to sources of natural radionuclides for industrial or medical applications. Neither uranium nor thorium are particularly rare elements with average crustal abundance around 3 ppm (similar to that of tin) and 9 ppm (slightly less than that of lead), respectively.

doi:10.1088/978-0-7503-2424-3ch11

Concentrations vary with lithology and tend to be higher in acid, igneous rocks, often substantially so, for example, in granite pegmatites and sedimentary deposits derived from them. The majority of ores from which uranium, thorium and radium were, and still are, extracted are geologically 'old' and consequently, the natural ^{238}U, ^{235}U and ^{232}Th series nuclides are close to secular equilibrium. Given that any anthropogenic input is always superimposed on the natural inventory already present, both the variability with underlying geology and the assumption of secular equilibrium are extremely important when attempting to differentiate potential sources of contamination, as discussed in the next section.

Health problems associated with the exploitation of uranium and thorium-bearing minerals were known well before the discovery of radioactivity itself, usually from exposure to radon gas. Nevertheless, it is only in recent years that the full environmental and health impacts of industrial radioactivity or 'NORM' (Naturally Occurring Radioactive Material) have become apparent. Indeed, for most of the population they far exceed those arising from the nuclear industry owing to a combination of high concentrations of $^{226/228}$Ra, ^{210}Pb and ^{210}Po together with historically less stringent controls on discharges and disposals. NORM impacts many sectors; the most important in terms of waste volumes being the extractive and chemical processing industries (Read *et al* 2013). Some of the scales affecting production equipment contain activity levels corresponding to those of intermediate level nuclear waste, defined in the United Kingdom as >4 GBq t^{-1} alpha and/or >12GBq t^{-1} beta-gamma. These materials require specialised treatment. The bulk of industrial wastes are much less active but still require careful handling and disposal to limit exposures. This has now been recognised in the Basic Safety Standards published by the European Commission (EC 2014), International Atomic Energy Agency (IAEA 2014) and national regulators (e.g. HMSO 2016).

Some anthropogenic radionuclides are also now ubiquitous owing to fallout from nuclear weapons tests and major nuclear accidents, notably Chornobyl in 1986 and the Fukushima Daiichi incident in 2011. Most of the inventory released during these events will have decayed before, or shortly after, deposition, but those nuclides with longer half-lives (e.g. ^{90}Sr and ^{137}Cs) persist and provide vital data for constructing atmospheric circulation models. Inhomogeneity in their deposition at more local scales has been closely correlated to precipitation patterns and, as with naturally occurring nuclides, effectively represents 'background' against which more recent events can be compared. One such release occurred in September 2017 when approximately 250 TBq of ^{106}Ru (half-life 373.6 days) was dispersed over a large area of Central and Southern Europe. Attempts to pinpoint its source provide an excellent example of a well-coordinated international sampling and analysis campaign combining state-of-the-art sampling protocols with advanced isotope forensics (Hopp *et al* 2020).

11.3 Sampling and sample preparation

Radiochemical analysis does not begin with receipt of a sample in a laboratory but in establishing key objectives and the design of an integrated (and often iterative) sampling and analysis protocol. The number, type and location of samples will obviously vary depending on the context of the study and how the results are to be used. Indicative and/or random samples may well be sufficient to confirm the presence of a contaminant following monitoring or to verify the provenance of a diagnostic 'fugitive' radionuclide in the wider environment. They will almost certainly not be adequate for determining the extent and distribution of contamination, for estimating doses, demonstrating compliance and providing suitable information for decision-makers concerned with occupational and public safety. As a rule of thumb, the potential for error at the sampling stage is at least an order of magnitude greater than that introduced during sample preparation, which in turn exceeds that likely to be caused by analysis with a modern and properly calibrated instrument. Unfortunately, and despite the excellent guidance available in the literature (e.g. EPA 2012, Popek 2018), too many site investigations are still conducted with little contact between the sampling team, over whose procedures the analyst has no control, and the receiving laboratory. From that point onwards, errors may be introduced during storage, sample preparation, the analysis itself or subsequent data processing and evaluation. Irrespective of how thorough and fault-free these steps may be, there is nothing the analyst can do to compensate for poorly chosen and collected samples.

For compliance investigations, samples must be representative of the medium (e.g. soil) from which they were obtained. How much information is known beforehand concerning the contamination incident(s) and the history of the site will influence initial hypotheses on regions of enhanced concern and hence, a suitable sampling strategy. For example, national soil surveys typically follow a regular grid or stratified random sampling scheme to ensure consistency over large areas; neither would be suitable for an ellipsoidal plume resulting from atmospheric stack or aqueous pipe discharges. In such cases, a graduated approach with sampling density decreasing with increasing distance from the source might well constitute the most efficient use of available resources (figure 11.1).

Spatial heterogeneity also needs to be considered carefully at the scale of individual samples. Larger samples are always required where the grain size of the matrix increases from clay through silt and sand to gravel, and well-known empirical relationships exist. Additionally, radioactive contamination is often characterised by the presence of 'hot particles' in an otherwise uncontaminated matrix; in this respect

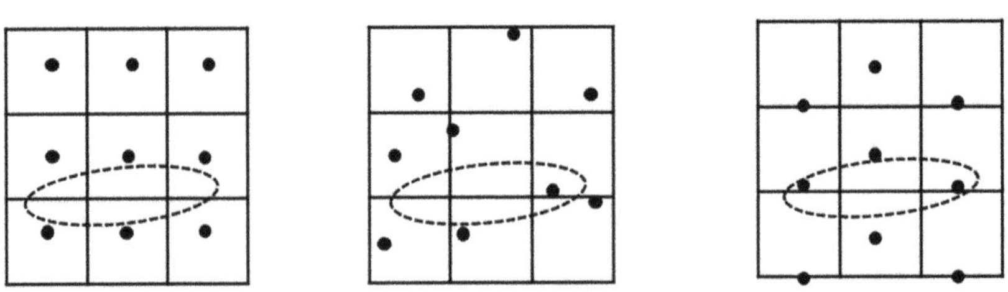

Figure 11.1. Regular, stratified random and herringbone sampling patterns applied to an ellipsoidal contamination plume.

they resemble precious metals in commercial ore deposits and the same statistical approaches are valid. As hot particles are easily detected by hand-held instruments there is a tendency to focus on them during sampling, which can lead to substantial bias. In one unpublished investigation of a former military site in south-west England, estimates of ^{226}Ra in soil supplied by several accredited laboratories ranged from 1 to >1000 Bq g^{-1}, with the highest values corresponding to 'more sensitive' analysis of small (0.5 g) samples using a well-type broad-energy Ge detector. An independent audit demonstrated that the differences were entirely attributable to selective sampling.

Finally, some mention should be made of indirect sampling whereby an attempt is made to remove surface contamination using swabs, which are then analysed. Wipe tests of this type are notoriously difficult to replicate as their effectiveness will vary with the swab (cotton bud, cloth, paper), solvent (water, organic), nature of the surface (metal, glass, laminate) form of radionuclide and the time since deposition. Nevertheless, they are standard practice in forensic investigations and during laboratory decontamination programmes.

Having outlined potential pitfalls in sampling, we will briefly cover some of the key points to bear in mind during preparation of the samples for radiochemical analysis.

Sample digestion is required for accurate determination of radionuclide concentrations by both mass spectrometry and radiochemical methods, including alpha spectrometry. Methods vary depending on the sample matrix, but most rely on aggressive reagents, typically hydrofluoric acid, to dissolve silicates, and a powerful oxidant, such as perchloric or nitric acid, to remove organics. Where organic content is particularly high, there may also be a need for prior ashing, which also serves as a pre-concentration technique. Conventional heating in a furnace risks loss of volatile components and is increasingly being eschewed in favour of microwave digestion, though the latter is limited in terms of sample size.

Complete dissolution of a sample is an unavoidable requirement, unless it can be proven that the analyte can be quantitatively leached. This is unlikely to be the case with the refractory minerals that host radium (barite), thorium (oxides and silicates) and lead (metallic or oxide), leading to reduced recovery and underestimation of radionuclide levels. In a recent inter-laboratory comparison study, > 80% of participating US laboratories significantly under-reported ^{238}U and ^{234}U in soil samples due to incomplete acid dissolution of refractory phases (Maxwell *et al* 2015).

Fusion using a suitable flux, followed by mild acid treatment, is a routine method in the geosciences owing to the preponderance of silicate minerals. Several fluxes have been employed, including sodium hydroxide, sodium peroxide (Na_2O_2) and ammonium bifluoride (NH_4HF_2); the most common is borate fusion using lithium borates ($LiBO_2$ and $Li_2B_4O_7$). The process is highly efficient with recovery close to 100% and can be readily automated (e.g. Braysher *et al* 2019). As with all methods, it is not without problems. The high proportion of flux compared to the sample may cause interferences during further separation and lead to unwanted matrix effects in the final analysis.

In an ideal world, many of the difficulties outlined above would be circumvented by direct sampling. Advances in mass spectrometry (MS), allowing hitherto inaccessible quantities of radionuclides to be measured, have proceeded in parallel with the development of increasingly sensitive *in situ* sampling methods, thereby minimising the need for sample preparation. Volatilisation and ionisation are achieved using a laser (e.g. laser ablation, LA-MS (Horstwood *et al* 2016); matrix assisted laser desorption, MALDI (Dreisewerd 2014), electrospray (e.g. desorption electrospray ionisation, DESI) (Morelato *et al* 2013) or solvent extraction. These methods offer great potential for highly sensitive mass spectrometric analysis but require specialised and expensive facilities, usually ultra-high vacuum conditions, and are unsuited to rapid throughput of large sample batches. Consequently, their use is likely to be restricted to high profile forensic cases rather than routine environmental analysis for the foreseeable future.

11.4 Analytical techniques

11.4.1 General principles

A vast array of analytical techniques is available to the modern radiochemist, each with its own advocates. Choosing one method over another is governed by exactly the same criteria as those in other branches of analytical chemistry; speed, simplicity, capacity, required limits of detection and quantification, cost and sample characteristics. In addition to matrix composition and homogeneity, the latter include factors such as whether there is a need to preserve the sample, as is often the case in forensic investigations. Essentially, all of the above questions reduce to the purpose of the analysis and the intended use of the results.

11.4.2 Non-destructive analysis and its limitations

Non-destructive or minimally destructive techniques are favoured for routine surveys and research or forensic work on samples of high intrinsic value. X-ray fluorescence (XRF) is the standard method for geochemical surveys as it is rapid, requires very little sample preparation and is highly scalable, with portable, transportable and automated, laboratory systems widely available. Although, in the current context, it is largely limited to determination of U and Th, it provides multi-elemental compositional information on sample matrices down to a few parts per million (ppm), which can prove invaluable in follow-up studies. Laser-induced breakdown spectroscopy (LIBS) is typically marketed as an alternative to XRF; however, the two methods are complementary, with LIBS providing information on the lightest elements that are inaccessible to XRF. Another laser-based method, Raman spectroscopy, can also be deployed in field settings and shows promise for identification of uranium phases in hazardous environments where direct sampling is impracticable (Frankland *et al* 2021).

Gamma spectrometry can also be considered a non-destructive technique as samples require little if any preparation prior to analysis. Moreover, intrinsic electric cooling systems for high resolution germanium detectors have facilitated fully portable systems for field measurements. Although detection levels do not match those achievable in the laboratory, for reasons discussed in the next section, direct field measurement of gamma-emitting radionuclides overcomes the constraints of selective sampling noted above and may provide a more representative guide to activity levels *in situ*.

At the other end of the spectrum are the highly specialised analytical techniques used to obtain isotopic information for geochronology or forensic purposes. In laser ablation inductively coupled mass spectrometry (LA-ICPMS), a high-energy laser is focussed on the sample generating a plasma and a (usually acceptable) degree of micro-pitting on the surface, which represents the only damage caused to the sample. The method is rapid, has excellent spatial resolution, low detection limits and requires very little sample preparation other than mounting. Consequently, it is used routinely for uranium-series dating of resistate minerals, such as zircon, where U, Th and Pb isotopic ratios are used to construct isochrons (Horstwood *et al* 2016). The main disadvantage is the high cost of the equipment and the relatively low precision owing to geometrical effects at the surface.

Ion beam analysis involves focussing an energetic ion beam (MeV) onto the surface of a sample to determine its composition, together with other chemical and structural information, via techniques such as particle-induced x-ray emission (PIXE), Rutherford backscattering (RBS), elastic backscattering (EBS), secondary ion mass spectrometry (SIMS) and nuclear reaction analysis (NRA). The suite of ion beam methods is extremely sensitive and particularly well suited to characterisation of thin films or 'hot particles' with depth resolution approaching the atomic scale; however, relatively few facilities are available owing to the cost and extensive infrastructure required.

Finally, mention is made of the synchrotron-based methods, micro-XRF, micro-x-ray diffraction and x-ray adsorption spectroscopy (XAS), incorporating x-ray adsorption near-edge (XANES) and extended x-ray adsorption fine structure (EXAFS). These are powerful tools for determining composition and crystal structure of solid phases and the oxidation state, coordination states and characteristic bond lengths of key components. As with ion beam analysis, the main constraint is the scarcity of synchrotron sources and the high cost associated with maintaining them.

11.4.3 Radiometric methods

The method of choice for routine analysis of radionuclides during decommissioning or remediation is normally radiometric counting; thus, alpha spectrometry, liquid scintillation counting (LSC) or gamma spectrometry is deployed depending on the decay mode. Longer-lived radionuclides may also be measured by mass spectrometry and this route would be favoured where decay counting times are considered too long.

A variety of solid-state scintillation detectors are used in gamma spectrometry, including thallium doped NaI, CsI, bismuth germanate ($Bi_3Ge_4O_{12}$), $LaBr_3$ and cadmium zinc telluride (CZT). All have excellent detection efficiencies and are relatively cheap to produce with the added advantage that they do not require cooling, making them a good choice for portable instruments. The main disadvantage is their poor energy resolution, which renders them unsuitable for characterising the complex spectra encountered in many nuclear and NORM situations. For this reason, they have been superseded by germanium semiconductor detectors for almost all laboratory and even some field applications.

A Ge detector provides far superior energy resolution across a wide energy range, albeit with lower efficiency. Replacing the conventional aluminium window with beryllium or carbon allows low energy gamma-emitting radionuclides to be quantified, notably ^{210}Pb (46 keV), an important component in NORM, and ^{241}Am (60 keV). Although the analysis is straightforward, it is not without its pitfalls. A recent paper highlighted major discrepancies amongst accredited laboratories when measuring ^{210}Pb in NORM scales owing to difficulties in accounting for matrix attenuation (Dal Molin *et al* 2018).

Composite lead and copper shields are used to reduce the contribution from background radiation. It is important therefore, that the shielding material itself does not contribute to low-background radiation measurements. Standard lead shielding contains ^{210}Pb, typically up to 500 Bq kg^{-1}. Employing aged lead can reduce this to 25 Bq kg^{-1} or less (Britton *et al* 2013). The often-complex spectra obtained for environmental samples can be resolved by increasingly sophisticated peak-fitting algorithms incorporating fuzzy logic, neural networks and wavelet analysis to discriminate low signal to noise ratios (Kamuda *et al* 2017). However, and notwithstanding advances in Ge crystal manufacture, electric cooling and software, the basic characteristics of Ge detectors have not changed for more than 30 years. Gilmore (2008) provides a comprehensive overview of the technique.

Liquid scintillation counting is the most widely used method for beta-emitting radionuclides, including ^{3}H, ^{63}Ni and ^{90}Sr. It can also be used (semi-quantitatively) for alpha determination, measuring nuclides that decay by electron capture (e.g. ^{55}Fe) using Auger electrons and Cherenkov counting with highly energetic emissions. Samples are shaken with a scintillation cocktail comprising a fluorescing aromatic species in an organic solvent, typically, toluene. The radiation-induced light pulses are then detected by two photomultiplier tubes (PMTs) in coincidence mode. As LSC is based on generating light pulses, usually in a liquid, it is susceptible to quenching from both the sample matrix and the measuring medium. A variety of techniques are available to overcome this constraint, such as 'triple-to-double-coincidence ratio' or TDCR involving three rather than two PMTs and two different coincident outputs. The advantages of LSC are its high counting efficiency and its simplicity, being easily automated for rapid sample throughput. The main disadvantage is poor energy resolution; only two beta-emitting radionuclides can be measured simultaneously and even then, only where the energies are markedly different. Therefore, radiochemical separation is a pre-requisite for samples containing several beta emitters. Hou (2018) provides a review of recent developments, focussing on environmental applications and nuclear decommissioning.

Although alpha emitters can be measured by LSC at very high efficiency ($\leqslant$100%), energy resolution is insufficient for quantification where more than one alpha source is present. Thus, alpha spectrometry is required unless the nuclide has a half-life long enough to make measurement by mass a viable option, as discussed in the next section.

In contrast to the techniques above, alpha spectrometry places significant demands on the analyst with regards to sample preparation. Solid samples must first be dissolved, contaminants likely to interfere with measurement of the target analyte removed by separation and the target source deposited as a thin film to avoid self-absorption of the particles emitted. Each of these steps can lead to loss of source activity and errors in the final analysis. Consequently, yield tracers, for example ^{209}Po in the determination of ^{210}Po, are used to monitor recovery throughout the procedure. Since the desired radionuclide activity is determined relative to the monitor, it is essential that the tracer displays near identical chemical behaviour and is added at the earliest possible stage. Issues with sample dissolution were addressed previously. Radiochemical separation serves two purposes: first, it removes other alpha-emitting nuclides with energies close to those of the target(s); second, it removes components of the matrix that would otherwise absorb the alpha particles and reduce the signal intensity. A range of separation schemes has been developed over the years for the principal radionuclides of interest in environmental studies (^{238}U, ^{232}Th natural series), nuclear decommissioning (^{236}U, ^{237}Np, $^{238/239/240/241}$Pu, ^{241}Am, ^{242}Cm) and both (^{235}U, ^{231}Pa). The reader is referred to Lehto and Hou (2011) for a comprehensive coverage.

The deposition stage also poses challenges; it can be achieved by various routes including solvent evaporation, co-precipitation of an insoluble salt, direct deposition (the method employed for ^{210}Po) and electrodeposition. With the latter, the nuclide and tracer are taken up from an electrolyte solution by a platinum or steel disc, which acts as a cathode, with a rotating platinum rod as the anode. Precise control of the current and other operating conditions is required to ensure a homogeneous layer for measurement. The counting of alphas relies on semiconductor detectors using ion-implanted silicon rather than Ge and operating in a vacuum. As with the sample itself, the detector is made very thin owing to the limited depth penetration of the emissions. Typical counting efficiencies are up to 30% across a wide energy range (2–8 MeV). Once obtained, alpha spectra are generally much easier to interpret than those obtained by gamma spectrometry or, indeed, LSC on mixed sources, as the peaks are fewer and usually well-defined.

11.4.4 Mass spectrometry

While there are several types of mass spectrometry (MS) available including thermal ionisation (TIMS), accelerator (AMS), glow discharge (GD-MS) and resonance ionisation (RIMS), the most frequently used instrument design is inductively coupled plasma mass spectrometry (ICP-MS), owing to its relatively low cost, ease of use and analytical flexibility. The capabilities of mass spectrometry for radionuclide measurement have been outlined in several extensive review papers (e.g. Larivière *et al* 2006, Hou and Roos 2008, Croudace *et al* 2017).

ICP-MS has been used for radionuclide measurement since the late 1980s, initially applied to long-lived actinides with limited interferences, e.g. ^{237}Np and ^{238}U

Table 11.1. Radionuclides now measurable by ICP-MS.

Example radionuclides	Half-life (y)	Occurrence
$^{238/235/234}U$, ^{232}Th, ^{231}Pa, ^{226}Ra, ^{210}Pb	10^2–10^9	Natural series, U mining, NORM
$^{240/239}Pu$, $^{243/241}Am$, ^{237}Np, ^{236}U	10^2–10^7	Fuel reprocessing, decommissioning
^{151}Sm, ^{137}Cs, ^{129}I, ^{99}Tc, ^{93}Zr, ^{90}Sr, ^{63}Ni	10^2–10^7	Waste management, nuclear forensics

(Brown *et al* 1988). As the technique has advanced in terms of sensitivity and interference removal capability, the number of radionuclides accessed has expanded, with around 20 radioisotopes measurable at present (table 11.1). Whereas most applications of mass spectrometry are geared towards low-level measurement of long-lived radionuclides as a rapid and more sensitive alternative to decay counting techniques, application to radionuclides with half-lives on the order of tens of years, including ^{90}Sr and ^{210}Pb, is now possible following chemical separation (Amr *et al* 2016, Baskaran *et al* 2018).

Mass spectrometry (most commonly ICP-MS) has been applied increasingly to radionuclide measurement to meet measurement challenges around nuclear decommissioning and environmental monitoring. Additionally, the ability to distinguish between single mass units allows for measurement of isotopic ratios to determine the source of nuclear contamination, with examples including $^{239}Pu/^{240}Pu$, $^{135}Cs/^{137}Cs$ and $^{127}I/^{129}I$. Such measurements have been valuable in monitoring environmental fallout on land and in the seas around Japan following the accident at the Fukushima Daiichi Nuclear Power Plant (Zheng *et al* 2014, Men *et al* 2019). Isotope ratio measurements are also increasingly being applied to atom counting of long-lived radionuclides using isotope dilution mass spectrometry (IDMS) to improve nuclear decay data (Essex *et al* 2018).

Regardless of the application, accurate radionuclide measurement by mass spectrometry requires the removal of multiple interferences. Firstly, an isobaric interference is an isotope at a similar mass to the analyte that cannot be resolved by the detector, such as ^{90}Zr on ^{90}Sr or ^{135}Ba on ^{135}Cs. Secondly, polyatomic interferences are formed by isotopes combining with gases in the high temperature plasma of an ICP-MS, again resulting in a product with a similar mass to the analyte. Examples include $^{89}Y^1H$, $^{74}Ge^{16}O$ and $^{50}Ti^{40}Ar$ on ^{90}Sr. Both isobaric and polyatomic interferences can be removed using a combination of offline chemical separation and/or online instrument separation depending on the instrument used. Finally, tailing interferences are caused by high concentrations of an isotope one or two mass units either side of the analyte. This tailing may be from a different element e.g. ^{238}U on ^{237}Np and ^{239}Pu but is more commonly from an isotope of the same element as the analyte; for example, ^{88}Sr on ^{90}Sr or ^{133}Cs on ^{135}Cs. In these cases, the tailing removal (termed the abundance sensitivity) is a function of the instrument used and, depending on the application, an analyte: tailing ratio of $\sim 10^{-9}$ may be required.

ICP-MS in particular, has a range of instrument designs with different advantages and limitations in terms of radionuclide measurement. It is also a flexible technique with regards to sample introduction. In some cases, this can change the nature of the sample measured, e.g. using laser ablation for direct measurement of solid materials, as discussed above. Solution-based sample introduction is the most common; however, within this, there is a range of sample introduction designs, some offering higher sample introduction efficiency, reduced sample uptake rate, increased total dissolved solid tolerance or interference removal, e.g. desolvating systems that reduce the water-loading into the plasma, decreasing hydride- and oxide-based polyatomic interference formation.

Quadrupole ICP-MS is the most straightforward design. Although it suffers from limited online interference removal capability and sensitivity compared to other designs, the latter can be improved by the addition of a collision (e.g. He) and/or reaction (e.g. O_2) gas into a cell. Collision gases act to reduce the energy of polyatomic interferences, preventing them from reaching the detector whilst maintaining the analyte signal. If the appropriate reaction gas is chosen, it will be thermodynamically favourable for it to react with either the analyte or the isobaric interference, thereby removing the interference. For example, ^{90}Zr reacts with O_2 to form $^{90}Zr^{16}O$, whereas the reaction with ^{90}Sr is endothermic and does not proceed (Amr *et al* 2016). In the case of ^{151}Sm, this reacts with O_2 to form $^{151}Sm^{16}O$, whilst the reaction with isobaric ^{151}Eu does not proceed. This offers a rapid support or potential replacement to offline chemical separation depending on the application; however, the impact on analyte sensitivity must be considered, particularly for relatively short-lived radionuclides such as ^{90}Sr.

Sector field ICP-MS offers superior analyte sensitivity to quadrupole designs, whilst the variable mass resolution can effectively remove some polyatomic interferences. Nevertheless, even high mass resolution is not capable of separating isobaric interferences from any of the radionuclides measurable by ICP-MS. Offline separation prior to measurement is required therefore, to achieve the low (fg/g) detection limits achievable by sector field instruments. Multi-collector ICP-MS has similar sample preparation requirements but has the advantage of being equipped with multiple detectors. This is beneficial for high precision isotope ratio measurements, as the ion beam is 'split' rather than using peak-hopping mode in single detector instruments. This is a powerful design for nuclear forensics and determining the source of the sample being measured, with isotopes of U and Pu commonly measured using this technique.

The last 10 years has seen the commercial availability of tandem mass spectrometry. This design consists of an additional quadrupole mass filter positioned before the reaction cell. The additional mass filter offers abundance sensitivity values on the order of 10^{-10}, which has been beneficial in tailing removal for several radionuclides including ^{88}Sr from ^{90}Sr, ^{135}Cs from ^{137}Cs and ^{127}I from ^{129}I (Warwick *et al* 2019). Positioning the mass filter before the collision-reaction cell allows the ion beam to be filtered prior to entering the cell, giving the user greater understanding of the reaction cell products formed. This is particularly useful for cell gases where multiple products can be formed such as NH_3 or CH_3F.

11.5 Quality control

To ensure confidence in measurement, sample preparation and measurement procedures must be validated through laboratory accreditation and the use of reference standards. Quality control for radionuclide measurement in the United Kingdom is managed through the United Kingdom Accreditation Service (UKAS), with Quality Management System ISO 17025:2005 specifying the competency requirements for general tests, calibrations and sampling. Validation and traceability throughout the process is vital to maintain compliance.

Solid, aqueous and gaseous reference materials support multiple applications of radionuclide measurement, including environmental monitoring, nuclear decommissioning, ensuring public and worker safety and emergency response. A number of organisations offer reference materials including, but not limited to, National Measurement Institutes (e.g. NIST, NPL), MAPEP, IAEA and Procorad.

Reference materials are defined as a '*material, sufficiently homogeneous and stable with respect to one or more specified properties, which has been established to be fit for its intended use in a measurement process*' (ISO17034). They are a valuable resource in ensuring measurement traceability, instrument calibration and validation of analytical techniques, ensuring accurate results over time and between laboratories. A certified reference material (CRM) is defined as a '*reference material characterised by a metrologically valid procedure for one or more specified properties, accompanied by a reference material certificate that provides the value of the specified property, its associated uncertainty, and a statement of metrological traceability*' (ISO 17034). The materials are developed in line with ISO Guides (31, 34 and 35), again often led by National Measurement Institutes.

The lack of reference materials for radioactivity measurements in environmental samples has been highlighted by several authors (Harms and Gilligan 2010, Parry 2012, Dal Molin *et al* 2018). Provision of reference materials for nuclear decommissioning is particularly challenging, as this industry presents a heterogenous and complex range of matrices and radionuclides. For example, the composition of contaminated concretes will vary widely between sites; therefore, validating a method with one concrete reference material does not guarantee that the method is capable of general concrete analysis. Decommissioning materials may also have a poorly-known history in terms of both matrix components and radionuclide content.

Only one CRM for NORM is currently available; a phosphogypsum material designed for the phosphoric acid production industry and made available via the IAEA (Shakhashiro *et al* 2011). It is by no means representative of either the industries affected by NORM nor the radioactive materials most commonly encountered. Development of a reference blast furnace slag applicable to both steel production and the cement industry (Braysher *et al* 2019) offers promise as does a produced water standard for the oil and gas sector (Al-Masri *et al* 2020) but many more initiatives of this type are needed.

Different approaches have been followed when developing reference materials. Synthetic versions start from known reagents allowing good control over the composition, albeit that simulant matrices cannot completely represent real materials. A similar issue is faced when spiking an inactive material with a known activity of one or more radionuclides. The approach ensures good control of radionuclide content, but spiking does not necessarily represent the same measurement challenge if, for example, the radionuclides are tightly held within a solid sample matrix. There is also a requirement for reference materials to be homogenous, which is again not necessarily representative of real materials.

Regardless of the approach used, robust, traceable reference materials are required and sample matrices with associated radionuclides must reflect evolving measurement techniques in order to remain relevant. Inter-laboratory comparison exercises or 'proficiency testing' help to improve measurement capability and can go some way towards building confidence that all potential sources of variability have been captured (e.g. Braysher *et al* 2021). The results generated will vary in absolute terms and in the level of (un)certainty depending on factors such as the mass analysed, sampling method, sample preparation, type of tracer and the measurement technique. Open publication and discussion of these differences raises awareness of the inherent limitations in radiochemical analysis among end-users, regulators and, importantly, the wider scientific community.

References

Al-Masri M, Hasan M and Al-Khateeb Y 2020 Preparation of produced water reference material for analysing ^{226}Ra by gamma spectrometry *J. Radioanal. Nucl. Chem.* **324** 225–31

Amr M, Helal A-F, Al-Kinani A and Balakrishnan P 2016 Ultra-trace determination of (90)Sr, (137)Cs, (238)Pu, (239)Pu and (240)Pu by triple quadruple collision/reaction cell-ICP-MS/MS: establishing a baseline for global fallout in Qatar soil and sediments *J. Environ. Rad.* **153** 73–87

Baskaran K, Blanchet-Chouinard G and Larivière D 2018 Attogram measurement of ^{210}Pb in drinking water by ICP-MS/MS *J. Anal. At. Spectrom.* **33** 603–12

Braysher E, Russell B, Woods S, Garcia-Miranda M, Ivanov P, Bouchard B and Read D 2019 Complete dissolution of solid matrices using automated borate fusion in support of nuclear decommissioning and production of reference materials *J. Radioanal. Nucl. Chem.* **321** 183–96

Braysher E *et al* 2021 Development of a reference material for analysing naturally occurring radioactive material from the steel industry *Anal. Chim. Acta.* **1141** 221–9

Britton R, Burnett J, Davies A and Regan P 2013 Determining the efficiency of a broad-energy HPGe detector using Monte Carlo simulations *J. Radioanal. Nucl. Chem.* **295** 2035–41

Brown R, Long S and Pickford C 1988 The measurement of long-lived radionuclides by non-radiometric methods *Sci. Total Environ.* **70** 265–74

Croudace I, Russell B and Warwick P 2017 Plasma source mass spectrometry for radioactive waste characterisation in support of nuclear decommissioning: a review *J. Anal. At. Spectrom.* **32** 494–526

Dal Molin F, Warwick P and Read D 2018 Under-estimation of ^{210}Pb in industrial radioactive scales *Anal. Chim. Acta.* **1000** 67–74

Dreisewerd K 2014 Recent methodological advances in MALDI mass spectrometry *Anal. Bioanal. Chem.* **406** 2261–78

EC 2014 Council Directive 2013/59/Euratom of 5 December 2013 laying down basic safety standards for protection against the dangers arising from exposure to ionising radiation, and repealing Directives 89/618/Euratom, 90/641/Euratom, 96/29/Euratom, 97/43/Euratom and 2003/122/Euratom *Off. J. Eur. Comm.* L 13/1 https://eur-lex.europa.eu/eli/dir/2013/59/oj/eng

EPA 2012 Sample Collection Procedures for Radiochemical Analytes in Environmental Matrices. Office of Research and Development, National Homeland Security Research Center. EPA/600/R-12/566

Essex R *et al* 2018 New determination of the ^{229}Th half-life *J. Radioanal. Nucl. Chem.* **318** 515–25

Frankland V, Milodowski A, Bright J and Read D 2021 The use of Raman and TRLF spectroscopy for differentiating early stage alteration products of spent nuclear fuel *Appl. Geochem.* **130** 104934

Gilmore G 2008 *Practical Gamma-Ray Spectroscopy* 2nd edn (New York: Wiley)

Harms A and Gilligan C 2010 Development of a neutron-activated concrete powder reference material *Appl. Radiat. Isot.* **68** 1471–6

HMSO 2016 *The Environmental Permitting (England and Wales) Regulations* (London: Her Majesty's Stationery Office) UK Statutory Instruments, No. 1154

Hopp T, Zok D, Kleine T and Steinhauser G 2020 Non-natural ruthenium isotope ratios of the undeclared 2017 atmospheric release consistent with civilian nuclear activities *Nat. Comms* **11** 2744

Horstwood M *et al* 2016 Community-derived standards for LA-ICP-MS U-(Th-)Pb geochronology—uncertainty propagation, age interpretation and data reporting *Geostand. Geoanal. Res.* **40** 311–32

Hou X 2018 Liquid scintillation counting for determination of radionuclides in environmental and nuclear application *J. Radioanal. Nucl. Chem.* **318** 1597–628

Hou X and Roos P 2008 Critical comparison of radiometric and mass spectrometric methods for the determination of radionuclides in environmental, biological and nuclear waste samples *Anal. Chim. Acta.* **608** 105–39

IAEA 2014 Safety standards for protecting people and the environment *Radiation Protection and the Safety of Radiation Sources: Int. Basic Safety Standards, General Safety Requirements Part 3* (Vienna: IAEA)

Kamuda M, Stinnett J and Sullivan C 2017 Automated isotope identification algorithm using artificial neural networks *IEEE Trans. Nucl. Sci.* **64** 1858–64

Larivière D, Taylor V, Evans D and Cornett R 2006 Radionuclide determination in environmental samples by inductively coupled plasma mass spectrometry *Spectrochim. Acta.* B **61** 877–904

Lehto J and Hou X 2011 *Chemistry and Analysis of Radionuclides: Laboratory Techniques and Methodology* (New York: Wiley-VCH.)

Maxwell S, Hutchison J and McAlister D 2015 Rapid fusion method for the determination of refractory thorium and uranium isotopes in soil samples *J. Radioanal. Nucl. Chem.* **305** 631–41

Men W, Zheng J, Wang H, Ni Y, Kumamoto Y, Yamada M and Uchida S 2019 Pu isotopes in the seawater off Fukushima Daiichi Nuclear Power Plant site within two months after the severe nuclear accident *Environ. Pollut.* **246** 303–10

Morelato M, Beavis A, Kirkbride P and Roux C 2013 Forensic applications of desorption electrospray ionisation mass spectrometry (DESI-MS) *Forensic Sci. Int.* **226** 10–21

Parry S 2012 Quality assurance in the nuclear sector *Radiochim. Acta* **100** 495–501

Popek E P 2018 *Sampling and Analysis of Environmental Chemical Pollutants: A Complete Guide* 2nd edn (Amsterdam: Elsevier)

Read D, Read G and Thorne M 2013 Background in the context of NORM-contaminated land *J. Rad. Prot.* **33** 367–80

Shakhashiro A *et al* 2011 The new IAEA reference material: IAEA-434 technologically enhanced naturally occurring radioactive materials (TENORM) in phosphogypsum *Appl. Radiat. Isot.* **69** 231–6

Warwick P, Russell B, Croudace I and Zacharauskas Z 2019 Evaluation of inductively coupled plasma tandem mass spectrometry for radionuclide assay in nuclear waste characterisation *J. Anal. At. Spectrom.* **34** 1810–21

Zheng J, Tagami K, Wenting B, Uchida S, Watanabe Y, Kubota Y, Fuma S and Ihara S 2014 $^{135}Cs/^{137}Cs$ isotopic ratio as a new tracer of radiocesium released from the Fukushima nuclear accident *Environ. Sci. Technol.* **48** 5433–8

IOP Publishing

Principles and Techniques of Radiological Protection

Michael Charles Thorne

Chapter 12

Effects of ionising radiations on the environment

N Horemans and N A Beresford

12.1 Introduction

The system for ensuring the protection of humans from ionising radiations began development early in the twentieth century (International X-ray and Radium Protection Committee 1929) and is now well established internationally (e.g. IAEA 2006a; ICRP 2007). However, the system for ensuring that the environment, or wildlife, is protected has only developed since about the year 2000. Previously there was an assumption that the control (or regulation) required to protect humans would ensure that other species were not put at risk. The 1990 Recommendations of the ICRP (International Commission on Radiological Protection) stated: '*The Commission believes that the standard of environmental control needed to protect man to the degree currently thought desirable will ensure that other species are not put at risk. Occasionally, individual members of non-human species might be harmed, but not to the extent of endangering whole species or creating imbalance between species. At the present time, the Commission concerns itself with mankind's environment only with regard to the transfer of radionuclides through the environment, since this directly affects the radiological protection of man.*' (ICRP 1991). National authorities responsible for radiation protection generally adopted this recommendation. Therefore, there was no need to assess the impact of authorised releases of radioactivity on wildlife and consequently there were no established approaches for conducting radiological environmental impact assessments.

However, the statement of the ICRP regarding protection of the environment was increasingly questioned. Criticisms included: a lack of supporting data for the statement (Thompson 1988, Linsley 1990); potential scenarios where wildlife may be more exposed than humans (Linsley 1990); and the need to demonstrate protection of the environment from any human activity (Pentreath 1999, Oughton and Strand 2004). Concurrently some countries began to establish requirements and guidelines for the protection of wildlife (often termed 'non-human biota' or 'non-human species') (e.g. United Kingdom (Copplestone *et al* 2001), Sweden (Holm *et al* 2002), USA (USDOE 2002), Canada (Environmental Canada and Health Canada 2003)).

doi:10.1088/978-0-7503-2424-3ch12

In response to these national requirements, approaches and models to assess the potential impact of radionuclides on wildlife began to be developed (e.g. USDOE 2002, 2004, Copplestone *et al* 2003, Beresford *et al* 2008a).

In 2007, the revised Recommendations of the ICRP (ICRP 2007) included a recognition of the need to consider radiological protection of the environment with the objective '*to maintain biological diversity, conservation of species, protection of the health and status of natural habitats, communities and ecosystems, with targets related to populations or higher organisational levels rather than individual organisms*'.

Following on from the revised Recommendations, the ICRP has begun the development of a framework to assess the radiation exposure of non-human species and consequent risks (ICRP 2008, 2009, 2014, 2017). The International Atomic Energy Agency (IAEA) has a role in transforming the ICRP Recommendations into practical guidance for application in regulatory frameworks. The IAEA Safety Fundamentals published in 2006 acknowledge the need to consider the environment within radiological assessment: '*People and the environment, present and in the future, must be protected against radiation risks*' (IAEA 2006b). In response to the changes in international recommendations, radiological environmental impact assessments are now being conducted in many countries for a wide range of facility types (e.g. see Brown *et al* 2016).

12.2 Overview of approaches used to assess risk to wildlife from ionising radiations

Since about the year 2000, several approaches along with supporting models and tools for the assessment of radiological risk to wildlife have been developed (see Beresford *et al* 2008a, 2022a, IAEA 2012, 2021). Some assessment tools implement tiered (or graded) screening assessment approaches beginning with highly conservative initial screening tiers (requiring minimal input) and progressing to more realistic assessments (requiring more input) if required. Other models have greater functionality for specific elements of the assessment (e.g. non-equilibrium modelling (Vives i Batlle *et al* 2008), dosimetry (Beaugelin-Seiller *et al* 2006), radon (Vives i Batlle *et al* 2017)). RESRAD-BIOTA (USDOE 2004), developed by the United Stated Department of Energy, and the ERICA Tool, initially developed by a European collaboration (Brown *et al* 2008, 2016), both implement a tiered assessment approach and are freely available (https://resrad.evs.anl.gov/codes/resrad-biota/; https://erica-tool.com/). Approaches, such as those implemented in the ERICA Tool and RESRAD-BIOTA, which are used in, or proposed for, regulatory assessment include common elements as discussed below.

12.2.1 Simplification of ecosystems and biodiversity

Generic screening level models cannot represent every potential species or ecosystem for which assessment might be required individually. Therefore, simplifications are required with ecosystems typically represented as 'freshwater' 'marine' or 'terrestrial' (USDOE (2004) also includes 'riparian'). To encompass the broad range of potential species a set of hypothetical organisms are considered; these may be termed reference organisms (Brown *et al* 2008) or reference animals and plants (RAPs) (ICRP 2008). Such hypothetical organisms are generally typical of broad wildlife types (e.g. bird, tree, mammal, crustacean, plant) rather than specific species, although the ICRP define their RAPs at the taxonomic level of family. To enable absorbed dose rates to be estimated, organisms are given a geometry (dimensions and mass) typical of a species falling within the broad organism category (Vives i Batlle *et al* 2007); organism geometries are commonly assumed to be ellipsoids. Data on radionuclide transfers and radiation effects are collated at these broad levels.

12.2.2 Radionuclide transfer

If the activity concentrations of organisms at the assessment site are not known, models for application in regulatory assessment use equilibrium whole-organism concentration ratios ($CR_{\text{wo-media}}$), where:

$$CR_{\text{wo}-\text{media}} = \frac{\text{Whole}-\text{organism activity concentration (Bq kg}^{-1}\text{ fresh mass)}}{\text{Activity concentration in unit mass or volume of a medium}}$$

The medium varies depending upon the ecosystem type:

- aquatic ecosystems—filtered water activity concentration (Bq l^{-1})
- terrestrial ecosystems—soil activity concentration (Bq kg^{-1} dry mass) or for some radionuclides (usually ^{3}H and ^{14}C) air activity concentration (Bq m^{-3}).

This approach has the advantage that it is simple to apply and that some data are available. The IAEA has published a handbook of $CR_{\text{wo-media}}$ values (IAEA 2014) for wildlife and the underlying database on which the handbook was based continues to be updated (see Copplestone *et al* 2013, Brown *et al* 2016). Regional $CR_{\text{wo-media}}$ databases have also started to be published (Hirth *et al* 2017).

However, the $CR_{\text{wo-media}}$ approach has some limitations. Firstly, whilst there are considerable amounts of data available (Copplestone *et al* 2013) there are many organism-radionuclide combinations for which there are no data. For instance, Brown *et al* (2016) reported that data were available for only approximately 50% of the about 1500 $CR_{\text{wo-media}}$ values required to populate the ERICA Tool database (version 1.2). Consequently, default values are often provided using 'extrapolation' approaches (e.g. by assuming a $CR_{\text{wo-media}}$ value for a similar organism or radionuclide with a similar behaviour) (Copplestone *et al* 2003, Brown *et al* 2013).

The second limitation of the $CR_{\text{wo-media}}$ approach is that for a given radionuclide-organism combination $CR_{\text{wo-media}}$ values can be highly variable with values ranging up to five orders of magnitude (see IAEA 2014). This variation leads to uncertainty in model predictions and is generally the major contributor to differences in predictions between different models (e.g. Beresford *et al* 2008a, 2010a, Wood *et al* 2009, Yankovich *et al* 2010, Johansen *et al* 2012, Stark *et al* 2015). Much of this variability will be the consequence of site-specific factors together with the pooling of data for many species to provide summarised $CR_{\text{wo-media}}$ values for broad wildlife groups. However, variability together with biases and other limitations in the available data limit the potential to disaggregate the summarised dataset to provide more refined $CR_{wo\text{-}media}$ values (e.g. for specific ecosystem types, wildlife groups by feeding strategy etc). An alternative to the $CR_{wo\text{-}media}$ model which is based on taxonomic groupings and takes account of inter-site variation has been proposed (Beresford *et al* 2013b, 2016a). This is based upon an approach proposed by plant scientists to relate plant nutrient uptake to phylogeny (or evolutionary history) (Willey 2014). Such phylogenetic relationships for plants have been demonstrated for a number of elements with radionuclides (e.g. Broadley *et al* 1999, Willey 2010, 2014). The first application of the approach as a potential method of predicting radionuclide uptake by wildlife was for caesium and freshwater fish (Beresford *et al* 2013b). The approach requires data being available for multiple species at multiple sites and uses residual maximum likelihood (REML) fitting of a linear mixed effects model. The output of the REML fitting is a mean value (or relative scaling value) for each taxon on a common scale. To apply the approach, measurements, or robust model predictions, are required for the site of interest. Using these data and the output of the REML fitting, predictions can be made for other taxa at that site. To date, in addition to caesium and freshwater fish, such taxonomic models have been fitted for terrestrial wildlife and lead, strontium, caesium selenium and uranium (Søvik *et al* 2017, Beresford and Willey 2019), and also marine wildlife and caesium (Brown *et al* 2019). The terrestrial and freshwater models give reasonable predictions when compared with blind test data and for those tested to date (Beresford *et al* 2013b, Søvik *et al* 2017, Beresford and Willey 2019). However, although a REML model could be fitted to data for marine organisms and caesium, comparisons with blind test data were poor (Brown *et al* 2019). In part this may be because marine waters have relatively similar chemistry, and this may reduce the influence of site compared with terrestrial and freshwater systems. As well as potentially providing an approach accounting for inter-site variation, such taxonomic based REML models also offer a scientifically based approach for estimating radionuclide transfers to species for which data are lacking.

At higher assessment levels, simple food chains can be modelled within RESRAD-BIOTA (USDOE 2004) which also has the advantage that the exposure of terrestrial organisms via contaminated freshwater can be predicted (this is not possible in the commonly used ERICA Tool). To enable simple food chains to be modelled, allometric relationships are used for many parameters (e.g. food and water ingestion rates) including radionuclide biological half-lives (USDOE 2002, Higley *et al* 2003). Allometry, or biological scaling, relates body size (often mass) to parameters most commonly as:

$$Y = aM^b$$

where Y is a biological variable (e.g. food ingestion rate, radionuclide biological half-life), M is body mass, and a and b are constants. The exponent (b) most often scales to quartile values with allometric relationships for radionuclide biological half-life tending to have b values approximating to 0.25 (Galeriu *et al* 2003, Higley *et al* 2003, Beresford and Vives i Batlle 2013a). Beresford and Vives i Batlle (2013a) demonstrated that b values approximating to 0.25 for radionuclide biological half-life can be explained by Kleiber's Law for metabolic rate (Kleiber 1932). RESRAD-BIOTA includes allometric biological half-life relationships for 16 radionuclides derived from fitting experimental data for different species. With the assumption that b equals 0.25, Beresford and Vives i Batlle (2013a) and Beresford and Wood (2014) proposed an approach to derive allometric radionuclide biological half-life relationships which does not require measurements of half-life for species across a large mass range, thereby making it possible to model a greater range of radionuclides. However, USDOE (2002) suggests that for a few radionuclides (Am, Ce, Eu, Pu and Th) the value of b is closer to 0.8, but the reasons for this are currently unclear (see Beresford and Vives i Batlle 2013a).

The $CR_{wo\text{-}media}$, approach as implemented in assessment models, assumes equilibrium between organism and media activity concentrations. Although this assumption is likely appropriate for contaminated sites and long-term planned releases, for unplanned releases assuming organism and media activity concentration equilibrium may significantly under- or over-predict exposure depending upon time since release (Vives i Batlle *et al* 2016). After the 2011 Fukushima accident, there was a desire to determine the potential impact on wildlife (Garnier-Laplace *et al* 2011, Strand *et al* 2014, Vives i Batlle 2016) and ideally such assessments require dynamic modelling. Existing models for aquatic ecosystems which may have been developed to assess human exposure can relatively easily be adapted to consider the exposure of wildlife, as was done following the Fukushima accident for marine ecosystems (Vives i Batlle *et al* 2016). On the other hand, dynamic models for the potential range of radionuclides of interest across the diversity of natural terrestrial ecosystems require development.

For aquatic ecosystems media activity concentrations are required for both water and sediment. If only one of these is available then the other is generally estimated using a distribution coefficient (K_d) value (Boyer *et al* 2018, Tomczak *et al* 2019, Kelleher *et al* 2022), where:

$$K_d(l\ kg^{-1}) = \frac{\text{activity concentration in sediment (Bq kg}^{-1}\text{ dry mass)}}{\text{activity concentration in (filtered) water (Bq l}^{-1}\text{)}}$$

12.2.3 Dosimetry

Absorbed dose to wildlife is defined as the amount of energy absorbed per unit mass in units of Gray (Gy). In the tools used for regulatory assessment, organisms are represented as homogenous in composition and typically as an ellipsoid. Organisms can be modelled in different exposure geometries, typically: in soil, on soil, in air, in sediment, on sediment, in the water column and on the water surface. To maximise the estimated dose rate, in the initial screening level of RESRAD-BIOTA the organism is to be both the smallest geometry for estimating external exposure and the largest geometry for estimating internal exposure. Radionuclide-specific dose coefficients (DCs) are used to calculate internal or external dose rates from organism or media activity concentrations, respectively (i.e. mGy d^{-1} per Bq kg^{-1} fresh mass for internal dose rate and mGy d^{-1} per Bq kg^{-1} fresh mass soil/sediment or Bq l^{-1} water for external dose rate). In the derivation of DCs it has often been assumed that some progeny radionuclides are included within the parent radionuclides DCs. This in effect assumes full equilibrium between the parent and its progeny including equilibrium in the transfer from environmental media to the organism. The approach used in early versions of the ERICA Tool (Brown *et al* 2008) and in ICRP (ICRP 2008) truncates a given decay chain, at the first radionuclide with a half-life greater than 10 days; all the members of the truncated decay chain are assumed to be in secular equilibrium and their activities are normalised to the activity of the parent radionuclide. In contrast, the USDOE truncates the decay chain at radionuclides with a physical half-life of 100 years or 180 days depending upon the assessment level (USDOE 2002). The effects of these different assumptions on estimated dose rates are discussed in Vives i Batlle *et al* (2011). A revised approach was presented in ICRP (2017), which allows for the consideration of the full decay chain without truncation and includes the effect of radioactive progeny into the dose coefficient of the parent in a way compatible to the assessment scenario (e.g. average doses during a year or lifetime). Dosimetry methods based upon this revised approach are now beginning to be used within assessment tools (e.g. Version 2.0 of the ERICA Tool uses an approach based upon ICRP (2017) to calculate DC values (ERICA Assessment Tool 2021)).

To account for the relative biological effectiveness of different radiation types, weighting factors are applied to estimate weighted absorbed dose rates. Although there is currently no international agreement on the values of weighing factors, commonly used assessment approaches use similar values. The ERICA Tool uses default radiation weighting factors of 10 for alpha radiation, 1 for (high-energy) beta and gamma radiation, and 3 for low-energy beta radiation (Brown *et al* 2008), whilst RESRAD-BIOTA (USDOE 2004) uses default values of 20 for alpha radiation and 1 for gamma and all beta radiation.

The assumption of simple homogenous geometry is generally considered sufficient for regulatory purposes (Beresford *et al* submitted, Ruedig *et al* 2015). However, more complex dosimetric methodologies are available if required. Using imaging technologies, voxel phantoms have been created for a number of organisms (Stabin *et al* 2006, Dogdas *et al* 2007, Kinase 2008, Caffrey and Higley 2013, Caffrey *et al* 2015a, 2015b, Ruedig *et al* 2014, Higley *et al* 2015). Voxel phantoms enable doses to be calculated to individual organs (modelling realistic heterogeneous distributions) including the consideration of organ-to-organ irradiation. Although not required for routine regulatory assessments, voxel phantoms could be useful in interpreting the results of dose-effects studies. Other models (e.g. Beaugelin-Seiller *et al* 2006) allow dose rates from heterogeneous radionuclide distributions in soil and sediment profiles to be modelled.

12.2.4 Estimating risk

For an assessment an estimated dose rate needs to be put into context, i.e. is this dose rate 'safe' for the organism/ecosystem under consideration?

In the 1990s the NCRP (1991), IAEA (1992) and UNSCEAR (1996) released reports with suggested dose rates below which populations of wildlife should not be adversely impacted by ionising radiation. These dose rates, 1 mGy d^{-1} for terrestrial animals, 10 mGy d^{-1} for plants and 10 mGy d^{-1} for aquatic organisms, were derived by 'expert review', with the process of derivation not always being transparently described (see Howard *et al* 2010). Although not formally proposed as benchmark values for use in assessment RESRAD BIOTA compares estimated dose rates for organisms against 'dose limits' (USDOE 2002) of 1 mGy d^{-1} for terrestrial and riparian animals, and 10 mGy d^{-1} for aquatic organisms and terrestrial plants; the dose limits were based on IAEA (1992) and NCRP (1991) (Howard *et al* 2010).

The ICRP (2008) proposed 'Derived Consideration Reference Levels' (DCRLs) for their RAPs using a documented expert judgment approach. The DCRLs are order of magnitude dose rate bands, using the definition 'a band of dose rate within which there is likely to be some chance of deleterious effects of ionising radiation occurring to individuals of that type of Reference Animal or Plant, derived from a knowledge of defined expected biological effects for that type of organism that, when considered with other relevant information, can be used as a point of reference to optimise level of effort expended on environmental protection, dependent upon the overall management objectives and the exposure situation'. The DCRLs are: 0.1–1 mGy d^{-1} (Reference Rat, Deer, Duck and Pine Tree); 1–10 mGy d^{-1} (Reference Frog, Trout, Flatfish and Wild Grass); 10–100 mGy d^{-1} (Reference Bee, Earthworm, Crab, Brown Seaweed) (ICRP 2008). At the time of writing, we are aware of developing environmental protection approaches which are adopting the lower end of the DCRL band as a screening value for use in assessments.

To derive a screening dose rate value for application in radiological environmental assessments Garnier-Laplace *et al* (2006) and Andersson *et al* (2009) adapted the species sensitivity distribution approach described in the European Union Technical Guidance Document (EC 2003) for deriving predicted no effects concentrations for chemicals. The species sensitivity distribution is a cumulative probability distribution based upon radiation effects data for different species. For the derivation of the screening dose rate both Garnier-Laplace *et al* and Andersson *et al* used EDR_{10} values (i.e. the dose rate giving rise to a 10% effect) as the input data to estimate a Hazardous Dose Rate 5% (HDR_5) value (i.e. the predicted dose rate at which 95% of species will be affected by less than 10%). To account for uncertainties, an assessment or safety factor of 5 was applied by Garnier-Laplace *et al* (2006), whilst Andersson *et al* (2009) applied a value of 2 to the resultant HDR_5 value to derive the screening dose rate. Although the input data and data treatment differed between Garnier-Laplace *et al* and Andersson *et al* they both derived a generic screening dose rate of 10 μGy h^{-1} (0.24 mGy d^{-1}) applicable to all organism types. The 10 μGy h^{-1} screening dose rate value is used as the default within the ERICA Tool (Brown *et al* 2008). A screening dose rate is not a 'limit', it is for application in conservative assessments with the aim that sites can be excluded from the need for further assessment. The screening dose rate is compared with the estimated incremental dose rate for the source(s) being assessed, i.e. it is not compared to the total dose rate including natural background exposure.

A disadvantage of a generic screening dose rate applied to all organisms is that it tends to identify the most exposed organism which will often be relatively radio-insensitive (Beresford *et al* 2010b). However, at the time the screening dose rate values were derived there were insufficient data to apply the species sensitivity approach to estimate organism type specific values (Andersson *et al* 2009).

It may also be useful to compare assessment outputs (and any suggested benchmark values) to natural background exposure of wildlife. Hosseini *et al* (2010) suggested that weighted background absorbed dose rates to aquatic organisms are in the range 0.4–1.9 $\mu Gy\ h^{-1}$ and Beresford *et al* (2008b) estimated a typical range of 0.07–0.6 $\mu Gy\ h^{-1}$ for terrestrial organisms (in the UK). These values for terrestrial organisms do not include estimates of the contribution to dose rate from inhaled ^{222}Rn and its progeny which for burrowing terrestrial mammals will be at least an order of magnitude higher than the dose rates from other naturally occurring radionuclides (i.e. ^{40}K and ^{238}U and ^{232}Th series radionuclides) (Macdonald and Laverock 1998, Beresford *et al* 2012).

12.3 Ionising radiation effects on different organisms

The degree to which ionising radiation (IR) will interact with an organism is dependent on the nature of the radiation: its energy and its linear energy transfer (LET). The LET describes the way in which particles lose energy when they travel through matter. Hence, heavy particles such as alpha particles have high-LET values, whereas beta particles can be considered high-LET or low-LET, depending on their kinetic energy. High-energy β-particles interact less often with matter along their path, while low-energy β-particles have more interactions. The concept of LET does not, in principle, apply to photons (x-ray and gamma radiation) because they are uncharged, but does apply to the secondary charged particles generated when they are scattered. Taking such scattering into account, their behaviour in matter can be described as low-LET.

A consequence of the interaction of ionising radiation with biological cells and tissues is the potential damage that might subsequently lead to overall damage to the organism. At a cellular and subcellular level, the effects induced by ionising radiation in any species are in essence similar to those described for humans, i.e. the induction of direct or indirect damage to different biomolecules. Direct damage occurs when particles directly ionise biomolecules along their track (Desouky *et al* 2019). Indirect damage comes from the subsequent production of hazardous radicals or breakdown products like reactive oxygen species (ROS). These radicals add to the endogenous pool of cellular ROS produced as by-products of cellular metabolism like respiration and photosynthesis (Bae *et al* 2011, Karuppanapandian *et al* 2011). When there is an imbalance between ROS production and the antioxidative scavenging capacity in a cell there is so-called oxidative stress that will lead to different adverse effects. Although oxidative stress has been observed to occur in many abiotic stress conditions including after exposure to gamma radiation (Real *et al* 2004, Horemans *et al* 2011, Xie *et al* 2019, Ludovici *et al* 2020, Song *et al* 2020) ROS production is not always detrimental to a cell or organism. In fact, ROS plays a pivotal role in cellular metabolism and in cell signalling in all organisms (Bae *et al* 2011, Mittler and Blumwald 2015, Dickson and Zhou 2020, Fichman and Mittler 2020). Additionally, although evidence has shown that organisms exposed to high acute radiation exposure have increased ROS production, it is disputable whether radiolysis of water can account for ROS production and subsequently induce oxidative stress at chronic low dose rates (i.e. such as those found in the Chornobyl Exclusion Zone or Fukushima impacted areas of Japan) (Smith *et al* 2012).

Whether directly or indirectly, ionising radiation can cause extensive modifications of biomolecules in the cell, such as bond breaking and crosslink formation in lipids, proteins and carbohydrates. These modifications lead to structural as well as functional damage. As such, DNA, being one of the main biomolecules in a cell, can be directly affected after exposure to any type of ionising radiation leading to single or double stranded breaks. There is ample evidence for the induction of DNA damage after exposure to radiation in different organisms (Real *et al* 2004, UNSCEAR 2008, Horemans *et al* 2019, Xie *et al* 2019). Whether or not this leads to permanent damage or changes in DNA and hence to adverse or long-term effects is highly dependent on whether adequate DNA repair mechanisms are induced (Hunt *et al* 2013, Horemans *et al* 2019). DNA damage is, however, critical as DNA carries the information necessary to build and maintain cellular and organismal stability. Any modification or reshuffling that results from DNA damage is therefore a possible threat to the integrity of genetic information (UNSCEAR 2008).

Research over many years has led to a reasonably good understanding of the impact that exposure to radiation has on organisms when exposed to a relatively high dose rate generally applied under laboratory studies (Real *et al* 2004). Within the framework of radiation protection, research was mainly conducted to see if exposure resulted in changes to individual overarching effects such as morbidity, mortality, reproduction and fecundity. This research directly fed into the screening values and different organism-dependent benchmarks that have been suggested for application within environmental radiological assessments to date (see 12.2.2). However, to fully understand the impact of radiation on the environment it is recognised that more environmentally relevant conditions need to be studied and research is required to link the effects between different levels of biological complexity from molecular to individual and from individual to the ecosystem (Gilbin *et al* 2021). In this section, we give an overview of the current state of understanding of the impact of radiation on the environment.

12.3.1 Learning from radiologically contaminated areas

Globally there are several areas continuously exposed to comparatively high levels of radiation due to contamination originating from routine or accidental releases of different radionuclides. These include (former) uranium mine and milling sites, and areas affected by nuclear accidents (Vandenhove 2002, Howard *et al* 2017). These legacy sites offer an opportunity to study the effects of radiation on different organisms and the interactions of these organisms with the environment (Smodis *et al* 2014, Lourenco *et al* 2016, Lourenco *et al* 2017, Beresford *et al* 2020c). Here we give an overview of the effects of radiation observed in areas impacted by nuclear accidents. Although effects have been reported in areas impacted by uranium mines and industries (Pyle *et al* 2002, Favas and Pratas 2014, Lourenco *et al* 2017, Trenfield *et al* 2019) we do not consider these as it is difficult to separate impacts of radiation from those of chemical toxicity.

Two INES 7 (major accident) and one INES 6 (serious accident) level classified nuclear accidents[1] have occurred at Chornobyl (1986) and Fukushima Dai-ichi (2011) nuclear power plants (NPPs) and the Mayak plutonium production site at Kyshtym (1957), respectively. In addition to a difference in magnitude in the released amounts of radionuclides following these accidents (UNSCEAR 1996, Steinhauser *et al* 2014) the major long-lived radionuclides also differ between the three contaminated areas. Strontium-90 is the major contaminant around the Kyshtym facility. Both Sr-90 and Cs-137 are dominant together with lower activity concentrations of Pu-isotopes and Am-241 in the Chornobyl Exclusion Zone. In Japan, Cs-137 is the predominant radionuclide present in the environment ten years after the Fukushima Dai-ichi accident. The temporal pattern of the releases differed between the two NPP accidents with releases in Japan being more protracted (UNSCEAR 2020). Due to political situations and a focus on human protection, environmental, studies of the impacts of radiation on the environment conducted shortly after these incidents are lacking and fragmentary. This hampers the distinction between effects due to the initial acute high doses of radiation and the current chronic low dose and low dose rate exposures (Fesenko 2019, Beresford *et al* 2020b). However, in the longer-term, more extensive research has been carried out in the different areas to understand the effects of long-term exposure on wildlife (see reviews by e.g. IAEA 2006a, 2015, Hinton *et al* 2007, Geras'kin *et al* 2008, UNSCEAR 2013, Beresford *et al* 2016b, Fesenko 2019). Here we focus on literature for the terrestrial or freshwater environments as these ecosystem types are present in the Chornobyl, Fukushima and Mayak areas. However, we acknowledge that the Fukushima accident represented the largest ever acute releases of radioactivity into the marine environment. An overview of the impact of the Fukushima Dai-ichi accident on marine non-human biota can be found in Vives i Batlle (2016), Vives i Batlle *et al* (2018).

12.3.1.1 Mayak

On September 29, 1957, an explosion occurred due to equipment failure in the cooling water system in a radioactive liquid waste storage tank at the Mayak Production Association (PA) located in the Russian Southern Urals. This accident led to an atmospheric release of 74 PBq of fission products (UNSCEAR 1996). The accident is often called the 'Kyshtym accident' referring to the town of Kyshtym which is close to the production site. Most of this release (~95%) comprised relatively short-lived radionuclides with a half-life of one year or less. Strontium-90 is the major

[1] See https://www.iaea.org/resources/databases/international-nuclear-and-radiological-event-scale for a description of the various categories of incident and accident.

radionuclide still present in the contaminated area. Almost 90% of the released radioactivity was deposited within the perimeter of the Mayak PA site, whereas the rest was, due to the prevailing wind, predominantly dispersed over a narrow trace of about 300 km that is known as the East Urals Radioactive Trace (UNSCEAR 1996, Vasilenko *et al* 2020, Akleyev and Degteva 2021). The paper of Akleyev and Degteva (2021) gives a good overview of the contamination levels over time including the management and protective and rehabilitation measurements taken for the human population.

In the East Urals both coniferous and deciduous trees were affected in the years after the accident. Depending on the dose rate, pine trees died off or showed parts with desiccated needles (Fesenko 2019). Birch trees appeared less sensitive but delayed bud formation as well as an early senescence were present for at least the first five years after which the trees seemed to recover (UNSCEAR 1996). It needs to be noted that there has been a habitat shift in the species composition of the forest that is now present from mainly pine to more deciduous trees with few pines, like the effect observed in the Red Forest in the Chornobyl Exclusion Zone (see further). This makes comparison with neighbouring, unimpacted sites difficult in ecological studies. Herbaceous plants were also affected in the most contaminated areas; it was shown that plants with buds at or near the soil surface disappeared from the communities in the most contaminated areas (Sr-90 levels above 18 MBq m^{-2}) in favour of short-lived plants or plants with dormant buds in more areal plant parts (see references within UNSCEAR 1996).

Soil invertebrates were among the most exposed animals due to them living in the litter layer and top soil with dose rates between 3 and 15 μGy d^{-1} during the initial acute phase (the definition of the acute phase varies in Fesenko (2019) from up to six months to up to one year after the accident). For some radiosensitive or highly exposed invertebrate species, the number dropped to almost zero in the first years after the accident. In 1969–71, soil-dwelling invertebrate numbers were on average two-fold lower in contaminated compared with control sites (Krivolutsky 1996b). Although 30 years after the accident recovery in the number of invertebrates was observed, the populations had not reached the same density as in comparable control sites (Fesenko 2019). This observation of continuing reduced numbers of invertebrates may have been due to the low mobility of soil invertebrates and hence a lack of migration into the area (Fesenko 2019).

In small mammals, lethal effects were found in the affected areas in the acute phase after the accident when doses accumulated above 10–20 Gy (for references see Fesenko 2019). In more recent years, pygmy wood mice (*Apodemus uralensis*) chronically exposed to ^{90}Sr and ^{137}Cs (~0.04–0.5 mGy d^{-1}) were captured from the East Ural Radioactive Trace between 2010 and 2014 to study potential metabolic changes as well as population density in these small rodents (Orekhova and Modorov 2017). Compared with a control population the exposed animals were characterised by a reduction in lipid metabolism, enhanced mitochondrial oxidation and increased antioxidative metabolites and enzymes, indicating that these mice were experiencing stress that leads to a depletion in cell energy. This metabolic response was correlated both with whole body radiation dose as well as with population size indicating possible harmful effects accumulating both from radiation and non-radiological stressors (Orekhova and Modorov 2017). These data confirm that in addition to radiation, population size and density and hence availability of resources are non-radiation stressors that can affect an organism's metabolism (Orekhova and Modorov 2017) and stress the difficulty in isolating the potential impact of radiation from other confounders under field conditions.

Larger herbivorous animals (both wildlife and livestock) were exposed to the high ^{90}Sr levels predominantly through uptake of contaminated food. Such large, domesticated grazing animals were highly exposed at the time of the accident because they were left in the field. Estimated external doses were 1.4–3.0 Gy in the days after the accident and doses to the gastrointestinal tract (GIT) of 4–24 Gy were estimated (Alexakhin *et al* 1994). The high initial doses and dose rates led to the development of symptoms of acute radiation sickness and death in cattle and sheep. Animals grazing in the less contaminated areas (estimated GIT dose of 1–2 Gy) were observed over a period of six months after the accident and showed some signs of radiation exposure (e.g. in production of blood cells) (Fesenko 2019) although any longer-term impact on animals in these less contaminated areas is unknown (Fesenko, personal communication).

12.3.1.2 Chornobyl

The Chornobyl accident on 26th of April 1986 was the largest accidental release of anthropogenic radionuclides to the environment (IAEA 1986); releases of radioactivity from the damaged reactor continued over 10 days. The accident, and the many studies looking at the consequences of it for man and environment have been summarised in a number of reports of the International Atomic Energy Agency (IAEA 1986, 2006a) and United Nations Scientific Committee on the Effects of Atomic Radiation (UNSCEAR 1996, 2008) as well as in numerous scientific papers and reviews (e.g. Kryshev *et al* 2005, Chesser and Baker 2006, Hinton *et al* 2007, Beresford and Copplestone 2011, Steinhauser *et al* 2014, Beresford *et al* 2016b, Geras'kin, 2016, Fuller *et al* 2017, Horemans *et al* 2019, Beresford *et al* 2020c).

Highest deposition of radionuclides was to the west of the reactor in a trace that included a pine plantation now known as the 'Red Forest'. In this forest, a massive death of pine trees was observed shortly after the nuclear accident (Arkhipov *et al* 1994). In addition, trees in close proximity to the reactor showed signs of injury 2–3 weeks after the accident as yellowing and needle death leading to early senescence, a growth arrest of pine trees, a reduced number of seeds, damaged reproductive buds and necrosis of meristems and young shoots, as reviewed by Geras'kin *et al* (2008). The effect of the accident is also clearly visible in the thickness in the yearly radial growth of the tree stems resembling the effect of years of extreme drought (Holiaka *et al* 2020). At a sub-organismal level, studies of aberrant cell frequencies in the root meristem of plant seedlings (Geras'kin *et al* 2011), increased levels of single- and double-strand breaks (Georgieva *et al* 2017) and gene deregulation have been reported (Zelena *et al* 2005).

In this acute phase, defined as the first month to a maximum of three months after the Chornobyl accident, the collapse of specific groups of soil biota and a significant decrease in the population density of the forest litter community were observed in the most contaminated areas (Krivolutzkii and Pokarzhevskii 1992, Krivolutsky 1996b). The accident happened at the end of April in the middle of the reproduction cycle of most soil organisms and it was observed that young earthworms (*Nicodrilus caliginosus* S., *Dendrobaena octaedra* S.) did not hatch from cocoons (Krivolutzkii and Pokarzhevskii 1992). This led to a reduction by a factor of 30 in the number of invertebrates in the litter layer of forests 3–7 km from the nuclear reactor (Krivolutzkii and Pokarzhevskii 1992). In the year after the accident, populations were reported to have recovered. After 2.5 years, the immature/adult ratio and total biomass were not different from less or uncontaminated sites, but the species diversity remained markedly lower (Krivolutsky 1996a). Even in 2001, Jackson *et al* (2005) suggested a loss of diversity in the terrestrial invertebrate population in higher contaminated areas in the Chornobyl Exclusion Zone. However, in 2016, Beresford *et al* (2022b) did not find a relationship between radiation exposure and soil invertebrate biological activity instead reporting significant relationships between soil invertebrate biological activity and soil parameters. This study did show that soil biological activity was lower in the Red Forest, the area where pine trees died due to radiation in 1986, though the authors speculated that this was possibly the residual consequence of an acute high radiation exposure in 1986.

During the first phase, negative effects were also observed in vertebrates. For example, in 1987 close to the Chornobyl NPP, an increase in infertile eggs of 33% was found in brown frogs (*Rana arvalis*) shortly after the accident (see references in Sazykina and Kryshev 2006). In the next two years, this increase in infertility was still present though lower: 17% in 1988 and 3% in 1989 compared with a level of 0.01% found on average in uncontaminated areas. In addition, some subcellular effects were observed in frogs from contaminated areas such as an increased level of chromosomal aberrations (Eliseeva *et al* 1994) and an increase in the number of micronuclei (Voitovich and Afonin 2000). Animals in the contaminated areas had an increased initial level of aberrant bone marrow cells and had a different rate of apoptosis after an additional radiation exposure (Voitovich and Afonin 2000, Voitovich 2001). Finally, Vinogradov and Chubinishvili (1999) found a decrease in genome size in frog species with increasing dose rates.

Whereas the acute phase of the first months was characterised by high levels of fast decaying radionuclides and a likely high bioavailability of radionuclides, the chronic phase started after the first year with the overall dose rate reduced due to the decay of the short-lived radionuclides. As such, the dose rates are nowadays mainly dominated by Cs-137 and Sr-90 and to a minor extent also by Pu-isotopes and Am-241 (Kashparov *et al* 2018, Beresford *et al* 2020a). The long-term evolutionary responses and population effects in vertebrates exposed to lower more chronic radiation exposures currently present in the Chornobyl Exclusion Zone have been investigated for different species ranging from bank voles (*Myodes glareolus*), frogs (e.g. common tree frog *Hyla orientalis*), birds (different species including great tit (*Parus major)* and barn swallows (*Hirundo rustica*)) and large mammals such as brown bear (*Ursus arctos*) and bison (*Bison bonasus*) (Meeks *et al* 2007, Galván *et al* 2014, Deryabina *et al* 2015, Gashchak *et al* 2016, Baker *et al* 2017, Gashchak *et al*

2017, Omar-Nazir *et al* 2018, Cannon and Kiang 2020, Car *et al* 2021). However, the results are not always in agreement leading to some controversy on the long-term effects of the Chornobyl accident on wildlife populations (Beresford *et al* 2020b). The comparison of the oxidative stress and DNA damage levels in different bird species even indicated that specimens sampled in higher contaminated areas showed a higher level of antioxidative stress defense coupled to lower DNA damage and oxidative stress levels (Galván *et al* 2014). This was taken by the authors as proof of potential adaptation of the birds to the chronic dose levels.

Some of the potential interfering processes that makes field population studies on long-term effects to low dose radiation difficult to interpret are the potential for changes in genetic diversity due to induced mutations and DNA damage leading to adaptation and selection as well as the migration between exposed and non-exposed populations (Vives i Batlle *et al* 2020, Car *et al* 2021). Using nuclear and mitochondrial markers Car *et al* (2021) investigated the genetic diversity in 19 populations of common tree frog (*Hyla orientalis*) sampled across and near the Chornobyl Exclusion Zone. The authors found a higher mitochondrial diversity and absence of genetic erosion and based on their studies concluded that the tree frog populations are small and under a high mitochondrial mutation rate possibly leading to deleterious individual effects although the population as a whole is able to cope; the study species has a high egg production and the authors suggested for future studies species with smaller egg production/offspring sizes be used. As frogs have a very specific habitat occupation depending on their life stage and time of year they are exposed to different (terrestrial and aquatic) radiation sources (Burraco and Orizaola 2022) with some exposure scenarios (e.g. in sediment) maximizing exposure.

It has been shown that more than 30 years after the accident, strongly affected populations seemed to have recovered, as reviewed by Geras'kin *et al* (2008). For plants and non-migratory animals, this is possibly due to reduced dose rates and hence reduced exposure after the initial phase of the accident (Arkhipov *et al* 1994, Zelena *et al* 2005) while for most animals there is a potential migration of animals from less contaminated areas to the more contaminated (Jackson *et al* 2005). However, the estimated doses to different organisms remain high enough to anticipate that effects at some level of biological complexity can occur (Kashparov *et al* 2018, Beresford *et al* 2020a, Vives i Batlle *et al* 2020) though it is still far from understood how long-term exposure to a low/very low dose rate might affect individuals and populations (Adam-Guillermin *et al* 2018, Horemans *et al* 2019) or if and how local populations might become more adapted over time (see 16.3.3). Furthermore, there are contradicting reports on the present impact of radiation on different wildlife in the Chornobyl affected area (Beresford *et al* 2020b, Mothersill *et al* 2020). Whereas some studies have reported no or little effect of the current low chronic dose rates present in Chornobyl, e.g. for mammals, birds, soil invertebrates and *Arabidopsis* plants (Chesser and Baker 2006, Galván *et al* 2014, Deryabina *et al* 2015, Bonzom *et al* 2016, Horemans *et al* 2018) others have indicated a high impact on different populations (insects, mammals and soil invertebrates) at dose rates of below 1 μGy h^{-1} (Moller and Mousseau 2009,

2013, 2018) which is comparable to the natural background exposure rate of organisms. A review of the potential explanations and suggestions of a way forward for research to address this contradiction is given in Beresford *et al* (2020b). A confounding factor when considering wildlife populations in the Chornobyl Exclusion Zone is the 'positive' impact of the removal of humans and cessation of agriculture etc in 1986.

12.3.1.3 Fukushima

The accident at the Fukushima Dai-ichi NPP was the consequence of a failure in the cooling system due to loss of electrical power after a major 9.0 magnitude earthquake and subsequent tsunami that occurred on March 11, 2011. The chain of events as outlined in the UNSCEAR (2013) report had led to a shut down and a loss of all electrical power including the emergency diesel generators which subsequently led to a meltdown in the three operational reactor units as well as hydrogen build-up which led to the explosions. Atmospheric as well as direct releases of contaminated cooling water into the sea occurred. Approximately 80% of the atmospheric release of different radionuclides was deposited over the ocean due to the prevailing wind (Vives i Batlle 2015). The remaining 20% was deposited in the terrestrial and freshwater environment largely within the Fukushima prefecture (Hirose 2020). Total (internal plus external) dose rates to biota were measured using media (soil or water) and biota activity concentration measurements as well as estimations through the ERICA Tool for many plants, insects, invertebrates and vertebrates. As for Chornobyl, environmental exposure depends on the time after the accident and different phases can be distinguished depending on the dose rates (Strand *et al* 2014). A first initial phase, the first week after the tsunami, is characterised by high and variable releases throughout the week depending on the prevalent wind direction as well as the ongoing events at the site, of which more than 80% was transported towards the Pacific Ocean (Steinhauser *et al* 2014). This initial phase was followed by an intermediate phase from the beginning of April 2011 until mid-June 2011 when acute releases had ceased but short-lived radionuclides such as I-131 were still present, and finally a late phase starting from 3 months onwards when short-lived radionuclides had largely decayed (Strand *et al* 2014).

For the initial and intermediate phase (the first 90 days) an international assessment concluded that within the wildlife species considered only for microalgae would the accumulated dose (7 Gy) exceed the generic benchmark of 4.8 Gy for acute exposures, mainly due to the peak in exposure to I-131 (Strand *et al* 2014). From the start of the late phase (mid-June 2011) the same evaluation compared estimated dose rates to the ICRP Derived Consideration Levels (DCRLs; see above). It was concluded that dose rates to deer and large mammals (71 μGy h^{-1}) and grass (26 μGy h^{-1}) fell within these DCRLs (Strand *et al* 2014, Fuma *et al* 2017). However, the dose rates were below the DCRLs for most herbaceous plants, amphibians, insects and earthworms in the first year after the accident even in the most contaminated areas. The study concluded that although for some radio-sensitive organisms or sedentary species living in highly contaminated areas sublethal and individual level effects have been reported (cytogenic, physiological, morphological), an overall negative impact on the whole population following exposure to the 2011 nuclear accident is unlikely (Strand *et al* 2014, UNSCEAR 2020).

It remains to be seen if chronic long-term exposures to elevated dose rates will lead to long-term effects. However, several studies have reported detrimental effects on individual organisms including induction of cytogenetic physiological and morphological changes. In Japan, in contrast to Chornobyl and Mayak, the pine trees did not die in the highest contaminated areas. In a study that ran from 2014 to 2016 at eight different localities in the Fukushima Exclusion Zone (FEZ), similar morphological differences, such as loss of apical dominance, were recently found in young Japanese red pine (*Pinus densiflora*) in the FEZ, whereas no morphological abnormalities were found in mature trees of Japanese red pine (Yoschenko *et al* 2016). The authors report a low frequency of abnormalities (11%–14%) in less contaminated regions (measured ambient dose rate on the day of sampling: 0.33–1.1 μGy h^{-1}), while in the highest contaminated regions (measured ambient dose rate: up to 9 μGy h^{-1}) a higher frequency of abnormalities was observed, up to 67%.

Studying the pale blue grass butterfly *Zizeeria maha* as an indicator species to evaluate the environmental impact of ionizing radiation in the Fukushima affected area led to similar conclusions (Hiyama *et al* 2012, Hiyama *et al* 2013). Mild morphological abnormalities were observed in some individuals of adult butterflies collected one month after the accident but an increase of the severity of these abnormalities occurred in the F1 generation, which were inherited in the F2 generation. These abnormalities were suggested to be explained by random mutation on important genes or by epigenetic mechanisms. However, for the pale blue grass butterfly none of these underlying mechanisms were studied by the authors nor was the link to radiation exposure or other confounding factors acknowledged in the paper (UNSCEAR 2020). After publication of the original paper other researchers tried to estimate the dose rate to the larvae reported, using the ERICA Tool, and making conservative assumptions estimated a maximum dose rate of 8 μGy h^{-1} assuming all ingested caesium was retained by the larvae over the course of their 12-day development (Copplestone and Beresford 2014). As this would indicate an LD_{50} near background dose rates it is very unlikely that radiation exposure is the only or a major contributor to the observed abnormalities. Furthermore, in a follow-up laboratory study butterflies developing from larvae on a diet of Cs-137 contaminated leaves collected in Fukushima contaminated areas (43–450 kBq kg^{-1}) showed morphological abnormalities, whereas feeding the larvae an artificial diet with high levels of Cs-137 (139 MBq kg^{-1}) did not result in any detectable abnormalities in the developing butterflies (Gurung *et al* 2019). It was concluded by the authors that the butterflies were relatively tolerant of the radioactive exposure and, hence, the field-observations could not originate solely from the radiation exposure. In a follow-up study to address this apparent laboratory-field controversy it was shown that the observed abnormalities in the field might be the consequence of changes in nutrient composition in the leaves of the larval host plants rather than a direct consequence of the radiation exposure to the plants. It was shown that the sodium but not the lipid concentration in the leaves of the host plants significantly negatively correlated with increasing radiation levels leading to a potential nutrient imbalanced diet for the butterflies (Sakauchi *et al* 2021). The sodium concentration ranged from 31 mg/100 mg to only 1 mg/100 mg in the most contaminated sampling plot. However, Sakauchi *et al* did not suggest a causal relationship between plant Na concentrations and radiation dose, nor did they discuss confounding factors which may have resulted in the observation. Whereas a low sodium concentration does not pose physiological problems to the plants it might be a crucial deficit in the larvae's diet contributing to the observed abnormalities and high mortality rate (Sakauchi *et al* 2021).

To study the potential effect of the Fukushima accident in small mammals, a number of studies have focused on the Japanese field mouse (*Apodemus speciosus*) (Okano *et al* 2016, Ariyoshi *et al* 2020, Fujishima *et al* 2020, Anderson *et al* 2021). Whereas Fujishima *et al* (2020) showed some structural aberrations in the chromosomes occurring in the mice collected in the highest contaminated areas, other studies failed to find a consistent link between the radiation exposure and the estimated endpoint. For example no significant difference in the frequency of apoptotic cells nor abnormalities in the sperm were found in male field-collected mice (Okano *et al* 2016) nor in the occurrence of DNA damage, inflammation markers or genome instability in the bone marrow of the same mouse species (Ariyoshi *et al* 2020). In the latter study, however, a decrease in the ability of hematopoietic progenitor cells to form colonies was found which is taken as an indication for a changed ability of the bone marrow cells to differentiate into different blood cells. Studying larger mammals, Sato *et al* (2020) examined DNA damage in blood samples of cattle (estimated absorbed dose rate $<16\ \mu Gy\ h^{-1}$) living in the 'difficult-to-return zone' in the Fukushima prefecture with two different markers (8-deoxyguanosine, COMET and micronucleus assays) and concluded that based on their assays DNA damage did not significantly increase in the radiation exposed animals.

12.3.2 Studying species sensitivities

To be able to estimate the impact of radiation on the environment, it is essential to take the differences in response between different organisms into account. That differences in sensitivity to radiation between different organisms exist has been recognised for several decades. It ranges from an acute dose of 10 Gy being lethal for most vertebrates up to the bacterium *Deinococcus radiodurans* withstanding 12 kGy (Daly *et al* 1994, Harrison and Anderson 1996, Adam-Guillermin *et al* 2018). Hence it seems that procaryotes are more radioresistant than vertebrates and also than invertebrates and plants. However, even within one phylum the sensitivity to radiation can differ across orders of magnitude (Harrison and Anderson 1996). Sensitivity of species is additionally dependent on the developmental stage and or time of the year. Annual plants for example, in early spring, have rapidly dividing cells more prone to damage than in late autumn or winter when most plants are in a physiologically dormant stage (Adam-Guillermin *et al* 2018, Fesenko 2019). Hence, the immediate impact on the environment of the nuclear accidents of Fukushima (early spring, March), Chornobyl (spring, April) or Mayak (autumn, October) will have differed. The Chornobyl accident impacted plants at their most sensitive period which is May until early June (in Ukraine), whereas from October to March most annual plants are in a physiological dormancy making them less sensitive (Fesenko 2019). Additionally, there are indications that a past exposure might trigger changes in a species altering its response to future radiation exposures (Day *et al* 2006, Adam-Guillermin *et al* 2018, Laanen *et al* 2021). The fact that radiation exposure can potentially result in observable effects in the progeny leading to trans- or multi-generational effects has recently led to discussions as well as concerns for both humans and the environment (Horemans *et al* 2019, Mothersill *et al* 2020).

An understanding of the mechanisms resulting in organisms being more radio-sensitive or resistant to radiation is lacking. In the early 1960s Sparrow and Miksche (1961) suggested that genome size, number of chromosomes ploidy status and intervals between cell divisions were determining factors. Studying the difference between human and plant radiation sensitivity, Arena *et al* (2014) added a number of other factors such as presence of increased DNA damage repair mechanisms and a strong antioxidative response and specifically for plant cells having a cell wall and often displaying polyploidy would add to their resistance. In contrast, the study of Sharma *et al* (2017) suggested that the presence of symmetrical H-Mn^{2+} complexes in a cell rather than its capacity to repair DNA damage or scavenge ROS, are the drivers of radiation resistance within and across different taxonomic groups.

In defining benchmark dose rates for application in assessments the difference in sensitivity between species/organisms has been considered by setting up sensitivity distributions (SSDs). As described in section 12.2.2 the SSD model is in essence dose rates at which specific species are affected to a certain extent, e.g. EDR_{10} or the dose rate giving rise to a 10 % effect. According to the Technical Guidance Document (TGD) (EC 2003), the SSD can be used to define the dose rate which is hazardous for only a small fraction of the species in the ecosystem. This approach which was initially used for chemicals, was applied to derive radiological benchmark values (Garnier-Laplace *et al* 2006, Andersson *et al* 2009) and led to a generic screening value of 10 μGy h^{-1} which could be used in risk assessments to screen out scenarios that are of no concern (see section 12.2.2). A comparison between an SSD purely based on laboratory-based experiments and one on field-based studies in the Chornobyl Exclusion Zone suggested that the median HDR_{50} value differed between the two scenarios by a factor of 8.5 (Garnier-Laplace *et al* 2013). However, this comparison was based on a relatively limited set of data and should be repeated using larger datasets as well as comparable species in both the laboratory and field studies. However, most of the available radiation effects data, including those used in the setting of the screening dose rate benchmark by SSD, are from laboratory-based studies looking at responses in individual organisms on overarching endpoints such as reduced growth, morbidity, morphology, reproduction and fertility. From a radiation protection point of view, it is, however, important to gather information on whether the radiation impact is different under environmentally relevant conditions. This includes: (i) considering chronic low dose exposure conditions preferably spanning the lifespan of the organism or even several generations; (ii) environmentally relevant surroundings including unfavourable growth conditions and/or exposures to other stressors; (iii) studying the overall impact on the community or ecosystem.

Different issues have been identified that can reduce the transferability of laboratory collected toxicity data to real-life field exposures situations (van Straalen and Denneman 1989, Beresford *et al* 2020c, Spurgeon 2020). These comprise factors inherent to the exposure as well as to the exposed organism. Field exposures are almost always heterogeneous both in time and space and more complex than laboratory exposures including the presence of different stressors, factors influencing bioavailability or sub-optimal growth conditions. Additionally compared with 'model' and genetically uniform organisms often used in laboratory testing, field organisms are generally more diverse.

12.3.3 Adaptation and epigenetic regulation of radiation responses

Radionuclide contaminated ecosystems in, for instance, nuclear accident affected regions experience persistent elevated levels of ionising radiation that may cause adaptive changes in plant, vertebrate and invertebrate populations. Organisms living in such ecosystems have been shown to adapt to optimise their fitness. For comparison in areas heavily polluted with toxic chemicals, rapid evolutionary change in wild populations has been observed (Whitehead *et al* 2012, Saaristo *et al* 2018). Organisms have different mechanisms that enable them to respond and adapt to such environmental stress. Immediate adaptive responses that facilitate acclimation include energy-demanding defense mechanisms like detoxification and repair. The added costs of adaptation might affect the individual's overall fitness or reproductive success, thus imposing a selective pressure on successive generations. Consequently, populations of organisms may adapt to long-term environmental stresses, either via permanent genetic change (Kovalchuk *et al* 2004), or possibly via transgenerational epigenetic inheritance (Mirouze and Paszkowski 2011, Lind and Spagopoulou 2018).

Long-term exposures to low levels of environmental stressors including radionuclides have been linked to lasting responses in organisms within, but also over multiple exposed generations (Mirbahai and Chipman 2014, Jimenez-Chillaron *et al* 2015, Hanson and Skinner 2016, Marczylo *et al* 2016, Schultz *et al* 2016, Horemans *et al* 2019). Yet, these responses seem to vary in magnitude as well as in outcome. In some cases it has been shown to lead to more adapted phenotypes (Walker *et al* 2006, Antonova *et al* 2014, Singer *et al* 2016) or to adaptation in population phenotypes (Coors *et al* 2009, Costa *et al* 2012). In contrast, there is also evidence suggesting that long-term exposures to chronic low dose radiation can lead to an increased population sensitivity and even to population declines (Beasley *et al* 2012, Parisot *et al* 2015). Hence, within the scientific literature there is an active debate on whether organisms have the ability to adapt when chronically exposed to enhanced radiation levels and radionuclides (Kovalchuk *et al* 2004, Galván *et al* 2014, Boubriak *et al* 2016, Moller and Mousseau 2016).

The potential presence of long-term and/or transgenerational effects raises concerns about whether our existing risk assessment approaches, which are largely based on laboratory effects data within one generation, are sufficient to estimate if the environment is protected from radiation in the long-term. For wildlife an overview of existing studies looking at possible adaptive responses performed in the Chornobyl Exclusion Zone was published in 2016 which suggested that the species diversity studied so far, as well as the experimental setups with small sample sizes, is too limited to conclusively determine whether adaptive responses are present (Boubriak *et al* 2016). This review identified that relevant studies were limited to one on birds, two on voles, two on bacteria and nine on plants. A pilot study on soybean seedlings collected in the Chornobyl Exclusion Zone showed that levels of DNA damage correlated with the radiation gradient, though the recovery of DNA damage was faster in the more radioactively contaminated samples (Georgieva *et al* 2017). Boubriak *et al* (2016) reported an enhanced DNA repair mechanism for birch (*Betula* spp.) pollen and primrose (*Primula* spp.) seed embryos found in contaminated areas of Chornobyl. Doing a meta-analysis of their own work, the Boubriak research team concluded that there are indications, at least in plants, for adaptive responses in Chornobyl but that mechanistic understanding is still limited (Boubriak *et al* 2016).

Scots pine trees (*Pinus sylvestris*) in the Chornobyl Exclusion Zone were followed in the period 1986–1991 and showed radiation dependent injuries. As early as 1987, however, signs of recovery and adaptation were observed in all but the highest contaminated areas in the standing trees and new progeny emerged (Arkhipov *et al* 1994). The molecular mechanisms involved in these adaptive processes may include favourable mutations in essential genes. However, such an event would be rare. By assuming a germline mutation rate in plants of about 10^{-5} to 10^{-6} per gamete, only one mutation in 500 000 plants would be expected (Kovalchuk *et al* 2003). So Kovalchuk and co-authors suggested that mechanisms explaining this rapid adaptation are more likely to be linked to epigenetic processes.

Epigenetic marks are hereditary traits which do not involve a change in the DNA sequence itself, but in the DNA environment, i.e. DNA methylation, micro-RNA interaction and histone modifications (Berger 2003, Goldberg *et al* 2007, Allis and Jenuwein 2016). Some epigenetic changes have been proven to be stable and can lead to transgenerational heritable changes. Epigenetic processes can be induced by stressors (like radiation) and may lead to altered/less fit phenotypes or offspring. Epigenetic changes may, however, also lead to increased resistance to adverse conditions. Recent literature has provided evidence for epigenetic marks induced by adverse conditions encountered by parents to be partly stable across generations leading to transgenerational heritable changes in offspring (Verhoeven *et al* 2010, Guerrero-Bosagna *et al* 2012, McCarrey 2012, Dutilleul *et al* 2014, Guerrero-Bosagna and Jensen 2015, Gombeau *et al* 2017, Trijau *et al* 2018, Guirandy *et al* 2019). In particular, evidence has been accumulated for the transfer of DNA methylation patterns in the germline (Verhoeven *et al* 2010, Verhoeven *et al* 2016). For example, in the Brassicaceae plant *Arabidopsis*, nickel chloride caused a change in DNA methylation patterns some of which was inherited by the following generation (Li *et al* 2015). Similarly dandelion (*Taraxacum officinale*) showed altered DNA methylation when exposed to different stressors which was largely inherited by the next generation of this asexually reproducing plant (Verhoeven and van Gurp 2012, Verhoeven *et al* 2016). There is accumulating evidence indicating that exposure to radiation leads to alterations in epigenetic marks including DNA methylation in different organisms (Horemans *et al* 2018, Hurem *et al* 2018, Kamstra *et al* 2018, Horemans *et al* 2019, Lindeman *et al* 2019, Murat El Houdigui *et al* 2020, Thaulow *et al* 2020, Laanen *et al* 2021). When used in conjunction with macroscopic responses, epigenetic analysis could enhance the ecological relevance of molecular biomarkers and this may be a promising way forward to establishing the effects of chronic low-level radiation exposure.

12.3.4 The way forward: towards an ecosystem approach

In order to study the overall effect of radiation on people and the environment it is not sufficient to look at the individual or population level but to go one level higher to consider the ecosystem as a whole (Chesser and Baker 2006, Beresford *et al* 2016b, Brechignac 2016, Mothersill *et al* 2018, Cannon and Kiang 2020, Rhodes *et al* 2020). Adaptation and transgenerational effects on populations of individual species to radiation together with the effects of species interactions and additional environmental drivers will all contribute to the response of the ecosystem. Most studies to date on radiation effects have studied individuals not considering species interactions and the sustainability and complexity of the ecosystem (Mothersill *et al* 2018). Adaptation and transgenerational effects are therefore two (of several) potentially important factors contributing to overall ecosystem function at a contaminated site. As shown by Sakauchi *et al* (2021) studying the interrelation between the pale blue grass butterfly and its host plant in Fukushima affected areas it is important to better understand the ecological and functional relationships between different nodes of an ecosystem in order to be able to estimate the impact of long-term low dose radiation on the environment. Ecosystem function may be assessed by measuring relative abundance of organisms present (e.g. biodiversity) and processes such as nutrient cycling and pollination. Ecosystem structure and function underpin the goods and services that people gain from the environment. An overview of the research needs to address these questions has recently been published by Gilbin *et al* (2021).

References

Adam-Guillermin C, Hertal-Aas T, Oughton D, Blanchard L, Alonzo F, Armant O and Horemans N 2018 Radiosensitivity and transgenerational effects in non-human species *Ann. ICRP* **47** 327–41

Akleyev A V and Degteva M O 2021 Radioecological consequences of radioactive releases due to weapons-grade plutonium production at the 'Mayak' facility in the Russian Federation *J. Radiol. Prot.* **41** 12

Alexakhin R M, Karaban R T, Prister B S, Spirin D A, Romanov G N, Mishenkov N N, Spiridonov S I, Fesenko S V, Fyodorov Y A and Tikhomirov F A 1994 The effects of acute irradiation on a forest biogeocenosis—experimental data, model and practical applications for accidental cases *Sci. Total Environ.* **157** 357–69

Allis C D and Jenuwein T 2016 The molecular hallmarks of epigenetic control *Nat. Rev. Genet.* **17** 487–500

Anderson D, Beresford N A, Ishiniwa H, Onuma M, Nanba K and Hinton T G 2021. Radiocesium concentration ratios and radiation dose to wild rodents in Fukushima Prefecture *J. Environ. Radioact.* **226** 106457

Andersson P, Garnier-Laplace J, Beresford N A, Copplestone D, Howard B J, Howe P, Oughton D and Whitehouse P 2009 Protection of the environment from ionising radiation in a regulatory context (PROTECT): proposed numerical benchmark values *J. Environ. Radioact.* **100** 1100–8

Antonova E V, Pozolotina V N and Karimullina E M 2014 Variation in the seed progeny of smooth brome grass, *Bromus inermis Leyss.*, under conditions of chronic irradiation in the zone of the Eastern Ural Radioactive Trace *Russ. J. Ecol.* **45** 508–16

Arena C, De Micco V, Macaeva E and Quintens R 2014 Space radiation effects on plant and mammalian cells *Acta Astronaut.* **104** 419–31

Ariyoshi K *et al* 2020 Environmental radiation on large Japanese field mice in Fukushima reduced colony forming potential in hematopoietic progenitor cells without inducing genomic instability *Int. J. Radiat. Biol.* **98** 1147–58

Arkhipov N P, Kuchma N D, Askbrant S, Pasternak P S and Musica V V 1994 Acute and long-term effects of irradiation on Pine (*Pinus silvestris*) stands post-Chernobyl *Sci. Total Environ.* **157** 383–6

Bae Y S, Oh H, Rhee S G and Do Yoo Y 2011 Regulation of reactive oxygen species generation in cell signaling *Mol. Cells* **32** 491–509

Baker R J, Dickins B, Wickliffe J K, Khan F A A, Gaschak S, Makova K D and Phillips C D 2017 Elevated mitochondrial genome variation after 50 generations of radiation exposure in a wild rodent *Evol. Appl.* **10** 784–91

Beaugelin-Seiller K, Jasserand F, Garnier-Laplace J and Gariel J C 2006 Modelling the radiological dose in non-human species: principles, computerization and application *Health. Phys.* **90** 485–93

Beasley D E, Bonisoli-Alquati A, Welch S M, Moller A P and Mousseau T A 2012 Effects of parental radiation exposure on developmental instability in grasshoppers *J. Evol. Biol.* **25** 1149–62

Beresford N A *et al* 2008a An international comparison of models and approaches for the estimation of the radiological exposure of non-human biota *Appl. Radiat. Isot.* **66** 1745–9

Beresford N A, Barnett C L, Jones D G, Wood M D, Appleton J D, Breward N and Copplestone D 2008b Background exposure rates of terrestrial wildlife in England and Wales *J. Environ. Radioact.* **99** 1430–9

Beresford N A *et al* 2010a Predicting the radiation exposure of terrestrial wildlife in the Chernobyl exclusion zone: an international comparison of approaches *J. Radiol. Prot.* **30** 341–73

Beresford N A, Hosseini A, Brown J E, Cailes C, Beaugelin-Seiller K, Barnett C L and Copplestone D 2010b Assessment of risk to wildlife from ionising radiation: can initial screening tiers be used with a high level of confidence? *J. Radiol. Prot.* **30** 265–84

Beresford N A and Copplestone D 2011 Effects of ionizing radiation on wildlife: what knowledge have we gained between the Chernobyl and Fukushima accidents? *Integr. Environ. Assess. Manag.* **7** 371–3

Beresford N A, Barnett C L, Vives i Batlle J, Potter E D, Ibrahimi Z-F, Barlow T S, Schieb C, Jones D G and Copplestone D 2012 Exposure of burrowing mammals to 222Rn *Sci. Total Environ.* **431** 252–61

Beresford N A and Vives i Batlle J 2013a Estimating the biological half-life for radionuclides in homoeothermic vertebrates: a simplified allometric approach *Rad. Environ. Biophys.* **52** 505–11

Beresford N A, Yankovich T L, Wood M D, Fesenko S, Andersson P, Muikku M and Willey N J 2013b A new approach to predicting environmental transfer of radionuclides to wildlife taking account of inter-site variation using residual maximum likelihood mixed-model regression: a demonstration for freshwater fish and caesium *Sci. Total Environ.* **463–464** 284–92

Beresford N A and Wood M D 2014 A new simplified allometric approach for predicting the biological half-life of radionuclides in reptiles *J. Environ. Radioact.* **138** 116–21

Beresford N A, Wood M D, Vives i Batlle J, Yankovich T L, Bradshaw C and Willey N 2016a Making the most of what we have: application of extrapolation approaches in radio-ecological wildlife transfer models *J. Environ. Radioact.* **151** 373–86

Beresford N A, Fesenko S, Konoplev A, Skuterud L, Smith J T and Voigt G 2016b Thirty years after the Chernobyl accident: what lessons have we learnt? *J. Environ. Radioact.* **157** 77–89

Beresford N A and Willey N 2019 Moving radiation protection on from the limitations of empirical concentration ratios *J. Environ. Radioact.* **208–209** 106020

Beresford N A, Barnett C L, Gashchak S, Maksimenko A, Guliaichenko E, Wood M D and Izquierdo M 2020a Radionuclide transfer to wildlife at a 'Reference site' in the Chernobyl Exclusion Zone and resultant radiation exposures *J. Environ. Radioact.* **211** 105661

Beresford N A *et al* 2020b Towards solving a scientific controversy—the effects of ionising radiation on the environment *J. Environ. Radioact.* **211** 106033

Beresford N A, Scott E M and Copplestone D 2020c Field effects studies in the Chernobyl Exclusion Zone: lessons to be learnt *J. Environ. Radioact.* **211** 105893

Beresford N A, Beaugelin-Seiller K *et al* 2022a Ensuring robust radiological risk assessment for wildlife: insights from the International Atomic Energy Agency EMRAS and MODARIA programmes *J. Radiol. Prot.* **3** 42

Beresford N A, Wood M D, Gashchak S and Barnett C L 2022b Current ionising radiation doses in the Chernobyl Exclusion Zone do not directly impact on soil biological activity *PLoS One* **17** e263600

Berger S L 2003 Chromatin and gene regulation—molecular mechanisms in epigenetics *Science* **300** 252–4

Bonzom J M *et al* 2016 Effects of radionuclide contamination on leaf litter decomposition in the Chernobyl exclusion zone *Sci. Total Environ.* **562** 596–603

Boubriak I, Akimkina T, Polischuk V, Dmitriev A, McCready S and Grodzinsky D 2016 Long term effects of Chernobyl contamination on DNA repair function and plant resistance to different biotic and abiotic stress factors *Cytol. Genet.* **50** 381–99

Boyer P, Wells C and Howard B J 2018 Extended Kd distributions for freshwater environment *J. Environ. Radioact.* **192** 128–42

Brechignac F 2016 The need to integrate laboratory- and ecosystem-level research for assessment of the ecological impact of radiation *Integr. Environ. Assess. Manag.* **12** 673–6

Broadley M R, Willey N J and Mead A 1999 A method to assess taxonomic variation in shoot caesium concentration among flowering plants *Environ. Pollut.* **106** 341–9

Brown J E, Alfonso B, Avila R, Beresford N A, Copplestone D, Pröhl G and Ulanovsky A 2008 The ERICA tool *J. Environ. Radioact.* **99** 1371–83

Brown J E, Beresford N A and Hosseini A 2013 Approaches to providing missing transfer parameter values in the ERICA Tool—how well do they work? *J. Environ. Radioact.* **126** 399–411

Brown J E, Alfonso B, Avila R, Beresford N A, Copplestone D and Hosseini A 2016 A new version of the ERICA tool to facilitate impact assessments of radioactivity on wild plants and animals *J. Environ. Radioact.* **153** 141–8

Brown J E, Beresford N A and Hevrøy T 2019 Exploring taxonomic and phylogenetic relationships to predict radiocaesium transfer to marine biota *Sci. Tot. Environ.* **649** 916–28

Burraco P and Orizaola G 2022 Ionizing radiation and melanism in Chornobyl tree frogs *Evol. Appl.* **15** 1469–79

Caffrey E A and Higley K A 2013 Creation of a voxel phantom of the ICRP reference crab *J. Environ. Radioact.* **120** 14–8

Caffrey E A, Johansen M P and Higley K A 2015a Voxel modeling of rabbits for use in radiological dose rate calculations *J. Environ. Radioact.* **151** 480–6

Caffrey E A, Johansen M P and Higley K A 2015b Organ dose rate calculations for small mammals at Maralinga, the Nevada test site, Hanford, and Fukushima: a comparison of ellipsoidal and voxelized dosimetric methodologies *Radiat. Res.* **184** 433–41

Cannon G and Kiang J G 2020 A review of the impact on the ecosystem after ionizing irradiation: wildlife population *Int. J. Radiat. Biol.* **98** 1054–62

Car C *et al* 2021 Unusual evolution of tree frog populations in the Chernobyl exclusion zone *Evol. Appl.* **15** 203–19

Chesser R K and Baker R J 2006 Growing up with Chernobyl *Am. Scientist* **94** 542–9

Coors A, Vanoverbeke J, De Bie T and De Meester L 2009 Land use, genetic diversity and toxicant tolerance in natural populations of Daphnia magna *Aquat. Toxicol.* **95** 71–9

Copplestone D, Bielby S, Jones S R, Patton D, Daniel C P and Gize I 2001 Impact assessment of ionising radiation on wildlife *R&D Publication 128* (Bristol: Environment Agency) https://assets.publishing.service.gov.uk/government/uploads/system/uploads/attachment_data/file/290300/sr-dpub-128-e-e.pdf

Copplestone D, Wood M D, Bielby S, Jones S R, Vives J and Beresford N A 2003 *Habitat regulations for Stage 3 assessments: radioactive substances authorisations* Environment Agency R&D Technical Report https://wiki.ceh.ac.uk/pages/viewpage.action?pageId=115016440&preview=/115016440/115212618/SP3-101-SP1A-E-E.pdf

Copplestone D, Beresford N A, Brown J E and Yankovich T 2013 An international database of radionuclide concentration ratios for wildlife: development and uses *J. Environ. Radioact.* **126** 288–98

Copplestone D and Beresford N A 2014 Questionable Studies Won't Help Identify Fukushima's Effects https://theconversation.com/questionable-studies-wont-help-identify-fukushimas-effects-26772

Costa D, Marien J, Janssens T K S, van Gestel C A M, Driessen G, Sousa J P, van Straalen N M and Roelofs D 2012 Influence of adaptive evolution of cadmium tolerance on neutral and functional genetic variation in *Orchesella cincta Ecotoxicology* **21** 2078–87

Daly M J, Ouyang L, Fuchs P and Minton K W 1994 In vivo damage and recA-dependent repair of plasmid and chromosomal DNA in the radiation-resistant bacterium *Deinococcus radiodurans J. Bacteriol.* **176** 3508–17

Day T K, Zeng G X, Hooker A M, Bhat M, Scott B R, Turner D R and Sykes P J 2006 Extremely low priming doses of X-radiation induce an adaptive response for chromosomal inversions in pKZ1 mouse prostate *Radiat. Res.* **166** 757–66

Deryabina T G, Kuchmel S V, Nagorskaya L L, Hinton T G, Beasley J C, Lerebours A and Smith J T 2015 Long-term census data reveal abundant wildlife populations at Chernobyl *Curr. Biol.* **25** R824–826

Desouky O, Ding N and Zhou G 2019 Targeted and non-targeted effects of ionizing radiation *J. Radiat. Res. Appl. Sci.* **8** 247–54

Dickson K B and Zhou J 2020 Role of reactive oxygen species and iron in host defense against *Front. Biosci.* **25** 1600–16

Dogdas B, Stout D, Chatziioannou A F and Leahy R M 2007 Digimouse: a 3D whole body mouse atlas from CT and cryosection data *Phys. Med. Biol.* **52** 577–87

Dutilleul M, Bonzom J M, Lecomte C, Goussen B, Daian F, Galas S and Reale D 2014 Rapid evolutionary responses of life history traits to different experimentally-induced pollutions in *Caenorhabditis elegans BMC Evol. Biol.* **14** 14

European Commission (EC) 2003 Technical Guidance Document on Risk Assessment in Support of Commission Directive 93/67/EEC on Risk Assessment for New Notified Substances and Commission Regulation (EC) No. 1488/94 on Risk Assessment for Existing Substances. Office for Official Publication of the European Communities: Luxembourg

Eliseeva K G, Vojtovich A M, Ploskaya M V and Smal S E 1994 Genetic monitoring of brown frog populations inhabiting radiocontaminated areas of Belarus *Radiats. Biol. Radioekol.* **34** 838–46

Environmental Canada and Health Canada 2003 Canadian Environmental Protection Act 1999. Priority Substances List Assessment Report releases of Radionuclides from Nuclear Facilities (Impact on Non-Human Biota). Environmental Canada, Health Canada

ERICA Assessment Tool 2021 ERICA Assessment Tool: Version 2.0. Overview of changes. October 2021, Version 1. https://erica-tool.com/wp-content/uploads/2021/11/ERICA-Comparison-report-Version-2-Nov-2021.pdf

Favas P J C and Pratas J 2014 Uranium bioavailability and environmental risk assessment in soils contaminated by mining *Int. Conf. on Environment Systems Science and Engineering (ESSE) (Los Angeles, CA)* pp 43–6

Fesenko S 2019 Review of radiation effects in non-human species in areas affected by the Kyshtym accident *J. Radiol. Prot.* **39** R1–R17

Fichman Y and Mittler R 2020 Rapid systemic signaling during abiotic and biotic stresses: is the ROS wave master of all trades? *Plant J. Cell Mol. Biol.* **102** 887–96

Fujishima Y *et al* 2020 Assessment of chromosome aberrations in large Japanese field mice (*Apodemus speciosus*) in Namie Town, Fukushima *Int. J. Radiat. Biol.* **98** 1159–67

Fuller N, Smith J T, Nagorskaya L L, Gudkov D I and Ford A T 2017 Does Chernobyl-derived radiation impact the developmental stability of *Asellus aquaticus* 30 years on? *Sci. Total Environ.* **576** 242–50

Fuma S *et al* 2017 Radiocaesium contamination and dose rate estimation of terrestrial and freshwater wildlife in the exclusion zone of the Fukushima Dai-ichi Nuclear Power Plant accident *J. Environ. Radioact.* **171** 176–88

Galeriu D, Beresford N A, Takeda H, Melintescu A and Crout N M J 2003 Towards a model for the dynamic transfer of tritium and carbon in mammals *Radiat. Prot. Dosim.* **105** 387–90

Galván I, Bonisoli-Alquati A, Jenkinson S, Ghanem G, Wakamatsu K, Mousseau T A and Møller A P 2014 Chronic exposure to low-dose radiation at Chernobyl favors adaptation to oxidative stress in birds *Funct. Ecol.* **28** 1387–403

Garnier-Laplace J, Della-Vedova C, Gilbin R, Copplestone D, Hingston J and Ciffroy P 2006 First derivation of predicted-no-effect values for freshwater and terrestrial ecosystems exposed to radioactive substances *Environ. Sci. Technol.* **40** 6498–505

Garnier-Laplace J, Beaugelin-Seiller K and Hinton T G 2011 Fukushima wildlife dose reconstruction signals ecological consequences *Environ. Sci. Technol.* **45** 5077–8

Garnier-Laplace J, Geras'kin S, Della-Vedova C, Beaugelin-Seiller K, Hinton T G, Real A and Oudalova A 2013 Are radiosensitivity data derived from natural field conditions consistent with data from controlled exposures? A case study of Chernobyl wildlife chronically exposed to low dose rates *J. Environ. Radioact.* **121** 12–21

Gashchak S, Gulyaichenko Y, Beresford N and Wood M 2017 European bison (*Bison bonasus*) in the Chornobyl Exclusion Zone (Ukraine) and prospects for its revival *Proc. Theriol. Sch.* **2017** 58–66

Gashchak S, Gulyaichenko Y, Beresford N A and Wood M D 2016 Brown bear (*Ursus arctos L.*) in the Chornobyl Exclusion Zone *Proc. Theriol. Sch.* **2016** 71–84

Georgieva M, Rashydov N M and Hajduch M 2017 DNA damage, repair monitoring and epigenetic DNA methylation changes in seedlings of Chernobyl soybeans *DNA Repair* **50** 14–21

Geras'kin S, Oudalova A, Dikareva N, Spiridonov S, Hinton T, Chernonog E and Garnier-Laplace J 2011 Effects of radioactive contamination on Scots pines in the remote period after the Chernobyl accident *Ecotoxicology* **20** 1195–208

Geras'kin S A 2016 Ecological effects of exposure to enhanced levels of ionizing radiation *J. Environ.Radioact.* **162** 347–57

Geras'kin S A, Fesenko S V and Alexakhin R M 2008 Effects of non-human species irradiation after the Chernobyl NPP accident *Environ. Int.* **34** 880–97

Gilbin R *et al* 2021 An updated strategic research agenda for the integration of radioecology in the European radiation protection research *J. Environ. Radioact.* **237** 106697

Goldberg A D, Allis C D and Bernstein E 2007 Epigenetics: a landscape takes shape *Cell* **128** 635–8

Gombeau K, Bourdineaud J-P, Ravanat J-L, Armant O, Camilleri V, Cavalie I, Floriani M and Adam-Guillermin C 2017 Epigenetic, histopathological and transcriptomic effects following exposure to depleted uranium in adult zebrafish and their progeny *Aquat. Toxicol.* **184** 14–25

Guerrero-Bosagna C, Covert T R, Haque M M, Settles M, Nilsson E E, Anway M D and Skinner M K 2012 Epigenetic transgenerational inheritance of vinclozolin induced mouse adult onset disease and associated sperm epigenome biomarkers *Reprod. Toxicol.* **34** 694–707

Guerrero-Bosagna C and Jensen P 2015 Globalization, climate change, and transgenerational epigenetic inheritance: will our descendants be at risk? *Clin. Epigenetics* **7** 3

Guirandy N *et al* 2019 Adverse effects induced by chronic gamma irradiation in progeny of adult fish not affecting parental reproductive performance *Environ. Toxicol. Chem.* **38** 2556–67

Gurung R D, Taira W, Sakauchi K, Iwata M, Hiyama A and Otaki J M 2019 Tolerance of high oral doses of nonradioactive and radioactive caesium chloride in the Pale Grass Blue Butterfly *Zizeeria maha Insects* **10** 290

Hanson M A and Skinner M K 2016 Developmental origins of epigenetic transgenerational inheritance *Environ. Epigenetics* **2** dvw002

Harrison F L and Anderson S L 1996 *Taxonomic and Developmental Aspects of Radiosensitivity* (Livermore, CA: Lawrence Livermore National Lab. (LLNL))

Higley K A, Domotor S L and Antonio E J 2003 A kinetic–allometric approach to predicting tissue radionuclide concentrations for biota *J. Environ. Radioact.* **66** 61–74

Higley K, Ruedig E, Gomez-Fernandez M, Caffrey E, Jia J, Comolli M and Hess C 2015 Creation and application of voxelised dosimetric models, and a comparison with the current methodology as used for the International Commission on Radiological Protection's Reference Animals and Plants *Anal. ICRP.* **44** 313–30

Hinton T G, Alexakhin R, Balonov M, Gentner N, Hendry J, Prister B, Strand P and Woodhead D 2007 Radiation-induced effects on plants and animals: findings of the United Nations Chernobyl forum *Health Phys.* **93** 427–40

Hirose K 2020 Atmospheric effects of Fukushima nuclear accident: a review from a sight of atmospheric monitoring *J. Environ. Radioact.* **218** 106240

Hirth G A, Johansen M P, Carpenter J G, Bollhöfer A and Beresford N A 2017 Whole-organism concentration ratios in wildlife inhabiting Australian uranium mining environments *J. Environ. Radioact.* **178–179** 385–93

Hiyama A, Nohara C, Kinjo S, Taira W, Gima S, Tanahara A and Otaki J M 2012 The biological impacts of the Fukushima nuclear accident on the pale grass blue butterfly *Sci. Rep.* **2** 570

Hiyama A, Nohara C, Taira W, Kinjo S, Iwata M and Otaki J M 2013 The Fukushima nuclear accident and the pale grass blue butterfly: evaluating biological effects of long-term low-dose exposures *BMC Evol. Biol.* **13** 25

Holiaka D, Fesenko S, Kashparov V, Protsak V, Levchuk S and Holiaka M 2020 Effects of radiation on radial growth of Scots pine in areas highly affected by the Chernobyl accident *J. Environ. Radioact.* **222** 13

Holm L-E, Hubbard L, Larsson C-M and Sundell-Bergman S 2002 Radiological protection of the environment from the Swedish point of view *J. Radiol. Prot.* **22** 235–47

Horemans N, Nauts R, Vives i Batlle J, Van Hees M, Jacobs G, Voorspoels S, Gaschak S, Nanba K and Saenen E 2018 Genome-wide DNA methylation changes in two Brassicaceae species sampled alongside a radiation gradient in Chernobyl and Fukushima *J. Environ. Radioact.* **192** 405–16

Horemans N, Spurgeon D J, Lecomte-Pradines C, Saenen E, Bradshaw C, Oughton D, Rasnaca I, Kamstra J H and Adam-Guillermin C 2019 Current evidence for a role of epigenetic mechanisms in response to ionizing radiation in an ecotoxicological context *Environ. Pollut.* **251** 469–83

Horemans N, Vanhoudt N, Janssens M, Van Chaze B, Wannijn J, Van Hees M and Vandenhove H 2011 On the nature and timing of oxygen radical production following exposure of *Arabidopsis thaliana* leaves to uranium, cadmium or a combination of both stressors *Radioprotection* **46** S491–6

Hosseini A, Beresford N A, Brown J E, Jones D G, Phaneuf M, Thørring H and Yankovich T 2010 Background dose-rates to reference animals and plants arising from exposure to naturally occurring radionuclides in aquatic environments *J. Radiol. Prot.* **30** 235–64

Howard B J, Beresford N A, Andersson P, Brown J E, Copplestone D, Beaugelin-Seiller K, Garnier-Laplace J, Howe P, Oughton D and Whitehouse P 2010 Protection of the environment from ionising radiation in a regulatory context—an overview of the PROTECT coordinated action project *J. Radiol. Prot.* **30** 195–214

Howard B J, Fesenko S, Balonov M, Prohl G and Nakayama S 2017 A comparison of remediation after the Chernobyl and Fukushima Daiichi accidents *Radiat. Prot. Dosimetry.* **173** 170–6

Hunt C R, Ramnarain D, Horikoshi N, Iyengar P, Pandita R K, Shay J W and Pandita T K 2013 Histone modifications and DNA double-strand break repair after exposure to ionizing radiations *Radiat. Res.* **179** 383–92

Hurem S, Martin L M, Lindeman L, Brede D A, Salbu B, Lyche J L, Alestrom P and Kamstra J H 2018 Parental exposure to gamma radiation causes progressively altered transcriptomes linked to adverse effects in zebrafish offspring *Environ. Pollut.* **234** 855–63

International Atomic Energy Agency (IAEA) 1986 *Summary Report on the Post-accident Review Meeting on the Chernobyl Accident* (Vienna: International Atomic Energy Agency)

International Atomic Energy Agency (IAEA) 1992 *Effects of Ionizing Radiation on Plants and Animals at Levels Implied by Current Radiation Protection Standards Technical Reports Series No. 332* (Vienna: International Atomic Energy Agency)

International Atomic Energy Agency (IAEA) 2006a *Environmental Consequences of the Chernobyl Accident and their Remediation: Twenty Years of Experience* (Vienna: International Atomic Energy Agency)

International Atomic Energy Agency (IAEA) 2006b *Fundamental Safety Principles—Safety Fundamentals. IAEA Safety Standards No. SF-1* (Vienna: International Atomic Energy Agency) https://www-pub.iaea.org/MTCD/publications/PDF/Pub1273_web.pdf

International Atomic Energy Agency (IAEA) 2012 Modelling radiation exposure and radionuclide transfer for non-human species *Report of the Biota Eorking Hroup of EMRAS Theme 3* (Vienna: International Atomic Energy Agency) Environmental modelling for radiation safety (EMRAS) programme. IAEA-TECDOC-1678 http://www-ns.iaea.org/downloads/rw/projects/emras/final-reports/biota-report.pdf

International Atomic Energy Agency (IAEA) 2014 *Handbook of Parameter Values for the Prediction of Radionuclide Transfer to Wildlife* (Vienna: International Atomic Energy Agency) IAEA-TRS-479 https://www-pub.iaea.org/MTCD/Publications/PDF/Trs479_web.pdf

International Atomic Energy Agency (IAEA) 2015 *The Fukushima Daiichi Accident* Non-serial Publications (Vienna: International Atomic Energy Agency) https://iaea.org/publications/10962/the-fukushima-daiichi-accident

International Atomic Energy Agency (IAEA) 2021 Radiological environmental protection of wildlife: modelling the exposure and effects *A Joint Summary Report by MODARIA I Working Groups 8 & 9, and MODARIA II Working Group 5* IAEA-TECDOC-1986 (Vienna: International Atomic Energy Agency)

International Commission on Radiological Protection (ICRP) 2007 The 2007 recommendations of the International Commission on Radiological Protection, ICRP Publication 103 *Ann. ICRP* **37(2–4)** https://icrp.org/publication.asp?id=ICRP%20Publication%20103

International Commission on Radiological Protection (ICRP) 2008 Environmental protection: the concept and use of Reference Animals and Plants, ICRP Publication 108 *Ann. ICRP* **38(4–6)** https://icrp.org/publication.asp?id=ICRP%20Publication%20108

International Commission on Radiological Protection (ICRP) 2009 Environmental protection: transfer parameters for Reference Animals and Plants, ICRP Publication 114 *Ann. ICRP* **39(6)** https://icrp.org/publication.asp?id=ICRP%20Publication%20114

International Commission on Radiological Protection (ICRP) 2014 Protection of the environment under different exposure situations, ICRP Publication 124 *Ann. ICRP* **43(1)** https://icrp.org/publication.asp?id=ICRP%20Publication%20124

International Commission on Radiological Protection (ICRP) 2017 Dose coefficients for non-human biota environmentally exposed to radiation, ICRP Publication 136 *Ann. ICRP* **46(2)** https://doi.org/10.1177%2F0146645317728022

International Commission on Radiological Protection (ICRP) 1991 1990 Recommendations of the International Commission on Radiological Protection, ICRP Publication 60 *Ann. ICRP* **21(1-3)** https://journals.sagepub.com/doi/pdf/10.1177/ANIB_21_1-3

International X-ray and Radium Protection Committee 1929 International recommendations for x-ray and radium protection *Ann. ICRP* **OS–1** 62–73

Jackson D, Copplestone D, Stone D M and Smith G M 2005 Terrestrial invertebrate population studies in the Chernobyl exclusion zone, Ukraine *Radioprotection* **1** 857–63

Jimenez-Chillaron J C, Nijland M J, Ascensao A A, Sardao V A, Magalhaes J, Hitchler M J, Domann F E and Oliveira P J 2015 Back to the future: transgenerational transmission of xenobiotic-induced epigenetic remodeling *Epigenetics* **10** 259–73

Johansen M P *et al* 2012 Assessing doses to terrestrial wildlife at a radioactive waste disposal site: inter-comparison of modelling approaches *Sci. Tot. Environ.* **427–428** 238–46

Kamstra J H, Hurem S, Martin L, Lindeman L C, Legler J, Oughton D, Salbu B, Brede D A, Lyche J L and Alestrom P 2018 Ionizing radiation induces transgenerational effects of DNA methylation in zebrafish *Sci. Rep.* **8** 13

Karuppanapandian T, Moon J C, Kim C, Manoharan K and Kim W 2011 Reactive oxygen species in plants: their generation, signal transduction, and scavenging mechanisms *Aust. J. Crop Sci.* **5** 709–25

Kashparov V, Levchuk S, Zhurba M, Protsak V, Khomutinin Y, Beresford N A and Chaplow J S 2018 Spatial datasets of radionuclide contamination in the Ukrainian Chernobyl Exclusion Zone *Earth Syst. Sci. Data* **10** 339–53

Kelleher K, McGinnity P, Howard B J, Boyer P, Vidal M and Bildstein O 2022 The use of the IAEA MARIS database in determining the variability of sediment distribution coefficients in the marine environment and potential implications for marine dispersion modelling *J. Radiol. Prot.* **42** 030501

Kinase S 2008 Voxel-based frog phantom for internal dose evaluation *J. Nucl. Sci. Technol.* **45** 1049–52

Kleiber M 1932 Body size and metabolism *Hilgardia* **6** 315–53

Kovalchuk I, Abramov V, Pogribny I and Kovalchuk O 2004 Molecular aspects of plant adaptation to life in the Chernobyl zone *Plant Physiol.* **135** 357–63

Kovalchuk O, Burke P, Arkhipov A, Kuchma N, James S J, Kovalchuk I and Pogribny I 2003 Genome hypermethylation in *Pinus silvestris* of Chernobyl—a mechanism for radiation adaptation? *Mutat. Res.—Fundam. Mol. Mech. Mutag.* **529** 13–20

Krivolutsky D 1996a Dynamics of biodiversity and ecosystems under conditions of radioactive contamination *Dokl. Bolg. Akad. Nauk.* **347** 166–8

Krivolutsky D A 1996b Soil fauna as bioindicator of radioactive pollution *Biodiversity of the Four Corners Area* (Biodiversity Foundation for Africa)

Krivolutzkii D A and Pokarzhevskii A D 1992 Effects of radioactive fallout on soil animal populations in the 30-km zone of the Chernobyl atomic power-station *Sci. Total Environ.* **112** 69–77

Kryshev I I, Sazykina T G and Beresford N A 2005 Effects on wildlife ed J T Smith and N A Beresford *Chernobyl Catastrophe and Consequences* (Chichester: Praxis Publisher/Springer) pp 267–87

Laanen P *et al* 2021 Changes in DNA methylation in *Arabidopsis thaliana* plants exposed over multiple generations to gamma radiation *Front. Plant Sci.* **12** 611783

Li Z G, Chen X, Li S P and Wang Z C 2015 Effect of nickel chloride on *Arabidopsis* genomic DNA and methylation of 18S rDNA *Electron. J. Biotechnol.* **18** 51–7

Lind M I and Spagopoulou F 2018 Evolutionary consequences of epigenetic inheritance *Heredity* **121** 205–9

Lindeman L C *et al* 2019 Gamma radiation induces locus specific changes to histone modification enrichment in zebrafish and Atlantic salmon *PLoS One* **14** 15

Linsley G 1990 Protection of natural ecosystems. The impact of radiation from waste disposal practices: a collation of recent review studies and assessments *J. Radiol. Prot.* **10** 153–6

Lourenco J, Marques S, Carvalho F P, Oliveira J, Malta M, Santos M, Goncalves F, Pereira R and Mendo S 2017 Uranium mining wastes: the use of the Fish Embryo Acute Toxicity Test (FET) test to evaluate toxicity and risk of environmental discharge *Sci. Total Environ.* **605** 391–404

Lourenco J, Mendo S and Pereira R 2016 Radioactively contaminated areas: bioindicator species and biomarkers of effect in an early warning scheme for a preliminary risk assessment *J. Hazard. Mater.* **317** 503–42

Ludovici G M, Oliveira de Souza S, Chierici A, Cascone M G, d'Errico F and Malizia A 2020 Adaptation to ionizing radiation of higher plants: from environmental radioactivity to Chernobyl disaster *J. Environ. Radioact.* **222** 106375

MacDonald C R and Laverock M J 1998 Radiation exposure and dose to small mammals in radon-rich soils *Arch. Environ. Contam. Toxicol.* **35** 109–20

Marczylo E L, Jacobs M N and Gant T W 2016 Environmentally induced epigenetic toxicity: potential public health concerns *Crit. Rev. Toxicol.* **46** 676–700

McCarrey J R 2012 The epigenome as a target for heritable environmental disruptions of cellular function *Mol. Cell. Endocrinol.* **354** 9–15

Meeks H N, Wickliffe J K, Hoofer S R, Chesser R K, Rodgers B E and Baker R J 2007 Mitochondrial control region variation in bank voles (*Clethrionomys glareolus*) is not related to Chernobyl radiation exposure *Environ. Toxicol. Chem.* **26** 361–9

Mirbahai L and Chipman J K 2014 Epigenetic memory of environmental organisms: a reflection of lifetime stressor exposures *Mutat. Res. Genet. Toxicol. Environ. Mutagen.* **764** 10–7

Mirouze M and Paszkowski J 2011 Epigenetic contribution to stress adaptation in plants *Curr. Opin. Plant Biol.* **14** 267–74

Mittler R and Blumwald E 2015 The roles of ROS and ABA in systemic acquired acclimation *Plant Cell* **27** 64–70

Moller A P and Mousseau T A 2009 Reduced abundance of insects and spiders linked to radiation at Chernobyl 20 years after the accident *Biol. Lett.* **5** 356–9

Moller A P and Mousseau T A 2016 Are organisms adapting to ionizing radiation at Chernobyl? *Trends Ecol. Evol.* **31** 281–9

Moller A P and Mousseau T A 2018 Reduced colonization by soil invertebrates to irradiated decomposing wood in Chernobyl *Sci. Total Environ.* **645** 773–9

Møller A P and Mousseau T A 2013 Assessing effects of radiation on abundance of mammals and predator–prey interactions in Chernobyl using tracks in the snow *Ecol. Indic.* **26** 112–6

Mothersill C *et al* 2018 When a duck is not a duck; a new interdisciplinary synthesis for environmental radiation protection *Environ. Res.* **162** 318–24

Mothersill C E *et al* 2020 From tangled banks to toxic bunnies; a reflection on the issues involved in developing an ecosystem approach for environmental radiation protection *Int. J. Radiat. Biol.* **98** 1185–200

Murat El Houdigui S, Adam-Guillermin C and Armant O 2020 Ionising radiation induces promoter DNA hypomethylation and perturbs transcriptional activity of genes involved in morphogenesis during gastrulation in Zebrafish *Int. J. Mol. Sci.* **21** 4014

National Council on Radiological Protection and Measurements (NCRP) 1991 Effects of ionising radiation on aquatic organisms *NCRP Report 109* (Bethesda, MD: NCRP)

Okano T, Ishiniwa H, Onuma M, Shindo J, Yokohata Y and Tamaoki M 2016 Effects of environmental radiation on testes and spermatogenesis in wild large Japanese field mice (*Apodemus speciosus*) from Fukushima *Sci. Rep.* **6** 8

Omar-Nazir L, Shi X P, Moller A, Mousseau T, Byun S, Hancock S, Seymour C and Mothersill C 2018 Long-term effects of ionizing radiation after the Chernobyl accident: possible contribution of historic dose *Environ. Res.* **165** 55–62

Orekhova N A and Modorov M V 2017 East Urals Radioactive Trace: dose-dependent functional-metabolic effects in the myocardium of the pygmy wood mouse (*Apodemus uralensis*) taking into account population size *J. Environ. Radioact.* **175** 15–24

Oughton D H and Strand P 2004 The Oslo consensus conference on protection of the environment *J. Environ. Radioact.* **74** 7–17

Parisot F, Bourdineaud J P, Plaire D, Adam-Guillermin C and Alonzo F 2015 DNA alterations and effects on growth and reproduction in *Daphnia magna* during chronic exposure to gamma radiation over three successive generations *Aquat. Toxicol.* **163** 27–36

Pentreath R J 1999 A system for radiological protection of the environment: some initial thoughts and ideas *J. Radiol. Prot.* **19** 117–28

Pyle G, Swanson S and Lehmkuhl D 2002 Toxicity of uranium mine receiving waters to early life stage fathead minnows (*Pimephales promelas*) in the laboratory *Environ. Pollut.* **116** 243–55

Real A, Sundell-Bergman S, Knowles J F, Woodhead D S and Zinger I 2004 Effects of ionising radiation exposure on plants, fish and mammals: relevant data for environmental radiation protection *J. Radiol. Prot.* **24** A123–37

Rhodes O E Jr. *et al* 2020 Integration of ecosystem science into radioecology: a consensus perspective *Sci. Tot. Environ.* **740** 140031

Ruedig E, Caffrey E A, Hess C and Higley K A 2014 Monte Carlo derived absorbed fractions for a voxelized model of *Oncorhynchus Mykiss*, a Rainbow Trout *Radiat. Environ. Biophys.* **53** 581–7

Ruedig E, Beresford N A, Gomez Fernandez M E and Higley K 2015 A comparison of the ellipsoidal and voxelized dosimetric methodologies for internal, heterogeneous radionuclide sources *J. Environ. Radioact.* **140** 70–7

Saaristo M *et al* 2018 Direct and indirect effects of chemical contaminants on the behaviour, ecology and evolution of wildlife *Proc. Biol. Sci.* **285** 20181297

Sakauchi K, Taira W, Toki M, Tsuhako M, Umetsu K and Otaki J M 2021 Nutrient imbalance of the host plant for larvae of the Pale Grass Blue Butterfly may mediate the field effect of low-dose radiation exposure in Fukushima: dose-dependent changes in the sodium content *Insects* **12** 23

Sato I, Sasaki J, Satoh H, Natsuhori M, Murata T and Okada K 2020 Assessments of DNA damage and radiation exposure dose in cattle living in the contaminated area caused by the Fukushima nuclear accident *Bull. Environ. Contam. Toxicol.* **105** 496–501

Sazykina T and Kryshev I I 2006 Radiation effects in wild terrestrial vertebrates—the EPIC collection *J. Environ. Radioact.* **88** 11–48

Schultz C, Unrine J, Tsyusko O and Svendsen C 2016 The transgenerational toxicity of Ag ions, nanopaticles and sulphadised nanoparticles across 10 *Caenorhabditis elegans* generations *Proc. R. Soc. B-Biol. Sci.* **283** 20152911

Sharma A *et al* 2017 Across the tree of life, radiation resistance is governed by antioxidant Mn^{2+}, gauged by paramagnetic resonance *Proc. Natl. Acad. Sci. U. S. A.* **114** E9253–60

Singer A C, Shaw H, Rhodes V and Hart A 2016 Review of antimicrobial resistance in the environment and its relevance to environmental regulators *Front. Microbiol.* **7** 1728

Søvik A *et al* 2017 Final Report of WP3 Activities. COMET Deliverable. D-N 3.7. https://radioecology-exchange.org/sites/default/files/files/COMET%20Deliverable%20D3_7%20WP3%20Final%20report_PU%20version.pdf

Smith J T, Willey N J and Hancock J T 2012 Low dose ionizing radiation produces too few reactive oxygen species to directly affect antioxidant concentrations in cells *Biol. Lett.* **8** 594–7

Smodis B, Strok M and Cerne M 2014 Radioecology around a closed uranium mine *J. Radioanal. Nucl. Chem.* **299** 765–71

Song Y, Xie L, Lee Y, Brede D A, Lyne F, Kassaye Y, Thaulow J, Caldwell G, Salbu B and Tollefsen K E 2020 Integrative assessment of low-dose gamma radiation effects on *Daphnia magna* reproduction: toxicity pathway assembly and AOP development *Sci. Total Environ.* **705** 13

Sparrow A H and Miksche J P 1961 Correlation of nuclear volume and DNA content with higher plant tolerance to chronic radiation *Science* **134** 282–3

Spurgeon D J 2020 Higher than … or lower than …? Evidence for the validity of the extrapolation of laboratory toxicity test results to predict the effects of chemicals and ionising radiation in the field *J. Environ. Radioact.* **211** 14

Stabin M G, Peterson T E, Holburn G E and Emmons M A 2006 Voxel-based mouse and rat models for internal dose calculations *J. Nucl. Med.* **47** 655–9 https://pubmed.ncbi.nlm.nih.gov/16595500/

Stark K *et al* 2015 Predicting exposure of wildlife in radionuclide contaminated wetland ecosystems *Env. Poll.* **196** 201–13

Steinhauser G, Brandl A and Johnson T E 2014 Comparison of the Chernobyl and Fukushima nuclear accidents: a review of the environmental impacts *Sci. Total Environ.* **470** 800–17

Strand P, Aono T, Brown J E, Garnier-Laplace J, Hosseini A, Sazykina T, Steenhuisen F and Batlle J V I 2014 Assessment of Fukushima-derived radiation doses and effects on wildlife in Japan *Environ. Sci. Technol. Lett.* **1** 198–203

Thaulow J, Song Y, Lindeman L C, Kamstra J H, Lee Y, Xie L, Alestrom P, Salbu B and Tollefsen K E 2020 Epigenetic, transcriptional and phenotypic responses in *Daphnia magna* exposed to low-level ionizing radiation *Environ. Res.* **190** 12

Thompson P M 1988 Environmental monitoring for radionuclides in marine ecosystems: are species other than man protected adequately? *J. Environ. Radioact.* **7** 275–83

Tomczak W, Boyer P, Krimissa M and Radakovitch O 2019 Kd distributions in freshwater systems as a function of material type, mass-volume ratio, dissolved organic carbon and pH *J. Appl. Geochem.* **105** 68–77

Trenfield M A, Harford A J, Mooney T, Ellis M, Humphrey C and van Dam R A 2019 Integrating laboratory and field studies to assess impacts of discharge from a uranium mine and validate a water quality guideline value for magnesium *Integr. Environ. Assess Manag.* **15** 64–76

Trijau M, Asselman J, Armant O, Adam-Guillermin C, De Schamphelaere K A C and Alonzo F 2018 Transgenerational DNA methylation changes in *Daphnia magna* exposed to chronic γ irradiation *Environ. Sci. Technol.* **52** 4331–9

United Nations Scientific Committee on the Effects of Atomic Radiation (UNSCEAR) 1996 *Sources and Effects of Ionizing Radiation Report to the General Assembly with Scientific Annex* (New York: United Nations)

United Nations Scientific Committee on the Effects of Atomic Radiation (UNSCEAR) 2008 Annex E, Effects of Ionizing Radiation on Non-human Biota https://unscear.org/unscear/en/publications/2008_2.html

United Nations Scientific Committee on the Effects of Atomic Radiation (UNSCEAR) 2013 Sources, Effects and Risks of Ionizing Radiation, vol 1 https://unscear.org/unscear/en/publications/2013_1.html

United Nations Scientific Committee on the Effects of Atomic Radiation (UNSCEAR) 2020 2020/2021 Report, Sources, Effects and Risks of Ionizing Radiations, 4 vol https://unscear.org/unscear/en/publications/2020_1.html

United States Department of Energy (USDOE) 2002 A graded approach for evaluating radiation doses to aquatic and terrestrial biota *Technical Standard DOE-STD-1153–2002* (Washington DC: United States Department of the Environment) https://standards.doe.gov/standards-documents/1100/1153-AStd-2002

United States Department of Energy (USDOE) 2004 RESRAD-BIOTA: a tool for implementing a graded approach to biota dose evaluation *User's Guide, Version 1. U.S.* (Springfield VA: Department of Commerce, Technology Administration, National Technical Information Service) https://resrad.evs.anl.gov/docs/RESRAD-BIOTA_Manual_Version_1.pdf

van Straalen N M and Denneman C A J 1989 Ecotoxicological evaluation of soil quality criteria *Ecotox. Environ. Safe.* **18** 241–51

Vandenhove H 2002 European sites contaminated by residues from the ore-extracting and -processing industries *Int. Congr. Ser.* **1225** 307–15

Vasilenko E K, Aladova E E, Gorelov M V, Knyazev V A, Kolupaev D V and Romanov S A 2020 The radiological environment at the Mayak PA site and radiation doses to individuals involved in emergency and remediation operations after the 'Kyshtym Accident' in 1957 *J. Radiol. Prot.* **40** R23–45

Verhoeven K J F, Jansen J J, van Dijk P J and Biere A 2010 Stress-induced DNA methylation changes and their heritability in asexual dandelions *New Phytol.* **185** 1108–18

Verhoeven K J F and van Gurp T P 2012 Transgenerational effects of stress exposure on offspring phenotypes in apomictic dandelion *PLoS One* **7** e38605

Verhoeven K J F, Vonholdt B M and Sork V L 2016 Epigenetics in ecology and evolution: what we know and what we need to know *Mol. Ecol.* **25** 1631–8

Vinogradov A E and Chubinishvili A T 1999 Genome reduction in a hemiclonal frog *Rana esculenta* from radioactively contaminated areas *Genetics* **151** 1123–5

Vives i Batlle J *et al* 2007 Inter-comparison of absorbed dose rates for non-human biota *Radiat. Environ. Biophys.* **46** 349–73

Vives i Batlle J, Wilson R C, Watts S J, Jones S R, McDonald P and Vives-Lynch S 2008 Dynamic model for the assessment of radiological exposure to marine biota *J. Environ. Radioact.* **99** 1711–30

Vives i Batlle J *et al* 2011 The estimation of absorbed dose rates for non-human biota: an extended intercomparison *Radiat. Environ. Biophys.* **50** 231–51

Vives i Batlle J 2015 Exposures and effects in the marine environment after the Fukushima accident *Ann. ICRP* **44** 331–46

Vives i Batlle J 2016 Dynamic modelling of radionuclide uptake by marine biota: application to Fukushima assessment *J. Environ. Radioact.* **151** 502–11

Vives i Batlle J *et al* 2016 Inter-comparison of dynamic models for radionuclide transfer to marine biota in a Fukushima accident scenario *J. Environ. Radioact.* **153** 31–50

Vives i Batlle J 2016 Impact of the Fukushima accident on marine biota, five years later *Integr. Environ. Asses.* **12** 654–8

Vives i Batlle J, Ulanovsky A V and Copplestone D 2017 A method for assessing exposure of terrestrial wildlife to environmental radon (^{222}Rn) and thoron (^{220}Rn) *Sci. Tot. Environ.* **605–606** 569–77

Vives i Batlle J *et al* 2018 Marine radioecology after the Fukushima Dai-ichi nuclear accident: are we better positioned to understand the impact of radionuclides in marine ecosystems? *Sci. Total Environ.* **618** 80–92

Vives i Batlle J, Sazykina T, Kryshev A, Wood M D, Smith K, Copplestone D and Biermans G 2020 Modelling the effects of ionising radiation on a vole population from the Chernobyl Red Forest in an ecological context *Ecol. Model.* **438** 109306

Voitovich A 2001 Bone tumor in frog (*Rana temporaria* L.) in radionuclide contaminated environment (in Russian) *Proc. Natl. Acad. Sci. Belarus.* **45** 91–4

Voitovich A and Afonin V 2000 Natural populations of small vertebrates in the ecologic-genetic monitoring system (in Russian) *Ecol. Ration. Land Use e Centuries Interface* **2** 35–6

Walker C H, Hopkin S P, Sibly R M and Peakall D B 2006 *Principles of Ecotoxicology* 3rd edn (London: Taylor & Francis)

Whitehead A, Pilcher W, Champlin D and Nacci D 2012 Common mechanism underlies repeated evolution of extreme pollution tolerance *Proc. Biol. Sci.* **279** 427–33

Willey N J 2014 Soil to plant transfer of radionuclides: predicting the fate of multiple radioisotopes in plants *J. Environ. Radioact.* **133** 31–4

Willey N J 2010 Phylogeny can be used to make useful predictions of soil-to-plant transfer factors for radionuclides *Radiat. Environ. Biophys.* **49** 613–23

Wood M D, Beresford N A, Barnett C L, Copplestone D and Leah R T 2009 Assessing radiation impact at a protected coastal sand dune site: an intercomparison of models for estimating the radiological exposure of non-human biota *J. Environ. Radioact.* **100** 1034–52

Xie L, Solhaug K A, Song Y, Brede D A, Lind O C, Salbu B and Tollefsen K E 2019 Modes of action and adverse effects of gamma radiation in an aquatic macrophyte *Lemna minor Sci. Total Environ.* **680** 23–34

Yankovich T L *et al* 2010 An international model validation exercise on radionuclide transfer and doses to freshwater biota *J. Radiol. Prot.* **30** 299–340

Yoschenko V, Nanba K, Yoshida S, Watanabe Y, Takase T, Sato N and Keitoku K 2016 Morphological abnormalities in Japanese red pine (*Pinus densiflora*) at the territories contaminated as a result of the accident at Fukushima Dai-Ichi Nuclear Power Plant *J. Environ. Radioact.* **165** 60–7

Zelena L, Sorochinsky B, von Arnold S, van Zyl L and Clapham D H 2005 Indications of limited altered gene expression in *Pinus sylvestris* trees from the Chernobyl region *J. Environ. Radioact.* **84** 363–73

Chapter 13

International organisations influencing radiological protection and recommendations on protection against ionising radiation

C Clement

This chapter begins with a brief history of radiological protection (section 13.1), leading to a presentation of the many international organisations that develop and influence radiological protection today (section 13.2). This is background for the second half (section 13.3) that introduces the System of Radiological Protection now in place that forms the basis of standards, legislation, and practice of radiological protection worldwide. The epilogue (section 13.4) describes the evolution of radiological protection.

13.1 History of radiological protection

Radiological protection began not long after the discovery of x-rays by Wilhelm Röntgen in November 1895 (Röntgen 1895). Revolutionary, practical implications were immediately obvious. From Röntgen (translated from the original German to English):

> *If one holds the hand between the discharge apparatus and the screen, one sees the darker shadows of the hand bones in the only slightly dark silhouette of the hand.*

There is little doubt as to why there was a rapid worldwide spread of the use of x-rays in medicine: for the first time it was possible to look inside the human body without surgical intervention. Experiments in radiation therapy also began almost immediately.

Radiation therapy was tested as early as 1896 (Belot 1904), and x-ray units were used in military field hospitals as early as 1897 (Churchill 1898). As one might imagine, these very early units, especially used in field hospitals, were not always reliable. From Churchill:

doi:10.1088/978-0-7503-2424-3ch13

> *On the 12th of August, Lieut.-Colonel J. Lamb, who had been wounded on the night of the 26th of July, died. An early amputation might have saved his life; but this was postponed in the expectation that the Rontgen Rays would enable the bullet to be extracted. The Rays arrived from India after some delay. When they reached Malakand, the experiment was at once made. It was found, however, that the apparatus had been damaged in coming up, and no result was obtained.*

With Curie's discovery of radium in 1898 (Curie 1898) use of radiation in medicine continued to increase, as did reports of radiation-induced damage in practitioners and patients.

Within months of the discovery of x-rays, early investigators were observing damage to the skin including erythema, dermatitis, and ulceration (Drury 1896, Leppin 1896, Grubbé 1933). From Drury:

> *… he was again exposed, six days after—namely, May 28th, this time for an hour and a-half … After the patient left the laboratory he again felt great nausea, and soon felt so weak that he had to get brandy on his way home. Next evening he noticed the abdomen slightly red, as if sun-burned; a patch about 1| inch square, just below the umbilicus, being redder than the rest … eighteen days after the second exposure, I saw the sore for the first time. It had begun to cause a great deal of discomfort from the abundant discharge pouring from it. I found a patch 7| inches across by 8¼ inches long … It had all the appearance of an acute irritative eczema, with exfoliated epidermis, and a profuse sero-purulent discharge … in about ten days the exudation began to lessen: the epithelium began to grow in from the edge, exactly as in a burn …*

The first published advice on radiological protection appeared in an article by Wolfram Fuchs in the Western Electrician on 12 December 1896 (Fuchs 1896), just one year after Röntgen's accidental discovery. From Fuchs:

> *In the last nine months we have applied the X-ray to all parts of the body, and out of 1,400 cases under our supervision we have had but four instances of the slow-healing burns which have lately attracted considerable attention through the columns of the press.*

Fuchs saw the value in the use of x-rays in medicine despite the harmful effects and said as much in his article. However, his next sentence made it clear that he was one of the pioneers of radiological protection:

> *… the injury may be regarded as slight in comparison with the benefits resulting from this wonderful discovery. However, it is desirable, of course, to prevent the inconvenience and pain of these 'sunburns' …*

To this end, Fuchs made three specific suggestions:

> *First—Shorten the exposure to a minimum.*
> *Second—Place the tube not nearer than 12 inches from the body.*
> *Third—Rub vaseline well into the skin and leave a coating on the part of the body to be exposed; also, use glass plates for the protection of the parts not to be exposed.*

The ideas behind these practical suggestions are still used today to optimise protection: reduce time exposed, increase distance from the source, and use shielding.

Despite the rapid acknowledgement of harmful effects and initial suggestions to improve practice, it was three decades before international bodies were formed to address radiological protection. At the first International Congress of Radiology (ICR), held in London in 1925, the International X-ray Unit Committee, now the International Commission on Radiation Units and Measurements (ICRU), was formed. Discussions were also held regarding creation of an international radiological protection committee, and the International X-ray and Radium Protection Committee (IXRPC) was created in 1928 at the second ICR in Stockholm. The IXRPC became the International Commission on Radiological Protection (ICRP) in 1950.

Initially, radiological protection recommendations (ICR 1929) were aimed primarily at the protection of 'x-ray and radium workers' in medical facilities. At that time, the main concern was 'injuries to superficial tissues', although 'derangements of internal organs and changes in the blood' were also mentioned. The focus was on keeping doses below specified levels to avoid these effects.

Over the years, protection expanded to include other workers, patients, the public, and the environment. In parallel, protection became more complex as the understanding of radiation effects grew. By the 1950s and 1960s, it was clear that radiation exposure could have other negative consequences including cancer and genetic effects. There were concerns that any level of exposure may involve some degree or risk, and, due to incomplete knowledge especially at low levels, any unnecessary exposure should be avoided.

For interested readers, more detailed accounts of the history of radiological protection are available. The essential reference is Bo Lindell's *The History of Radiation, Radioactivity, and Radiological Protection* in four volumes recently translated from Swedish to English (Lindell 2019a, 2019b, 2020a, 2020b). Less complete, but a quicker read, are papers by Bo Lindell (Lindell 1996), and Clarke and Valentin (2009).

Today, nearly a century after it was established, ICRP continues to manage the System of Radiological Protection that is used as the basis of standards, regulations, and practice worldwide, as described further in section 13.3. However, before delving into the current System, it is instructive to understand the international framework in which it has evolved during recent decades.

13.2 The international framework of science, recommendations, guidance, and standards

The prevailing view of how international organisations interact to develop science, recommendations, guidance, and standards is simpler than the reality. Nonetheless, the simplified view is not entirely incorrect and is a reasonable entry point for understanding the more complex international framework of the many radiological protection organisations, some of which are listed in table 13.1. Furthermore, it is reasonably representative of how some of the international organisations work together to bring scientific knowledge, ethical values, practical factors, and regulatory considerations into the developing international recommendations, guidance, and standards.

Table 13.1. International organisations referred to in this chapter with their abbreviations.

ALLIANCE	European Radioecology Alliance
AOARP	Asian and Oceanic Association for Radiation Protection
EAN	European ALARA Network
EC	European Commission
EURADOS	European Radiation Dosimetry Group
EURAMED	European Alliance for Medical Radiation Protection Research
FAO	Food and Agriculture Organisation
FORO	Ibero-American Forum of Radiological and Nuclear Regulatory Agencies
HERCA	Heads of the European Radiological Protection Competent Authorities
IAEA	**International Atomic Energy Agency**
IARC	International Agency for Research on Cancer
ICRP	**International Commission on Radiological Protection**
ICRU	International Commission on Radiation Units and Measurements
IEC	International Electrotechnical Commission
ILO	International Labour Organisation
IOMP	International Organization for Medical Physics
IRPA	**International Radiation Protection Association**
ISO	International Organisation for Standardization
ISOE	Information System on Occupational Exposure Network
ISR	International Society of Radiology
MELODI	Multidisciplinary European Low Dose Initiative
NEA	Nuclear Energy Agency
NERIS	European Platform on Preparedness for Nuclear and Radiological Emergency Response and Recovery
PAHO	Pan American Health Organization
UNEP	United Nations Environment Programme
UNSCEAR	**United Nations Scientific Committee on the Effects of Atomic Radiation**
WENRA	Western European Nuclear Regulators Association
WHO	World Health Organization
WNA	World Nuclear Association

NB: Highlighted organisations are those shown as the central four in figure 13.1.

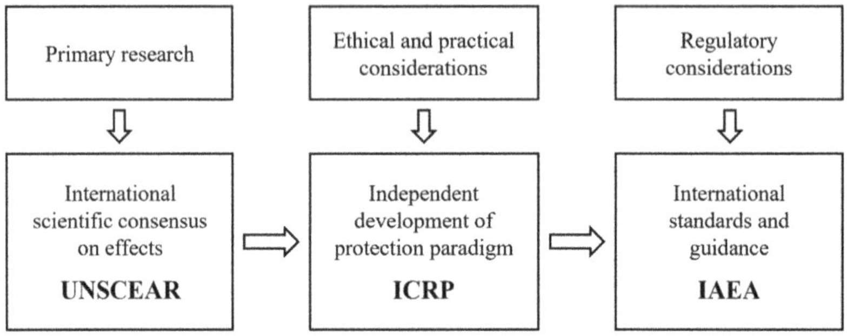

Figure 13.1. A classical, simplified view of interactions between international organisations leading to international standards and guidance on radiological protection.

Figure 13.1 is a schematic of this simplified view. It shows only three organisations, with UNSCEAR developing an international scientific consensus on the effects of radiation, ICRP using the results of UNSCEAR along with ethical and practical considerations to develop the System of Radiological Protection, and IAEA using the results of ICRP along with regulatory considerations to prepare internationally agreed standards and guidance. (For Europe, the last actor in this simplified view would be the EC rather than IAEA).

As one might expect, the reality is not quite so simple. Today there are dozens of international organisations, global and regional, inter-governmental and non-governmental, with an interest in radiological protection, and many more at the national level. While the basic path of UNSCEAR—ICRP—IAEA persists, the process is never quite this linear and there are more players acting at various points. Trying to describe all these organisations and their interactions is practically impossible, both because many of the interactions are fluid, adapting to changing circumstances and levels of interest, and because the number of organisations continues to grow. While the four central organisations, ICRP (established 1928), UNSCEAR (established 1955), IAEA (established 1957), and IRPA (established 1965), have been around for more than half a century, other organisations such as HERCA (established 2007), ALLIANCE (established 2009), MELODI (established 2009), and EURAMED (established 2017) have appeared much more recently.

Figure 13.2 is one way to envisage this cloud of international organisations. At its centre are the four organisations highlighted in table 13.1: UNSCEAR, ICRP, IAEA, and IRPA. A broad structure is implied by a loose grouping of these and several other representative organisations around four general domains: science, principles, standards, and practice. Even this somewhat nebulous representation should be viewed with caution; the mandates of many of these and other organisations encompass several of these domains. The main point of the figure is not to classify the organisations, but to demonstrate the number of organisations involved and reflect the complexity of the interactions, in contrast to the simple view presented earlier.

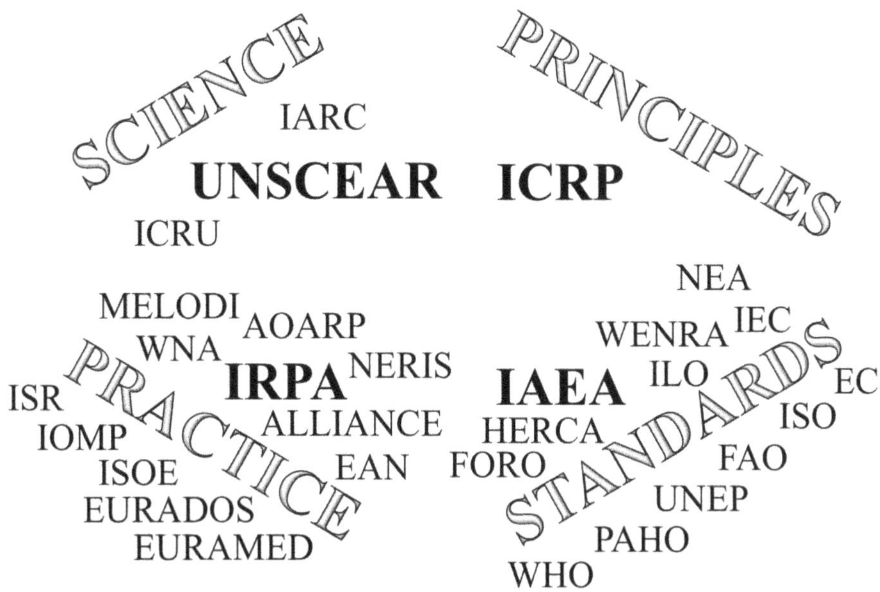

Figure 13.2. A more realistic representation of the cloud of international organisations influencing developments in science, principles, standards, and practice of radiological protection worldwide, loosely grouped around these four general domains. The central organisation in each domain, UNSCEAR, ICRP, IAEA, and IRPA, is highlighted. The set of organisations shown is representative, not exhaustive, corresponding with table 13.1.

Based on this, one might conclude that effective and efficient interactions are a significant challenge. There is no doubt that the situation is complex, but there are mechanisms in place to aid collaboration. However, several regional organisations aid coordination, e.g., AOARP, FORO, HERCA, and several European research platforms. An important mechanism at the international level is the Inter-Agency Committee on Radiation Safety (IACRS). Formed in 1990 to aid in the development of the first International Basic Safety Standards for Protection against Ionizing Radiation and for the Safety of Radiation Sources (IAEA 1996), IACRS consists of eight inter-governmental (EC, FAO, IAEA, ILO, NEA, PAHO, UNSCEAR, and WHO) organisations and five non-governmental bodies (ICRP, ICRU, IEC, IRPA, and ISO). IACRS promotes consistency and coordination of policies in radiation safety, primarily through regular exchanges between its organisations and agencies, and through information updates released through its website.

Nonetheless, each organisation has its own role to play within the international framework of organisations with an interest in radiological protection. The remainder of this section is dedicated to examining a selection of these.

13.2.1 United Nations Scientific Committee on the Effects of Atomic Radiation (UNSCEAR)

In 1955, the General Assembly of the United Nations established UNSCEAR to assess the levels, effects, and risks of exposure to ionising radiation from all sources. The Committee reports the results of its assessments to the UN General Assembly annually and makes its reports publicly available. These reports are regarded by governments and the scientific community as authoritative and independent and are used by national and international organisations as the basis for radiation risk assessment and for developing national and international policies, standards, and programmes on radiological protection.

The main task of UNSCEAR is to collect, analyse and evaluate the up-to-date scientific data on sources, effects, and mechanisms (including risk) of ionising radiation that underpins the basis for decision making and protection of public, workers, and the environment worldwide.

13.2.2 International Commission on Radiological Protection (ICRP)

Founded in 1928, ICRP is an independent, international organisation, formally a charity registered with the Charity Commission of England and Wales. ICRP works to advance radiological protection for the public benefit, protecting patients, workers, the public and the environment from harmful effects of radiation. Almost all the approximately 300 members are part-time volunteer experts working, e.g., in research institutions, government agencies, medical facilities, universities, and industry.

The central work of ICRP is the development and promotion of the System of Radiological Protection, which forms the basis of all radiological protection standards, guidance, regulations, and practice worldwide. This System is based on the best scientific knowledge, sound ethical principles, and nearly a century of experience. See section 13.3 'An overview of the System of Radiological Protection developed by ICRP'.

In preparing its recommendations, ICRP considers the fundamental principles and quantitative bases upon which appropriate radiological protection measures can be established, while leaving to the various national protection bodies the responsibility of formulating the specific advice, codes of practice, or regulations that are best suited to the needs of their individual countries.

The System of Radiological Protection is described at a high level in the so-called general recommendations, most recently in ICRP *Publication 103* The 2007 Recommendations of the International Commission on Radiological Protection (ICRP 2007). General recommendations are published only rarely (see table 17.2); additional recommendations and guidance from ICRP on more specific topics are published in its dedicated journal *Annals of the ICRP*.

The recommendations of ICRP have no legal standing of their own but may become binding by being incorporated in other instruments, e.g., the IAEA Basic Safety Standards for Protection against Ionizing Radiation and for the Safety of Radiation Sources (IAEA 2014), the ILO Radiation Protection Convention 1960

(No. 115) (ILO 1960a) and its accompanying Recommendation (No. 114) (ILO 1960b), and national legislation.

13.2.3 International Atomic Energy Agency (IAEA)

The IAEA's role in radiological protection is to assist its Member States and the radiological protection professional community in general, in regulation and application of its standards, and by contributing to further improvement of the System of Radiological Protection through collecting feedback from its Member States. This entails identifying and effectively addressing conceptual, scientific, policy, regulatory, operational, and societal issues and developing relevant standards and guidance documents.

According to the IAEA statute, Article 6, the IAEA shall '... *establish or adopt, in consultation and, where appropriate, in collaboration with the competent organs of the United Nations and with the specialised agencies concerned, standards of safety for protection of health and minimisation of danger to life and property (including such standards for labour conditions), and to provide for the application of these standards to its own operations as well as to the operations making use of materials, services, equipment, facilities, and information made available by the Agency or at its request or under its control or supervision; and to provide for the application of these standards, at the request of the parties, to operations under any bilateral or multilateral arrangement, or, at the request of a State, to any of that State's activities in the field of atomic energy*'.

The IAEA General Safety Requirements Part 3, Radiation Protection and Safety of Radiation Sources: International Basic Safety Standards (IAEA 2014), is one of the key standards addressing radiation safety in all situations where people are exposed. These Standards, as well as its preceding editions, reflect ongoing efforts over several decades to prepare relevant standards to contribute to the safe use of radiation sources and to provide for their application.

To ensure that sound and up-to-date scientific knowledge is adequately and appropriately reflected in its standards, the IAEA takes into account the findings of UNSCEAR and the recommendations of international expert bodies, notably ICRP.

Whereas the ICRP recommendations establish fundamental radiological protection principles that can be subsequently translated into a more regulatory context for implementation in the regulatory and legal frameworks of individual countries to ensure a satisfactory level of radiological protection, the IAEA standards may serve as a benchmark that may be directly used, transferred, or adopted into regulatory systems of individual states. As such, the IAEA standards are important instruments for practical implementation of ICRP recommendations and a framework for the development of national regulations.

Although the regulation of nuclear and radiation safety is a national responsibility, many IAEA Member States have decided to adopt the IAEA's safety standards for use in their national regulations. In this way, the harmonisation of radiological protection approaches, and criteria are strengthened, and their application is enhanced.

In preparing its standards, the IAEA liaises with other international organisations, professional societies, and regulatory and public health authorities. For example, it liaises with WHO when preparing standards relevant to radiological protection of patients, and with ILO when addressing radiological protection of workers. International organisations, in many cases, become formal co-sponsors of selected standards or guidance documents.

13.2.4 International Radiation Protection Association (IRPA)

IRPA is the international professional association for radiation protection. Through national and regional Associate Societies and radiological protection professionals, IRPA promotes excellence in radiological protection by providing benchmarks of good practice and enhancing professional competence and networking. IRPA encourages the highest standards of professional conduct, skills, and knowledge for the benefit of individuals and society.

IRPA has over 18 000 radiological protection professionals as members, coming from 68 countries. Its membership covers all fields of radiological protection, from fundamental science, policy, and regulation to practical application in industry, healthcare, and research facilities. The experience of IRPA members ranges from countries with well-developed radiological protection systems and infrastructure, to situations where there is little local support or infrastructure. This enormous range of practical expertise across the whole radiological protection spectrum gives IRPA its unique position to represent the views of the profession.

Based on this platform of wide practical experience, IRPA acts as the international voice of the radiological protection profession in its interactions with the major international bodies in the radiological protection field. IRPA has formal relationships with the key players involved in the development of principles, standards, and legislation, such as ICRP, IAEA, WHO/PAHO, NEA, EC, and many others. During extensive interactions with these bodies, IRPA's aim is to feedback practical experience from the profession to help identify areas for improvement and ensure that proposals are suitable for practical implementation.

IRPA also has the objective of developing and sharing good practices across the Associate Societies and individual members. This often takes the form of publishing Guidance Documents in key areas of practice. This encourages and supports the application of best practice locally, including providing a sound basis for national radiological protection societies to engage on these topics with national regulatory authorities in the development of legislation and regulatory practice.

13.2.5 European Commission (EC)

The Euratom Treaty (EU 2010a), one of the founding treaties of the European Union in force since 1958, empowers the European Atomic Energy Community to establish uniform basic safety standards to protect the health of workers and the public against dangers arising from ionising radiation. The first Euratom Basic

Safety Standards Directive—the BSS Directive—was adopted in February 1959 taking account of the, at that time latest, scientific knowledge in radiological protection and relevant international recommendations, such as ICRP *Publication 1* (ICRP 1959). Subsequently, with increasing scientific knowledge in radiological protection, the BSS Directive has been regularly amended in the years 1962, 1966, 1976, 1980, 1984 and 1996. Major revisions took place in the light of new ICRP publications on the System of Radiological Protection; the 1976 BSS Directive was based on ICRP *Publication 26*, the 1996 BSS (EC 1996) on ICRP *Publication 60* (ICRP 1991), supplemented in 1997 by a Directive for the protection of patients (EC 1997), which makes explicit reference to ICRP recommendations in the area of medical exposures, and the 2013 revision of the BSS Directive was triggered by ICRP *Publication 103* (ICRP 2007).

The 2013 BSS Directive (EC 2014) follows, to the extent possible, the radiological protection philosophy introduced in ICRP *Publication 103*. Following ICRP's recommendation to consistently integrate exposure to natural radiation, the 2013 BSS Directive introduces binding requirements on protection from exposure to natural radiation sources, in particular radon, cosmic rays and naturally occurring radioactive material (NORM). The Directive also refers to the latest ICRP recommendations when it comes to scientifically established values and relationships for the calculation of doses from measurable quantities. The BSS Directive is supplemented by a Directive laying down requirements for the protection of the health of the public regarding radioactive substances in water intended for human consumption (EC 2013). This is in line with the recommendations developed and published by the WHO.

European Union Member States are bound to comply with the obligations laid down in the legally binding Directives. The EC supports the practical implementation of its Directives through the publication of guidelines, recommendations, and information material, e.g., on radon in workplaces (EU 2019), on monitoring individuals subject to external and internal exposure (EU 2010b, 2018), on emergency preparedness and response, and in the medical area, published in the Radiation Protection Series of the European Commission. All these supporting materials are based on recommendations and guidelines published by, e.g., ICRP, ICRU, and WHO.

Based on international guidelines and recommendations, Europe has developed and implemented the most advanced set of legislation aiming at the highest level of protection of individuals—members of the public, workers, and patients—exposed to any source of radioactivity or ionising radiation. The EC is committed to maintaining this high level of protection while closely following scientific developments in radiological protection research, which may have an impact on future radiological protection policy.

13.2.6 European Nuclear Installations Safety Standards Initiative (ENISS)

The European Nuclear Installations Safety Standards Initiative (ENISS), established in 2005, brings together European license holders with nuclear power units, fuel reprocessing plants, or large waste storage facilities. Its ambition is to work together with regulators to produce high level, practicable, and harmonised safety targets and rules in Europe. As of 2021, there are companies from 17 European countries, EU-States, Switzerland, and Ukraine, who share their experience within ENISS.

The focus of ENISS is the development and application of harmonized safety standards for the design and operation of nuclear power plants and other nuclear installations, and in this regard ENISS cooperates closely with the European authorities, namely the Western European Nuclear Regulators' Association (WENRA). WENRA has issued several so-called Safety Reference Levels and ENISS is a major stakeholder in the development of these safety guidelines.

Based on its wealth of experience of many workers involved in various aspects of practical radiological protection daily, ENISS is able to influence changes to radiological protection recommendations, guidance, standards, and legislation to ensure improved safety or protection without unjustified increases in cost or other resources through its relationships with key organisations like ICRP and IAEA.

ENISS has observer status in the IAEA Safety Standards Committees, enabling it to comment on drafts of IAEA safety standards and bring in practical experience when commenting. It also has an Expert Group on Radiation Protection that regularly reviews draft ICRP documents, providing valuable feedback to help ensure the publications reflect the practices and experience of operators.

13.2.7 Heads of the European Radiological Protection Competent Authorities (HERCA)

HERCA was created in 2007 to initiate an exchange of knowledge and experience to facilitate practical and harmonized solutions to important regulatory issues in radiological protection within Europe.

It has existed for more than 10 years as a voluntary association of the Radiation Protection Authorities (RPA) in Europe enabling them to work together to identify common significant radiological protection issues and propose harmonisation and/or practical solutions towards a common approach for these issues, whenever possible. The goal of HERCA is to contribute to a high level of radiological protection throughout Europe. HERCA members are the European Radiation Protection Authorities (RPA) represented by the Heads of the authorities.

HERCA is a forum for the RPA to share information and experience, and acts as a stakeholder forum on radiological protection matters able to react, e.g., to an external request for advice on radiological protection and regulatory matters.

A key player and pillar in the radiological protection community in Europe, HERCA has relationships with several relevant stakeholders such as ICRP, IAEA, EC, and IRPA, and approaches stakeholder involvement with a view to maximising the efficiency and efficacy of its efforts. Positions approved and recommendations agreed by HERCA are expected to result in concrete actions having an impact on national policies and practices in radiological protection.

13.2.8 International Electrotechnical Commission (IEC)

The IEC is the world's leading organisation preparing and publishing international standards for all electric and electronic devices and systems. It brings together 86 countries (62 Member and 24 Affiliate countries), representing 98% of the world population and 96% of world energy generation. Close to 20 000 experts cooperate on the global IEC platform.

All IEC international standards are consensus-based and represent the needs of key stakeholders of every nation participating in IEC work. Worldwide, the IEC standards serve as a basis for national standardisation as well as references when drafting international tenders and contracts and for conformity evaluation of instrumentation.

IEC Subcommittee 45B 'Radiation protection instrumentation', one of the 207 Technical Committees (TC) and Subcommittees (SC), prepares standards addressing instrumentation used for the measurement of ionising radiation in the workplace, to the public, and in the environment for radiological protection purposes; illicit trafficking detection of radioactive material and identification of radionuclides; and radiation-based security screening.

TC 62 'Electrical Equipment in Medical Practice' develops international standards and other publications concerning electrical equipment, electrical systems and software used in healthcare and their effects on patients, operators, other persons, and the environment. TC 62 has 4 subcommittees:

SC 62 A Common aspects of electrical equipment used in medical practice
SC 62B Diagnostic imaging equipment
SC 62 C Equipment for radiotherapy, nuclear medicine and radiation dosimetry
SC 62D Electromedical equipment

TC 62's IEC 60 601 series of publications deals with the protection of patients, operators and the environment against hazards related to medical equipment and its application. Such hazards include, among others, ionising and non-ionising radiation hazards, foreseeable misuse, usability, and electromagnetic compatibility hazards. The series includes requirements for the risk management process of manufacturers.

SC 62B and SC 62 C publish standards dedicated to radiological equipment and software when using ionising radiation, such as x-ray, computed tomography, and radiotherapy equipment. Software, IT networks, and artificial intelligence are significant aspects to be considered. Personal protective equipment against ionising radiation, such as aprons, is also addressed in standards of SC 62B.

13.2.9 International Labour Organization (ILO)

The ILO was established in 1919 by the Treaty of Versailles to bring governments, employers, and trades unions together for united action in the cause of social justice and better living conditions everywhere. It is a tripartite organization, with representatives of workers and employers taking part in its work on equal status with representatives of governments. The ILO became the first specialized agency of the United Nations in 1946. One of the main features of the ILO, in addition to its tripartite structure, is its standard-setting activity. Some 60 international conventions and recommendations concern the protection of workers against occupational hazards.

Radiological protection is part of the ILO's action on the protection of workers against sickness, disease and injury arising out of employment, as mandated by the Constitution of the International Labour Organization (ILO 1974). In 1949, the ILO published a set of practical international standards on radiological protection which were revised and considerably extended in 1957 and were incorporated into the ILO Manual of Industrial Radiation Protection. In June 1960, the International Labour Conference adopted the Radiation Protection Convention 1960 (No. 115) (ILO 1960a), and its accompanying Recommendation (No. 114). Convention No. 115 (ILO 1960b) has been ratified by fifty countries.

ILO standards have exerted considerable influence on the laws and regulations of member States. Many texts have been modelled on the relevant provisions of ILO instruments. Drafts of new legislation or amendments are often prepared with ILO standards in mind to ensure compliance with ratified Conventions or to permit the ratification of other Conventions. Trade unions use ILO standards to support arguments in bargaining and in promoting legislation. Governments frequently consult the ILO, both formally and informally, about the compatibility of proposed legislative texts with international labour standards.

The ILO Convention No 115 applies to all activities involving the exposure of workers to ionising radiations in the course of their work and provides that each Member of the ILO which ratifies it shall give effect to its provisions by means of laws or regulations, codes of practice or other appropriate means.

Under Article 3(1) of the ILO Convention No. 115, 'In the light of knowledge available at the time, all appropriate steps shall be taken to ensure effective protection of workers, as regards their health and safety, against ionising radiations.' Article 3(2) of the Convention states that 'Rules and measures necessary for this purpose shall be adopted', and Paragraph 3 of the Radiation Protection Recommendation 1960 (No. 114), states that for the purpose of giving effect to Article 3(2), 'every Member should have due regard to the recommendations made from time to time by the International Commission on Radiological Protection and standards adopted by other competent organisations.' In addition, among the protective steps that have to be taken under Article 3(1), Article 6(1) provides for the fixing, for various categories of workers, of 'maximum permissible doses of ionising radiations which may be received from sources external to or internal to the body and maximum permissible amounts of radioactive substances which can be taken into the body', and Article 6(2) specifies that 'such maximum permissible doses and amounts shall be kept under constant review in the light of current knowledge'. In assessing compliance with these requirements, it has been the practice of the ILO Committee of Experts on the Application of Conventions and Recommendations to refer to the current knowledge as embodied in the recommendations of the ICRP and other international reference sources based on the same recommendations, such as the BSS (co-sponsored by a number of international organizations, including the ILO from 1982 onwards), as well as the ILO Code of Practice on radiation protection of workers (ionising radiations) approved by the Governing Body of the ILO at its 234th Session (November 1986) (ILO 2016), which gives practical guidance on the implementation of a radiological protection programme at the enterprise level.

13.2.10 OECD Nuclear Energy Agency (NEA)

The NEA is an inter-governmental agency operating under the framework of the Organisation for Economic Co-operation and Development (OECD). The NEA Strategic Plan provides the direction for their programme of work in all areas. Based on this foundation, the NEA's overarching goal in radiological protection of human health and the environment is to assist member countries in the implementation and further development of the system of radiological protection and related science, regulation, and operations. This NEA goal is implemented by the Agency's Committee on Radiological Protection and Public Health (CRPPH).

This standing technical committee, which has been in existence for more than fifty years, addresses evolving radiological protection issues including underlying science and concepts, policy and regulation, and practice. The work produced by the CRPPH supports radiological protection development and implementation in member countries by providing timely identification of significant topics and emerging issues, analysing their possible implications for radiological protection practices and regulation, and contributing to the development of approaches for their resolution. The Committee is an authoritative source of information for national authorities, policy makers, practitioners, and interested stakeholders, providing advice, guidance, framework, feedback from regulation, and operational approaches, on multidisciplinary areas comprising radiological protection, public health, and the environment. The Committee's participative working methods facilitate rapid progress in those areas where international consensus on radiological protection concepts, regulatory issues and practices is sought.

For example, through the organisation of experience exchange meetings, workshops, expert/working groups, joint undertakings/services, and emergency exercises, the CRPPH has interpreted the science from UNSCEAR, the policy from ICRP, and the standards from IAEA to pragmatically develop paths forward and good practice when faced with new and emerging challenges. This supports national governments and international organisations to help them maintain a state-of-the-art level and to use the best practices.

13.2.11 World Health Organization (WHO)

As the agency within the United Nations system with a specific mandate to coordinate international public health work, the main objective of WHO is the attainment by all peoples of the highest possible level of health. The WHO has a decentralized structure with 150 country offices, six regional offices and it has its headquarters in Geneva. Its governing body is the World Health Assembly (WHA), made up of delegates from the 194 WHO Member States.

The engagement on radiological protection and health has a long history within WHO and this has been reflected in several WHA resolutions. In line with its core function of 'setting norms and standards and promoting and monitoring their implementation', WHO has contributed to the revision of the International Radiation Basic Safety Standards (BSS) as one of the eight BSS cosponsoring organizations.

The WHO has a programme on radiation and health addressing protection against both ionising and non-ionising radiation exposures. With regards to ionising radiation, WHO is conducting a programme on protection of patients, workers, and the public under planned, existing, and emergency exposure situations. Addressing several activities in the areas of risk assessment, risk management and risk communication, this programme brings together relevant stakeholders to support the implementation of the BSS worldwide.

13.3 An overview of the system of radiological protection developed by ICRP

ICRP *Publication 103* The 2007 Recommendations of the International Commission on Radiological Protection (ICRP 2007) is the current set of general recommendations at the time of writing (2021). This publication describes the entire System of Radiological Protection ('the System') at a relatively high level.

Figure 13.3 is a schematic representation of the broad parts of the System. The organisation of this section is based largely on this scheme.

13.3.1 Science, ethics, and experience

The System is based on three pillars: scientific knowledge, ethical values, and practical experience.

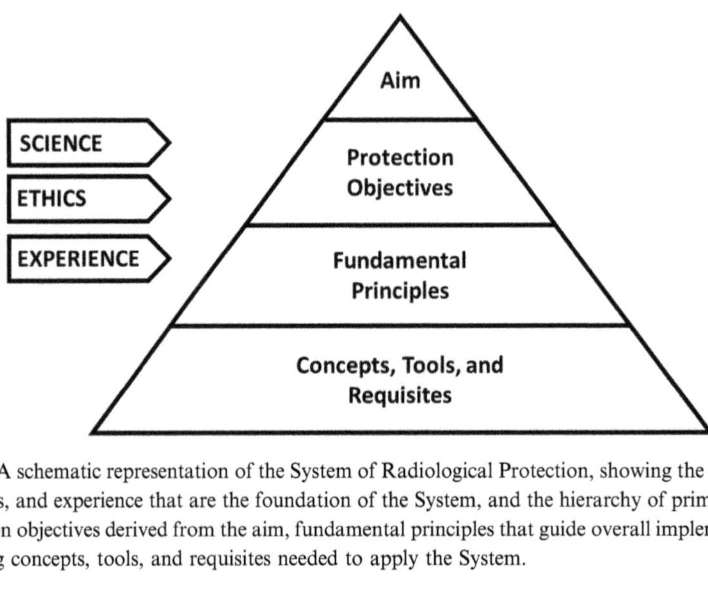

Figure 13.3. A schematic representation of the System of Radiological Protection, showing the three pillars of science, ethics, and experience that are the foundation of the System, and the hierarchy of primary aim at the top, protection objectives derived from the aim, fundamental principles that guide overall implementation, and the remaining concepts, tools, and requisites needed to apply the System.

Scientific knowledge is necessary to understand effects and risks of radiation exposure. Many branches of science are applied in the System: physics to understand the way radiation interacts with matter, biokinetics to understand how substances move through the human body, oncology to understand the mechanism of radiogenic cancer induction, ecology to understand ecosystem impacts, and many more.

However, scientific knowledge alone cannot answer questions of how to behave to ensure the protection of people and the environment. Science can tell us 'what is' but we need ethics to help us decide 'what ought to be'. This was most famously articulated by the Scottish philosopher David Hume in his *Treatise of Human Nature* (Hume 1739), and often referred to as Hume's Law or the is–ought problem. For example, through scientific knowledge we have some understanding of the relationship between exposure to a dose of radiation and the possible consequences, but we need to apply ethical values to recommend what level of exposure is tolerable, acceptable, or neither in each circumstance.

Practical experience is also crucial. There is no point in developing a System of Radiological Protection, no matter how perfectly it matches our best scientific understanding and how precisely it follows the most appropriate ethical values, if it is not practicable. For the System to be beneficial, it must be used.

13.3.2 Scope

In principle, the scope of the System encompasses all exposures of all living things to ionising radiation from any source, regardless of origin or magnitude. This includes natural and artificially produced sources. However, the System can only be applied fully when the source of exposure or the pathways leading to an exposure can reasonably be controlled.

13.3.3 Scientific basis

Even more than other aspects, the scientific basis of the System is given a very cursory treatment here. Other chapters, especially those on dose, effects, and risks, cover this in more depth. Here, the discussion is limited to high-level concepts of effects and dose essential to understanding the System.

For the purposes of protection, radiation-induced effects are grouped into two broad categories: cancer and heritable disease, and harmful tissue reactions. This categorisation is both biological and practical. It is also an important simplification specifically for protection and should not be used in place of more complex descriptions necessary to describe specific effects in a clinical or research setting.

Cancer and heritable disease (aka stochastic effects) are the result of mutation of individual cells, where the probability of occurrence increases with increasing dose, with the assumption that there is no dose threshold below which this probability is zero.

Protection against these effects is management of dose to keep (assumed) risks as lows as reasonably achievable.

Harmful tissue reactions (aka deterministic effects) are the result of the injury of populations of cells, where the severity of the effect increases with increasing dose and characterised by a dose threshold below which there is no clinical harm. Protection against these effects is simpler, as it is possible, in principle, to avoid the effect if doses are kept below the threshold.

Radiation dose is a proxy for risk from exposure to radiation. However, the two are closely related, as equivalent and effective dose are calculated based on risk.

Absorbed dose, D, is the starting point, being a physical quantity that can be measured directly. It is the mean energy ($d\bar{\varepsilon}$) imparted to matter by ionising radiation divided by the mass (dm). For radiological protection purposes, the absorbed dose is typically averaged over an organ, tissue, or sensitive region of an organ or tissue. The SI unit of absorbed dose is joule per kilogram (J kg^{-1}), and it is given the special name of gray (Gy).

$$D = \frac{d\bar{\varepsilon}}{dm}$$

Equivalent dose in an organ or tissue, H_T, is equal to the absorbed dose, $D_{T,R}$ delivered by radiation type R averaged over the tissue or organ T, multiplied by a radiation weighing factor, w_R. The weighting factor reflects the relative effectiveness of different types of radiation (e.g., alpha, beta, photon, neutron) to cause harm. The SI unit of equivalent dose is joule per kilogram (J kg^{-1}), and it is given the special name of sievert (Sv).

$$H_T = \sum_R w_R D_{T,R}$$

Effective dose, E, is a weighted sum of equivalent doses to all organs and tissues. Tissue weighting factors, w_T, reflect the relative radiosensitivity of each organ to radiation. Like equivalent dose, the SI unit of effective dose is joule per kilogram (J kg^{-1}), and it is given the special name of sievert (Sv).

$$E = \sum_T w_T H_T$$

Effective dose is another simplification in the System to ensure it remains practicable. The radiation and tissue weighting factors are based on scientific evidence, but the values are binned to keep the number of different values manageable.

Effective dose is a central concept for radiological protection of humans. It is a single quantity that accounts for exposures from all types of ionising radiation, and all sources whether they originate from inside or outside the body. Most dose limits, dose constraints, and reference levels are in effective dose.

13.3.4 Ethical basis

The fundamental question in radiological protection is what action, if any, we should take when faced with existing sources of radiation, or when considering introducing new ones. Scientific facts are essential to understanding, but, alone, are not enough. Ethical values are necessary for making recommendations on how to behave in light of our scientific knowledge.

Ethics have been a crucial part of the System since the beginning, although rarely was this aspect described in depth in ICRP publications before ICRP *Publication 138* 'Ethical Foundations of the System of Radiological Protection' (ICRP 2018). This was the first ICRP publication dedicated to describing the ethical foundation of the System. It developed a common language that facilitates discussions of the ethical basis, and identified four core ethical values inherent in the System (ICRP 2018):

> ***Beneficence/non-maleficence:*** *promoting or doing good and avoiding doing harm. This is reflected, for example, in the primary aim of the system of radiological protection: … an appropriate level of protection … without unduly limiting … desirable human actions.*
>
> ***Prudence:*** *making informed and carefully considered choices without full knowledge of the scope and consequences of an action. Prudence is reflected, for example, in the consideration of uncertainty of radiation risks for both humans and the environment.*
>
> ***Justice:*** *fairness in the distribution of advantages and disadvantages. Justice is a key value underlying, for example, individual dose restrictions that aim to prevent any individual from receiving an unfair burden of risk.*
>
> ***Dignity:*** *the unconditional respect that every person deserves, irrespective of personal attributes or circumstances. Personal autonomy is a corollary of human dignity. This underlies, for example, the importance placed on stakeholder participation and the empowerment of individuals to make their own informed decisions.*

Three procedural values were also identified to aid in the practical implementation of the System: accountability, transparency, and inclusiveness.

13.3.5 Aims and objectives

The primary aim of the system of radiological protection is 'to contribute to an appropriate level of protection for people and the environment against the detrimental effects of radiation exposure without unduly limiting the desirable human actions that may be associated with such exposure.' (ICRP 2007). This is an embodiment of the values of beneficence (promoting good) and non-maleficence (avoiding doing harm). It also reveals a central characteristic of the System: the need to find the best balance of a complex mix of pros and cons in each circumstance. There is an emphasis on the radiological aspect, since this is the purview of the System, but recognises that many other considerations must also be taken into account.

The human health objectives of the System are 'to manage and control exposures to ionizing radiation so that deterministic effects are prevented, and the risks of stochastic effects are reduced to the extent reasonably achievable.' (ICRP 2007)

The structure of the human health objectives relates directly to the classification of effects into harmful tissue reactions (deterministic effects) that have thresholds, and cancer and heritable effects (stochastic effects) that, in principle, do not, as described in the section on 'Scientific Basis'. In short, the human health objectives are to prevent effects that are, in principle, preventable, and to reduce the risk of effects that cannot be completely avoided.

The objectives of protection of the animals and plants in their natural environmental setting are: 'preventing or reducing the frequency of deleterious radiation effects to a level where they would have a negligible impact on the maintenance of biological diversity, the conservation of species, or the health and status of natural habitats, communities and ecosystems' (ICRP 2007).

Here the emphasis is on the protection of groups of animals and plants, not individuals.

13.3.6 Fundamental principles

The System includes three fundamental principles, derived from the primary aim, based on the core ethical values (although developed before these were concretely identified in *Publication 138*), and which guide overall implementation of the System:

- Justification;
- Optimisation of protection;
- Application of dose limits.

All three fundamental principles are connected to all four core ethical values, although some values are more prominently linked to some principles.

Justification: 'The principle of justification: Any decision that alters the radiation exposure situation should do more good than harm' (ICRP 2007).

When introducing a new source of radiation or changing or removing a risk of actual or potential exposure, the benefits must outweigh the detriments. This principle applies under all circumstances.

There are clear and direct links between the principle of justification and the core ethical values of beneficence (promoting good) and non-maleficence (avoiding harm).

Optimisation of protection: 'The principle of optimisation of protection: the likelihood of incurring exposures, the number of people exposed, and the magnitude of their individual doses should all be kept as low as reasonably achievable, taking into account economic and societal factors' (ICRP 2007).

Optimisation of protection requires that the margin of benefits over detriments (the net benefit) to society and individuals be maximized. This principle applies under all circumstances.

A core ethical value of prudence is central to optimisation of protection. In most circumstances, radiological protection decisions are made for levels of exposure where our knowledge of radiation risks is incomplete. This calls for carefully considered choices based as much on ethical considerations as scientific ones.

Justice (fairness) is also central to optimisation of protection, as optimisation tools like dose constraints and reference levels help to avoid severely inequitable outcomes for individuals.

Application of dose limits: 'The principle of application of dose limits: The total dose to any individual from regulated sources in planned exposure situations other than medical exposure of patients should not exceed the appropriate limits recommended by the Commission' (ICRP 2007).

This principle applies only for planned exposure situations, as the strict application of dose limits in other circumstances, where there is less *a priori* control over radiation exposures, could lead to grossly non-optimised solutions. In these circumstances, constraints and reference levels serve a similar purpose as described under 'Optimisation of protection'.

The core ethical values of justice and dignity are central here, assuring no individual bears an unfair burden of risk, regardless of their personal attributes or circumstances.

13.3.7 Concepts, tools, and requisites

Going much beyond the aims, protection objectives, and fundamental principles of the System is beyond the scope of this chapter. This section serves only as an introduction to some of the higher-level concepts including the categories of exposure and exposure situations, and criteria such as dose limits, constraints, and reference levels.

This is likely to be unsatisfactory for readers with a keen interest in radiological protection. The best reference for a comprehensive description is ICRP *Publication 103* 'The 2007 Recommendations of the International Commission on Radiological Protection' (ICRP 2007) itself.

ICRP defines three **categories of exposure** (ICRP 2007):

> ***Occupational exposure*** *is defined by the Commission as all radiation exposure of workers incurred as a result of their work … that can reasonably be regarded as being the responsibility of the operating management.*
>
> ***Medical exposure*** *[is] the exposure of individuals for diagnostic, interventional, and therapeutic purposes, … exposures (other than occupational) incurred knowingly and willingly by individuals such as family and close friends helping either in hospital or at home in the support and comfort of patients … [and] exposures incurred by volunteers as part of a programme of biomedical research that provides no direct benefit to the volunteers.*
>
> ***Public exposure*** *encompasses all exposures of the public other than occupational exposures and medical exposures.*

ICRP also defines three **exposure situations** (ICRP 2007):

> ***Planned exposure situations*** *are situations involving the deliberate introduction and operation of sources. Planned exposure situations may*

give rise both to exposures that are anticipated to occur (normal exposures) and to exposures that are not anticipated to occur.

Emergency exposure situations *are situations that may occur during the operation of a planned situation, or from a malicious act, or from any other unexpected situation, and require urgent action in order to avoid or reduce undesirable consequences.*

Existing exposure situations *are exposure situations that already exist when a decision on control has to be taken, including prolonged exposure situations after emergencies.*

Both the categories of exposure and exposure situations are used to aid decision-making in radiological protection. Although every situation is unique, the general approach is similar within a particular category of exposure and exposure situation. For example, as described under 'Fundamental principles', justification and optimisation of protection apply in all exposure situations, but the application of individual dose limits applies only in planned exposure situations.

Moreover, the radiological protection criteria used vary depending on the exposure situation.

Dose limits are perhaps the most well-known of the criteria, although they are applied only in limited circumstances. They apply to the total dose to any individual from regulated sources in planned exposure situations other than medical exposure of patients.

Dose limits are regarded as hard restrictions on doses to individuals as they are typically introduced into radiological protection legislation, with penalties associated with causing an individual to exceed these limits.

Dose constraints are also only used in planned exposure situations, in conjunction with optimisation of protection to restrict individual doses. They relate to the dose to an individual from a specific source and serve as an upper bound on the predicted dose in the optimisation of protection for that source. They are not used in the medical exposure of patients, but can be applied to restrict exposures to comforters, carers, and volunteers in biomedical research.

Reference levels are used in emergency or existing exposure situations. They represent the level of dose or risk above which it is judged to be inappropriate to plan to allow exposures to occur.

Diagnostic reference levels (DRLs) are used exclusively for protection of patients and are 'a form of investigation level used to aid in optimisation of protection in the medical exposure of patients for diagnostic and interventional procedures' (ICRP 2017). DRLs are used to assess optimisation of protection in a facility by comparing the distribution of doses to patients undergoing a specific procedure to DRLs set based on good practice. They do not apply to individual patients, and do not represent a line between good and bad practice.

Derived consideration reference levels (DCRLs) are used in the protection of animals and plants. They are a band of dose rate within which there is likely to be some chance of deleterious effects of ionising radiation occurring to individuals of that type of reference animal or plant (derived from a knowledge of defined expected biological effects for that type of organism) that, when considered together with other relevant information, can be used as a point of reference to optimise the level of effort expended on environmental protection, dependent upon the overall management objectives and the relevant exposure situation (ICRP 2008). When doses to reference organisms are below the DCRLs there is likely little need for concern. Doses within the DCRL band may warrant further consideration, and doses above the DCRL band may warrant action depending on the broader circumstances.

13.4 Evolution of the system of radiological protection

The System of Radiological Protection must continue to evolve to remain fit for purpose. There have been major changes in the past, for example related to advances in knowledge, especially the discovery in the 1950s that radiation exposure could cause cancer. There have also been major changes based more on philosophy and practicability, for example in ICRP *Publication 26* (ICRP 1977) which introduced the foundation of the System of Radiological Protection still in use today.

ICRP *Publication 103* (ICRP 2007) contains the general recommendations of ICRP current as of writing this in 2021 and described in the previous section of this chapter. It is the backbone of the System of Radiological Protection. In 2018, about two decades since work began to develop what became the 2007 Recommendations, ICRP began a review of the System with the intention to prepare new general recommendations for around the end of the decade to supersede ICRP *Publication 103*. The beginnings of the review were modest, comprising a few internal discussions and interactions with some other organisations to get initial feedback on what areas might need attention.

The driver is to ensure that the System of Radiological Protection remains fit for purpose for the next generation. It must keep up with advances in scientific knowledge, and reflect exposures not previously covered such as those in innovative uses of radiation in medical diagnosis and treatment, and in human exploration of space.

Despite the constant need for evolution, there are also benefits to stability. Regulatory systems and radiological protection practices need to adjust to new recommendations. This is not always easy and inexpensive.

ICRP balances these two pressures by continually refining elements of the System and the principles of its application, while keeping major updates to the general recommendations to a minimum. Table 13.2 lists each of the general recommendations of ICRP, showing the number of years since the previous version was published. The long pause from 1937 to 1950 is due to World War II.

The trend towards more time between updates of the general recommendations points to the growing stability of the System of Radiological Protection. From 1928 to 1937 changes came rapidly as the pioneers of radiological protection created and refined their work. Immediately after World War II, the discovery that radiation exposure can increase the risk of cancer challenged some of the basic concepts of the System, demanding another series of rapid revisions in the 1950s and 1960s. Since

Table 13.2. General recommendations of ICRP.

General recommendations	Years since previous version
ICRP *Publication 103* 2007 Recommendations (ICRP 2007)	17
ICRP *Publication 60* 1990 Recommendations (ICRP 1991)	13
ICRP *Publication 26* 1977 Recommendations (ICRP 1977)	11
ICRP *Publication 9* 1966 Recommendations (ICRP 1966)	2
ICRP *Publication 6* 1964 Recommendations (ICRP 1964)	5
ICRP *Publication 1* 1959 Recommendations (ICRP 1959)	3
1956 Recommendations (ICRP 1958)	2
1954 Recommendations (ICRP 1955)	4
1950 Recommendations (ICRP 1951)	13
1937 Recommendations (IXRPC 1938)	3
1934 Recommendations (IXRPC 1934)	3
1931 Recommendations (ICR 1931)	3
1928 Recommendations (ICR 1929)	—

Publication 26 (ICRP 1977), the time between updates has steadily increased as the System matures, and our understanding of the underlying science steadily improves without major revolution. The next general recommendations will be published no less than 20 years after the current general recommendations (ICRP 2007) were released.

It seems unlikely that work on the System of Radiological Protection will end in the foreseeable future. Advances in techniques and technologies that use ionising radiation continue to be made, and protection must keep pace. In addition, our understanding of radiation effects continues to evolve, and the next big discovery that challenges the way people and the environment should be protected from radiation may come at any time.

Radiological protection is a noble profession, dedicated to keeping people and the environment safe. It is also an exciting one, both highly multidisciplinary and ever evolving. The challenges are great, but this only makes the successes more satisfying.

Acknowledgments

Many thanks to the individuals who provided information about their organisations in section 13.2. Any errors or omissions are entirely the responsibility of this chapter's author.

References

Belot J 1904 *Traité de Radiothérapie* (Paris: G. Steinheil) English edn

Belot J 1905 *Radiotherapy in Skin Disease* (New York: Rebman)

Churchill W S 1898 *The Story of the Malakind Field Force* (London: Longman's Green & Co.)

Clarke R and Valentin J 2009 The history of ICRP and the evolution of its policies, ICRP Publication 109 *Ann. ICRP* **39(1)**

Curie M 1898 Rayons emis par les composes de l'uranium et du thorium *C. R. Acad. Sci., Paris* **126** 1101

Drury H C 1896 Dermatitis caused by Roentgen x-rays *Br. J. Med.* **2** 1377

EC 1996 Council Directive 96/29/Euratom of 13 May 1996 Laying down Basic Safety Standards for the Protection of the Health of Workers and the General Public Against the Dangers Arising from Ionizing Radiation, Official Journal of the European Communities, Series L, No. 159

EC 1997 Council Directive 97/43/Euratom of 30 June 1997 on Health Protection If individuals Against the Danger of Ionizing Radiation in Relation to Medical Exposure, Official Journal of the European Communities, Series L, No. 180, 1997

EC 2013 Council Directive 2013/51/Euratom of 22 October 2013 Laying Down Requirements for the Protection of the Health of the General Public with Regard to Radioactive Substances in Water Intended for Human Consumption, Official Journal of the European Union OJ L296, 7.11.2013, pp. 11–20

EC 2014 Council Directive 2013/59/Euratom of 5 December 2013 Laying Down Basic Safety Standards for Protection Against the Dangers Arising from Exposure to Ionising Radiation, and Repealing Directives 89/618/Euratom, 90/641/Euratom, 96/29/Euratom, 97/43/Euratom and 2003/122/Euratom, Official Journal of the European Union OJ L13, 17.1.2014, pp. 1–73

EU 2010a *European Union, Euratom Treaty—Consolidated Version* (Luxemburg: Publications Office of the European Union)

EU 2010b *RP N° 160: Technical Recommendations for Monitoring Individuals Occupationally Exposed to External Radiation* (Publications Office of the European Union) https://op.europa.eu/en/publication-detail/-/publication/9e018fd7–15c1–4679-8960-b9aaa8ffa099/language-en/format-PDF/source-193113083 (accessed March 2021)

EU 2018 *RP N° 188 Technical Recommendations for Monitoring Individuals for Occupational Intakes of Radionuclides* (Publications Office of the European Union) https://op.europa.eu/en/publication-detail/-/publication/ed6ac2a8-b7cf-11e8–99ee-01aa75ed71a1/language-en/format-PDF/source-193112819 (accessed March 2021)

EU 2019 *RP N° 193 Radon in Workplaces—Implementing the Requirements in Council Directive 2013/59/Euratom.* (Publications Office of the European Union) https://op.europa.eu/en/publication-detail/-/publication/93cc4aff-47c5–11ea-b81b-01aa75ed71a1/language-en/format-PDF/source-193112630 (accessed March 2021)

Fuchs W C 1896 Effect of the Röntgen rays on the skin *Western Electrician* **1896** 291

Grubbé E H 1933 Priority in the therapeutic use of x-rays *Radiology* **XXI** 156–62

Hume D 1739 *A Treatise of Human Nature* (London: John Noon)

IAEA 1996 International Basic Safety Standards for Protection Against Ionizing Radiation and for the Safety of Radiation Sources. IAEA Safety Series No. 115

IAEA 2014 Radiation Protection and Safety of Radiation Sources: International Basic Safety Standards. IAEA Safety Standards Series No. GSR Part 3

ICR 1929 International Recommendations for X-ray and Radium Protection. Second International Congress of Radiology, Stockholm, 1928. Acta Radiologica Supplementum III Pars I, 62–73. Kungliga Boktryckeriet. P.A. Nordstedt & Söner, Stockholm

ICR 1931 The work of the International X-ray Unit Committee and the International X-ray and Radium Protection Commission. Third International Congress of Radiology, Paris, 1931 *Acta Radiol.* **12** 586–94

ICRP 1951 International Recommendations on Radiological Orotection. Revised by the International Commission on Radiological Protection at the Sixth International Congress of Radiology, London, 1950 *Br. J. Radiol.* **24** 46–53

ICRP 1955 Recommendations of the International Commission on Radiological Protection *Br. J. Radiol.* **Suppl. 6** 1–92

ICRP 1958 Report on amendments during 1956 to the Recommendations of the International Commission on Radiological Protection (ICRP) *Radiat. Res.* **8** 539–42

ICRP 1959 *Recommendations of the International Commission on Radiological Protection* (New York: Pergamon) Now known as ICRP Publication 1

ICRP 1964 Recommendations of the International Commission on Radiological Protection *ICRP Publication 6* (Oxford: Pergamon)

ICRP 1966 Recommendations of the International Commission on Radiological Protection *ICRP Publication 9* (Oxford: Pergamon)

ICRP 1977 Recommendations of the ICRP, ICRP Publication 26 *Ann. ICRP* **1(3)**

ICRP 1991 1990 Recommendations of the International Commission on Radiological Protection, ICRP Publication 60 *Ann. ICRP* **21(1–3)**

ICRP 2007 The 2007 Recommendations of the International Commission on Radiological Protection, ICRP Publication 103 *Ann. ICRP* **37(2–4)**

ICRP 2008 Environmental protection: the concept and use of reference animals and plants, ICRP Publication 108 *Ann. ICRP* **38(4–6)**

ICRP 2017 Diagnostic reference levels in medical imaging, ICRP Publication 135 *Ann. ICRP* **46(1)**

ICRP 2018 Ethical foundations of the system of radiological protection, ICRP Publication 138 *Ann. ICRP* **47(1)**

ILO 1960a *Radiation Protection Convention, 1960 (No. 115)* (International Labour Organisation)

ILO 1960b *Radiation Protection Recommendation, 1960 (No. 114)* (International Labour Organisation)

ILO 1974 Constitution of the International Labour Organization, Geneva (1919), as amended up to 1974

ILO 2016 ILO. General Observation of the Committee of Experts on the Application of Conventions and Recommendations (CEACR) adopted in 2015 and published at the 105th Session of the International Labour Conference (2016) on the Radiation Protection Convention, 1960 (No. 115)

IXRPC 1938 International recommendations for x-ray and radium protection. Revised by the International X-ray and Radium Protection Commission at the Fifth International Congress of Radiology, Chicago, September 1937 *Br. Inst. Radiol.* **(leaflet)** 1–6

Leppin O 1896 Aus kleine Mitteilungen. Wirkung der Röntgenstrahlen auf die Haut *Dtsch. Med. Wochenschr.* **28** 454

Lindell B 1996 The history of radiation protection *Radiat. Prot. Dosim.* **68** 83–95

Lindell B 2019a The History of Radiation, Radioactivity, and Radiological Protection. Part 1. The Time Before World War II: Pandora's Box (http://nsfs.org/wp-content/uploads/2019/06/Pandoras-Box-2019-05-31.pdf)

Lindell B 2019b The History of Radiation, Radioactivity, and Radiological Protection. Part 2. The 1940s: The Sword of Damocles (http://nsfs.org/wp-content/uploads/2019/08/Sword-of-Damocles-2019-07-31.pdf)

Lindell B 2020a The History of Radiation, Radioactivity, and Radiological Protection. Part 3. 1950–1966: The Labours of Hercules (http://nsfs.org/wp-content/uploads/2020/03/Hercules_Downloadable_V1.pdf)

Lindell B 2020b The History of Radiation, Radioactivity, and Radiological Protection. Part 4. 1967–1999 + : The Toil of Sisyphus (http://nsfs.org/wp-content/uploads/2020/06/The-Toil-of-Sisyphus-2020-06-28.pdf)

Röntgen W C 1895 Über eine neue Art von Strahlen *Sitzungsberichte d. Phys. Med. Ges. Würzburg* **9** 132

IOP Publishing

Principles and Techniques of Radiological Protection

Michael Charles Thorne

Chapter 14

Radiological protection in the nuclear industry

M C Thorne and J K Humphries

14.1 Introduction

The nuclear fuel cycle includes activities related to the mining and milling of uranium ores, the enrichment of uranium, fuel fabrication, the operation of reactors used for research and power production, the reprocessing of spent fuel, and the disposal of low-intermediate- and high-level wastes, and of spent fuel if reprocessing is not adopted. It could be argued that the development, production and maintenance of nuclear weapons should also be included. Although, issues related to nuclear weapons are not discussed explicitly in this chapter, some of the information presented is also of relevance in that context. Issues related to the transport of radioactive materials are also excluded from consideration, because they are addressed in chapter 18.

14.2 The mining and milling of uranium

14.2.1 The mining and milling process and the history of uranium exploitation

Radiological protection issues related to the mining and milling of uranium are addressed in IAEA (2020a, 2020b). Although thorium can be used as a source of fuel for nuclear reactors, uranium is, and is likely to remain, the principal material used in both power-production and research reactors. Mining and milling of uranium cover both the initial extraction process and the subsequent treatment of the ore (or leachant in some extraction processes) to provide supplies of well-defined chemical forms of the element for subsequent enrichment (as required) and use.

Uranium can be recovered by underground mining, open-pit mining and *in situ* leaching (ISL). The ISL process involves dissolving the uranium within the ore body by circulating groundwater fortified with oxygen and a slightly acid or slightly alkaline chemical additive through injection wells and then extracting the uranium-containing solution through recovery wells. A variant technique is heap leaching, in which the ore is mined conventionally, but then placed on surface pads where extractive acid or alkaline liquors are pumped over and through the material.

doi:10.1088/978-0-7503-2424-3ch14

Uranium can also be recovered as a by-product from phosphate fertiliser production and from the mining of other metals, e.g. copper and gold. In such contexts, the treatment process to recover uranium can be more complex than in cases where uranium is the primary product. However, waste heaps arising from mining activities may be readily available for heap leaching.

The processing of uranium ore typically involves crushing, grinding or milling, followed by extraction of uranium by acid or alkaline leaching and separation of the uranium-bearing liquid. Uranium-bearing leachate solutions obtained from ISL, heap leaching and ore processing are then concentrated, purified, precipitated and dried. The final products include U_3O_8, UO_4, UO_2 and ammonium diuranate (IAEA 2020a, 2020b). The primary precipitate is usually ammonium diuranate $\{(NH_4)_2U_2O_7\}$. This is commonly known as 'yellow cake' due to the colour of the product from the early production facilities. However, it is now commonly oxidised to produce U_3O_8 or UO_3 and varies in colour from brown to black (IAEA 2023).

A refining stage is used to produce uranium metal. This involves dissolution of yellow cake in nitric acid, with extraction of the uranium from the acid feed using tributyl phosphate (TBP) diluted with kerosene (or another suitable hydrocarbon mixture). The uranium is then stripped into acidified water to yield highly purified uranyl nitrate. $UO_2(NO_3)_2$ which is calcined to UO_3 and reduced with hydrogen to UO_2. Powdered UO_2 can be treated with gaseous HF to produce UF_4. The UF_4 can then be reduced to uranium metal using magnesium (the Ames process) or fluorinated using fluorine gas to produce UF_6, which is the feedstock for isotopic enrichment (IAEA 2023).

Early uranium production was largely as a by-product from mines in Saxony, Bohemia and Cornwall in the first part of the 19th century. At that time, uranium was used mainly to colour glass and ceramics and the first uranium colour-production factory was opened at Jáchymov, Czech Republic (then known as St. Joachimsthal, Bohemia) in 1855 (IAEA 2023).

The total production of uranium during the 19th century was about 1000 tonnes. However, production accelerated rapidly in the first few decades of the 20th century spurred on by the high price of radium. New mining areas developed, e.g. in Kyrgistan, Sweden, Australia, the USA, Madagascar and Portugal. The USA dominated the world uranium market from 1912 to 1922, which is when high-grade ore from the Belgian Congo also entered the market. In 1930, high-grade ore was discovered at Great Bear Lake in northwest Canada with a radium-refining plant commencing operations at Port Hope on Lake Ontario in 1933. The discovery of nuclear fission in the late 1930s and its exploitation after World War II resulted in the rapid growth of large-scale uranium prospecting and mining activities during the 1950s (IAEA 2023). As of 2013, the major producers of uranium, expressed as metal content, were Kazakhstan (22 567 tonnes), Canada (9331 tonnes) and Australia (6370 tonnes). Identified reasonably assured resources of uranium are located mainly in Australia (29%), Kazakhstan (12%), Russian Federation (9%), Canada (8%), Niger (7%), South Africa (6%), Namibia (6%), Brazil (5%) and the USA (4%) (IAEA 2023).

14.2.2 Main routes of occupational exposure

The IAEA (2020a) discusses occupational exposures in the context of the following broad categories of activity:

- Exploration;
- Underground mining;
- Surface mining;
- *In situ* leach mining and processing;
- Heap leaching;
- Processing facilities;
- Non-conventional uranium extraction as a by-product;
- Uranium-tailings facilities;
- Material in transport;
- Decommissioning.

For each of these categories, the IAEA (2020a) describes the associated process, discusses issues of design and operation, identifies the principal pathways of radiological exposure, addresses approaches to the control of those exposures, and finally provides guidance on monitoring of those pathways and on dose assessment. These principal pathways comprise external gamma irradiation, inhalation of Rn-222 and its short-lived progeny and the inhalation of dusts containing long-lived radionuclides. The relative importance of these pathways depends on the environmental context, e.g. inhalation of Rn-222 and its progeny tends to be of greater significance in poorly ventilated underground spaces and the inhalation of dusts where dry drilling and blasting are undertaken. Therefore, wet drilling and mining techniques are preferred. For both Rn-222 and its progeny and aerosols, mitigation by an effective ventilation regime is an important consideration, with workers located preferentially upstream of areas of contamination. In situations where ventilation control cannot be achieved in the short term, e.g. because of a failure of the ventilation system, the use of respiratory protection may be warranted.

In monitoring for external exposure, personal, electronic direct-reading dosimeters can be useful for monitoring near strong gamma sources in high-grade mines. However, such dosimeters are not generally used for official purposes, for which the standard practice is to issue passive integrating dosimeters, e.g. thermoluminescent or optically stimulated luminescence dosimeters (TLDs or OSLs).

In the case of Rn-222 and its progeny, underground concentrations can vary rapidly both spatially and temporally. This means that real-time monitoring and control measures may be required. Continuous monitors for radon progeny are important in monitoring locations with the potential for high concentrations and in alerting workers to changing conditions. Also, the variability in concentrations and the practical difficulties of tracking workers in underground conditions mean that it is often preferable to use personal monitors to assess exposures to radon progeny.

Monitoring of dust incorporating long-lived radionuclides can be based on a combination of personal and area monitoring, with the emphasis on area sampling with high-volume air samplers in those areas with a lower potential for significant exposures. Drilling operations can generate high dust concentrations and drillers are prone to bypass control mechanisms owing to a perceived need to 'feel' the drill to maximise performance. Exposure to dust during the maintenance of equipment may be controlled by ensuring that equipment is cleaned before it is maintained.

In general, radiological exposures arising from surface mining are lower than those arising from underground mining. However, the principal exposure mechanisms, and monitoring and control procedures, are similar.

In all mining operations, although the focus should be on the principal pathways outlined above, there is also a need to insist on good hygiene practices to limit ingestion and to be aware of the potential for contaminated wounds, e.g. limiting the potential for exposure to sharp metal objects and ensuring that existing wounds are properly dressed prior to work.

In *in situ* leaching processes, enhanced external exposures may arise from concentrates of radium in some process components and in areas where large amounts of yellow cake concentrates are produced, packaged and stored. Enhanced exposures to radon and its progeny can arise where dissolved radon is brought to the surface in the leaching solution. Although radon concentrations can be high in areas where ventilation is limited, radon progeny concentrations may be low because of the limited amount of time available for their ingrowth. Radiological impacts of aerosols incorporating long-lived radionuclides are almost exclusively associated with yellow cake drying areas, because this is the first stage at which dry materials are handled. However, there should be an awareness that the drying of spills of liquor can result in aerosols in other areas. Similarly, in heap leaching, exposure to radioactive aerosols can be limited by keeping the heap wet. Whereas external exposure from heap materials may be significant, radon is less of an issue than in underground mining because of its rapid dispersion in the open air.

Processing involves initial stages of ore blending, crushing or grinding, and the production of slurries for subsequent treatment by acid or alkali leaching, with the residual solids directed to the tailings processing route and the uranium-rich solution directed to the purification and concentration route, and thence to drying and packaging of the final product. For most workers on a processing plant, external irradiation is the main route of exposure, because operations on slurries and solutions predominate, while the arising of dust from initial ore crushing can be mitigated by water sprays, as well as control of ventilation and exclusion of workers from the crushing area. These measures are also of importance in mitigating exposure to radon and its progeny because radon tends to be liberated from the ore during the crushing process. Maintenance of process equipment can give rise to enhanced external exposures because of the build-up of uranium-series radionuclides in processing tanks and pipework. Emptying such vessels can reduce self-shielding effects and result in increased dose rates. Enclosed vessels or areas can have enhanced radon concentrations arising from radium-rich residues and it is good practice to flush such vessels or areas with clean air before the entry of maintenance personnel. Negative pressurisation of process vessels may be used to prevent the escape of radon and its progeny if the tank hatches are opened for inspection or sampling. Throughout, water-based clean-up procedures are preferred to reduce dust generation.

In most process operations, potential radiation exposures in accident situations would not be sufficiently large as to prevent emergency responders from performing their duties. However, a detailed analysis of potential accidents may be required for facilities involved in the processing of high-grade uranium. Such facilities may also require the installation of shielding to limit external exposures during routine working. Control of airborne dust is also of specific importance when mining or processing high-grade ores.

Non-conventional uranium extraction generally involves small production quantities and low-grade ores. Therefore, it does not give rise to any specific radiological protection issues.

Uranium processing can result in the production of large quantities of waste material, known as 'tailings'. These contain up to 85% of the initial radioactivity of the ore, as well as heavy metals and other toxic contaminants. Furthermore, the physical and chemical processing of the ore means that these radioactive and other contaminants are more readily available in the tailings than in the original ore. Tailings are managed by various strategies including thickening, dewatering, backfilling underground, desulphuring or blending with waste rock. However, characteristically, they are transferred to tailings ponds, where they are progressively dewatered and subsequently capped with an engineered structure in which the top layer is vegetated soil. Underlying layers may include coarse material to prevent biointrusion and enhance drainage, and geotextiles or clay layers to limit water infiltration. Similar techniques are used in capping near-surface facilities for the disposal of solid radioactive wastes. Before such a cap has been installed, there is the potential for significant external exposure from the exposed material. Also, radon can be exhaled, particularly as the tailings dry out, though it tends to rapidly disperse after release. Coverage of the tailings with about one metre of soil is sufficient to substantially reduce dose rates from external exposure, and to strongly attenuate the exhalation of radon, because of its radioactive decay while it is being transported through the cap by a combination of advection and diffusion.

For routine transport of uranium products, external exposure is the main consideration and the normal control practices of time, distance and shielding apply. Details of requirements governing the transport of radioactive materials, including uranium products, are provided in chapter 18.

The decommissioning of a uranium operation is a combination of plant demolition and disposal, large-scale Earth moving, stabilisation of the surface waste facilities and rehabilitation of the surface areas. Issues can arise with the internal contamination of process equipment and with the generation of aerosols if the capping on waste heaps or disposals of tailings is disturbed. Unfortunately, many uranium operations throughout the world have been terminated or abandoned without the performance of an adequate programme of decommissioning. These uranium liabilities constitute environmental hazards that will require remediation in the future. In addition to the radiological exposure pathways described above, the reflooding of mines after active drainage has ceased may result in groundwater that is highly contaminated with uranium salts entering the surface environment as seeps, springs and streams. In addition, these liabilities can constitute conventional hazards due to subsidence and slope instabilities of piles of waste rock. Issues relating to uranium liabilities are discussed in detail in IAEA (2023).

14.3 Reprocessing

When the fuel is identified as spent, it is removed from the reactor and stored for a short period of time in the local reactor fuel storage pond, to allow the short-lived nuclides to decay and minimize their impact during transport. After a few years, the fuel is removed from the local storage pond using a flask and sent to another location.

In the UK, all spent nuclear fuel is sent to the Sellafield site, where there are two reprocessing facilities. Oxide fuel (such as Advanced Gas-cooled Reactor Fuel, and Pressurized Water Reactor Fuel) was reprocessed in the Thermal Oxide Reprocessing Plant (ThORP) until November 2018 when reprocessing was stopped at the request of the UK Government. All un-reprocessed oxide fuel will be stored in fuel storage ponds at Sellafield awaiting the definition of a final disposal route, assumed to be to a Geological Disposal Facility.

Magnox Fuel from the now shutdown fleet of Magnox Nuclear Power stations including Calder Hall and Chapelcross, is reprocessed in the Magnox Reprocessing Plant, which is delivering the Magnox Operating Plant objective of reprocessing as much Magnox fuel as possible. This is because the fuel cladding is not suitable for long-term pond storage as it can degrade during wet storage if the conditions are not carefully managed and therefore it is both the Best Available Technique (BAT) and As Low As Reasonably Practical (ALARP) to reprocess this fuel into well-defined products that can be safely stored.

Reprocessing of spent nuclear fuel into products is a hazardous activity and a nuclear safety case is required to ensure that all hazards are considered, and engineering controls are implemented to minimize human interactions with the process and reduce dose uptake from routine and non-routine tasks. In the UK, nuclear safety cases are overseen by the regulator, The Office for Nuclear Regulation (ONR).

The radiological protection hazards from reprocessing are different at the various stages of the process in which the radionuclides are routed into the differing streams to allow the chemical separation to occur. It is always wise, therefore, to ask the question where in the process has an event occurred prior to giving any radiation protection advice and to confirm the radionuclides present.

14.3.1 Stages of reprocessing and their radiological protection issues

Dissolution: In both the Sellafield reprocessing plants, the first step is to dissolve the fuel in hot nitric acid. In ThORP, a gadolinium compound was added as a neutron poison and the fuel was sheared by a large weight into small sections to allow the nitric acid access to the uranium bar inside the cladding. In Magnox reprocessing, the cladding of the fuel is removed prior to transport to the reprocessing plant and the entire bar of uranium fuel is fed into the dissolver. No active neutron poisoning is added as the fuel is not enriched in U-235. In both facilities, heat is applied to dissolve the fuel.

Chemical separation: Both the Sellafield plants use the same theory of solvent separation using pressure and temperature to change valency states of radionuclides between solvent and aqueous phases to separate different species into the aqueous and solvent streams. This is completed through separate cycles, described below.

- Highly Active Cycle (HA cycle): This is the first cycle of the chemical separation and removes the fission products which are sent to another plant for initial storage and eventual vitrification into glass blocks. For foreign fuel the HA waste has been transported back to the country of origin for them to store.
- Plutonium Purification Cycle (PP Cycle): The plutonium is removed from the dissolved fuel stream and fed into the next cycle.
- Plutonium Finishing Cycle (PF Cycle): The plutonium is concentrated and then powdered. The powder is then packed into cans for storage (this is an offshoot of the PP cycle).
- Uranium Purification Cycle (UP Cycle): This removes the uranium from the solvent phase and allows solvent to be recycled back to the HA cycle to continue within the facility.
- Uranium Finishing Cycle (UF cycle): Uranium-bearing aqueous liquor is concentrated and sprayed into a powder. The powder is then stored in drums for reuse.

Product finishing: Once each stream is purified, it is then finished and turned into powder, dried, and put into storage vessels, usually cans for plutonium and drums for uranium. The radiological risk here is high, as the powders tend to be highly mobile and contamination control needs continual reinforcement. For plutonium, gloveboxes give engineered control, although gloveboxes are only as good as the weakest part of the box, the glove, with potential internal dose from breaches in the gloves. For uranium, human access to the processing areas is manageable and visible yellow powder means contamination control can become lax, due to a perception any radioactive contamination will be easily seen.

Waste processing: There are solid, liquid, and gaseous byproducts of nuclear reprocessing. Solid materials include fuel cladding (which is encapsulated in cement then stored) as well soft waste (such as used PPE) and contaminated equipment from the process. Liquids such as raffinates are treated to minimize activity and control temperature and pH prior to final dispatch to sea. Gaseous waste is treated by Highly Efficient Particulate (HEPA) filtration to remove the radioactive particles and wet treated (such as with a scrubber) to remove chemical hazards prior to discharge to the air. The liquid and gaseous releases are under appropriate environmental legislation and permits.

14.3.2 Radiological hazards from nuclear reprocessing

Criticality: During processing, the fuel is dissolved and therefore the radionuclides are mobile in the solution, and this gives the potential for a critical mass to form and a criticality to occur, which would release huge amounts of radiation in a small period of time. Safeguards to prevent this include use of geometrically safe vessels, neutron poisons (such as gadolinium) and criticality detection systems to ensure prompt evacuation and minimize dose uptake if a criticality were to occur.

Radiation from the process: The process of cutting and dissolving spent nuclear fuel leads to high dose rates (of the order of hundreds mSv per hour), so these processes are remote, within cells or caves where no human access is permitted. Penetrations are required to allow monitoring to take place and ensure services (such as lighting or control systems for in cave machinery) are maintained. There is potential, therefore, for shine from these penetrations.

Contamination from the process streams: Although people are excluded from the process, the machines that deliver the work of reprocessing do require human interactions for maintenance or repair, and these items are heavily contaminated with very mobile fuel dust. The items are usually moved to a maintenance area following in-cave cleaning and personnel entries into the area allow assessment of the risk from both contamination and radiation present on the area of interest. Plugs are used to access cell equipment, and these sometimes are required to be removed, which has potential to lead to a release of contamination. Engineered controls (such as cascade ventilation) help to minimize this, but contamination controls such as temporary prepared areas, monitoring of individuals and items, and activity measured in air monitors would usually be deployed to identify any activity and promptly react to it.

Sampling: Although instruments are present that indicate the state of each process, samples of the process liquors are required to ensure product streams conform to specification. Some of these samples are highly active, especially from the HA cycle. In ThORP, there was a high level of engineered control, with remote sampling and transfer to laboratory for processing. In Magnox reprocessing, the controls are more hands-on with an increased risk of contamination spread and potentially high radiation samples.

The products arising from reprocessing are:

- Highly active waste—which is stored and then vitrified into glass for long-term storage
- Uranium oxide (UO_3) powder that can be recycled into new fuel
- Plutonium oxide which is stored in cans and has potential to be reused as a mixed oxide (MOX) fuel.

At Sellafield, all these products are stored awaiting reuse.

14.4 Waste disposal

14.1.1 Preferred approaches to the disposal of solid radioactive wastes

The nuclear fuel cycle gives rise to a wide diversity of solid radioactive wastes. Setting aside the wastes arising from uranium mining and milling that are discussed in section 14.2, wastes are typically distinguished into low-level waste (LLW), intermediate-level waste (ILW) and high-level waste (HLW). In addition, other materials may also be recognised as specific waste classes. These include spent nuclear fuel (SF), i.e. fuel elements that have been subject to irradiation in a reactor, depleted uranium (DU) arising from enrichment operations, and plutonium that has been separated from other elements in reprocessing but is subsequently found not to be required for MOX manufacture or for military applications.

Although the focus here is on wastes arising from the nuclear fuel cycle, in practice, solid radioactive wastes arising from other sources, e.g. other industries, research activities and nuclear medicine, are often handled by the same waste management organisation. Therefore, the following discussion is relevant to the handling and disposal of all types of solid radioactive wastes.

In the UK, very low-level waste (VLLW) is distinguished from LLW. VLLW is defined by reference to restrictive limits on activity concentrations. Sites that produce VLLW can dispose of the waste with regular household or industrial waste at permitted landfill facilities. The major components of VLLW from nuclear sites are building rubble, soil and steel items. These arise from the dismantling and demolition of nuclear reactors and facilities. LLW contains relatively low levels of radioactivity, in the UK not exceeding 4 GBq per tonne of alpha activity, or 12 GBq per tonne of beta/gamma activity.

Most LLW comes from the operation and decommissioning of nuclear facilities. The waste includes items such as scrap metal, paper and plastics. Some smaller amounts of LLW also come from hospitals and universities. About 94% of all UK radioactive wastes (by volume) are in the LLW category.

Some LLW near the radioactive content threshold is managed as higher activity waste (HAW). This may be because it is too difficult to separate from ILW or does not meet the Waste Acceptance Criteria for current LLW disposal facilities. HAW includes HLW, ILW and some LLW that is unsuitable for disposal in the UK Low Level Waste Repository (LLWR). HAW arises from activities such as reactor operations, reprocessing of spent nuclear fuel and decommissioning.

HLW is waste where the temperature may rise significantly because of its radioactivity. The design of waste storage or disposal facilities must take this into consideration. Heat output is also an important consideration when considering the disposal of SF. HLW is a by-product from reprocessing spent fuel from nuclear reactors. It typically occurs in liquid form and vitrification is used to convert it into a solid product suitable for disposal.

A wide variety of approaches have been proposed for the disposal of solid radioactive wastes and some, such as the injection of liquid waste into deep boreholes, have been implemented. However, there is now a consensus that geological disposal, in which the wastes are emplaced in engineered facilities within suitable geological formations is the preferred option, though above-surface disposal of short-lived wastes in above-ground engineered structures may also be adopted.

The IAEA (2020b) has provided a general typography of types of facility that have been developed or proposed for the geological disposal of radioactive wastes. This is illustrated in figure 14.1.

In general, geological disposal relies on the use of multiple barriers to prevent radionuclide release from the disposal facility and to limit the transport of any radionuclides that are released to the accessible environment. Thus, the host rock may be considered to have two broad functions. The first is to provide a passive and

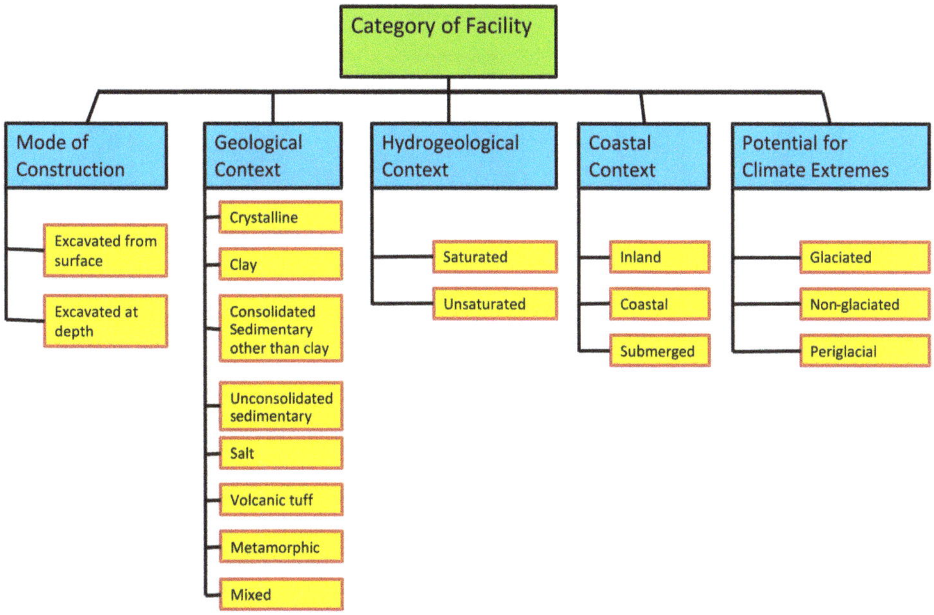

Figure 14.1. Types of geological disposal facilities and the contexts in which they may be developed. IAEA (2020b), © IAEA, 2020, p. 18, ISBN 978-92-0-105020-5

stable environment for the engineered facility. The second is to ensure that radionuclides released from the engineered facility are transported only slowly to the accessible environment, allowing time for substantial radioactive decay to occur during their transport. The relative importance of these two functions differs depending upon the type of waste to be disposed of, the design of the engineered system, and the strategy adopted for siting the facility (including location, preferred host rock and depth).

Achieving a stable environment, requires consideration of both the potential for seismic events and surface processes that could adversely affect the disposal facility. For example, in the KBS-3 concept adopted for spent fuel disposal in Sweden and Finland, the spent fuel is packaged into 0.05 m thick copper canisters with a cast-iron insert that gives them mechanical strength under external loading. These canisters are to be emplaced in deposition holes at a depth of about 450 m in ancient basement rocks (see figure 14.2).

The copper canisters are projected to be resistant to penetration by corrosion on a timescale of hundreds of thousands of years or more. However, they could fail because of seismic events causing movements on faults passing through the deposition holes. Therefore, the overall layout of the repository is such that zones on which substantial fault movements could occur are avoided by applying respect distances. Furthermore, at a smaller scale, individual deposition holes are excluded from use if they are found to be intersected by seismically significant secondary faults.

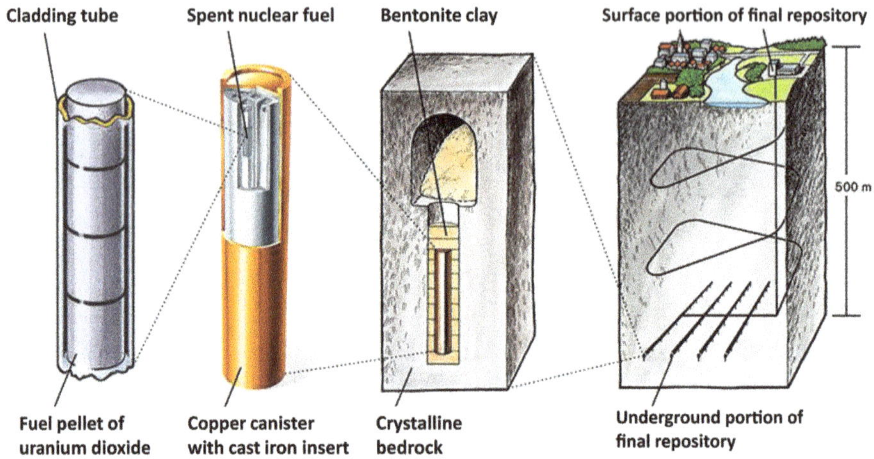

Figure 14.2. The KBS-3 concept for the disposal of spent nuclear fuel. Image courtesy of © Svensk Kärnbränslehantering AB

An example of a surface event that could adversely affect a KBS-3 repository is the passage of an ice sheet over the site during a future glacial episode. This would impose a substantial additional mechanical load on the containers. However, the canister and insert are designed to support this load together with loads arising from the overburden of rock present and the swelling pressure of the bentonite buffer surrounding the canister.

Achieving a passive environment requires consideration of the chemical characteristics of the groundwaters that could contact the various engineered components of the facility. In the context of a KBS-3 facility, there is a need to avoid groundwaters that could be aggressive either in terms of eroding the bentonite buffer or by delivering corrosive agents to the canister surface. Again, external events need to be considered when determining whether the passive environment will be maintained in the long term. For example, the passage of an ice sheet could force oxygenated waters containing low concentrations of solutes to repository depth, and such waters could adversely affect both buffer stability and container integrity. In this context, the low hydraulic conductivity of the rock at depth is an important factor in mitigating the penetration of such waters.

If radionuclides are released from an engineered facility, their passage to the accessible environment can be limited by the slow movement of groundwater, by diffusion of the radionuclides into the rock matrix and by sorption of those radionuclides to fracture surfaces, the minerals deposited in fractures and to surfaces within the rock matrix. In addition, within the engineered facility and in its immediate vicinity radionuclide migration may be precluded by their co-precipitation, e.g. Ra-226 may be co-precipitated in barium-radium sulphate, which is highly insoluble. However, in practice, co-precipitation effects are seldom invoked in safety arguments related to the geological disposal of radioactive wastes.

Disposal in salt formations comprises a special case, in that bulk groundwaters are not present. The only water that exists within such a formation is located within small, isolated voids within the salt. Thus, the safety of disposal in a salt formation is mainly determined by other considerations, such as the long-term stability of the salt (bearing in mind its tendency to deform by creep under mechanical loads) and its potential for future exploitation as a resource.

In argillaceous formations, notably plastic clays, the water content may be high (several tens of percent by volume), but the small sizes of the pores in the material and its homogeneity at larger spatial scales, mean that the water is immobile or flows to only a very limited degree under imposed hydraulic gradients. In these conditions, transport of radionuclides is mainly by diffusion and timescales of transport for unabsorbed radionuclides across a few tens of metres of the material can be hundreds of thousands of years. Therefore, with such host rocks, attention is properly focused on the potential occurrence of geological anomalies (e.g. sand lenses) or human perturbations (e.g. boreholes) that could provide short circuits for water flows and hence for radionuclide transport.

In hard-rock formations, fractional water contents can be low, e.g. less than one percent in some igneous rocks. However, this is because the water is present in an array of fractures with only a few fractures of modest aperture present per unit volume. In such rocks, the volumetric flow of groundwater may be limited, but the velocity of flow can be high, resulting in short transport times for unabsorbed radionuclides from the repository to the accessible environment. In such rock types, absorption can play a significant role in limiting the transport of radionuclides through the host rock, as can diffusion into the rock matrix because the water within the matrix is stationary, meaning that a radionuclide that has diffused into the matrix has to diffuse out again before it can be transported onward.

Absorption is strongly element-dependent and is generally much greater for cationic species than for anions. This means that anions, such as Cl-36 as chloride, I-129 as iodide and Tc-99 as pertechnetate (TcO_4^-), which are typically only present as minor constituents in waste streams can be the dominant radionuclides determining the radiological impact of a disposal facility. In addition to limited sorption, the transport of anions can be enhanced because the negative charge on anions makes it difficult for them to penetrate through negatively charged pore throats in the rock, meaning that they do not diffuse as readily into the matrix as do cations.

The diversity of sites selected or under consideration for the disposal of solid radioactive wastes is illustrated by the examples selected by the IAEA (2020b). Their selected examples are summarised in table 14.1, conforming to the classification scheme shown in figure 14.1.

Table 14.1 illustrates the diversity of contexts adopted for disposal of solid radioactive wastes. However, irrespective of context, proposals to site, construct, operate, close and relinquish control on such facilities require radiological (and, in some cases, non-radiological) impact assessments. The development of such impact assessments and their role in the overall safety case for the facility are discussed below.

Table 14.1. Illustrative examples of different types of existing or proposed facilities for the disposal of solid radioactive wastes (from IAEA (2020b)).

Region or site	Mode of construction	Geological context	Hydrogeological context	Coastal context	Potential for glaciation
Mol, Belgium	Near surface	None	Unsaturated	Inland	Unglaciated
Mol, Belgium	Excavated at depth	Clay	Saturated	Inland	Periglacial
Switzerland	Excavated at depth	Clay	Saturated	Inland	Glaciated
Forsmark (SFR), Sweden	Excavated at depth	Crystalline	Saturated	Submerged	Glaciated
Forsmark (KBS-3), Sweden	Excavated at depth	Crystalline	Saturated	Coastal	Glaciated
El Cabril, Spain	Excavated from surface	Unconsolidated sedimentary	Unsaturated	Inland	Unglaciated
Lowland Britain Periglacial	Excavated at depth	Mixed	Saturated	Inland	Glaciated or
Olkiluoto, Finland	Excavated at depth	Crystalline	Saturated	Coastal	Glaciated
Canada, LLW/ILW	Excavated at depth	Consolidated sedimentary	Saturated	Inland but lakeside	Glaciated
Canada, spent fuel	Excavated at depth	Not selected	Saturated	Not selected	Glaciated
Yucca Mountain	Excavated at depth	Volcanic tuff	Unsaturated	Inland	Unglaciated

14.4.2 Radiological impact assessments for geological disposal facilities for solid radioactive wastes

Radiological impact assessments for disposal facilities for solid radioactive wastes may be characterised into evaluations of radiological impacts on:

- Workers occupationally exposed during construction, operation and closure of the facility;
- Members of the public and the environment during construction, operation and closure of the facility;
- Members of the public and the environment during the post-closure period.

Where significant amounts of chemically toxic materials are disposed of with the radioactive wastes (e.g. lead shielding materials) or where toxic materials could be released from materials used to condition and package the wastes or construct the facility, non-radiological impact assessments may be required in each of these contexts.

Radiological impact assessments applicable during the phases of construction, operation and closure will be like those undertaken for other types of facility handling radioactive materials and are not addressed in detail here. The construction period is included because many repository concepts are intended to receive wastes over several decades. Thus, construction of new sections of the facility may occur in parallel with waste emplacement in other sections, and with backfilling of sections where waste emplacement is already complete. Parallel working of this nature has implications for radiological control, e.g. in terms of area classification and the updating of ventilation controls in underground areas.

In respect of radiological impacts on members of the public and the environment during construction, operation and closure, radionuclide releases are likely to be mainly to the atmosphere, because only limited amounts of liquid radioactive wastes should arise (particularly if the wastes have been conditioned and packaged before transport to the disposal site). Releases to the atmosphere can arise because of a need to have vented packages for some waste types. This need arises where significant gas evolution may arise from the packaged wastes during storage or after disposal. Bulk gases may arise from the degradation of organic wastes or from corrosion of metals under anaerobic conditions. The likely bulk gases are hydrogen, methane and carbon dioxide, with radionuclide releases in the gas phase including H-3 (as hydrogen gas or tritiated water vapour), C-14 (as methane or carbon dioxide) and Rn-222. There could also be releases of radioisotopes of inert gases, such as krypton, though these are likely to have been substantially lost from the wastes during treatment. In addition, other volatile radioisotopes could be released, e.g. I-129 as methyl iodide.

Radiological impact assessments for the post-closure period give rise to issues that are not typically addressed in other contexts. The post-closure period when the site remains under administrative control is relatively straightforward. During this period, a monitoring programme is likely to be put in place (keeping in mind the need to ensure that invasive monitoring, e.g. by pumping water from boreholes to assay its radionuclide content, does not adversely affect the containment properties of the facility). In addition, administrative controls should prevent other human activities with the potential to adversely affect the facility. However, it has to be recognised that control of the site cannot be maintained indefinitely, so it is generally assumed that knowledge of the use of the site for disposing of radioactive wastes is lost within a few hundred years. This is several half-lives of some of the key radionuclides likely to be disposed of (e.g. Co-60, Sr-90, Cs-134 and Cs-137), but is much less than the half-lives of many others (e.g. C-14, Cl-36, Ni-59, Se-79, Tc-99, I-129, Np-237, Pu-239, Pu-240). In practice, information relating to the use of the site may be slowly degraded rather than being lost instantaneously. However, for assessment purposes, it is generally assumed that the loss of information occurs instantaneously.

Beyond the time when loss of knowledge of the site is assumed to occur, various scenarios need to be assessed to determine the radiological impacts that could arise.

The timescale of such assessments typically varies from a few thousand years for LLW, particularly if its radioactive content is dominated by shorter-lived radionuclides, through to about one million years for SF, by which time its residual radioactivity is comparable with that of the uranium mined to fabricate it. On these timescales it is impossible to predict in detail the future characteristics of the site or human activities, so it is conventional to base assessments on human habits and behaviour at and around the site at the present day, or at analogous sites if potential long-term changes in the climate and landscape are being considered.

Because knowledge of the site is assumed to be lost, one or more of the scenarios studied usually relates to inadvertent intrusion into the repository or its immediate vicinity. Types of intrusion studied often include boreholes for exploration or exploitation of resources that pass through the repository. In such scenarios routes of exposure typically include external irradiation or inhalation of dust from core samples extracted from a borehole, or various pathways arising from localised contamination of soil in the vicinity of the borehole. Some assessments also include larger-scale excavations into or in the vicinity of a repository, particularly where a near-surface facility is being assessed.

Concerns over the potential for inadvertent intrusion into a repository often lead to a regulatory requirement that the repository is sited at a location and at such a depth that there is no, or limited, incentive to explore or exploit the host rock. Igneous basement tocks at depth are an example of a host rock containing little in the way of exploitable resources.

Intrusion in the vicinity of a repository can include scenarios in which deep wells are sunk for groundwater abstraction. This is the reference scenario in the License Application for an SF and HLW disposal facility at Yucca Mountain, Nevada (US DOE 2008). In this case, radionuclides are envisaged as being transported downward from the repository through the unsaturated zone, then horizontally in a regional aquifer before being extracted from deep groundwater wells in Amargosa Valley. Because deep groundwater wells typically have large pumping rates, they can capture extensive plumes of groundwater pollution. Thus, it is usual to assume, as is done in the case of Yucca Mountain, that the entire flux of radionuclides released from a repository is captured by a deep groundwater well downstream of that repository.

In addition to intrusion, post-closure radiological impact assessments typically include scenarios in which radionuclides are released to groundwaters and travel downstream in those groundwaters until they discharge at the ground surface or are sampled by shallow wells or boreholes with small abstraction rates that result in only limited perturbations in groundwater flow. For such scenarios and for scenarios in which deep groundwater wells abstract contaminated water, a wide range of pathways for radionuclide transport in the biosphere need to be considered. These pathways are like those discussed in chapter 11 and formal methods, such as interaction matrices, have been developed to ensure that comprehensive conceptual models of those pathways are developed and implemented in mathematical models. In this context, the IAEA sponsored the development of a formal methodology for the specification of reference biospheres for use in assessments (IAEA 2003) and this has recently been updated (IAEA 2022).

In addition, in some contexts, radionuclides can be released in the post-closure period in association with bulk releases of gases arising either from the degradation of organic materials or the corrosion of metals in an anaerobic environment. Here the gases of greatest interest are carbon dioxide and methane labelled with C-14 (H-3 is too short-lived to be of relevance and Rn-222 will mainly arise from Ra-226 that has either been transported to the soil zone or produced there from its longer-lived precursors). Experimental studies have shown that methane is almost completely metabolised to carbon dioxide by microbial processes in soils, so the main pathway giving rise to a radiological impact is the uptake by plants of C-14 bearing carbon dioxide and their subsequent consumption by domestic animals and humans.

As mentioned above, post-closure radiological impact assessments of disposal facilities for solid radioactive wastes typically consider timescales ranging from a few thousand out to one million years. On such timescales both the climate and the landscape at a disposal site can change markedly. Thus, such assessments typically include consideration of scenarios involving various alternative projections of changes in the climate and the landscape. The envelope of changes in climate is often based on extrapolations of the scenarios proposed by the Intergovernmental Panel on Climate Change (IPCC). Issues arising in incorporating changes in climate and landscape into assessment studies are addressed in detail in IAEA (2020a, 2020b).

The above discussion shows that post-closure radiological impact assessments of disposal facilities for solid radioactive wastes involve the formulation of multiple scenarios. In some assessments all of these are treated on a common basis, whereas in others a distinction is made between a reference scenario (which may be the expected scenario or may simply be a scenario selected to provide a basis for comparison with other scenarios), variant scenarios (which may be less likely than the reference scenario, or merely alternatives) and ‘what if?’ scenarios that explore situations that are not expected to arise, but that illustrate the robustness of the disposal system, e.g. to loss of function of one of the barriers. Where scenarios are associated with different probabilities of occurrence, the issue arises as to how their results should be used to demonstrate compliance with regulatory guidance. In some regulatory regimes there is a requirement to aggregate either doses or risks over the various scenarios weighting each scenario by its probability of occurrence. The complexities of aggregation can be compounded where each scenario has been modelled as multiple calculation cases or by a Monte Carlo approach with sampling of input parameters for the scenario from probability density functions. Specific issues arise if doses or risks arising from scenarios, calculation cases or realisations in a Monte Carlo are differently distributed in time or apply to spatially distinct population groups. For example, if various realisations in a Monte Carlo simulation result in peak doses at different times, should the combined result be calculated by summing the individual time distributions of dose and finding the peak of the resultant curve, or should the peaks from the individual curves be summed even though they occur at different times? The former approach seems intuitively more satisfactory, but it can result in risk dilution if a lack of knowledge of parameter values results in peaks that should be closely clustered in time being unduly spread out. Such issues give rise to difficult philosophical considerations and can imply complex mathematical computations. These matters are not discussed further here, but the reader should be aware that they continue to be vigorously debated and are not easily resolved.

References

IAEA 2003 *'Reference Biospheres' for Solid Radioactive Waste Disposal, Report of BIOMASS Theme 1 of the BIOsphere Modelling and ASSessment Programme, IAEA-BIOMASS-6* (Vienna, Austria: International Atomic Energy Agency)

IAEA 2020a *Occupational Radiation Protection in the Uranium Mining and Processing Industry, IAEA Safety Reports Series No. 100* (Vienna, Austria: International Atomic Energy Agency)

IAEA 2020b *Development of a Common Framework for Addressing Climate Change in Post-Closure Radiological Assessment of Solid Radioactive Waste Disposal. IAEA-TECDOC-1904* (Vienna, Austria: International Atomic Energy Agency)

IAEA 2022 *Biosphere Modelling for Long Term Safety Assessments of Solid Radioactive Waste Disposal Facilities: The Enhanced BIOMASS Methodology* (Vienna, Austria: International Atomic Energy Agency) (in preparation)

IAEA 2023 *The Environmental Behaviour of Uranium* (Vienna, Austria: International Atomic Energy Agency) Technical Reports Series No. 488

US DOE 2008 Yucca Mountain Repository License Application, US Department of Energy, Office of Civilian Radioactive Waste Management, DOE/RW-0573, Rev. 0, June 2008

IOP Publishing

Principles and Techniques of Radiological Protection

Michael Charles Thorne

Chapter 15

Radiological protection in radiology

C J Martin

15.1 The early use of radiation in medicine

One of the first things that Wilhelm Conrad Röntgen did when he discovered x-rays in 1895 was to produce images on film of the bones in his wife's hand. These images stimulated interest from around the world in the use x-rays in medicine and a rapid growth occurred during the last few years of the nineteenth century. X-ray imaging is now a standard part of medical practice, and most hospitals have radiology departments. The ability of radiation to kill cells was also recognised at an early stage (see section 1.2 for references to early studies on the deleterious effects of radiation). Also, Marie and Pierre Curie isolated radium shortly after Henri Becquerel had identified radiation being emitted from salts of uranium linked to radioactivity in 1896 and it was relatively rapidly adopted for use in medical treatments (see chapter 2 of Meggitt (2008)). Radiotherapy uses both sources placed in or near diseased tissues (brachytherapy) and high intensity beams for external treatments. Radiotherapy is usually provided as a regional service through major hospitals specializing in oncology.

In the early days, professionals working with x-rays were unaware of any potential risks and used their own hands to check outputs of their x-ray tubes. Moreover, the first x-ray tubes had no shielding and long exposures were required because of low sensitivity of the photographic plates. As the use of medical x-ray imaging became widespread during subsequent years, evidence of radiation injuries began to appear, initially taking the form of radiation dermatitis and later more serious deformities to the hands of pioneering practitioners. During the early twentieth century some workers began to use protective devices such as shielding around x-ray tubes, aprons and gloves containing lead, and lead screens. These measures marked the beginning of formal radiation protection in radiology. Radiological protection for medical applications began to be developed more formally during the 1920s. Recommendations on radiological protection practice from the International Commission on Radiological Protection and the

doi:10.1088/978-0-7503-2424-3ch15

International Electrotechnical Commission are now implemented though approved codes of practice linked to national regulations (Martin 2020).

15.2 X-ray imaging equipment

X-ray equipment comprises an x-ray tube and a device to record the x-rays transmitted through the body. The x-ray tube is based on the principle described in section 2.4 and contains a heated cathode filament and an anode comprising a metal disk with a bevelled edge, which are held at a potential difference of between 20 and 150 kV. The dimension of the filament and the acuteness of the bevel angle (anode angle 10°–18°) determine the size of the focal spot forming the image and so the potential resolution of the x-ray images. The anode disk is rotated rapidly to avoid prolonged irradiation of one point and the x-ray tube, which is evacuated, is housed in a container filled with oil to help dissipate the heat and shielded to minimise leakage radiation. The beam of x-rays produced is filtered to remove low energy x-rays that would be absorbed in the patient, increasing their dose and contributing little to the image formed. Filtration equivalent to at least 2.5 mm of aluminium is required for all medical and dental x-ray equipment, apart from mammography of the breast, and 0.1–0.6 mm of copper may be added to reduce risks of skin injury for specialist applications such as interventional radiology and cardiology. The size of the x-ray beam is limited by collimators, which in radiographic units take the form of pairs of plates containing lead that are used to adjust the size of the x-ray field in the two dimensions, but in fluoroscopy equipment circular apertures may be used.

15.2.1 Radiography

Radiography is the simplest and most widely used form of imaging in which an image is formed of x-rays transmitted through the body. For most of the twentieth century the image receptors used x-ray film sandwiched between phosphor plates that emitted light when x-rays were absorbed. However, film has now been almost entirely replaced by digital technologies. There are two main types: computed radiography (CR) and direct digital radiography (DR). In CR the image is stored on a photostimulable phosphor plate and converted to digital form later using a reader in which the image plate is scanned by a laser. DR systems use flat panel detectors containing diode arrays in which the image is stored directly. They may contain phosphors that convert the x-ray energy into light that is recorded by the diodes or photodiodes that generate electrons when exposed to x-rays. Digital systems have wider dynamic ranges and provide facilities for processing images. Radiographs for most examinations are taken with the patient lying on a low attenuation couch, with the x-ray beam directed vertically downwards, towards a bucky, which is a device below the table that contains the digital detector behind a grid to remove scatter (figure 15.1(a)). However, chest radiography is performed using a horizontal beam with the patient standing with their chest against a vertical bucky. Sometimes other examinations are also performed with a horizontal beam across the x-ray table.

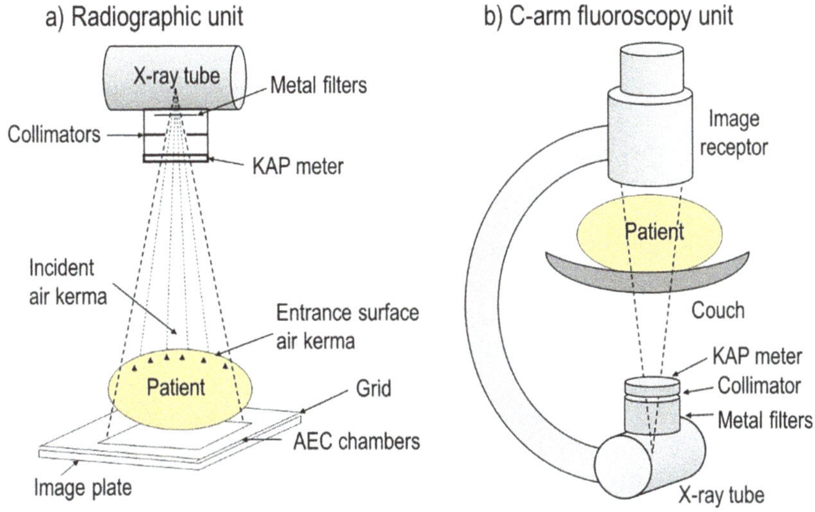

Figure 15.1. Arrangement of x-ray imaging systems for (a) radiography and (b) fluoroscopy. The KAP meter measures the Kerma Area Product for an examination (see text).

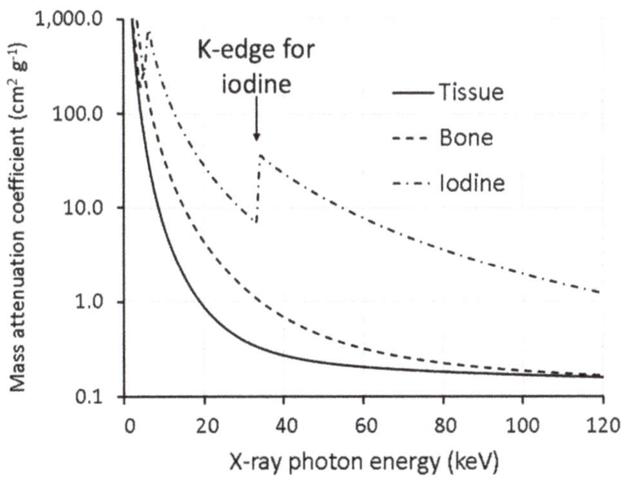

Figure 15.2. Variation of mass attenuation coefficients with photon energy for tissue, bone and iodine.

X-ray beams comprise photons with a wide range of energies up to the maximum determined by the tube potential (kV) (figure 15.2). The transmission of x-rays through tissue declines as photon energy decreases, therefore the choice of kV depends on the part of the body being imaged and is selected by the radiographer (Martin 2007a). Lower energy x-ray photons interact more strongly through the photoelectric effect and provide better contrast between tissues, but are more heavily absorbed, so give higher doses. Higher energy photons interact more through Compton scattering and give poorer contrast but penetrate more deeply through tissue (figure 15.2). Tube potentials of 70–90 kV are generally used for exposures of the trunk, except for chest radiographs where 120–140 kV might be used to reduce the difference in the air kerma transmitted through the low attenuation lungs and the higher attenuation heart and spine. 55–70 kV will be used for radiography on babies and infants, and dental radiography, 50–60 kV for extremities, and 20–35 kV for mammography of the breast which also uses different filters of molybdenum or rhodium instead of aluminium. The radiographer will increase the kV for heavier and obese patients in order to achieve sufficient x-ray penetration and avoid giving large patients unnecessarily high doses.

The number of photons is determined by the product of the tube current (mA) and exposure time (in seconds), usually termed the mAs. If a higher kV is used for a particular projection, the mAs can be reduced to give a similar exposure at the image plate with a lower dose to the surface of the patient (entrance surface dose). However, the contrast decreases and the amount of background scattered radiation increases as the kV is raised, so the choice of exposure factors (kV and mAs) is a compromise between the competing requirements (Martin 2007a). The quality of the image will, up to a point, tend to improve the more x-ray photons are used, so the level of image quality should be matched to the clinical task and the type of examination in terms of size, contrast, and noise level required for visualization of detail. Optimisation of exposures is an important aspect of all radiological imaging.

To improve consistency, radiographic exposures can be terminated automatically when the exposure of the image receptor reaches the required level to produce an image, using low attenuation detectors (usually three in number) positioned immediately in front of the image receptor. Most fixed radiographic units will be fitted with automatic exposure control (AEC) devices of this type. When film was being used for recording images, the choice of exposure factors was crucial to obtaining a usable image, but post-processing of digital images adjusts the greyscale for optimal viewing. This has limited the feedback of information on exposure levels to radiographers. What really determines the level of exposure required is the amount of noise in the image, and whether this allows features to be visualized that are required for making the clinical decisions required. Since images that have too high or too low a noise level cannot readily be identified at the time a radiograph is taken, numerical exposure indicators have been developed. These give a measure of the exposure relating to the image produced on digital image detector, and indicator levels should be monitored by radiographers.

A beam of x-rays transmitted through the body will contain significant numbers of scattered photons that will degrade the image and reduce contrast. Anti-scatter grids are fitted in front of the image receptor to remove this scattered radiation (figure 15.1). A grid comprises a plate, of similar size to the image plate, containing thin strips or lamella of lead that are perpendicular to the surface, and may have a number density of 60 cm^{-1}. These are sandwiched between layers of a low attenuation interspace material such as fibre or paper, formerly aluminium. Most of the x-ray photons that are transmitted through the body without any deviation will pass between the strips, whereas photons scattered at angles are more likely to hit the lead strips and be attenuated (Martin 2007b). How effective a grid is in removing scatter depends on the grid ratio between the depth of the lead strips parallel to the beam direction and the width of the fibre interspace perpendicular to the beam. Typical values for the grid ratio are between 18:1 and 8:1 with higher ratios being more effective in removing scatter. A device is incorporated into the bucky to move the grid (e.g. by oscillation or vibration), to avoid grid lines appearing in the image.

15.2.2 Fluoroscopy and interventional radiology

Fluoroscopy is a form of x-ray imaging that allows dynamic images of structures within the body to be viewed in real time. An x-ray tube aligned with an image receptor on the opposite side of the body produces images that can be used both in diagnosis and for interventional techniques to enable surgery to be performed with minimal intervention. Fluoroscopy is used to guide procedures such as minor repairs to vasculature using instruments inserted into the body via catheters. Fluoroscopic images were for decades formed using analogue image intensifiers, but these have largely been replaced by digital systems with flat panel detectors. Contrast for viewing soft tissues is often limited, but this can be improved by the introduction of highly absorbing material into accessible vessels within the body. These contrast agents contain elements with photoelectric K-edges at energies that enhance absorption of x-rays in the diagnostic range, barium (37 keV) for visualizing the gastro-intestinal system and iodine (33 keV) for blood vessels and highly vascular lesions (figure 15.2).

Original fluoroscopic units contained fluoroscopy and radiography facilities and consisted of a patient table that could be tilted to distribute contrast agent through the organs or structures of interest. The x-ray tube would be fixed under the table, with an image receptor above the table that would be positioned close to the patient by the operator located adjacent to the unit. These systems have been used for a wide variety of diagnostic examinations of the upper and lower gastro-intestinal (GI) tract, urinary tract, and a variety of other diagnostic procedures and therapeutic interventions. Conventional radiography/fluoroscopy systems do not allow multiple projections; so are not suitable for complex interventional procedures. The majority of fluoroscopy units now in operation use a 'C-arm' geometry with the x-ray tube and image receptor aligned and attached to a gantry in the shape of a 'C' (figure 15.1 (b)). The x-ray couch made from low attenuation material such as carbon fibre is supported from one end to allow great flexibility in the range of projection angles available that can be adapted to the clinical needs. Fluoroscopy systems use pulses of x-rays between about 2 and 15 ms long. The pulse rate may be about 30 pulses per second (pps), the same as the image frame rate, but lower rates of 3, 7.5, or 15 pps may be selected to reduce the dose with the gap between pulses on the display being filled with the last image recorded.

During dynamic observations, the x-ray tube will be moved to different parts of the body with varying attenuations, so the exposure level needs to be adjusted constantly. An automatic dose rate control sometimes referred to as automatic brightness control adjusts the exposure parameters to maintain a constant signal intensity at the image receptor. This is done by changing the exposure factors (kV

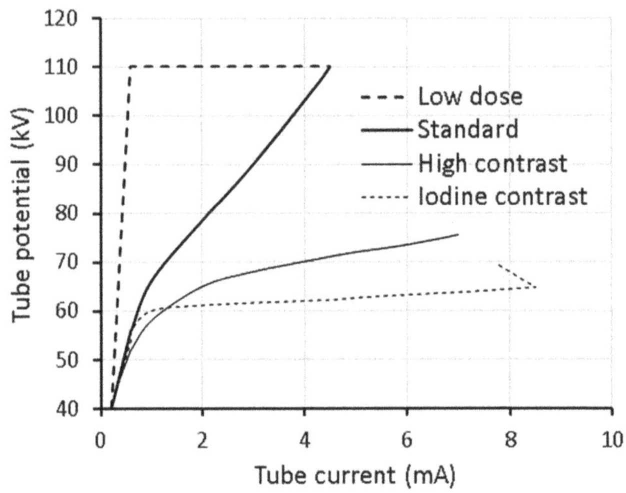

Figure 15.3. Graphs showing fluoroscopic kV versus mA options for different fluoroscopy modes. In the iodine contrast curve, the kV eventually rises steeply while the mA decreases to maintain the skin dose rate below the approved limit.

and mA) according to predetermined relationships (kV versus mA curves) stored in the x-ray generator. Different fluoroscopy modes can be selected by the operator to suit the patient and procedures being performed. Standard ones will raise both the kV and mA together at steady rates as the body attenuation increases, whereas ones designed for paediatric studies with lower attenuation and lower doses will raise the kV more rapidly (figure 15.3). Procedures that involve visualisation of vasculature using iodine contrast agents require a lower kV to be maintained while the mA is raised to take full advantage of the difference in contrast agent attenuation. The disadvantage is that the patient dose will rise more rapidly. Several series of digital radiographs (digital cine) may be recorded from different angles. The imaging of delicate blood vessels can be improved further using the digital subtraction angiography technique in which an image, recorded prior to the infiltration of iodine, is used as a mask that is subtracted from subsequent images recorded as iodine passes through the vessels, thereby removing the background tissue image.

The ability to visualise surgical manipulations inside the body through interventional x-ray techniques has revolutionised cardiac and other types of surgery. There are now interventional radiology and cardiology facilities in most major hospitals. Small surgical devices attached to guidewires are introduced and manipulated through catheters, and the development of new tools and techniques is allowing ever more complex procedures to be performed non-invasively. The procedures require powerful x-ray tubes with multiple fluoroscopy modes, with high spatial and temporal resolution image receptors and variable frame rates. The increase in complexity has led to a rise in doses to the skin with a risk of skin injury, and precautions must be taken to avoid giving high doses from prolonged exposures of localised areas (Jaschke *et al* 2020). Additional copper filters (0.1–0.6 mm) that attenuate low energy x-rays more heavily, are also introduced for some modes to reduce skin doses. The C-arms on advanced angiography systems can perform full rotations around the patient so that the digital image plate can be used to record a cone beam CT scan (see section 15.2.3). Bi-plane units with a pair of C-arms at right angles are used for specialist applications such as neuroradiology or paediatric cardiology. Fluoroscopy is no longer confined to the radiology and cardiology specialisms. Mobile C-arm fluoroscopy units are used in operating theatres. These are employed by a variety of specialties, including orthopaedic surgeons, gastro-enterologists, urologists, and vascular surgeons, as well as being used for pacemaker insertion.

15.2.3 Computed tomography scanners

A CT scanner comprises an x-ray tube and an array of detectors within a circular gantry and a patient couch that can move to different positions along the *z*-axis through the centre of rotation. Scanners use a collimated fan beam that is broad enough in the *x*–*y*-plane to intercept the entire body or head of the patient being imaged. CT scanners use tube potentials between 80 and 140 kV, with 100 or 120 kV being values used more frequently, and the beam is heavily filtered, by the equivalent of many mm of aluminium. A CT examination starts with a scan projection radiograph (SPR) recorded with the x-ray tube in a fixed angular position, the beam is collimated to give a fan beam, and the patient is moved through the gantry to build up an image similar to a radiograph. The range to be included in the CT scan in the longitudinal *z*-direction is then selected using this image and the patient positioned automatically to scan the regions selected (Samei and Pelc 2020).

During a scan, the x-ray tube and detectors rotate around the patient, collecting transmitted data from multiple angles or projections. In early CT scanners the detector consisted of a linear array and the beams were only 5–10 mm wide in the *z*-plane. The transmission data collected from one rotation were used to construct a computer-generated image of a section through the body and the couch was moved in steps through the gantry to produce a complete scan. The technology has progressed considerably since that time and now arrays of 64, 128 or 256 detectors along the *z*-axis are used to construct sectional mages from multiple slices recorded from a single rotation of the x-ray tube. In addition, the couch is moved continuously through the gantry so that the x-ray beam follows a helical path around the patient, and data from all the detectors are combined using weighted filters to generate a series of slices through the body, of a width chosen by the operator. The reconstructed CT images show cross-sections through the body, so overlying tissues are not superimposed. This provides a greater potential for identification of changes in tissue structure and abnormal pathology.

The speed that the couch travels through the gantry will be such that it moves a distance similar to the width of the detector array for each rotation. If the distance travelled and the beam width are the same, the pitch is said to be 1.0, but larger pitches resulting from more rapid movement of the couch or smaller ones with overlap of adjacent fields are used frequently for different applications. The most important criterion that determines whether an image is acceptable is the noise level. This relates to a combination of tube current and exposure time (mAs), which depends on the speed at which the x-ray tube rotates and the length of the scan. However, another factor is the width of the image slice that is reconstructed. If the slice width is reduced, fewer x-ray photons will contribute to the image, so higher values of mAs will be required to provide similar image quality levels for thinner slices. The tube current is selected by the radiographer, but in modern systems this can be modulated automatically according to the size and attenuation of the patient. The variation in attenuation along each patient in terms of a water equivalent diameter is determined from the SPR, and the tube current modulated to give a similar signal level at the image receptor at each position.

Image reconstruction has been performed using a technique called filtered back projection in which a high pass filter is applied to all the projections from multiple directions, which are then combined to form an image. There is potential to improve the image quality through a process where the pattern of attenuation from the initial image is used to generate a new data set of profiles that is then compared with the original. The differences are then used to make adjustments to reduce the image noise level and in this way image quality can gradually be improved through repeated iterations. This iterative reconstruction technique requires more computing power, and wider application has become possible only in the last decade.

CT scans provide a better overall visualization of body anatomy and the technique has the flexibility to provide images in different formats. However, this comes at the expense of a higher dose than older radiology techniques. Nevertheless, the use of CT is increasing and studies of radiation doses to patients from around the world indicate that where CT scanners are in use, 50%–70% of the dose from medical imaging is due to CT scans (Mettler *et al* 2008, Hart *et al* 2010, NCRP 2019). Optimisation of CT scanning techniques to achieve an acceptable balance between dose and image quality is of utmost importance.

15.3 Measurements of patient doses

15.3.1 Radiography and fluoroscopy

There is a need to know the dose levels for x-ray examinations if procedures are to be optimised, and surveys of patients are carried out at regular intervals. The output of x-ray equipment is measured either with ionisation chambers or semiconductor detectors calibrated in terms of air kerma in gray. The quantities used are particular to each technique. For radiographic and fluoroscopic procedures, measurements of equipment output are made in terms of air kerma, from which the air kerma incident on a patient's skin can be assessed using the inverse square law or the entrance surface air kerma (ESAK) including backscatter can be calculated. This quantity can also be measured using thermoluminescent dosemeters and is used in patient dose surveys for radiography. However, the kerma-area product (KAP), which is the product of the mean air kerma in an x-ray beam and the beam area is now used more frequently. It is usually measured with a large-area ionisation chamber fitted to the x-ray tube after the diaphragms that collimate the beam (figure 15.1) and the value is independent of the distance from the tube focus. A KAP meter records all the radiation that enters a patient, which relates to the overall exposure from the procedure, and is suitable for recording doses from fluoroscopic examinations in which the radiation is incident from different angles. KAP gives the operator an indication of the dose at the time the examination is performed. KAP meters are calibrated from measurements of air kerma in the x-ray beam multiplied by the area of the beam recorded at the same distance from the x-ray tube focus in Gy cm^2, but meters are often calibrated in different units (Martin and Sutton 2015).

To monitor the risk of skin injury in interventional procedures, another quantity, the cumulative air kerma at the interventional reference point is also measured. This is the total air kerma recorded during a fluoroscopic procedure on a C-arm unit at the centre of the x-ray beam at a position 15 cm in front of the isocentre. This is taken to approximate to the position of the skin, so the measurement represents the cumulative air kerma that would result if the x-ray beam were stationary throughout a procedure. Systems using the x-ray tube angulation and position are now being developed to provide maps of skin dose for interventional units to allow peak skin doses to be quantified and areas of skin receiving high exposures to be identified.

15.3.2 CT dosimetry

The methods used for CT dosimetry are based on a 100 mm long cylindrical (pencil) ionisation chamber. This is used to measure a dose length integral or summed dose profile along the z-axis of the CT scanner for a single slice axial CT scan with a typical beam width of 10 mm. The measurement is then divided by the nominal beam width (equal to the slice width multiplied by the number of slices) to give a CT Dose Index ($CTDI_{100}$) in mGy. The measurement is made in air along the axis of the CT scanner to measure output. The $CTDI_{100}$ is also measured in standard PMMA phantoms, typically 150 mm long, representing either the head (160 mm diameter) or body (320 mm) (Martin and Sutton 2015). Measurements are made in specially positioned holes at the centre and periphery of these phantoms and combined with a 1:2 weighting to give a weighted CTDI ($CTDI_w$). This is one of the dose quantities displayed on the consoles of CT scanners, but because the beam width (W) is often different from the table increment per tube rotation (I), it must be divided by the pitch (I/W) to give the volume averaged CTDI ($CTDI_{vol}$). This gives an indication of the dose level within the standard phantoms but is not a measure of the dose from an examination, so to provide a measure linked to this, the $CTDI_{vol}$ is multiplied by the scan length to give what is called the dose length product (DLP) and both the $CTDI_{vol}$ and DLP are used in patient dose surveys. Because the dimensions of the phantoms may be rather different from those of many patients, correction factors are now often used to derive size specific dose estimates (SSDE values) for phantoms of similar sizes to the patients to provide a better link to the doses to tissues of the patient.

15.3.3 Patient dose surveys

Radiologists and radiographers will be unaware of how well their units are performing in terms of dose, unless measurements are made of doses that patients

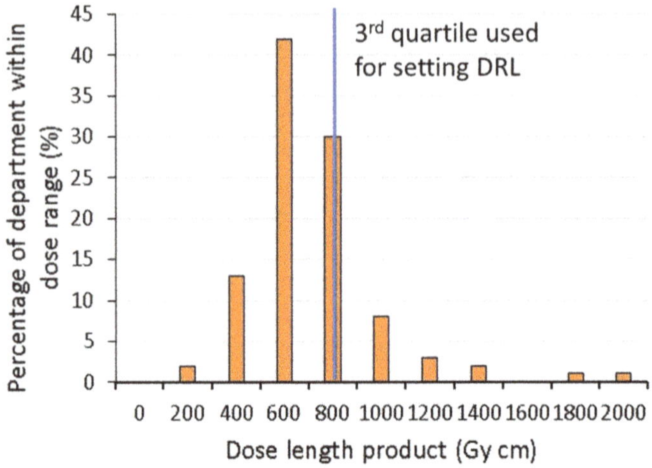

Figure 15.4. Distribution of median doses from a selection of hospitals for CT examinations of the chest, abdomen and pelvis, showing the 3rd quartile dose used for setting the DRL.

are receiving. However, even then, it is difficult to know what dose might be judged acceptable. The International Commission on Radiological Protection (ICRP) introduced the concept of diagnostic reference levels (DRLs) to aid in the standardization of doses for diagnostic procedures (ICRP 2017). DRLs are generally based on surveys of patient doses for specific examinations in many hospitals often undertaken at national level. The hospital surveys might include data from hundreds of patients downloaded from computer databases, or may be more limited, in the latter case a restriction in the range of patient weights to which the DRLs apply may be necessary, as weight can have a significant influence on the optimal dose. The distribution of the median patient doses from surveys performed by a representative selection of hospitals is then taken and the third quartile used to set a national DRL (figure 15.4), and this process is repeated periodically. This DRL, which three quarters of the hospitals considers provides an adequate level of image quality, is used as a reference. Patient dose surveys are undertaken routinely for local audit purposes to identify hospitals and x-ray equipment for which effort is required to reduce doses and optimise procedures. Repeat surveys should be undertaken following any optimisation, and the whole process should be repeated after an appropriate time interval.

15.4 Radiotherapy

Radiotherapy is the treatment of cancer through depositing energy from ionising radiations in body tissues to destroy cancer cells. Marie and Pierre Curie first used radium to treat disease and this marked the birth of radiotherapy. Initial treatments used either sources of radium or x-ray beams, to treat superficial lesions. X-ray units operating at tube potentials of 250–300 kV and various other devices were developed to produce higher energy radiation, but it was with the production of ^{60}Co sources emitting γ-rays with energies of 1.17 and 1.33 MeV, capable of much greater

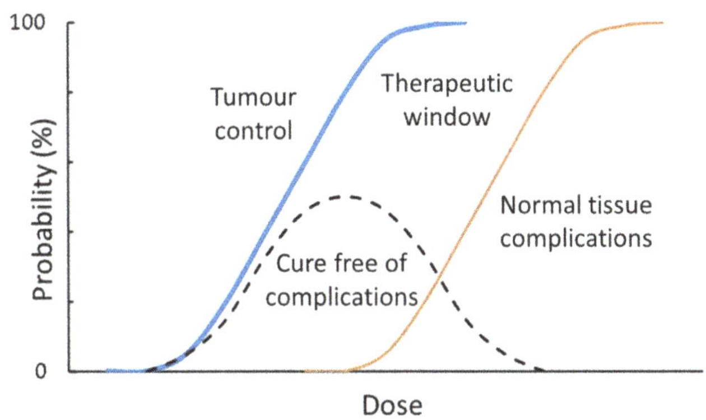

Figure 15.5. Dose dependence of the probabilities for tumour control and side effects in normal tissue, showing the narrow therapeutic dose window in which a cure free of complications can be achieved.

penetration into tissue, that treatment of tumours deep within the body with external beams became feasible. The number of ^{60}Co therapy machines still in use around the world is dwindling, being replaced by linear accelerators generating x-ray beams with accelerating voltages of 5–20 MV, commonly referred to as linacs. Therapies with particles such as protons and carbon ions have recently started to be used.

The principle of radiotherapy is to treat a volume of tissue containing a tumour and surrounding tissue that may have been infiltrated by malignant cells. Ideally the radiation dose should be high enough to kill the tumour cells, but that to normal healthy tissue should not be too high as to minimise risks of serious damage. Tumour cells that are more rapidly dividing tend to be more susceptible to radiation damage, but nevertheless giving a sufficiently high dose to the tumour while keeping the dose to surrounding tissues at a level that can be tolerated and give a cure free of complications is difficult (figure 15.5). Radiotherapy is used primarily to cure cancer (radical), but 10%–15% of cases are palliative to alleviate symptoms (pain) arising from the cancer or its metastatic deposits, because a cure is not possible. Tumour doses are usually in the range 30–75 Gy and treatment is divided into up to 30 daily fractions, to provide better tolerance of radiation effects in normal tissues (Mayles *et al* 2021).

15.4.1 Planning radiotherapy treatments

To maximise the chances of a successful outcome, the dose to tumour tissue should be as high as possible, while avoiding excessive damage to surrounding normal tissue. This requires careful planning. External beam therapy is a flexible treatment that can be used to treat tumours deep within different parts of the body. Linear accelerators used to deliver the treatment are mounted on a gantry and can be rotated in a vertical plane around an isocentre, allowing the beam to irradiate the patient from any angle. If the patient is positioned such that the tumour is at the isocentre, the tumour dose can be delivered additively by beams from predetermined directions, while keeping down the doses to surrounding tissues.

Planning of treatments is crucial to achieving success. Images recorded prior to the treatment are used to outline the shape of the target and define its position within the body to develop a treatment plan. Treatment planning systems (TPSs) incorporate sophisticated computers that produce plans for each individual patient, and these may be modified as they undergo treatment. The plan is based on CT scans of the region of interest and this may be coupled with other imaging techniques, such as magnetic resonance imaging and positron emission tomography (PET see chapter 16) that may be better able to identify the extent of diseased tissues. A clinical oncologist will confirm the extent of diseased tissue to be treated on each slice of the CT image, defining a clinical target volume that may include not only the tumour, but also metastases, lymph nodes that may be affected, and a margin for possible sub-clinical spread. This forms the basis for the planned target volume (PTV). Critical organs that are at risk of damage from radiation including the spinal cord, heart, rectum, and intestines are also identified. The TPS then plans the individual treatments to direct radiation beams from different directions with the aim of delivering a high, but uniform dose throughout the PTV, while keeping doses to critical structures as low as possible.

15.4.2 Linear accelerator treatments

Linear accelerators produce high energy x-ray beams by accelerating electrons to 5–20 MeV. The electrons are drawn from a tungsten filament located inside an electron gun by a voltage of about 50 kV. They are then accelerated to high energy and directed at the tungsten anode target, which is about 2 mm thick. The x-rays produced are transmitted through the target resulting in a high energy x-ray beam that diverges from the target. The linear accelerator is housed in the treatment head, which is heavily shielded, and the size of the beam is restricted using thick attenuating collimators to reduce leakage radiation. The x-ray beam then passes through a flattening filter to improve the uniformity, and a flat plate ionisation chamber, called a monitor chamber, to measure the amount of radiation delivered by the treatment beam. Wedges can also be placed in the beam to introduce a gradation of intensity across the beam for specific applications. A secondary collimator is used to set the size of the beam to be used for treatment. The treatment head is mounted on a C-arm gantry that can be moved in a 360° arc around the patient couch, allowing flexibility in the direction of delivery of the x-ray beam. In some systems, the angle of the couch in the horizontal plane can also be adjusted.

In early machines the x-ray fields were of standard shapes, but multi-leaf collimators, which consist of motorized high density tungsten alloy leaves positioned on opposite sides of the beam aperture that can be positioned independently, are now standard. These allow the beam to be shaped so that it conforms to the PTV, enabling more precise delivery of treatment doses. The multi-leaf collimators can also be moved during treatment delivery to modulate the distribution of dose across the field. This intensity modulated radiotherapy (IMRT) allows the generation of complex three-dimensional dose distributions around the PTV as it is moved around the patient, giving more uniform dose distributions across tumours, while sparing surrounding normal tissue in conformal radiotherapy. External beam radiotherapy doses for the treatment of cancer are generally delivered in fractions. The total dose prescribed is divided into smaller doses that are delivered over a period of several days or weeks. This reduces the toxic effects on healthy cells, allowing higher doses to be given to tumours. Various alternative treatment regimes are being developed based on radiobiological evidence, such as stereotactic body radiotherapy where high doses of radiation are delivered from many angles with less fractionation or further subdividing the dose into smaller fractions.

15.4.3 Image guided radiation therapy (IGRT)

More precise dose distributions conformed to the shape of a tumour require accurate positioning of the patient on the treatment couch. Immobilisation devices such as head shells, leg and ankle supports, and bags of polystyrene pellets are used in setting up each patient. Precise alignments are determined using optical devices such as lasers together with tattoos or reflectors on the patient's skin. However, positioning patients with an accuracy to the nearest mm is almost impossible. Errors can be reduced by imaging the patient immediately before treatment is administered, to allow corrections to be applied. IGRT is now in regular use in radiotherapy departments at the start of treatment and before delivery at some fractions. Thus, IGRT potentially allows better targeting of the tumour, increasing the dose delivered to give a higher probability of cure for the cancer, without a significant increase in dose to healthy normal tissues and thus increasing the therapeutic ratio.

A variety of techniques are used for IGRT. The radiation emitted by the treatment source, can be used to indicate the location of the beam with respect to bony landmarks, but the contrast given by MV x-rays for soft tissues is poor and the absorbed doses delivered to tissues are significant. Modern linear accelerators incorporate kV on-board imagers which can be used for planar radiographs or cone beam CT scans. These give good contrast in soft tissue for localisation and comparisons with reference images and provide the potential for the patient's position and the radiation beams to be adjusted to correct errors in alignment. Thus, IGRT provides additional accuracy in localisation of tumour volumes at the time of treatment, but there is a balance to be achieved, as frequent imaging exposures can give significant cumulative doses to much larger volumes within the patient than the PTV.

15.4.4 Particle therapy

So far only treatments with photon radiation have been considered. Beams of electrons can be obtained from a linac by removing the anode target, and these are suitable for treatment of superficial tumours, and protons and ions are also now used. X-rays penetrate deep into tissue, but the dose absorbed declines exponentially with depth. Protons and heavy ions, on the other hand, deposit more energy per unit length as the particles approach the end of their tracks through the body (figure 15.6). For a particle of a specific energy this is within a narrow range in depth (the so called

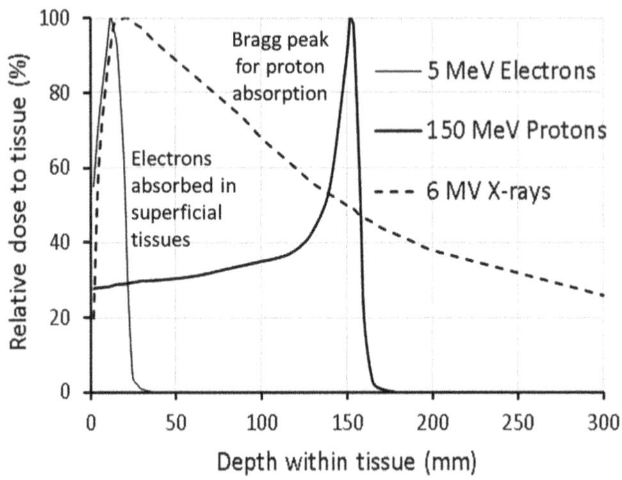

Figure 15.6. Energy deposition versus depth for different types of radiation.

'Bragg peak'). The dose increases towards the depth of the Bragg peak but declines to almost zero at greater depths. Therefore, particles have the potential to give greater doses to tumours at depth and help to minimise exposure of other normal tissues, potentially enabling higher doses to be delivered to tumour tissue.

The ions are first accelerated by means of a cyclotron, synchrotron or synchrocyclotron. The energy of the emerging particle beam determines the penetration, and so the depth at which the peak absorbed dose is delivered. The particles used in clinical practice are protons with energies of 100–250 MeV and carbon ions with energies of 320–430 MeV. Carbon ions have a higher linear energy transfer and relative biological effectiveness because they are heavier than protons. This produces many double-strand DNA breaks which are difficult for the tumour to repair. The range in depth delivered by a single energy beam may not cover the whole extent of a tumour, so particles with a range of energies may be required. A pencil beam of charged particles can be deflected using electro-magnets to scan in a raster pattern across the target area. In addition, the beam energy and hence the depth of penetration can be varied, so by varying the energy and position, an entire PTV can be covered in three dimensions, providing an irradiation pattern following the shape of the tumour. Early proton therapy systems used a broad beam produced using scattering devices, which was then shaped by collimators. These systems delivered a homogenous dose over the target area but did not allow the dose distributions to be conformed as precisely to the PTV. Many different possibilities for particle therapy are being investigated, including the use of neutrons combined with the introduction of boron into the tumour to take advantage of the boron capture reaction.

15.4.5 Other external beam radiotherapy devices

A variety of other types of radiotherapy machine are available for delivery of external treatments. Conventional x-ray units operating at energies up to 300 kV and mounted to give flexibility in beam direction are used for the treatment of skin and superficial lesions. The treatment fields are determined by applicators and lead inserts used to customize the field to the area on the patient's skin to be treated.

In tomotherapy a linear accelerator producing 6 MV x-rays is mounted on a gantry similar to a CT scanner that allows continuous rotation during treatments. and the radiation beam striking the patient is defined by a multi-leaf collimator. The combination of adjustments in the shape of the field, rotation of the x-ray source, and movement of the patient couch, enables complex dose distributions to be delivered. A linear array of detectors is mounted on the opposite side of the gantry from the accelerator to provide a cross-sectional image with MV x-rays that can be used to check patient position prior to treatment.

The CyberKnife employs a 6 MV linear accelerator with a smaller field size mounted on a computer-controlled robotic arm that allows the beam to be pointed in a range of directions. The target of the linear accelerator is typically about 80 cm from the body tissue being treated, and the fields are defined by cylindrical collimators less than 6 cm in diameter. Complex dose distributions can be built up over the tumour volume by moving the x-ray beam, so that it delivers 80–100 treatments from different orientations using variable dwell times.

The Gamma Knife uses about 200 ^{60}Co sources with activities totalling over 200 TBq mounted so that the radiation beams are focussed on a small volume in the brain to treat brain lesions. The position of the patient's head is fixed at the focus of the radiation beams. Gamma Knife treatments are a form of precision stereotactic radiosurgery that delivers high doses to small lesions with sub-mm accuracy with minimal radiation dose to healthy brain tissues surrounding the target. Treatments are usually completed in a single day.

15.4.6 Brachytherapy

Brachytherapy is a form of treatment in which a sealed radioactive source is placed next to the site to be treated, allowing high doses to be delivered to a small volume. The first brachytherapy treatments were performed in the early 20th century with radium, followed by ^{137}Cs, but ^{192}Ir is now the predominant radionuclide. The sources are encapsulated in non-radioactive capsules. Source applicators used for positioning the sources are placed in the patient in theatre. The patient is then returned to the ward and sources are inserted with the patient located in purpose-designed shielded rooms. The insertions were originally done by staff, but now remote after-loading systems are used for which the sources are driven into position by guidewires through flexible plastic catheters. They can be left in position for any length of time and can be inserted and withdrawn into a lead safe automatically, avoiding the need for staff to be present in the room when a patient is undergoing treatment. Early low dose rate systems used lower activity sources with the patient being left connected to the machine for several days, but with larger sources in the current high dose rate (HDR) machines, patients can be given treatment fractions in similar times to those required for external beam therapy. Many brachytherapy procedures are now performed on an outpatient basis. The treatments are planned using computer simulation. Specialised applicators linked via catheters to the HDR after-loader are used to position the sources for different parts of the body. Once staff have left the treatment room, the after-loader controls the delivery of the sources along the guiding catheters into the pre-specified positions within the applicators. The sources remain in place for a specified time determined by the treatment plan, following which they are returned along the catheters to a shielded safe within the after-loader.

There are two types of brachytherapy treatment, contact and interstitial, relating to the placement of the radioactive source. For contact brachytherapy, the radiation source is placed in a space next to the target tissue. This can be done through a body cavity (intracavitary brachytherapy) such as the cervix, uterus or vagina; a body lumen (intraluminal brachytherapy) such as the trachea or oesophagus; or externally (surface brachytherapy) such as the skin. Specialised applicators linked via catheters to an HDR after-loader are used to position the sources for different parts of the body. In interstitial brachytherapy, the sources are placed directly in the target tissue of the affected site. For prostate treatment, seeds of ^{125}I are inserted into an array in the prostate using needles, which are left permanently in place to deliver the planned treatment dose before they decay. For interstitial breast brachytherapy several catheters are positioned in the breast tissue. The catheters are connected to an HDR after-loader, programmed to deliver ^{192}Ir pellets to deliver the planned radiation dose to the treatment area, while sparing the surrounding breast tissue. Several options are available for breast treatment. These include intraoperative radiation therapy that is delivered directly to the site during surgery to remove the tumour using a mobile electronic device generating x-rays or electrons with an applicator (electronic brachytherapy). Intracavity therapy involves placement of an applicator in the breast cavity left after the removal of the tumour, that can then be connected to an HDR after-loader to deliver the radiation sources. Another option is to implant multiple radioactive ‘seeds’ into the breast, similar to permanent seed prostate brachytherapy. Non-melanoma skin cancer lesions of the eye can be treated with contact applicators, either linked to an HDR system or using ^{90}Sr or ^{106}Ru applicators for surface lesions.

15.5 Protection of staff and members of the public

15.5.1 General principles of x-ray installation design

Radiology and radiotherapy services use equipment that emits penetrating radiation. The units that emit significant levels of radiation are installed in fixed positions in rooms and these need to be shielded, although there are mobile radiographic units used on wards for taking chest x-rays and mobile fluoroscopic x-ray units that can be used in theatres. Restriction of exposure to radiation for staff and members of the public is achieved by using compromises between the three basic elements of shielding, distance, and time. The rooms in which equipment is installed will be controlled radiation areas with restricted access. They are designed to provide flexible and efficient working areas for different types of procedure, while keeping radiation exposure levels outside within set criteria linked to dose limits and providing protection for staff operating the equipment.

The radiation arises from three main sources—the primary beam, leakage and scatter. Leakage and scatter are referred to as secondary radiation. Protection is required that will reduce the dose or dose rate to the agreed acceptable criterion, usually defined in terms of annual dose or the equivalent. Protection for most x-ray facilities can be planned by taking account of time-averaged dose rates, although instantaneous dose rates will also need to be considered in some situations. Radiation protection requirements can be based on comparisons of the dose that would be accumulated from exposures over an agreed time, e.g. a week or year that is assessed from the radiation workload. The amount of scatter incident on the walls of an x-ray room will depend on placement of the equipment within the x-ray room, but the location of the x-ray room within the whole building is also important, as x-ray facilities are often adjacent to areas with relatively high occupancies. The shielding to these areas must be designed to provide an area behind which the dose rates are low so that members of staff do not need to wear additional personal protective equipment.

Shielded cubicles are provided for the majority of staff operating x-ray equipment, and during normal operation staff only need to be in the controlled area with equipment emitting x-rays if they are performing manipulations on a patient using fluoroscopic images to guide the procedure. The dose constraint used for members of the public is 0.3 mSv per annum in the UK and the same is usually recommended for members of staff who do not need to work in the same room in which radiation is being emitted. Entrances to rooms containing x-ray equipment require a sign to warn staff of the hazard. These are usually two-stage signs, the first stage being illuminated when the mains power is connected to the unit and the second stage lighting up when the x-ray set is being prepared or emitting x-rays.

15.5.2 Design of diagnostic and interventional radiology rooms

In radiology the primary beam will usually be directed downwards towards a patient and fall within the bounds of the bucky. However, horizontal beams are used in chest radiography for which the patient is standing in front of the bucky or when a radiograph is taken across the x-ray table. Leakage radiation from the x-ray tube is generally low and scattered radiation, produced when the x-ray beam is incident on the patient, is the predominant source against which protection is required. The technique factors for the range of procedures performed must be considered, particularly the average and maximum kV used. The air kerma level from scatter can be evaluated from the workload in terms of kerma-area product (KAP), as this contains all the x-radiation incident on the patient, so an assessment needs to be made of typical KAP values for the examinations performed and the numbers of patients. Scatter factors linked to KAP can be applied to calculate the level of scatter at 1 m from the patient and the inverse square law applied to calculate scatter levels at boundaries requiring protection (Sutton *et al* 2012, Martin 2015). Assessments may be needed for the primary beam directed horizontally for chest and cross-table exposures.

The primary beam for fluoroscopic equipment is absorbed by the image receptor, so only scattered radiation needs to be considered. There is a wide range in fluoroscopic equipment, with many different C-arm systems, but in normal practice the x-ray tube will be below the patient couch, but with some angulation to reduce exposure of the upper body of the operator. The throughput can vary markedly with interventional units often having high workloads. Both radiography and fluoroscopy rooms require protected control cubicles containing enough protection to ensure staff do not need to wear personal protective equipment, but these do not usually require a shielded door.

The whole of the primary beam in CT will be absorbed by the radiation detectors and equipment gantry. Thus, the only radiation to be considered in CT room design is again that scattered by the patient and imaging table. CT produces a well-defined pattern of scattered radiation that resembles an hourglass that is approximately symmetrical about the axis of rotation (figure 15.7). The scatter kerma from a CT scanner is related to dose length product (DLP) in a similar way to the relationship for conventional radiology to KAP (Sutton *et al* 2012) but can also be determined from manufacturer supplied isodose curves. Although most of the protection can be based on time-averaged dose rates, the instantaneous dose rates may need to be considered because periods of x-ray emission in CT can last considerably longer than exposures in radiography and fluoroscopy. In addition, shielding of the roof/floor is likely to be required, although because of attenuation by the gantry this will only be required for oblique angles above about 30°. The control cubicle will need a shielded door to protect against scatter from adjacent walls and the barriers to the control cubicle and other areas may need to be extended to the full height of the room to reduce scatter from the ceiling slab into these areas.

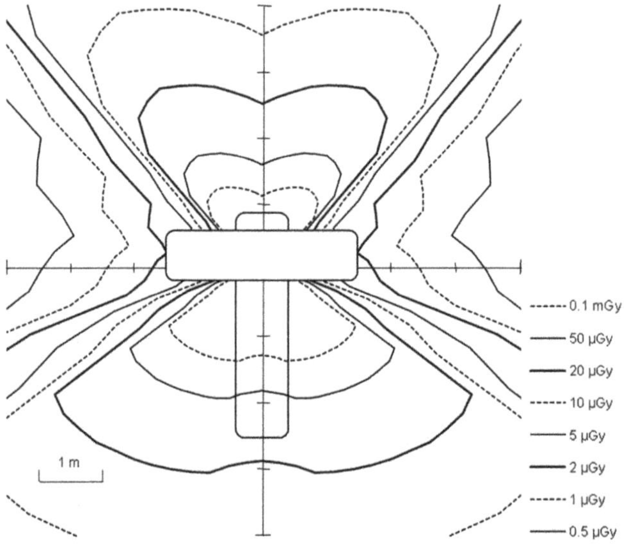

Figure 15.7. Contour map of the distribution of scattered radiation around a CT scanner for a body scan with a DLP of 1000 mGy cm.

15.5.3 Evaluation of protective barrier requirements for radiology

The best way of determining workload is from local dose audit data, but if such data are not available published data can be used. Most of the workload in a radiographic room generally arises from chest, spine and abdominal examinations, with chest radiographs being performed on a wall Bucky and the rest using table radiography. A range of kVs will be used, but an average towards the upper end of the range might be 85 kV. A typical weekly workload is a KAP of 150 Gy cm^2, while for a busy department it might be a KAP of 500 Gy cm^2. This provides enough information for calculation of the air kerma at the walls of radiographic room with adjustments for the inverse square law. The workload for fluoroscopic, interventional and CT rooms will be more varied and require more detailed assessment (Sutton *et al* 2012).

The workload per year can be used to calculate the annual air kerma level and this can then be compared with the dose constraint derived from the regulatory requirements used as the design criterion. The dose constraint is likely to be in terms of effective dose, but the ratio of effective dose to air kerma for the x-ray energies used in radiology is between 0.6 and 0.9, so this adds a degree of conservatism. The dose constraints can be raised by dividing by an occupancy factor if occupancies of adjacent rooms are limited, and for many areas this will be between 0.2 and 0.5, and lower for corridors, toilets, storerooms, etc. The ratio of the dose constraint to the air kerma level from the calculation will then give the transmission required for the barrier.

The transmissions required for x-ray room walls will typically be 10^{-2}–10^{-3}. A wide range of materials are available from plasterboard, which is often sufficient for dental radiography, to brick and concrete blocks. Lead incorporated into plywood or plasterboard is commonly added to increase protection. Sets of standard equations are available for calculation of x-ray transmissions and barrier thicknesses with coefficients linked to different materials and energies of x-ray beams (Sutton *et al* 2012). Since the kV used in CT is generally higher than in other radiology applications and the beam more heavily filtered, specific transmission factors are used for CT in the equation. Lead thicknesses of the order of 1 mm are generally sufficient for radiographic rooms, but 2 mm or more may be required for CT scanning rooms. Most purpose-designed x-ray rooms have windows positioned more than 2 m above floor level to avoid significant levels of scattered radiation at positions that might be occupied by staff or members of the public. Concrete slabs with a density of the order of 2350 kg m^{-3} and with a thickness of about 150 mm will generally provide sufficient protection against high-dose interventional procedures or even high-kV modalities such as CT, but lightweight concrete is often used and there may be variations in concrete thickness or even hollow sections in the concrete, so care is needed. Doors and cubicle screens are custom-made by specialist suppliers and incorporate sheet lead into the frames.

15.5.4 Design of radiotherapy bunkers

The rooms in which linear accelerators and other therapy devices using penetrating radiations are housed require substantial shielding usually of concrete to reduce the

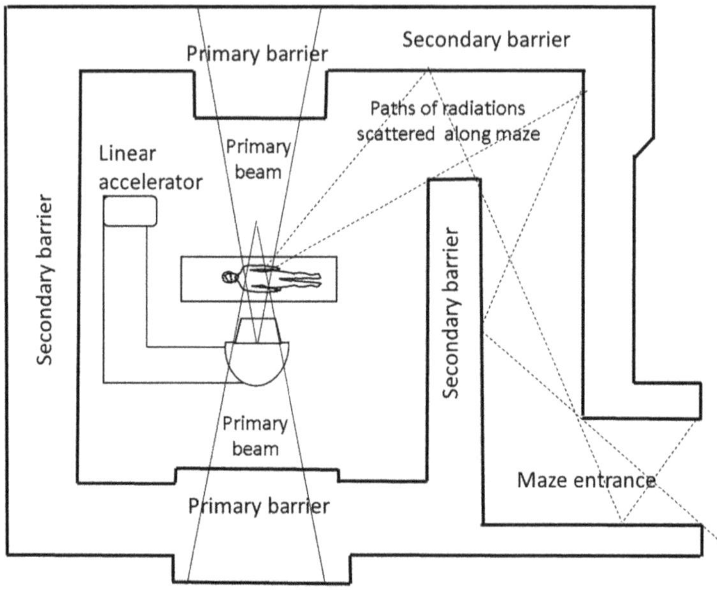

Figure 15.8. Plan of linear accelerator bunker with radiation paths of the primary beams pointing towards the barriers on opposing sides of the bunker, and x-rays scattered down the maze.

dose rates outside to acceptable levels. The rooms are commonly referred to as bunkers, and all people other than the patient being treated must be excluded during treatments. The greatest protection is required against the primary beam. Linear accelerators and some other types of radiotherapy equipment rotate the treatment beam around a point, called the isocentre, corresponding to the position of the irradiation target, so the primary barrier must extend around a band covering a 360° rotation (figure 15.8). Other walls in the bunker will be exposed to high levels of scatter, so will also need substantial shielding to achieve the required dose constraint (Horton and Eaton 2017).

The approach to calculation of shielding requirements is similar for all external beam radiotherapy, but relative contributions from the primary beam, leakage and scattered radiation will vary. Several x-ray beam energies, such as 6, 10, and 15 MV with a linear accelerator may be used and the workload for each energy may need to be considered separately. Calculations of barrier thicknesses require the number of patients in a given period, the dose delivered for each treatment, the fraction of the time that the primary beam points in different directions, and the proportions of time for which different treatment modes are used. Treatment doses are generally quoted in terms of the dose at the isocentre and dose rates at barriers can be calculated by applying the inverse square law. The resultant annual dose from each component of the primary beam incident on each wall of the barrier can then be summed for the calculation. Similar calculations are performed for leakage and for scattered radiation for which scatter factors are available. These are considered to be isotropic, so no orientation factor is involved. For more specialised types of radiotherapy equipment, where the primary radiation beam is intercepted, manufacturers give secondary dose rates from leakage and scatter in the form of polar plots or rectangular grids. Bunkers are also used for HDR brachytherapy with the design based on dose rates from the sources used and the numbers of treatments, while for low dose brachytherapy shielded rooms may be provided within a ward.

When dose levels are compared with dose constraints, attenuation factors of from 10^{-4}–10^{-5} are typically required, so barrier thickness requirements are often calculated from data in the form of tenth value layers. Because of the level of attenuation required and the high energy of the x-ray beams, concrete is the best option for providing the protection. Barriers of poured high density concrete (2350 kg m^{-3}) will typically be 1–2 m thick, and higher density concrete may be used to reduce this. The concrete density is critical for the shielding requirement, so quality-control tests are required to confirm the density of each batch. If block construction is used then positions of mortar joints that could provide shine-through paths need to be considered, and interlocking blocks are usually used. Radiotherapy treatment rooms require significant services including air conditioning, heating, lighting, and electrical power, as well as access for dosimetry cables, and penetrations must not compromise the radiation protection. If rooms above the treatment bunker are occupied, protection will be required to reduce dose rates to the appropriate levels. Higher dose rates from the upward-pointing primary beams than other beam orientations may be allowed if access to the roof is prohibited during linear accelerator operation and the roof is designated as a controlled area. But there is still a risk of skyshine due to scatter from air molecules, which can give measurable dose rates at ground level outside the building.

Because of the substantial shielding required, mazes are used frequently to provide entry to bunkers (figure 15.7). Mazes may allow easier entrance, because shielded doors are heavy and cumbersome and need to be power operated. Dose rates can be reduced to acceptable levels at the maze entrance by multiple scattering and absorption through careful planning of maze design. Linear accelerators operating at 10 MV and above produce neutrons and neutron dose must also be taken onto account in maze design and walls may be lined with neutron attenuators such as borated polyethylene. A shielded door for 15 MV x-rays would contain typically 6 mm of lead as well as neutron attenuator. A short maze or nib can be helpful even when a door is used, in order to reduce incident radiation and so the amount of shielding required.

15.5.5 Personal protective equipment and personnel monitoring

Staff who need to operate diagnostic x-ray equipment directly wear lead/rubber aprons that contain the equivalent of 0.25–0.5 mm of lead. Aprons protect the body but not the head, neck and limbs, to keep weight to a minimum (Hiles *et al* 2016). Staff working in interventional rooms will also normally wear a shielded collar to protect the thyroid. Overhead suspended lead acrylic shields are used by operators in angiographic and interventional procedures who need to remain close to the patient, and these provide the most effective way to reduce radiation dose to the head and upper body (ICRP 2018). There have been many reports of the observance of opacities in the eye lenses of interventional clinicians (ICRP 2012) and the ceiling suspended screens provide the best protection for the eyes, but it may be necessary for operators to wear protective eye wear containing 0.5–0.75 mm of lead in some cases. Table-mounted shields are useful for protecting the legs of interventional radiologists during C-arm fluoroscopy, as they may receive significant doses from scattered radiation from x-ray tubes underneath the couch. Mobile screens are sometimes also used to provide radiation protection in operating theatres.

Since staff wear lead rubber aprons protecting their bodies when working directly with x-rays, this presents issues of obtaining separate assessments of doses to their body, head, and eyes. Interventional clinicians and those who are more heavily exposed will wear two dosemeters, one on their body under their lead apron and the other either adjacent to their eyes or outside the collar of their coat (Martin *et al* 2018). Other staff who work with x-rays but are exposed at a lower level will usually only wear one dosemeter, but practices about dosemeter position vary across the world. The dose measured by the dosemeter under the apron assesses dose to the trunk, which can be regarded as a measure of effective dose, although it does not include unprotected organs and tissues. However, the sensitivity of a dosemeter in this position is low, so a dosemeter worn outside the collar can give a better indication of dose levels in the environment and that to the eyes. If the result is divided by ten, this will give an approximate value for the likely effective dose, provided the wearer has always worn a lead apron when exposed to x-rays (ICRP 2018).

In radiotherapy facilities, protection of personnel is based primarily on working remotely from the source. Staff should never be inside bunkers when a patient is being treated and, in addition to systems of work and local rules, interlocks and other engineering controls are used to prevent inadvertent access to bunkers when radiation is being emitted. Therefore, occupational doses should be at very low levels and personal protective equipment is not worn routinely apart from during fluoroscopy procedures. Most personnel dose monitors will not show any measurable doses. but dose monitoring is required in case of inadvertent exposures caused by a breakdown in engineering controls or systems of work, which could have significant consequences in radiotherapy. For example, with HDR brachytherapy, staff intervention would be required in the unlikely event that the treatment source was not returned to its shielded container.

There are a few situations in which radiotherapy staff may be exposed to lower levels of radiation. For example, linear accelerators operating at energies over 10 MV may induce radioactivity in the treatment head. The dose rates will decline within minutes, so exposure can be limited by restricting entry to the linac bunker for a short period following high energy beam operation. Personal monitoring will highlight changes in dose levels around facilities where there has been a change of equipment or usage. Extremity monitoring may also be required where there is a risk of the hands coming close to a source. Examples are in centres where brachytherapy using ^{192}Ir wire is still loaded manually and for clinicians who insert ^{125}I or other radioactive seeds directly into patients.

References

Hart D, Wall B F, Hillier M C and Shrimpton P C 2010 *Frequency and Collective Dose for Medical and Dental X-ray Examinations in the UK 2008 Report HPA-CRCE-012* (Chilton: HPA)

Hiles P A, Hughes H, Arthur D and Martin C J 2016 *Personal Protective Equipment* (London: British Institute of Radiology)

Horton P and Eaton D 2017 *Design and Shielding of Radiotherapy Treatment Facilities, IPEM Report 75* (Bristol: IOP Publishing)

ICRP 2012 ICRP statement on tissue reactions and early and late effects of radiation in normal tissues and organs—threshold doses for tissue reactions in a radiation protection context, ICRP Publication 118 *Ann. ICRP* **41(1/2)**

ICRP 2017 Diagnostic reference levels in medical imaging, ICRP Publication 135 *Ann. ICRP* **46(1)**

ICRP 2018 Occupational radiological protection in interventional procedures, ICRP Publication 139 *Ann. ICRP* **47(2)**

Jaschke W, Bartal G, Martin C J and Vano E 2020 Unintended and accidental exposures, significant dose events and trigger levels in interventional radiology *Cardiovasc. Intervent. Radiol.* **43** 1112–41

Martin C J 2007a The importance of radiation quality for optimisation in radiology *Biomed. Imaging Interv. J.* **3** e38 Invited review for special issue on Dose optimisation, http://biij.org/2007/2/e38

Martin C J 2007b Optimisation in general radiography *Biomed. Imaging Interv. J.* **2** e18 Invited review for special issue on Dose Optimisation, http://biij.org/2007/2/e18

Martin C J 2015 Radiation shielding for diagnostic radiology *Radiat. Prot. Dosim.* **165** 376–81

Martin C J 2020 The role of ICRP in medicine: past, present and future *Health Technol.* **10** 1427–35

Martin C J and Sutton D G (eds) 2015 *Practical Radiation Protection in Healthcare* 2nd edn (Oxford: Oxford University Press)

Martin C J, Temperton D H, Hughes A and Jupp T 2018 *Guidance on the Personal Monitoring Requirements for Personnel Working in Healthcare* (York: IPEM)

Mayles P, Nahum A E and Rosenwald J C 2021 *Handbook of Radiotherapy Physics: Theory and Practice (two volumes)* 2nd edn (Boca Raton, FL: CRC Press)

Meggitt G 2008 *Taming the Rays: A History of Radiation and Protection* (Lulu.com)

Mettler F A, Thomadsen B R, Bhargavan M, Gilley D B, Gray J E, Lipoti J A, McCrohan J, Yoshizumi T T and Mahesh M 2008 Medical radiation exposure in the US in 2006: preliminary results *Health Phys.* **95** 502–7

NCRP 2019 *Medical exposure of patients in the United States* NCRP Report No 184 Bethesda, MDNCRP

Samei E and Pelc N J 2020 *Computed Tomography Approaches, Applications, and Operations* (New York: Springer)

Sutton D G, Martin C J, Williams J R and Peet D 2012 *Radiation Shielding for Diagnostics Radiology* (London: British Institute of Radiology)

IOP Publishing

Principles and Techniques of Radiological Protection

Michael Charles Thorne

Chapter 16

Radiological protection in nuclear medicine

C J Martin

16.1 Introduction

Nuclear medicine is the use of radioactive substances for diagnosis and treatment of disease. Radioactive atoms are attached to biologically active molecules that accumulate in different body tissues. The radiation emitted can then be detected and either an image formed of the tissues that have taken up the radioactivity or measurements made to show the level of function in a tissue based on the uptake and retention and so demonstrate the effect of disease. Nuclear medicine is also used in treatment of some diseases with radiopharmaceuticals that accumulate within diseased tissue, irradiating the tissue from within. Imaging in nuclear medicine, often called radionuclide imaging, differs from radiological imaging in that it shows the physiological function of organs and tissues rather than their structure. Nuclear medicine images do not provide the same level of anatomical detail and so are often combined with CT scans in hybrid scanners.

Nuclear medicine started as a technique after World War II. The early radionuclides used were produced in cyclotrons, but subsequently nuclear reactors took on the dominant role. Radioiodine collects in the thyroid and was used in diagnosis, initially with hand-held detectors, and treatment of thyroid disease. Rectilinear scanners, in which a detector above the patient was moved in steps to build up an image, were developed during the 1950s. The invention by Hal Anger in 1957 of the gamma camera, which contained large flat crystal detectors optically coupled to arrays of photomultiplier tubes, combined with the adoption of technetium-99m (^{99m}Tc) as a radionuclide suitable for imaging, promoted a rapid development in radionuclide imaging during the 1960s.

16.2 Radionuclide imaging

16.2.1 Radiopharmaceuticals with radionuclides emitting photon radiation

The choice of radionuclides that will provide reasonable images while not giving patients too high a dose of radiation is crucial. For any radionuclide to be suitable for

doi:10.1088/978-0-7503-2424-3ch16

imaging, it must emit γ-rays with energies that are high enough for the photons to escape from the patient, but low enough for easy detection. Radionuclides that emit particle radiations at the same time as γ-rays should be avoided, as particle radiations will be absorbed within patient tissues and so increase the dose without contributing to image formation. In addition, the radionuclide should emit enough radiation to allow imaging to be carried out over a relatively short timescale of 10–40 min but not continue irradiating the body for a long time afterwards. Therefore, radionuclides with short half-lives are suitable, but if the half-life is too short this will create difficulties in transport and supply.

Dealing first with imaging based on photon-emitting radionuclides, ^{99m}Tc has been the workhorse of nuclear medicine departments for over half a century. It is a meta-stable state in the decay of ^{99}Mo with a half-life of 6 h. ^{99m}Tc decays to ^{99}Tc and emits only 140 keV γ-rays. The parent radionuclide ^{99}Mo, which has a 66 h half-life, can be transported in generators to hospitals where the decay product ^{99m}Tc is extracted. The ^{99}Mo in the form of molybdate, MoO_4^{2-} is adsorbed onto acid alumina (Al_2O_3) and forms pertechnetate TcO_4^- when it decays. The pertechnetate is not bound as tightly to the alumina and can be dissolved and removed preferentially by passing saline through the column (figure 16.1(a)). The ^{99m}Tc is eluted from the column by piercing the rubber cap of an evacuated vial with a needle connected to the column and drawing in the saline solution containing ^{99m}Tc in the form of $NaTcO_4$. This will be done in a radiopharmacy at the start of each working day to collect the ^{99m}Tc produced by decay of ^{99}Mo over the previous 24 h (figure 16.1(b)). The solution may then be added to the pharmaceutical kit to be used in an appropriate concentration (labelling), or the sodium pertechnetate used directly as the radiopharmaceutical to be injected into patients. The product of the decay is ^{99}Tc, which has a half-life of 2.1×10^5 y, so that the level of activity

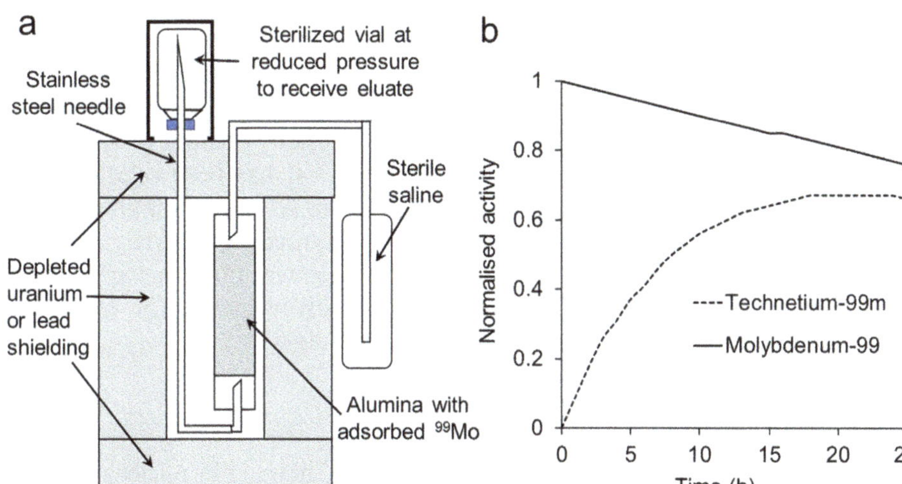

Figure 16.1. (a) ^{99m}Tc generator and (b) plots showing the decay of ^{99}Mo and the increase in ^{99m}Tc within the generator over a 24 h period.

Table 16.1. Properties of radionuclides used for imaging (Delacroix *et al* 2002, Smith and Stabin 2012).

Radionuclide	Mode of decay	Principal γ-emissions (keV)	Half-life (T_{phys})	First tenth value layer (mm Pb)
^{67}Ga	Electron capture	93, 185, 300	3.26 d	4.8
^{75}Se	Electron capture	136, 265, 280	119.8 d	4.32
^{81m}Kr	Metastable state	190	13 s	1.93
^{99m}Tc	Metastable state	141	6.02 h	0.905
^{111}In	Electron capture	171, 245	2.83 d	1.96
^{123}I	Electron capture	27, 159	13.2 h	1.12
^{201}Tl	Electron capture	x/γ 75, 167	73.1 h	0.887

remaining is low. The elution, preparation, and labelling are carried out in radiopharmacies linked to nuclear medicine departments and, because of the decay of ^{99}Mo, departments will purchase at least one generator per week. The generators incorporate shielding usually of depleted uranium.

To demonstrate physiological function a pharmaceutical is required that will be taken up by the relevant tissue and a radionuclide that can be bound to it. The resulting radiopharmaceutical should accumulate predominantly in the tissue of interest, although in practice most will also collect in other parts of the body. The period over which the radiopharmaceutical accumulates in the tissue affects the imaging procedure, and the time for which it remains in the body determines the dose that the patient receives. The dose depends on both the physical half-life (T_{phys}) and biological half-life (T_{biol}). The effective half-life (T_{eff}) is given by:

$$\frac{1}{T_{eff}} = \frac{1}{T_{phys}} + \frac{1}{T_{biol}}$$

Biological half-lives vary considerably, so wherever possible a radionuclide with as short a physical half-life as possible is used. Radionuclides used for imaging are shown in table 16.1 and decay predominantly via electron capture, so avoiding exposure from β-particles. Once a radiopharmaceutical has been selected for an examination, it is administered to patients, usually via an injection although other routes are used where appropriate. Images may be acquired immediately or delayed depending on the physiological processes being investigated. For some applications, images may also be acquired at several stages post-injection, for example in renography to view uptake and clearance from the kidneys.

16.2.2 Gamma cameras

The gamma camera is the main instrument used for imaging in nuclear medicine. The camera contains a large flat scintillation crystal usually of sodium iodide doped with thallium NaI[Tl], sealed within a light-tight housing. The crystal is optically coupled to an array of photomultiplier tubes (PMTs) all contained within the head of the camera. A gamma camera head is typically 60 cm × 45 cm with a close-packed array of about sixty PMTs. The camera head is mounted on a gantry so that it can be placed close to the patient and when γ-ray photons interact with the crystal a flash of light is produced that is detected by the PMTs.

A simple system with just a crystal could not differentiate between photons travelling in different directions and would not provide a useful image of anatomy. Therefore, a collimator is placed in front of the crystal/PMT array allowing only photons incident perpendicularly on the camera head to be detected (figure 16.2(a)). The collimator consists of a sheet of lead, 25–75 mm thick, containing thousands of closely spaced holes. In the common form, the holes are perpendicular to the face of the collimator and parallel to each other. This arrangement cuts out photons hitting the surface at oblique angles, so that only photons incident at angles close to the perpendicular are collected. The collimator properties determine the imaging performance, with smaller holes improving the resolution but decreasing the sensitivity. Thus, the choice of collimator is a compromise, and a typical resolution is about 7 mm while the sensitivity is about 0.1%. Several collimators may be available in a department for examinations requiring different resolutions or for imaging higher energy γ-rays. Collimators with converging holes are used to magnify images of small organs (figure 16.2(b)). The surface of the NaI[Tl] crystal that lies immediately behind the collimator and the sides of the crystal must be light-tight, and this is usually accomplished with a layer of thin aluminium, whereas the rear surface is connected via an optical plate to the bank of PMTs. The system accumulates counts from fluorescent light flashes emitted when γ-ray photons are absorbed by the NaI[Tl] crystal, which are converted into electrical signals by the PMTs, amplified, and transmitted on for processing.

The signals from the PMTs are digitised and used to produce the image. The energy is obtained by summing all the PMT outputs, and the signal position is

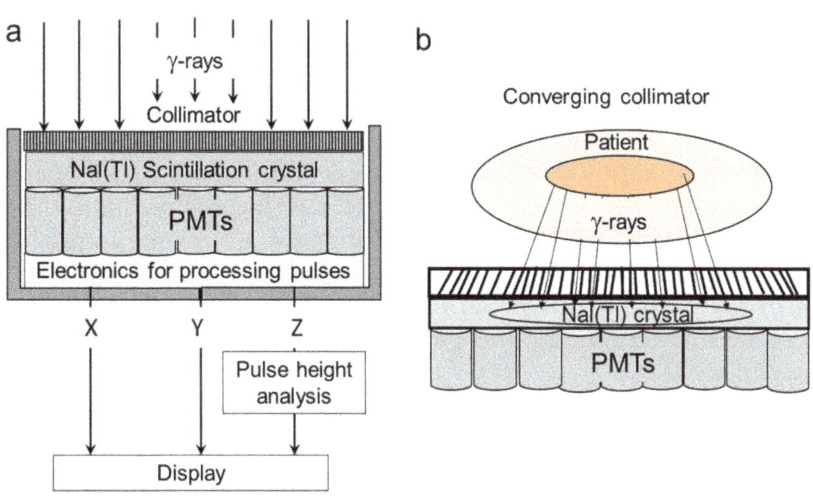

Figure 16.2. Section through (a) gamma camera head with scintillation crystal backed by an array of PMTs and (b) formation of a magnified image with a converging collimator.

computed from the centroid of the group of PMTs with the strongest signals. The γ-rays detected are a mixture of photons emitted directly and ones that have undergone Compton scattering within the patient that do not carry useful information. Pulse height analysis is used to select γ-rays within a window around the main photopeak, which for 140 keV ^{99m}Tc γ-rays is about 30–40 keV wide, and correction circuits adjust for non-uniformities and reduce loss of contrast due to scatter. The image is displayed typically as a 1024 × 1024 matrix, giving a map of physiological function (see figure 16.4 later).

16.2.3 Single-photon emission computed tomography (SPECT)

SPECT imaging is the tomographic form of single-photon emission radionuclide imaging and uses the same radiopharmaceuticals. SPECT -requires a gantry capable of rotating the gamma camera around the patient, and although this can be performed with a single-headed camera, a dual-headed version with cameras spaced by 180°, on opposite sides of the patient, would normally be used, as acquisition times need to be much longer than for x-ray CT scans. SPECT images are acquired by rotating the gamma camera(s) around the patient in steps of 3°–6° while acquiring an image at each projection, which might take 15–20 s, so that a full scan could take about 20 min. The signals are processed to form a series of planar images that are then used with tomographic reconstruction algorithms to yield a 3D data set that can be manipulated to show slices along any axis of the body. Patient movement can be an issue because of the long scan times and automated correction algorithms, in which shift projection images to align organs of interest are used to correct for small amounts of movement. The algorithms used for CT reconstruction are similar to those for x-ray CT, with filtered back projection being replaced by iterative reconstruction (section 15.2.3), which can correct for attenuation and depth-dependent blurring. The reconstructed images have lower resolution than planar images and are displayed as 64 × 64 or 128 × 128 matrices, with pixel sizes of 3–6 mm.

Photons originating from deep within the body are more likely to be absorbed or scattered by overlying tissues and this will reduce the apparent activity in these tissues. Transmission imaging is used to derive correction factors for tissue attenuation. Many SPECT systems now have an integrated x-ray CT scanner that can provide a map of tissue attenuation that can be incorporated into the SPECT reconstruction. Hybrid SPECT/CT systems use x-ray CT images to provide more detailed anatomical information that can be used to identify the locations of regions of high or low activity, as the SPECT image may not show other anatomical structures, making identification of precise locations with respect to other organs and tissues difficult.

16.2.4 Positron-emitting radiopharmaceuticals

PET uses radionuclides that emit positrons, detecting the pairs of 511 keV photons emitted when the positron interacts with an electron in an annihilation event. A positron will only travel a short distance within the body while losing its kinetic

Table 16.2. Properties of radionuclides used for PET imaging (Delacroix *et al* 2002).

Radionuclide	Mode of decay	Positron energy (keV)	Half-life (T_{phys})
^{11}C	β^+, Electron capture	960	20.4 m
^{13}N	β^+, Electron capture	1198	9.97 m
^{15}O	β^+, Electron capture	1732	2.03 m
^{18}F	β^+, Electron capture	633	110 m
^{64}Cu	β^+, Electron capture	653, β^- 0.578	12.7 h
^{68}Ga	β^+, β^-, Electron capture	1899	68 m
^{82}Ru	β^+, Electron capture	3379	1.27 m

energy, typically less than 1 mm, and will then interact with an electron in the surrounding tissue. The two particles annihilate in the interaction and all the mass is converted into two 511 keV photons that are emitted in opposing directions. The most widely used PET radionuclide is ^{18}F, which can be readily incorporated into glucose, as fluorodeoxyglucose [^{18}F]FDG. This is a marker for uptake of glucose, which is correlated with certain types of tissue metabolism, especially cancer, and it is now the standard radionuclide used for cancer patient management. Data on the radionuclides used for PET imaging are given in table 16.2.

The radionuclides used in PET scanning can be readily combined with molecules processed by the body such as water and ammonia or incorporated into molecules that bind to receptors or other sites of drug action. However, most have short half-lives and so have to be produced using a cyclotron in close proximity to the PET imaging facility. The half-life of ^{18}F is long enough to allow it to be manufactured commercially at another location and transported to other imaging centres, and ^{82}Ru generators are available containing ^{82}Sr, which has a 25-day half-life and decays by electron capture to^{82}Ru.

16.2.5 Positron emission tomography (PET)

PET images are formed with a ring of scintillation detectors. Positron annihilation produces a pair of photons moving in opposite directions, which are recorded by the ring of detectors (figure 16.3(a)). The technique depends on the coincident detection of the pair of photons from each disintegration. Flashes of light from the scintillators are registered by photomultiplier tubes or silicon avalanche photodiodes. Flashes that arrive within a timing-window of a few nanoseconds are recorded and the remainder are rejected. The location of the source will be along a straight line of coincidence (the line of response).

PET scanners comprise a large number of scintillation crystals placed in circular ring surrounding the patient. There may be up to 48 multiple rings stacked along the scanner axis and separated by retractable lead collimation septa. The detector crystals are large blocks of scintillation material that are relatively thick (e.g. 30 cm) to detect the high energy γ-rays and each is linked to several PMTs (figure 16.3(b)). The blocks are partially cut through in the direction of γ-ray travel to create an array

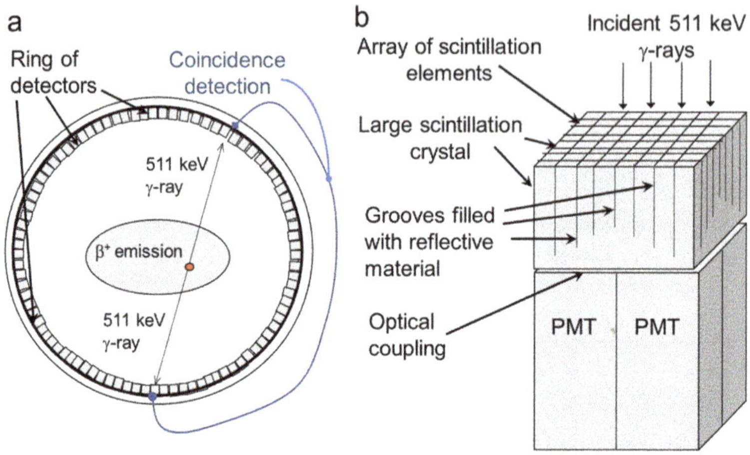

Figure 16.3. PET scanner showing (a) coincidence counting of annihilation photons and (b) detail of scintillation crystal, which in this example might be 50 cm × 50 cm × 30 cm thick from which light is collected by four PMTs.

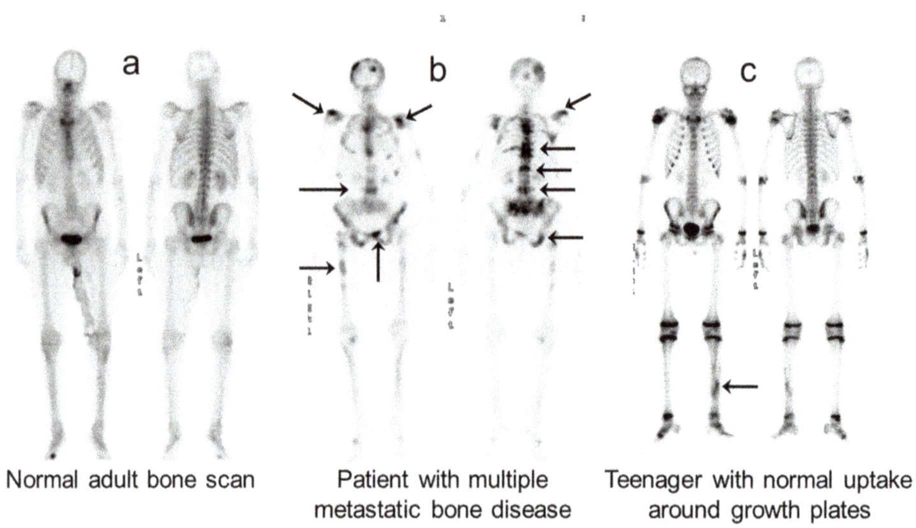

Figure 16.4. Bone scan images recorded on a gamma camera for (a) normal adult, (b) patient with metastatic bone deposits highlighted by arrows and (c) a normal teenager with uptake around growth plates and a fractured tibia indicted with an arrow. Image credit: Alison Bolster

of detectors and the cuts are filled with reflective material to prevent light formed at the top of the crystal from producing a broad spread at the PMTs 30 cm away. Pulse height analysis is used to select the signal from the 511 keV photons and the X–Y position relative to the centre of the block of four PMTs is determined from the relative magnitudes of the signals detected by each PMT. The scintillation material should have a high yield, a short decay time to allow short coincidences to be resolved, emit a wavelength near 400 nm to which PMTs are most sensitive, and be optically transparent. NaI[Tl] was used in early PET scanners, but crystals with higher atomic numbers such as bismuth germinate oxide BGO and lutetium orthosilicate doped with cerium LSO(Ce), which have a higher probability of photoelectric interaction, are now used.

Coincidence events are grouped into projection images, corresponding to the angle of γ-ray emission for each view. These are analogous to the projections in SPECT and an image is reconstructed in a similar manner. Pre-processing of the data is carried out to reduce the influence of scattered photons and random coincidences and correct for variations in detector sensitivity. Correction for attenuation of the emitted photons is performed either with a transmission scan using a ^{68}Ge rotating rod source or a CT scanner for hybrid units, and iterative reconstruction is used to generate the final image. PET scanner detectors with good time resolution could potentially localize an event to the segment of a chord, the length of which would depend on the time resolution. LSO(Ce) has a response time of 450 ps and measurements of the delay in photon detection can be used to estimate the time of flight. The current localization is only within about 10 cm but can provide some improvement in performance.

Modern PET scanners have multiple detector rings, so forming a cylinder of detectors placed around the patient. There are two approaches to data collection and reconstruction: 2D multi-slice in which the collimation septa are extended, and 3D in which the septa are retracted. In the 2D case each detector ring is treated as a separate entity, only coincidences within the ring are detected, and the image from each ring is reconstructed separately. In the 3D option coincidences are also detected between rings and a 3D data set for the entire volume reconstructed. The 3D reconstruction requires more computing power but has ten times the sensitivity and a higher signal-to-noise ratio. The 3D technique is more susceptible to random coincidences and scattered γ-rays, but is becoming more widely used, and improvements in detector timing resolution are providing better random coincidence rejection.

16.3 Preparation and administration of radiopharmaceuticals

16.3.1 The radiopharmacy

Preparation of radiopharmaceuticals is a key component of nuclear medicine services. Nuclear medicine departments may have their own radiopharmacy, or a large central radiopharmacy may serve several departments. The design of a radiopharmacy needs to take account of good manufacturing and aseptic dispensing, as well as radiation protection considerations, both of which are tightly regulated. Requirements include rooms for the preparation processes, a quality control laboratory, secure storage areas for radioactive materials and waste, a reception area for radiopharmaceutical deliveries, and a shower for decontamination. Facilities within laboratory areas and nuclear medicine in general for handling radioactive materials will require sealed benches, ventilation systems that do not recirculate radioactivity between rooms, hand washing facilities close to room exits, and a designated sink for disposal of aqueous waste. Aseptic manipulations in a radiopharmacy must be performed in a workstation with a controlled internal environment, which might be a laminar flow cabinet within a cleanroom or an isolator with filtered air. Entry and exit for the cleanroom should be through a separate room in which staff change into sterilized clothing and entry may be via a 'step over' barrier.

^{99m}Tc eluted from a ^{99}Mo generator is combined with a commercial kit taking the form of a freeze dried non-radioactive compound contained within a sealed sterile vial. Kits enable the $NaTcO_4$ to be transformed into a range of radiopharmaceuticals simply by adding the pertechnetate to the vial and dissolving the contents, sometimes with a short incubation period. The reconstituted kit vial is then subdivided into administrations for single patients, commonly referred to by nuclear medicine staff as 'patient doses'. In an alternative 'wet' method, the kit can be reconstituted with sterile saline, and subdivided into vials for individual patients and the $NaTcO_4$ added to each vial later. This reduces the number of manipulations performed with active syringes and so can help to reduce staff doses. Radiopharmaceuticals labelled with other radionuclides are supplied as products that are ready for use, in vials labelled with the radiopharmaceutical, activity, and date.

Because standard activities are administered to every patient, careful planning, recording, and measuring of activity levels is required at each stage. The aim is to ensure that the nuclear medicine department has sufficient activity to carry out each procedure at the designated time. Detailed documentation and paper labelling of all vials is essential to avoid errors that could result in patients receiving unnecessary doses of radiation (Martin *et al* 2019). Activity meters, calibrated for a range of radionuclides, are used for checking activity levels both in the radiopharmacy and in the nuclear medicine department to confirm that the correct activity has been prepared and is being administered to the patient. The meter calibration will be confirmed periodically and regular constancy checks are made with surrogate long half-life check sources.

16.3.2 PET radionuclide production with a cyclotron

PET radionuclides such as ^{11}C, ^{13}N, or ^{15}O, have short half-lives (table 16.2) and can only be used if there is a cyclotron on-site, so specialist PET facilities usually have a cyclotron incorporated into the same building. A cyclotron will require its own shielded vault, but self-shielded cyclotrons are available, although they will still require additional shielding (McJury and Tonge 2014). PET cyclotrons produce radionuclides through bombarding targets with ^{1}H (p) or ^{2}H (d), that produce nuclear reactions such as (p,n), (p,2n), (p,α), and (d,n). The primary beam of charged particles will be completely absorbed by the target, but shielding is required against γ-rays and neutrons produced in nuclear reactions. Concrete is usually the best compromise for shielding, containing sufficient hydrogenous material to attenuate neutrons, whilst also protecting against γ-rays. Typical wall thicknesses would be of the order of 0.5–0.6 m of concrete (density 2.3 g cm^{-3}) for a 10–11 MeV self-shielded cyclotron, and 2.0–2.4 m of concrete for a non-shielded 18 MeV cyclotron and the vault may have a shielded door or a maze. The pattern of air flow within the facility should be designed to control airborne contamination and avoid re-circulation of air. In general, the cyclotron vault should be at the lowest pressure, the hot laboratories at an intermediate pressure, and the surrounding public areas at a higher pressure. There should be systems for control of access to the bunker and staff should wear specific protective clothing.

The radiopharmacy should be close to the cyclotron, to allow rapid transfer of radionuclide from the target to shielded hot cells. Radionuclides in liquid or gaseous form from the targets, within the cyclotron bunker, should be delivered to the points of use through a shielded transport system, which might be through walls or in trenches under the vault floor. Release of radioactive gases from cyclotron operation can result from rupture of the target or a poor connection in a delivery line. The main discharge point of the cyclotron and PET laboratory should be on the roof of the building and exhaust air travelling through the extraction system must be monitored before release. The monitoring system can act as an alarm for detection of unexpected discharges, as well as providing a record of routine gaseous releases. Automated devices are available to perform synthesis, filling vials and dispensing in syringes, with minimal need for manual intervention. The synthesis modules for radiopharmaceutical production are located inside hot cells normally operated at a low pressure to reduce the risk of leaks into the laboratory. The final dispensing of radiopharmaceutical in a syringe for injection into patients can be performed either in the radiopharmacy or in a shielded clinical area.

Components of the cyclotron may be activated, if they are hit by the proton beam or in close proximity to it, or if they are irradiated by secondary neutrons produced in the target. The most activated components will be the foils used to seal the targets. These are often made of Havar, an alloy containing cobalt, chromium, iron, and tungsten and the radionuclides produced by activation include ^{51}Cr, ^{57}Co and ^{54}Mn. Dose rates at 1 m from targets after ^{18}F production can be several mSv h^{-1}, so after disconnecting a target, a period is required to allow for decay of radioactivity with dose rates taking several hours to decrease to hundreds of $\mu Sv.h^{-1}$.

16.3.3 Design of nuclear medicine facilities

In nuclear medicine departments the sources of radiation are patients, vials, and syringes containing radioactive material, which are not confined to a single room. The radiopharmacy, therapy rooms, and administration and scanning areas will all usually be designated as controlled areas, because of the potential for contamination as well as external radiation (McJury and Tonge 2014, Martin and Sutton 2015). The adjoining corridors could be supervised or controlled areas and patient waiting areas with their own toilets will be required. Shielding requirements will be based on assumptions about numbers of patients, the times they spend in the department after injection, and lengths of scans (Salehzahi and Tse 2020). Additional shielding of rooms for the radionuclides used in photon emission nuclear medicine imaging is often unnecessary, as the lead tenth value layer for ^{99m}Tc is only 0.9 mm and values for other radionuclides are only a few mm (table 16.1), so any solid walls will provide protection. Shielding may sometimes be required to prevent degradation of nuclear medicine images from radiation emitted from a patient in an adjoining room. Shielding will be required in PET facilities, as the tenth value layer for the 511 keV γ-rays is 15–16 mm of lead, and care will need to be taken in planning the movement of patients through a department in order to minimise direct exposure (Sutton *et al* 2012). Patients being imaged with [^{18}F]FDG are required to rest for a period following injection, so shielded bays that can take several patients will be required. Shine-through paths from bays where patients are resting to areas occupied by staff should be minimized and dedicated toilet facilities provided for patients.

16.4 Medical diagnostic procedures

16.4.1 Single-photon emission imaging

Nuclear medicine imaging studies tend to be organ-, tissue- or disease-specific, exploiting different ways in which the body processes substances when disease is present. The radiopharmaceutical will normally be taken up by specific tissues and there may need to be an interval after injection before images are recorded. Examples of different types of examination will be discussed here to illustrate the processes involved.

Bone scan: Nuclear medicine bone scans of the whole body are used for investigating bone pain and bone-related pathology. Methylene-diphosphonate (MDP) is preferentially taken up in regions of bone growth and by attaching ^{99m}Tc areas of bone development can be identified. (figure 16.4). A period of 2–3 h after injection is required to allow uptake of radiopharmaceutical before scans are performed. Regions of increased physiological function will have increased concentrations of radioactivity and show up as ‘hot spots’. Radionuclide uptake will occur for a variety of pathologies such as stress fractures, infections, and bone metastases associated with other types of cancer.

Heart scan: Myocardial perfusion scans play an important role in the evaluation of coronary artery disease. In multi-gated acquisition scans (MUGAs) separate image data sets, triggered by electrocardiograms, are collected for different phases of the heart cycle to provide a series of images showing the beating heart. Two injections of activity are given about 20 min apart, and scans may be taken both at rest and after exercise. Gated SPECT imaging can provide quantitative information about myocardial perfusion, and the thickness and contractility of the myocardium, and enable the left ventricular ejection fraction, stroke volume, and cardiac output to be calculated to provide prognostic information on the risk of an adverse cardiac event.

Renogram: Scans of the kidney with ^{99m}Tc diethylene triamine pentaacetic acid (DTPA) can show not only the shape of the functioning kidney tissue, but also changes with time. Count data for the two kidneys and the bladder can be collected separately and activity versus time curves generated to enable the function of each kidney to be assessed, and identify or exclude an obstruction to renal drainage, or diagnose kidney disease.

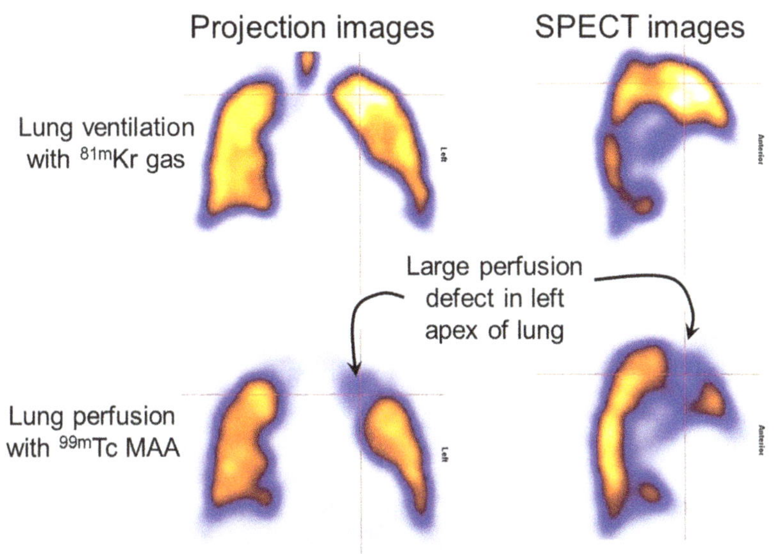

Figure 16.5. Lung scan images for ventilation and perfusion scans: on the left as gamma camera planar projection images and on the right as SPECT images, showing reduction in perfusion. Image credit: Alison Bolster

Lung scan: Pulmonary ventilation and perfusion (V/Q) scans of the lungs are performed in two parts to assess lung function. V/Q scans are useful in the evaluation of pulmonary embolism. In the ventilation scan the patient breathes in an aerosol labelled with ^{81m}Kr from a generator of ^{81}Rb that has a 4.6 h half-life, and this enables imaging of the pattern of air flow in the lungs and helps to identify regions where there is no air movement. For the perfusion scan, human albumin macro-aggregates 10–40 μm in diameter labelled with ^{99m}Tc are injected slowly, these collect throughout the lung tissue in which blood flow is normal and enable areas where flow is depleted to be identified (figure 16.5).

16.4.2 PET procedures

PET scanning with fluorodeoxyglucose (FDG) labelled with ^{18}F is widely used in clinical oncology. FDG is a glucose analogue for which the uptake is elevated in rapidly growing malignant tumour cells, where the FDG becomes trapped. After the radiopharmaceutical has been administered. the patient must rest for about an hour to obtain the optimal bio-distribution. The concentrations of FDG represent regions of tissue metabolic activity and images can be used to assess the possibility of the spread of cancer to other sites within the body (metastases). FDG-PET can be used for diagnosis, staging, and monitoring treatment of a wide range of cancers and has become a major clinical tool in cancer diagnosis and management.

Brain imaging with [^{18}F]FDG-PET can measure regional glucose uptake and is used in diagnosis of neuropathology. Brain pathologies such as Alzheimer's disease decrease brain metabolism of both glucose and oxygen in tandem. [^{18}F]FDG and other ^{18}F-labelled radiopharmaceuticals can differentiate Alzheimer's disease from other dementia processes for early diagnosis. PET imaging with FDG can also be used for localization of seizure focus that will appear as a region of decreased uptake, and other ^{18}F-labelled agents are used for the visualization of neuroreceptors in the investigation of various neuropsychiatric and neurologic illnesses. PET imaging with ^{15}O indirectly measures blood flow to the brain which is assumed to correlate with increased brain activity. Since the half-life is only 2 min, ^{15}O must be piped directly from a medical cyclotron.

16.4.3 *In vitro* nuclear medicine

Nuclear medicine includes non-imaging investigations involving the assay of low levels of radioactivity. Samples are placed adjacent to a NaI[Tl] scintillator often in the form of a cylinder into which the sample is placed, and light flashes counted with a PMT. Counting systems are designed to take large numbers of samples and record the counts automatically. Counting of β-emitters is carried out by dissolving samples in liquid scintillator and recording counts with PMTs. Examples of the tests carried out and the radionuclides used are given here:

- Blood plasma volume with ^{125}I-human serum albumin;
- Red cell volume with ^{51}Cr-sodium chromate;
- Breath tests for *Helicobacter pylori* linked to duodenal ulcers with ^{14}C-urea;
- Blood loss from the gastrointestinal tract with ^{51}Cr-sodium chromate;
- Protein loss from the gastrointestinal tract with ^{51}Cr-chromic chloride;
- Test for megaloblastic anaemia with ^{57}Co Schilling test;
- Studies of ferrokinetics with ^{59}Fe for haematological disorders;
- Renal function tests with ^{125}I-hippuran and ^{51}Cr-ethylenediaminetetraacetic acid;
- Thyroid uptake with ^{123}I-sodium chloride.

16.4.4 Optimization of radiation protection for the patient

The Medical Internal Radiation Dose Committee (MIRD) of the American Society of Nuclear Medicine created a method of estimating organ absorbed dose after radiopharmaceutical administration which is referred to as the MIRD Schema and this has been developed further by the International Commission on Radiological Protection (ICRP). This uses biokinetic models to evaluate radionuclide distribution and transit times through different compartments within the body. Activity-time curves are generated for all parts of the system and used to assess mean absorbed doses to radiosensitive organs and tissues throughout the body. Early simulations used phantoms with organs based on geometrical shapes, but recent calculations use more realistic phantoms based on CT scans of 'average individuals' (ICRP 2009). Organ equivalent doses and effective doses are available for a wide range of radiopharmaceuticals for adults and children aged 1, 5, 10 and 15 y in publications produced by the ICRP (ICRP 2015).

To enable consistent practices across each country, the amounts of activity that should be administered for all diagnostic procedures are set out in terms of

Table 16.3. Radiopharmaceuticals used for common nuclear medicine examinations for adult patients, together with UK DRLs and effective doses (ARSAC 2018).

Radiopharmaceutical	Imaging task	DRL (MBq)	Effective dose (mSv)
^{18}F-fluorodeoxyglucose	Tumour imaging	400	7.6
^{67}Ga-citrate	Infection/inflammation	150	15
^{81m}Kr gas	Renal imaging	6000	0.2
^{99m}Tc-colloid	Lymph node	20	0.09
^{99m}Tc-diphosphonates	Bone imaging	600	2.9
^{99m}Tc-DMSA	Renal imaging	80	0.7
^{99m}Tc-DTPA	Renal imaging	300	1.5
^{99m}Tc-erythrocytes	Blood pool imaging (MUGA)	800	5.6
^{99m}Tc-iminodiacetates	Liver biliary function	150	2.4
^{99m}Tc-human albumin, macro-aggregates	Lung perfusion	100 (planar) 200 (SPECT)	1.1 2.2
^{99m}Tc Technegas	Lung ventilation	40	0.6
^{99m}Tc-HMPAO	Brain imaging	500	5
^{99m}Tc-MAG3	Renal imaging	100	0.7
^{99m}Tc-MIBI	Myocardial imaging	800 (SPECT)	7.2
^{99m}Tc-leococytes	Infection/inflammation	200	2.2
^{99m}Tc-pertechnetate	Thyroid imaging	80	1
^{99m}Tc-tetrofosmin	Myocardial imaging	800 (SPECT)	6
^{111}In-leucocytes	Infection/inflammation	20	7.2
^{123}I-iodide	Thyroid imaging	20	6.1
^{123}I-MIBG	Neuroectodermal tumour	400	5.2
^{123}I ioflupane	Dementia and Parkinsons dis.	185	4.6
^{201}Tl chloride	Myocardial imaging	80	11

diagnostic reference levels (DRLs) recommended at a national level. A list of the adult DRLs and effective doses for some common procedures used in the UK is given in table 16.3. These are in terms of the total activity administered to a patient (ARSAC 2018). Standard adjustments are made to the administered activity based on weight for paediatric examinations, and in some countries adult DRLs are also defined in terms of the activity per unit weight. Every nuclear medicine exposure should be justified by a nuclear medicine practitioner, usually a radiologist or nuclear medicine physician, and the protocol chosen for the examination should be optimized for the patient. A larger amount of activity may be used if justified by the clinical conditions, perhaps to speed up the procedure for a patient who is in extreme pain or to obtain an image that is diagnostic for one who is obese.

Much of the radioactivity used in nuclear medicine is excreted through the bladder or colon and irradiation of these organs accounts for 20%–50% of the effective doses for many examinations (Martin 2011). After the activity has been administered to a patient, steps that can be taken to minimize the dose linked to excretion by advising good hydration through drinking water and frequent voiding to reduce the accumulation of activity in the bladder. For patients undergoing procedures involving ^{123}I, ^{125}I or ^{131}I administered as iodide or chemical forms that metabolise to iodide, the dose to the thyroid can be significant. Administration of a thyroid blocking agent, such as potassium iodide, potassium iodate or potassium perchlorate will reduce thyroid uptake if the thyroid is not being investigated.

CT scans that accompany SPECT and PET imaging can themselves give a considerable radiation dose to a patient, but the level of image quality required depends on the purpose for which the image is to be used and this has a significant impact on dose. Therefore, it is important to identify the purpose for which CT images are needed. This might be for attenuation correction, to provide anatomical positioning information, or there may be a need for a full diagnostic image quality CT investigation, and different protocols and DRLs should be used for each application (section 15.3.3).

In nuclear medicine there are potential problems when a radiopharmaceutical injection has been extravasated and there is a risk of deterministic effects such as pruritus, erythema and ulceration (Martin *et al* 2019). The dose will depend on the radiations emitted, the physical half-life, and the rate of biological clearance, and will be greater for radionuclides emitting β-particles.

The foetus is more susceptible to damage by radiation than is the adult, so precautions are taken when examining or treating pregnant patients, especially with radionuclides that have longer half-lives. The administration of a radiopharmaceutical to a pregnant woman will result in exposure of the foetus to radiation emitted from adjacent maternal organs and from any radioactivity transferred across the placental boundary. Therefore, patients are questioned about the possibility of pregnancy and the date of their last menstrual period checked. If a patient is found to be pregnant then the possibility of postponing the procedure will be considered. Extensive tabulated data on foetal dose levels following administration of a variety of radiopharmaceuticals is available (ICRP 2001). Following administration of ^{131}I, ^{32}P, ^{89}Sr and ^{90}Y for therapeutic purposes, patients may be advised to avoid conception for a period that could extend to several months.

When a radiopharmaceutical is administered to a mother who is breast feeding, radioactivity will be secreted in her milk and will be ingested by the infant if breast feeding is continued. Therefore, use of other imaging techniques or postponing the examination may be considered. If a nuclear medicine procedure is essential, then breast feeding may need to be interrupted but the radiation risk must be balanced against the long-term benefits of breast feeding (ICRP 2004). The strategy adopted will depend on the examination or treatment, but an option may be to express breast milk and store it in a refrigerator to feed the infant during the period of interruption.

16.5 Radionuclide therapy

16.5.1 Planning treatments

Radionuclide therapy involves the administration of radiopharmaceuticals that emit β- and occasionally α-particles that can deliver significant local doses to diseased

Table 16.4. Properties of radionuclides used in radionuclide therapy (Delacroix *et al* 2002, Smith and Stabin 2012).

Radionuclide	Decay mode	Principal emissions: maximum β energies and γ-ray energies (MeV)	Half-life	First tenth value layer in Perspex/lead and Range in air
^{32}P	β	E_β 1.71	14.3 d	6 mm Perspex, Range in air 6 m
^{89}Sr	β	E_β 1.463	50.5 d	5.5 mm Perspex, Range in air 6 m
^{90}Y	β	E_β 0.523 (<1%), 2.284	64.0 h	9.2 mm Perspex, Range in air 9 m
^{131}I	β	E_β 0.606, E_γ 0.364	8.04 d	9.9 mm lead
^{153}Sm	β	E_β 0.703, 0.634, E_γ 0.103	46.7 h	2.4 mm Perspex, 0.43 mm lead
^{169}Er	β	E_β 0.344, 0.352	9.4 d	0.8 mm Perspex
^{177}Lu	β	E_β 0.497, E_γ 0.208, 0.113	6.71 d	2.11 mm lead
^{186}Re	β	E_β 0.939, 1.077, E_γ 0.137	3.78 d	3.4 mm Perspex
^{223}Ra	α	E_α 5.5, 5.6, 5.7	11.4 d	Thin plastic or paper

tissues. The first radionuclide therapy treatments performed in the 1940s used radioiodine for treating thyroid disease and others using ^{32}P followed in the 1950s. New treatment options are now becoming available with different radionuclides and pharmaceuticals. The therapies encompass a range of radionuclides (table 16.4), targeting mechanisms, and methods of administration, and are used for both benign and malignant disease (ICRP 2019). Optimising the ability to treat cancer cells while minimizing potential adverse side effects and complications in normal tissues present many challenges.

The calculation of absorbed dose to internal organs, normal tissues, and tumours is important for achieving the optimum therapeutic result. Because most radiopharmaceuticals are administered systemically, radionuclide therapy will also irradiate some normal organs and tissues. Therefore, the use of radiopharmaceuticals for cancer treatment requires detailed, patient-specific dosimetry to evaluate the maximum radiation dose that can be delivered to cancerous tumours, while at the same time ensuring the radiation doses to critical normal organs and tissues are minimized and do not exceed tolerated levels.

16.5.2 Iodine therapies

^{131}I-iodide (radioiodine) is used routinely for treatment of hyperthyroidism and goitre. The thyroid produces hormones that regulate all aspects of the body's metabolism, and an overactive thyroid will accelerate body metabolism. Hyperthyroidism can be caused by Graves' disease, in which the entire thyroid gland is overactive, or by nodules within the gland that produce too much thyroid hormone. ^{131}I-iodide is administered orally, usually in the form of a capsule containing 300–800 MBq ^{131}I that accumulates in the thyroid gland, where the β-particles emitted destroy thyroid cells. Radioiodine is primarily excreted via urine, but also through faeces and perspiration, with the mean effective half-life in the body being about 5 days. The levels of activity used for benign thyroid conditions are such that patients are usually allowed to return home after treatment. The consent of carers who may be in close contact with the patient is required. Advice to carers will depend on the circumstances and detailed written instructions and guidance will be provided to the patient and their family.

Radioiodine is also the treatment of choice for the ablation of thyroid cancer and is used as an adjunct for patients who have undergone thyroidectomy, to ablate remnants of thyroid tissue and treat distant metastatic disease. The amounts administered are several GBq and treatments may be repeated for recurrent or persistent disease. As these patients are hospitalised, there are risks to nurses, technologists, medical physics staff, and physicians. Patients will be treated in single rooms with separate toilet and washing facilities, staff and visitors should use shielding when communicating with patients, and wards should be well ventilated because of the high level of volatility of ^{131}I. Advice should be given to carers when patients return home similar to that for treatment of hyperthyroidism. Unwitting administration of ^{131}I during early pregnancy will expose the embryo/foetus to external radiation from radioiodine in the mother's bladder. Radioiodine can cross the placental boundary and once the foetus reaches ten weeks gestational age, any free radioiodine can be taken up by the foetal thyroid with a risk of hypothyroidism (ICRP 2001). Ensuring that patients are not pregnant is therefore a major concern.

16.5.3 Other therapy radionuclides

Various other β-emitting radionuclides are used in therapies and initial trials are being performed with α-emitters to achieve very localised treatments while sparing normal tissues. A selection of examples covering the range of treatments carried out is given here.

Bone metastases: Metastases from prostate and breast cancers are found in the skeletons of up to 85% of patients and cause severe pain. Low-energy β-emitting radiopharmaceuticals such as ^{153}Sm-EDTMP (ethylenediamine tetramethylene phosphonate) and ^{89}Sr chloride can deliver high radiation doses to bone metastases and are used in palliative treatments. The bone marrow will receive a lower dose, but care is required in avoiding bone marrow suppression. The α-emitting radionuclide ^{223}Ra-dichloride is undergoing trials as a potentially curative approach for castration-resistant prostate cancer patients with bone metastases and has a theoretical advantage of sparing much of the marrow irradiation because of the short range of α-emissions.

Radioimmunotherapy: Radiolabelled antibodies that recognise tumour-specific antigens can deliver therapeutic radiation to tumour tissue. Most radionuclides are β-emitters such as ^{131}I, ^{90}Y, ^{186}Re, and ^{153}Sm, but α-emitters such as ^{225}Ac and ^{213}Bi are undergoing preliminary trials. Injection of Zevalin containing about 1 GBq of

^{90}Y is used to reduce the number of B-cells for treatment of non-Hodgkin's lymphoma and has a high response rate.

Liver cancer: Liver carcinoma and liver metastases can be treated by direct infusion of a radiotherapeutic substance into the hepatic artery, from where liver tumours preferentially take their blood supply, normal liver being fed predominantly by the portal vein. The radionuclide ^{90}Y is used to label microspheres that are trapped in tumour capillaries and the activities of 3–20 GBq are delivered by intravenous infusion. The treatment is palliative, but offers a potential for delivering high absorbed doses with precision targeting. However, it is a relatively invasive procedure and will irradiate normal tissues such as the lungs, gut, normal liver, and pancreas.

Neuroblastoma accounts for about 6% of childhood cancers. It is treated with ^{131}I-metaiodobenzylguanidine (mIBG), which is taken up by cells of the sympathetic nervous system. Neuroblastoma arises from the neural crest cells involved in the development of the nervous system. It occurs in the adrenal glands or in the nerve tissue and can spread to the bones and liver. The aim of treatment is predominantly palliative.

Neuroendocrine tumours express somatostatin receptors and analogues ^{90}Y-DOTATOC and ^{177}Lu-DOTATATE that target a somatostatin receptor are used for palliative treatments.

Polycythaemia vera: ^{32}P orthophosphate is used to treat polycythaemia vera and essential thrombocythaemia, which are chronic progressive myeloproliferative disorders characterised by an over-production of erythrocytes and thrombocytes, respectively. ^{32}P administered either orally or by intravenous injection is incorporated into the DNA of rapidly proliferating cells and the treatment supresses blood cell production by irradiation of the bone marrow.

Rheumatoid or osteoarthritis: Radiopharmaceuticals have provided a cost-effective treatment for rheumatoid or osteoarthritis and haemophilic synovitis for over 40 years. ^{90}Y and ^{32}P colloid are used for larger joints such as the knee, ^{186}Re-colloid for the elbow and ankle, and ^{169}Er-citrate for metatarsophalangeal joints. The aim of radiosynovectomy is to reduce swelling and provide pain relief. The limited range of the injected radionuclides ensures little irradiation of adjacent tissues and reported side effects are rare.

Prostate cancer: ^{177}Lu treatment for cancer utilizes a molecule that attaches itself to the prostate-specific membrane antigen (PSMA) receptors located on cancer cells. PSMA is found on the surface of prostate cells, and the expression of PSMA increases in patients with prostate cancer and is also found in metastases. The therapy reduces tumour size and controls growth.

16.6 Radiation protection of nuclear medicine staff

Radiation sources in nuclear medicine such as radionuclide generators and radionuclide storage facilities all require dedicated shielding for protection of nuclear medicine staff. But the patients are a source of exposure for a range of healthcare staff including nuclear medicine technicians, medical physicists, doctors, nurses, and porters.

16.6.1 Handling of radiopharmaceuticals

Radiopharmacists may manipulate significant amounts of radioactivity, so shielding will be required for the head and body during manipulations. The tenth value thickness for ^{99m}Tc γ-rays is 0.9 mm thick lead, that for ^{18}F is 5 mm of lead, and that for ^{131}I is 11 mm. In a small radiopharmacy within a nuclear medicine department, a benchtop shield containing a sufficient thickness of lead in the wall and base, which might amount to 8–14 mm for handling ^{18}F, may be used, with a lead glass window for viewing manipulations (figure 16.6(a)). In a regional radiopharmacy dispensing tens of GBq of radiopharmacejutical for several nuclear medicine departments the shielding may take the form of a thick lead wall built across the lower section of the dispensing cabinet. Manipulations should also be carried out over spill trays lined with absorbent material to reduce risks of spreading activity from any spillage.

Vials containing either $NaTcO_4$ eluted from a ^{99m}Tc generator or other radiopharmaceuticals for administration to patients, must all be appropriately shielded for as much of the time as possible and tongs or tweezers should be used to pick up vials and other sources of radioactivity. Lead pots with wall thicknesses of about 1 mm are used for single-patient vials, and pots with wall thicknesses of 2–4 mm for multi-patient vials. Special tungsten vial shields with screw tops with an aperture for withdrawal of liquid are generally used for manipulations in radiopharmacies. Since radiopharmacists carry out repeated manipulations of radioactive liquids using syringes, syringe shields should always be used and their consistent use has been shown to reduce staff finger doses by 85% (Whitby and Martin 2005). Syringe shields for γ-emitters tend to be made of tungsten with a lead glass window or made entirely of leaded glass to provide good visibility (figure 16.6(b)). For β-emitters, such as ^{14}C or ^{3}H, Perspex syringe shields are generally used to minimise the generation of

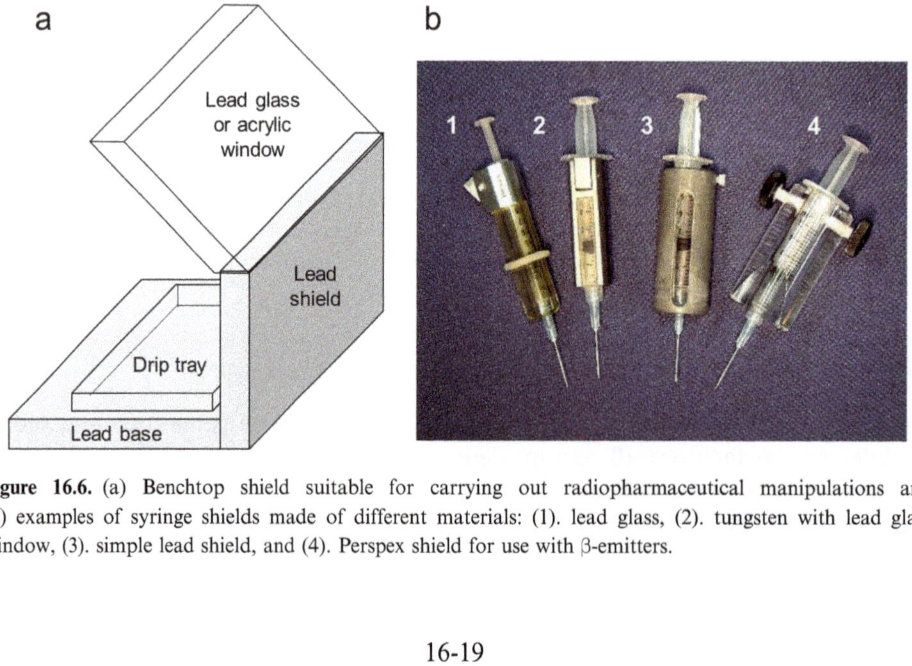

Figure 16.6. (a) Benchtop shield suitable for carrying out radiopharmaceutical manipulations and (b) examples of syringe shields made of different materials: (1). lead glass, (2). tungsten with lead glass window, (3). simple lead shield, and (4). Perspex shield for use with β-emitters.

bremsstrahlung radiation, and ^{90}Y and ^{32}P will require 10 mm of PMMA to completely shield against β-radiation in both benchtop shields and syringe shields.

The specific γ-ray constants for PET radionuclides are an order of magnitude higher than that for ^{99m}Tc and the tenth value layer considerably larger. The drawing up of PET radiopharmaceuticals can lead to significant exposure and should be carried out with appropriate shielding. The ranges of positrons from PET radionuclides can result in significant eye exposure, in addition to the contribution from annihilation γ-rays. Aspects of the handling and preparation of radiopharmaceuticals such as the filling of syringes can be automated, and a fully automatic dispensing system can virtually eliminate hand exposure (Madsen *et al* 2006), but such systems are expensive so may not be an option for smaller radiopharmacies. Nuclear medicine staff will potentially be exposed to activity in vials and syringes when injections are drawn up and this should be done behind a local shield and again syringe shields used to reduce the dose both to the fingers and the body when giving injections (figure 16.6).

16.6.2 Protection from contamination

Protective clothing, such as aprons and gloves, is worn to limit skin contamination and manipulations should be carried out over trays lined with absorbent materials to contain spills. However, contamination cannot be completely avoided and levels in radiopharmacies and nuclear medicine departments need to be kept as low as practicable. Contamination monitoring should be undertaken regularly, including not only benches and flooring, but items that might have been touched by a contaminated glove, such as telephones and door handles. In areas of high ambient background this may need to be accomplished with wipe testing. Facilities should be available for monitoring hands and clothing of staff when entering or leaving an area and this may take the form of dedicated computerised hand and foot monitors with stored personnel profiles that can check for contamination and record results to a file.

The body fluids of patients to whom radiopharmaceuticals have been administered also present a source of potential contamination (Martin *et al* 2019). The level of hazard will depend on the radiations emitted and the half-life. Oral ^{131}I therapy administrations are potentially serious, as vomit will be highly radioactive during the period immediately following administration, so it is good practice to have a dedicated radiopharmaceutical administration room, which might be used for injections and oral administrations. Risks from contamination during treatment of incontinent patients need to be considered and appropriate contingency plans put in place to deal with foreseeable incidents.

16.6.3 The patient journey though the imaging department

In nuclear medicine departments, the main source of whole-body exposure for most staff is from patients injected with radiopharmaceuticals. The exposure of staff working in a nuclear medicine department can be quasi-continuous and the energy and dose rates of the radiation emitted will have a significant impact on the doses that staff receive. Essential exchanges of information with patients attending for a scan, such as the identification check and passing on of instructions, should be carried out before radiopharmaceutical is administered.

The insertion of a syringe needle into a patent vein can sometimes be difficult and occasionally require use of the fingertip to guide the needle, while the vein for injection is punctured. Insertion of a short butterfly cannula prior to radiopharmaceutical administration before handling the radioactive syringe can avoid the need for contact with the needle while it contains radiopharmaceutical and aid in dose reduction. Use of syringe shields of appropriate design substantially reduces doses to the fingertips during injections.

From the time at which the radiopharmaceutical is injected, the patient becomes a source of external radiation, against which protection of staff is necessary. External dose rates after administration of ^{99m}Tc are 5–10 μSv h^{-1} at 1 m, but those for PET ^{18}F are of the order of 30–50 μSv h^{-1} at 1 m. A period of about 60 min is needed to obtain the optimal bio-distribution of [^{18}F]FDG, so 3–4 shielded bays are needed in a PET facility, in which patients can rest before having their scan, and the shielding and patient movement need to be carefully planned. Staff carrying out PET scans should be within a shielded cubicle, so the main periods when they will be exposed is when positioning the patient, setting-up the scanner, and in assisting the patient to and from the scanning room. Scan times may vary, but with modern scanners it can be reduced below 20 min.

A risk assessment is required when planning staff duties and protection measures for a facility and this requires information on dose levels to which individuals will be exposed. These can be estimated by adding separate doses received over time periods spent at different distances from patients. Tables of dose rates have been published in a variety of texts (Martin and Sutton 2015) and these can be used together with inverse square law adjustments to assess exposure for different staff groups. The method should provide sufficient information for a risk assessment and determining initial dose monitoring requirements.

16.6.4 Issues relating to radionuclide therapy

Administration of therapy radionuclides emitting β-particles, such as ^{131}I, ^{90}Y, and ^{32}P requires high standards of safety in technique and the laboratory environment and appropriate shielding (table 16.4). Therapy administrations may be given in the therapy room where the patient is to stay or in a separate administration room, after which the patient is transported to the isolation suite. Preparation and transport of the therapy doses and administration apparatus present risks in terms of spillages and staff exposure that need to be considered. Slow intravenous administration may require additional protection using lead or Perspex shields to reduce staff doses. The patient's state of health, and the risk of nausea, incontinence or catheterisation will also impact on the potential risks to staff during administration.

Particular precautions should be taken when handling radioisotopes of iodine. The annual limits for intake of ^{131}I and ^{125}I are 0.9 MBq and 1.3 MBq, far lower than any other radionuclide used in healthcare, because of uptake by the thyroid.

Manipulation of radioisotopes of iodine should be carried out inside fume cupboards if there is a possibility of volatility and two pairs of protective gloves should be worn, as iodine-labelled materials can penetrate gloves. Uptake of iodine in the thyroid after an episode of contamination can be reduced by blocking the thyroid using stable iodine. The thyroid of staff handling MBq quantities of radioiodine should be checked regularly with a scintillation contamination monitor.

^{90}Y and ^{177}Lu need to be manipulated using syringes shielded with Perspex for intravenous administration to ensure that extremity doses are kept below occupational dose constraints (figure 16.6(b)). Doses to the fingertips from preparation and administration are typically in the range 5–10 mSv from single administrations, even when protection is optimised, but can be over 100 mSv if precautions are inadequate, so doses to the fingers should be monitored (Vanhavere *et al* 2012, ICRP 2019).

Rooms or suites used for treatment of radioiodine therapy patients must meet standards similar to those for a general radionuclide laboratory. The shielding requirements for high activities of ^{131}I are likely to amount to several centimetres of lead or lead equivalent materials. Distance together with local shields can generally be used to limit doses to ward staff and visitors. When a patient is discharged, the room should be monitored and decontaminated, handling of linen and other items contaminated with ^{131}I should be limited, and waste disposed of in the radioactive stream. Patients treated with other radionuclides may only be kept for a single overnight stay before release and present less risk to staff. However, control still needs to be exercised on any fluids produced by the patient (blood, urine, vomiting etc) as these may contain high activities of the administered radionuclide. The route for disposal of liquid waste from radioiodine therapy patients needs to be considered in the design of any therapy suite. On occasions, the inclusion of delay tanks that allow decay of ^{131}I before it is released through the drainage system has been considered, although this is unlikely to be appropriate in most cases because of the doses that would be incurred by staff in maintaining the tanks.

16.6.5 Personnel dosimetry

The manipulation of small vials and syringes containing radiopharmaceuticals can lead to significant exposure of the hands as well as the whole body. ^{99m}Tc decays predominantly with γ-ray emission but also emits a low-energy electron that is heavily absorbed in superficial tissues. This contributes to the superficial $H_p(0.07)$ dose equivalent to the skin. The radiation exposure across the body and head is comparatively uniform, as attenuating personal protective equipment is not worn generally, so personal dosemeters worn on the chest or collar will give an indication of eye doses through the Hp(10) reading linked to effective dose. However, for radionuclide therapies with ^{90}Y or ^{32}P emitting higher energy β-particles (table 16.4) an eye dosemeter calibrated in terms of Hp(3) would be required, as the dosemeter response varies substantially with energy and direction (Behrens 2012). It is often necessary to determine dose levels received from particular procedures or to ensure that a staff member who is pregnant does not receive too high a dose. Small electronic active personal dosimeters (APDs) will provide immediate feedback on dose levels, and some also allow subsequent analysis of data to show when doses were received. Characteristics such as energy and angular response, and battery life may have an important bearing on the performance.

The hands of radiopharmacy staff preparing ^{99m}Tc radiopharmaceuticals and nuclear medicine staff who administer imaging radiopharmaceuticals regularly should be monitored. There may be high dose gradients across the hands (Whitby and Martin 2005, Vanhavere *et al* 2012, Martin *et al* 2018) and these complicate routine monitoring, since the maximum dose may be 3–6 times greater than that at the base of the finger for both ^{99m}Tc and ^{18}F. The dosemeters available are either finger stalls which take the form of a sheath fitted over the finger with a thermoluminescent dosemeter at the tip or rings incorporating a dosemeter. The former is a better option for dosimetric accuracy, but finger stalls are not favoured by some staff because of loss of touch sensation. The distribution of dose across the hands varies between departments and individuals, because different methods are used for drawing up radiopharmaceutical (figure 16.7). However, if exposures are received from manipulations that are repeated many times, the pattern of dose distribution may be consistent for an individual and then a scaling factor can be applied to a ring dosemeter dose, determined from a trial monitoring period. Generally, if doses measured with ring dosemeters worn at the bases of the index fingers with the detector elements located on the palmar sides are less than 2 mSv per month, then ring dosemeters should be sufficient, but if the annual dose to a fingertip is likely to be over 100 mSv, then a finger stall is likely to be needed.

Radionuclides used for therapy emit particle radiations which can deliver significant doses to the fingers. Finger doses from the use of ^{90}Y for therapeutic procedures of 2–50 mSv GBq^{-1} have been recorded (Vanhavere *et al* 2012), so there is the potential for the dose to the fingertip from a single procedure to exceed the dose limit if adequate protection measures are not in place. Finger stall dosemeters

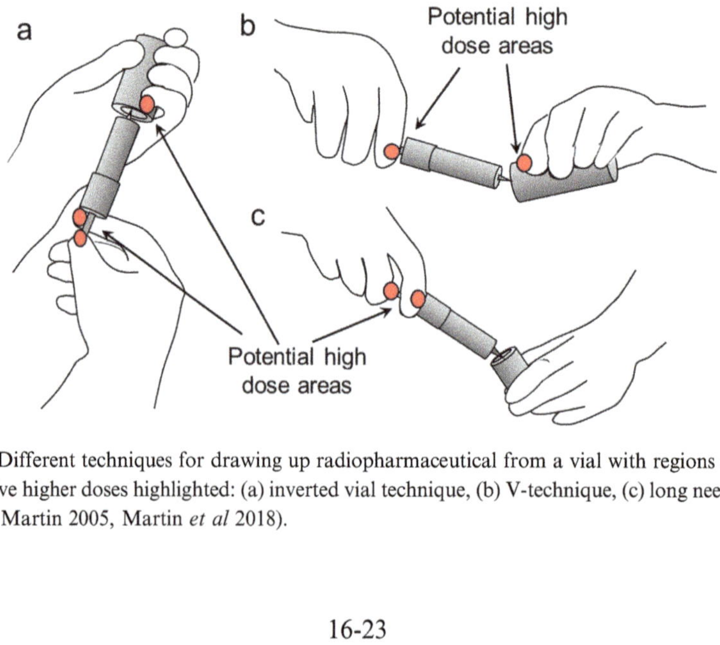

Figure 16.7. Different techniques for drawing up radiopharmaceutical from a vial with regions of the fingers likely to receive higher doses highlighted: (a) inverted vial technique, (b) V-technique, (c) long needle technique (Whitby and Martin 2005, Martin *et al* 2018).

should be worn for ^{90}Y therapies, but ring dosemeters may be adequate for therapies using the lower energy β-emitter ^{177}Lu.

References

ARSAC 2018 Notes for Guidance on the Clinical Administration of Radiopharmaceuticals and Use of Sealed Radioactive Sources *Administration of Radioactive Substances Advisory Committee (ARSAC)* (Public Health England: Chilton)

Behrens R 2012 On the operational quantity $H_p(3)$ for eye lens dosimetry *J. Radiol. Prot.* **32** 455–64

Delacroix D, Guerre J P, Leblanc P and Hickman C 2002 Radionuclide and radiation protection data handbook *Rad. Prot. Dosim.* **98** 1–168

ICRP 2001 Doses to the embryo and fetus from intakes of radionuclides by the mother, ICRP Publication 88 *Ann. ICRP* **31(1–3)**

ICRP 2004 Doses to infants from radionuclides ingested in mothers' milk, ICRP Publication 95 *Ann. ICRP* **34 (3–4)**

ICRP 2009 Adult reference computational phantoms, ICRP Publication 110 *Ann ICRP* **39 (2)**

ICRP 2015 Radiation dose to patients from radiopharmaceuticals: a compendium of current information related to frequently used substances, Publication 128 *Ann ICRP* **44 (2S)**

ICRP 2019 Radiological protection in therapy with radiopharmaceuticals, ICRP Publication 140 *Ann. ICRP* **48(1)**

Madsen M T, Anderson J A, Halama J R, Kleck J, Simpkin D J, Votaw J R, Wendt I I I, Williams R E, Yester L E and M V 2006 PET and PET/CT shielding requirements *AAPM Task Report 108, Med. Phys.* **33** 4–15

Martin C J 2011 Effective dose: practice, purpose and pitfalls for nuclear medicine *J. Radiol. Prot.* **31** 205–19

Martin C J and Sutton D G 2015 *Practical Radiation Protection in Healthcare* 2nd edn (Oxford: Oxford University Press)

Martin C J, Temperton D H, Hughes A and Jupp T 2018 *Guidance on the Personal Monitoring Requirements for Personnel Working in Healthcare* (New York: IPEM)

Martin C J, Marengo M, Vassileva J, Giammarile F, Poli G L and Marks P 2019 Guidance on prevention of unintended and accidental radiation exposures in nuclear medicine *J. Radiol. Prot.* **39** 665–95

McJury M and Tonge C 2014 Radiation protection in nuclear medicine *Institute of Physics in Medicine Report 109* (York: IPEM)

Salehzahi F and Tse J 2020 Shielding commissioning factors in nuclear medicine facilities *J. Radiol. Prot.* **40** 165–80

Smith D and Stabin M 2012 Exposure rate constants and lead shielding values for over 1100 radionuclides *Health Phys.* **102** 271–91

Sutton D G, Martin C J, Williams J R and Peet D 2012 *Radiation Shielding for Diagnostics Radiology* (London: British Institute of Radiology)

Vanhavere F 2012 ORAMED: Optimisation of Radiation Protection for Medical Staff. 7TH EURADOS Report 2012-02 http://eurados.org./en/Documents_Publications

Whitby M and Martin C J 2005 A multi-centre study of dispensing methods and hand doses in UK hospital radiopharmacies *Nuc. Med. Comm.* **26** 49–60

IOP Publishing

Principles and Techniques of Radiological Protection

Michael Charles Thorne

Chapter 17

Radiological protection and naturally occurring radioactive materials

D Read and M C Thorne

17.1 Introduction

Almost all materials contain traces of natural radionuclides, principally members of the ^{238}U and ^{232}Th decay chains, together with the less abundant ^{235}U chain[1]. Industrial processing may result in one or more of these radionuclides being concentrated to levels where they need to be regulated on health or environmental grounds, at which point they are classified as Naturally Occurring Radioactive Materials (NORM). The definition of NORM as *processed* material in which naturally occurring radionuclides are *concentrated* is important as it avoids regulatory issues caused by the inherent variation in raw materials due to geological factors. Thus, the term 'Technologically Enhanced NORM' or 'TENORM', used by some authors, is at best tautological; at worst, it implies that there is a category of NORM that is not technologically enhanced, raising the problematic issue of attempting to 'regulate Nature'.

Activities leading to generation of NORM are many and varied, including mining of metalliferous and non-metalliferous ores, ore processing, metal smelting, power production from coal, oil and gas extraction, exploitation of geothermal sources and water treatment (EC 2014). NORM also encompasses some manufactured products as well as residues and wastes from recycling and/or decontamination operations; for example, when dealing with legacy radium contamination from luminising works.

In the UK, the above are collectively termed Type 2 NORM activities to distinguish them from Type 1 activities in which uranium or thorium compounds are produced or deliberately added (HMSO 2016). Situations in which radionuclides are processed for their radioactive, fissile or fertile properties are not classified as

[1] Some regulations also include the ubiquitous ^{40}K.

doi:10.1088/978-0-7503-2424-3ch17

Table 17.1. Current classification of NORM activities in the UK.

TYPE 1	
Production of Th, Th compounds & where Th is deliberately added	Production of U, U compounds & where U is deliberately added
TYPE 2	
Extraction, production of REE & alloys	Mining & processing of ores except U ores
Oil and gas production	Removal of scales & precipitates
Utilising phosphate ore	TiO_2 pigment production
Extraction & refining of Zr & Zr compounds	Production of Sn, Cu, Al, Zn, Pb, Fe & steel
Dewatering coal mines	China clay extraction
Water treatment associated with drinking water	Remediation of any Type 1 or 2 activity

NORM activities and are considered within the ambit of nuclear regulation. Development of a framework for radiological protection in the case of NORM has evolved considerably in recent years. Both supranational (e.g. EC 2014, IAEA 2014) and national legislation (e.g. HMSO 2016) list a number of 'NORM industries' (table 17.1) for which materials containing low activity concentrations of natural radionuclides can either be excluded from regulatory control ('Out of Scope') or where the provisions can be relaxed subject to certain conditions being met.

Conversely, industrial processing can give rise to very high levels of radium isotopes (^{226}Ra, ^{228}Ra) in particular, owing to co-precipitation within insoluble barite ($BaSO_4$) scales. The other radionuclides to show substantial fractionation are ^{210}Pb and ^{210}Po, which occur not only as mineral deposits (e.g. galena, PbS) but also as scales composed of almost pure metallic lead or amalgams with volatile mercury (Ceccarello *et al* 2004). These materials require specialised treatment and are usually dealt with on a case-by-case basis. Most industrial wastes are considerably less active but nevertheless, require careful handling and disposal to limit exposures.

Radiobarite poses both external and internal radiological hazards during handling and maintenance of plant. The relatively long half-life of ^{226}Ra (1602 years) must be borne in mind when considering suitable disposal options, as must the potential for radon accumulation. Inhalation of dust and aerosols is the main radiological issue when dealing with ^{210}Pb and ^{210}Po; however, co-contaminants, such as mercury and arsenic, may prove to be of equal or even greater concern when present at high concentrations. In the past, significant quantities of NORM waste were consigned to nuclear facilities, whereas the current strategy is to facilitate the management of low-level waste from the NORM sector in line with the 'waste hierarchy' principle (DECC 2014).

Arisings from the diverse mineral processing sector are described in section 17.2, with examples from both metalliferous and non-metalliferous ore processing. The phosphate mineral industry is addressed in section 17.3.

17.2 Mineral processing and metal smelting

The main scale minerals present, and associated radionuclide enrichments are common to several key industries, of which the best-known example is co-precipitation of radium isotopes in barite. This almost insoluble phase is a major contributor to radiological dose in uranium mill tailings (e.g. Tuovinen *et al* 2016) and is the dominant scale mineral affecting the China Clay (kaolinite), rare earth, titanium and zircon industries (Read *et al* 2002, 2013, IAEA 2007, 2011, 2012).

The rare earth elements (REE) are indispensable to modern communication, being critical components of smart phones, computers, electronic displays and electric vehicles. They provide the basis for sustainable energy sources and advanced health care as the most powerful permanent magnets available for wind turbine generators and MRI scanners. Their use in lasers, semiconductors and fibre optics is well-established as is their application in satellite technology and the defence sector. China is the dominant producer and until very recently held an almost complete monopoly on global production (~120 000t per annum) based on its extensive deposits of bastnaesite ($LREE(CO_3)F$)[2]. Concerns surrounding security of supply have led to increased production from other countries, including monazite ($LREEPO_4$) deposits in Australia and other countries.

Exploitation of REE deposits is intrinsically linked to radioactivity and its management. The thorium content of bastnaesite typically ranges from 0.01% to 0.1% with uranium at ppm levels. In contrast, monazite ores contain much higher levels of both thorium (up to 20%) and uranium ($\leqslant$1%). Indeed, the highest grade REE deposit known, Steenkampskraal in the Western Cape, South Africa (Read *et al* 2002), was originally mined for thorium and is now being re-developed to extract REE. The ore body comprises up to 45 wt% REE oxides, 8% thorium and 600 ppm uranium. Underground mining of REE as well as storage of mined residues will need to carefully address accumulation of radon and thoron gas. Once extracted from primary ores and washed to remove gangue, harsh chemical separation is required on the concentrate and is expected to follow an alkali route at Steenkampskraal. This involves calcination, treatment with caustic soda and then acid dissolution to obtain soluble REE salts. The alternative acid route used elsewhere proceeds via high temperature chlorination to yield a mixture of REE chlorides. In either case, radioactive residues are generated in substantial quantities; with uranium and thorium arising as hydroxides following acid neutralisation and their progeny arising at the final separation step where barium chloride is added to remove radium as barite and sulphide is used to remove lead and polonium. The activity concentrations of thorium isotopes in the hydroxide fraction can reach 1 kBq g^{-1}, whereas radium, mainly as ^{228}Ra, in separated barite may be present at even higher concentrations (IAEA 2011). Consequently, all these residues require disposal in engineered facilities.

The world's largest producer, the open-cast Bayan Obo mine in China, taps a mainly bastnaesite ore with much lower thorium concentrations of around 400 ppm

[2] LREE = light rare earth elements.

(Li *et al* 2016). Historically, measures to reduce operational doses (estimated at $\geqslant$ 1 mSv y^{-1} for miners) were limited as indeed was the case at other REE operations in Australia, India and South America (IAEA 2011). Improved dust suppression, protective equipment and greater awareness of the radiological hazard has reduced airborne doses significantly in recent years. The concentrate is treated with H_2SO_4 at high temperatures leading to scaling of equipment and the same issues with residue disposal encountered elsewhere.

Those REE used in permanent magnets (neodymium, samarium, terbium and dysprosium) and which are driving commercial demand, are present at relatively low concentrations in most monazites. However, certain secondary, authigenic monazites contain an abundance of these elements and moreover, are notably low in thorium and uranium (Donnot *et al* 1973, Read *et al* 1987). Such deposits would be favoured provided they can be exploited economically. Until such time, processing of REE will continue to be subject to strict radiological protection practices.

China Clay results from low temperature, hydrothermal alteration of granite, a geochemical environment frequently associated with vein-type uranium deposits; thus, allowing radionuclides to become entrained with the clay during mining. The raw material removed from the pit face is first subjected to several physical separations to increase the kaolinite component and achieve the desired grain size. A dispersant and sulphuric acid are added at several stages releasing barium from the main constituents, feldspar and mica. Further dosing with acid and a reducing agent (sodium dithionite) to aid bleaching adds more sulphate to the system promoting barite precipitation. The huge volumes processed can generate a substantial inventory of even ultra-trace radium, which shows very high selectivity for the barite lattice, and accounts for activity concentrations in scale that can reach several thousand Bq g^{-1} (Read *et al* 2004). The deposits show marked secular disequilibrium and are typically enriched in ^{226}Ra over ^{228}Ra, the greater abundance of thorium in the source rocks being outweighed by the higher solubility of uranium.

Titanium dioxide pigments find widespread application as high-quality industrial coatings and have almost completely replaced earlier lead-based products owing to the latter's toxicity. The principal sources, ilmenite ($FeTiO_3$) and rutile (TiO_2), typically contain only traces of ^{238}U (ppm) and ^{232}Th (tens of ppm), but with annual production of several million tonnes, the potential radioactive inventory is substantial (IAEA 2012, Mantero *et al* 2013). Extraction of titanium is achieved by one of two methods; either the use of gaseous chlorine to generate $TiCl_4$ followed by oxidation or dissolution in concentrated sulphuric acid, depending on the feedstock. Unsurprisingly, it is with operations that follow the sulphate route where problems of NORM contamination are most prevalent.

The radioactive products from the TiO_2 industry are like those found with China Clay, above. High temperatures used in smelting and calcination operations lead to accumulation of volatile lead and polonium isotopes in fly ash and, if not captured, stack discharges. However, it is the co-precipitation of radium isotopes in sulphate (barite) scales that show the highest activity concentrations, commonly in the region of 100 Bq g^{-1}; in some cases, an order of magnitude higher (IAEA 2012). The scales tend to form in pipework and other equipment where there is turbulent flow, a sharp change in temperature or pressure, or during dewatering operations. The aggressive nature of the dissolution process combined with the greater abundance of thorium in the feedstock offsets, to a large extent, the higher mobility of uranium, with the result that ^{226}Ra and ^{228}Ra are present at similar levels.

Zirconium is used in a wide variety of applications including catalysts, fuel cells, ceramics, electronics, refractory products and advanced manufacturing components for extreme environments. By far the most important commercial resource is zircon ($ZrSiO_4$) with major deposits in Australia, South Africa and the USA producing almost a million tonnes per annum (IAEA 2007). Baddeleyite (ZrO_2) deposits make a minor contribution to global output, notably in Russia. Zircon is a markedly resistate mineral and, with a specific gravity of 4.7 g cm^{-3}, accumulates in heavy mineral placer deposits allowing it to be co-produced with titanium and, at some locations, the rare earths. Therefore, in addition to the radiological burden of zircon itself, one has also to consider the impact of processing associated minerals such as monazite, in which radionuclide concentrations may be much higher, as discussed above.

For several applications, the radiological impact is limited to inhalation of dust containing low ($\sim$1 Bq g^{-1}) activity concentrations of the parent isotopes during milling and transport. However, where the metal or pure oxide is required the situation changes substantially. Extraction of zirconium can be accomplished by high temperature ($>$ 2500 °C) fusion of the zircon sand feedstock leading to volatilisation of much the radionuclide inventory. With uranium and thorium concentrations in the raw material each typically $>$ 100 ppm, there is potential for significant release and subsequent concentration of radium, lead and polonium isotopes. Atmospheric stack discharge of volatiles, mainly ^{210}Pb and ^{210}Po, or handling and disposal of filters can pose a risk to the public and workers, respectively, as activity concentrations may reach several hundred Bq g^{-1} (IAEA 2007). On the other hand, manufacturing of zirconium compounds requires chemical dissolution using harsh reagents providing more scope for NORM accumulation at different stages of processing. Typically, the feedstock is first treated with concentrated NaOH at high temperature to generate sodium zirconate (Na_2ZrO_3), which is then dissolved in concentrated HCl to produce zirconium oxychloride ($ZrOCl_2.8H_2O$). The latter is a useful precursor for a range of other zirconium compounds and, if these include sulphates, co-precipitation of radium is inevitable. Alternatively, zircon (or baddeleyite) sand can be reacted directly with strong acid to yield a soluble salt; again, sulphuric acid will lead to radiobarite formation. Scales associated with the long-term production of high purity compounds have been found to contain exceptional levels of radium isotopes (Read 2015).

Iron ore used in steel-making is mainly in the form of hematite (Fe_2O_3), magnetite (Fe_3O_4), limonite ($FeO(OH).nH_2O$) and goethite ($FeO(OH)$), which require sintering with coke before being processed in the blast furnace. As the sintered product is collected, waste gases and fines are cooled and recovered by an electrostatic precipitator (ESP). Lead-210 and ^{210}Po, present at only trace amounts in the raw materials, are volatilised during sintering and concentrate in the ESP dusts, most of which are recycled back into the process. However, a proportion is released to the atmosphere via the main stack or escapes into the workplace (e.g. Dal Molin *et al* 2017). Activity concentrations for ^{210}Pb and ^{210}Po in ESP dusts are on the order of 1–100 Bq g^{-1}; consequently, protective measures are required for workers during both steel production and waste management. Because an iron ore sinter plant typically handles tens of thousands of tonnes per day, discharge permits for an integrated steel works may allow GBq quantities of these isotopes to be released each year.

17.3 Phosphate extraction

From the late 19th century to the present, the use of phosphate rock in the production of agricultural fertiliser has increased enormously and worldwide phosphate production is now about 1.4×10^8 tonnes per year, with the main reserves of phosphate located in Morocco, North America and Asia. During the 20th century, fertiliser production was mainly in the USA, Europe and China. The deposits that have been exploited lie within a few tens of metres of the surface and extraction has been by open-cast mining, either by drilling and blasting in hard rock or high-pressure water jetting in softer rock.

Uranium concentrations in phosphate rock range from 0.02 to 2 g kg^{-1}. Other metals, such as cadmium, arsenic and mercury are also often present. Radionuclide substitutions in the apatite structure of phosphate ore may include various cationic forms of uranium, thorium, radium, lead, polonium and bismuth, with U^{4+} substituting for Ca^{2+} in the apatite structure, because the two ions have almost identical ionic radii. Processes involved in the processing of phosphate rock for various purposes are shown in figure 17.1.

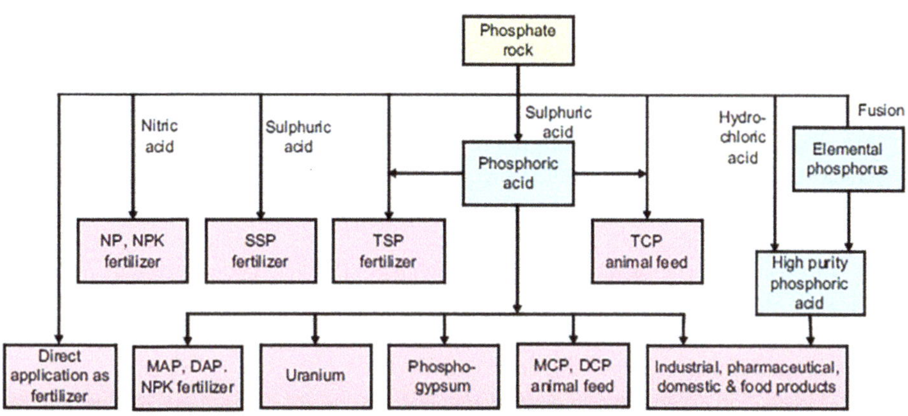

Figure 17.1. Phosphate rock product streams (NP: nitrogen–phosphorus; NPK: nitrogen–phosphorus–potassium; SSP: single super phosphate; TCP: tricalcium phosphate; TSP: triple superphosphate; MAP: monoammonium phosphate; DAP: diammonium phosphate; MCP: monocalcium phosphate; DCP: dicalcium phosphate). IAEA (2023), © IAEA, 2023, p. 351, ISBN 978-92-0-126722-1 978-92-0-105020-5

17.4 Mineral processing

Evaluation of radiological exposures to NORM during mineral processing is difficult owing to the diverse nature of the plants affected. The applicability of default dose coefficients and biokinetic models provided by the International Commission for Radiological Protection (ICRP) for converting chronic uptake of industrial NORM into dose remains unverified for many exposure scenarios. Indeed, most of the coefficients and models were derived from epidemiological studies following acute exposures and then usually only for a few relatively simple, chemical species. Most scales comprise phases of low to very low solubility such as barite, phosphogypsum, metallic lead and galena (Dal Molin *et al* 2018).

High occupational doses are most likely during maintenance, decommissioning and packaging waste for disposal. In routine operations, external gamma exposures would be largely attenuated by shielding except where intrusive work is being undertaken. Dust is much less of a problem in 'wet' areas but becomes the primary concern elsewhere, as highlighted in a recent study of the steel industry (Dal Molin *et al* 2017).

References

Ceccarello S, Black S, Read D, Weiss H, Schubert M, Kunze C and Grossmann J 2004 Radioactive scales from a natural gas production facility in the Altmark region, Germany *Appl. Mineral.* 395–8 Special edn.

Dal Molin F, Anderson D, Fisher R, Frost D and Read D 2017 Dose implications of chronic exposure to industrial NORM in iron ore processing *J. Rad. Prot.* **37** 402–21

Dal Molin F, Warwick P and Read D 2018 Under-estimation of ^{210}Pb in industrial radioactive scales *Anal. Chim. Acta.* **1000** 67–74

DECC 2014 Strategy for the management of Naturally Occurring Radioactive Material (NORM) waste in the United Kingdom https://assets.publishing.service.gov.uk/media/5a7d8e37e5274a6b89a50c38/Final_strategy_NORM.pdf

Donnot M, Guigues J, Lulzac Y, Magnien A, Parfenoff A and Picot P 1973 Un nouveau type de gisement d'europium: la monazite grise à europium en nodules dans les schistes paléozoiques de Bretagne *Miner. Depos.* **8** 7–18

EC 2014 Council Directive 2013/59/Euratom of 5 December 2013 laying down basic safety standards for protection against the dangers arising from exposure to ionising radiation, and repealing Directives 89/618/Euratom, 90/641/Euratom, 96/29/Euratom, 97/43/Euratom *Off. J. Eur. Comm.* **L** 13

HMSO 2016 *The Environmental Permitting (England and Wales) Regulations* (London: Her Majesty's Stationery Office) UK Statutory Instruments, No. 1154

IAEA 2007 *Radiation protection and NORM residue management in the zircon and zirconia industries* IAEA Safety Reports Series No. 51 IAEA

IAEA 2011 *Radiation protection and NORM residue management in the production of rare earths from thorium containing minerals* IAEA Safety Reports Series No. 68 IAEA

IAEA 2012 *Radiation protection and NORM residue management in the titanium dioxide and related industries* IAEA Safety Reports Series No. 76 IAEA

IAEA 2014 *Safety Standards for Protecting People and the Environment. Radiation Protection and the Safety of Radiation Sources: International Basic Safety Standards, General Safety Requirements Part 3* IAEA

IAEA 2023 *The Environmental Behaviour of Uranium* IAEA Technical Reports Series No. 488 International Atomic Energy Agency, Vienna

Li B, Wang N, Wan J, Xiong S, Liu H, Li S and Zhao R 2016 In-situ gamma-ray survey of rare-earth tailings dams—A case study in Baotou and Bayan Obo Districts, China *J. Env. Rad.* **151** 304–10

Mantero J, Gazques M, Bolivar J, Garcia-Tenorio R and Vaca F 2013 Radioactive characterization of the main materials involved in the titanium dioxide production process and their environmental radiological impact *J. Env. Rad.* **120** 26–32

Read D 2015 Analytical challenges posed by NORM contaminated sites *Environmental Radiochemical Analysis V* (Royal Society of Chemistry) pp 81–91

Read D, Cooper D and McArthur J 1987 The composition and distribution of nodular monazite in the Lower Palaeozoic rocks of Great Britain *Min. Mag.* **51** 271–80

Read D, Andreoli M, Knoper M, Williams C and Jarvis N 2002 The degradation of monazite: implications for the mobility of rare earth and actinide elements during low temperature alteration *Eur. J. Mineral.* **14** 487–98

Read D, Rabey B, Black S, Glasser F, Grigg C and Street A 2004 Implementation of a strategy for managing radioactive scale in the China Clay industry *Min. Eng.* **17** 293–304

Read D, Read G and Thorne M 2013 Background in the context of NORM-contaminated land *J. Rad. Prot.* **33** 367–80

Tuovinen H, Pohjolainen E, Vesterbacka D, Kaksonen K, Virkanen J, Solatie D, Lehto J and Read D 2016 Release of radionuclides from waste rock and mill tailings at a former pilot uranium mine in Eastern Finland *Boreal Env. Res.* **21** 471–80

IOP Publishing

Principles and Techniques of Radiological Protection

Michael Charles Thorne

Chapter 18

The transport of radioactive materials

M C Thorne

18.1 Introduction

Radioactive materials comprise one class of dangerous goods that can be transported by road, rail, sea and air. International transport of dangerous goods is governed by various agreements such as the European Agreement concerning the International Carriage of Dangerous Goods by Inland Waterway (AND), the International Maritime Dangerous Goods Code (IMDG Code), the European Agreement concerning the International Carriage of Dangerous Goods by Road (ADR), the Convention concerning International Carriage by Rail (COTIF) and the International Air Transport Association Dangerous Goods Regulation. Domestic transport of dangerous goods is typically covered by national regulations with similar provisions to these international agreements. In the case of radioactive materials, transport requirements generally trace back to the recommendations made by the International Atomic Energy Agency (see IAEA (2018), bearing in mind that these regulations are updated every two years). These recommendations are intended to ensure safety and protect people, property and the environment during the transport of radioactive material. This protection is to be achieved by containment of the radioactive contents, control of the external dose rate in the vicinity of a package, prevention of criticality for packages containing fissile material and prevention of damage caused by heat.

The transport regulations are necessarily complex, reflecting the wide variety of radioactive materials that may be transported, ranging from short-lived sources used in nuclear medicine through to high-level radioactive waste and spent nuclear fuel. This vast range of radioactive materials requires the use of a wide variety of different types of packages, with the robustness and shielding capacity of the various package types related to the characteristics of the radioactive materials that they may be used to transport.

A further consideration is that packages containing radioactive materials may be transported alone or with other packages that may also contain radioactive

doi:10.1088/978-0-7503-2424-3ch18

materials. Thus, additional rules govern the numbers of packages containing radioactive materials that may be transported in a single load.

In this chapter an outline is presented of the international recommendations governing the safe transport of radioactive materials. Section 18.2 sets out the various classes of radioactive materials that may be transported and the types of packages that are considered suitable for that transport. It also addresses the tests that are used to establish that these packages would maintain their integrity in accident conditions appropriate to the type of material being transported. Section 18.3 provides some illustrations of package designs that comply with the transport requirements. Section 18.4 sets out the dose rates that are acceptable in the vicinity of packages and introduces indices that are used to control external dose rates and risks of criticality, while section 18.5 deals with the conditions that need to be satisfied during transport, both to comply with radiological protection requirements and to ensure that criticality incidents do not occur. Section 18.6 comments briefly on security issues relating to the transport of radioactive materials, because attacks on radioactive materials in transit could give rise to incidents with significant radiological implications. Accidents and incidents during transport are then discussed in section 18.7.

18.2 Classes of radioactive materials and their packaging

Radioactive material that is to be transported is required to be assigned a specific United Nations (UN) number. The relevant numbers together with the shipping names and descriptions are given in table 18.1. Note that when used, the proper

Table 18.1. UN numbers and associated shipping names for radioactive materials.

UN number	Shipping name and (description)
UN 2908	Radioactive material, excepted package—empty packaging
UN 2909	Radioactive material, excepted package—articles manufactured from natural uranium *or* depleted uranium *or* natural thorium
UN 2910	Radioactive material, excepted package—limited quantity of material
UN 2911	Radioactive material—excepted package—instruments *or* articles
UN 3507	Uranium hexafluoride, radioactive material, excepted package (less than 0.1 kg per package, non-fissile or fissile-excepted)
UN 2912	Radioactive material, low specific activity (LSA-I) (non-fissile or fissile-excepted)
UN 3321	Radioactive material, low specific activity (LSA-II) (non-fissile or fissile-excepted)
UN 3322	Radioactive material, low specific activity (LSA-III) (non-fissile or fissile-excepted)
UN 3324	Radioactive material, low specific activity (LSA-II), fissile
UN 3325	Radioactive material, low specific activity (LSA-III), fissile
UN 2913	Radioactive material, surface-contaminated objects (SCO-I, SCO-II *or* SCO-III) (non-fissile or fissile-excepted)
UN 3326	Radioactive material, surface-contaminated objects (SCO-I or SCO-II), fissile
UN 2915	Radioactive material, type A package (non-special form, non-fissile or fissile-excepted)

UN 3327	Radioactive material, type A package, fissile (non-special form)
UN 3332	Radioactive material, type A package, special form (non-fissile or fissile-excepted)
UN 3333	Radioactive material, type A package, special form, fissile
UN 2916	Radioactive material, type B(U) package (non-fissile or fissile-excepted)
UN 3328	Radioactive material, type B(U) package, fissile
UN 2917	Radioactive material, type B(M) package (non-fissile or fissile-excepted)
UN 3329	Radioactive material, type B(M) package, fissile
UN 3323	Radioactive material, type C package (non-fissile or fissile-excepted)
UN 3330	Radioactive material, type C package, fissile
UN 2919	Radioactive material transported under special arrangement (non-fissile or fissile-excepted)
UN 3331	Radioactive material transported under special arrangement, fissile
UN 2977	Radioactive material, uranium hexafluoride, fissile
UN 2978	Radioactive material, uranium hexafluoride (non-fissile or fissile-excepted)

shipping names are required to be all in capitals (see table 1 of IAEA (2018)). Here only initial capitals are used. Alternatives are indicated by '*or*' and descriptive material is given in parentheses.

The first five categories relate to packages containing only small amounts of radioactive materials (see table 4 of IAEA (2018)) that also give low external dose rates in their immediate vicinity. In terms of package integrity, all that is required of an excepted package is that it should retain its radioactive contents under routine conditions of transport.

The second five categories relate to materials in which radionuclide concentrations are low. These are described as low specific activity (LSA) materials. LSA-I includes (but is not limited to) uranium and thorium ores, concentrates of such ores, other ores containing naturally occurring radionuclides, and compounds or mixtures of natural uranium, depleted uranium, natural thorium provided that they are unirradiated and in solid or liquid form. LSA-II includes solids, liquids and gases up to specified specific activity limits and water with a tritium concentration of up to 0.8 TBq l^{-1}. Finally, LSA-III relates only to solids up to a limiting specific activity and with a requirement that the activity is essentially uniformly distributed throughout that solid. LSA material can be transported unpackaged under specified conditions or in an industrial package (IP-1, IP-2 or IP-3). Packages of type IP-1 are subject only to conditions that apply to all packages for the transport of radioactive materials (except for excepted packages). These packages must be designed so that they can be safely transported, their external surfaces should be free of protruding features and easily decontaminated, should be capable of withstanding the effects of acceleration or vibration during transport, physically and chemically stable, and sufficiently well shielded to ensure that dose rates on their surfaces comply with specified limits. In addition, IP-2 packages must show no substantial loss of integrity when subject to a drop test from a height of 0.3–1.2 m, depending on mass, and a stacking test. IP-3 packages must satisfy the same requirements as Type A packages (discussed below). IP-2 and IP-3 packages can include tanks and freight containers that satisfy specific requirements.

For LSA materials and other materials discussed below, a distinction is made between non-fissile/fissile-excepted material and fissile material. Fissile material comprises material containing any of the fissile nuclides U-233, U-235, Pu-239 and Pu-241. However, the term 'fissile-excepted' provides some exclusions from this definition, essentially to address materials that contain fissile nuclides but do not pose a risk of criticality under transport conditions. Thus, fissile-excepted material includes unirradiated natural or depleted uranium, natural or depleted uranium that has been irradiated in thermal reactors only, material with a fissile nuclide content of less than a total of 0.25 g, and any combination of such materials. Some other fissile materials can be transported without the packages having to be identified as containing fissile material, but with specific conditions of carriage. Thus, for example, fissile nuclides with a mass of up to 2 g per package may be so transported, but only in a consignment that includes no more than 15 g of fissile nuclides (see paragraphs 417 and 570 of IAEA (2018)).

Two categories relate to surface-contaminated objects. These comprise three groups. SCO-I is characterised by low concentrations of fixed and non-fixed contamination on accessible surfaces, with different limits for the two types of contamination and for beta–gamma emitters plus low toxicity alpha emitters and for other alpha emitters. SCO-II has higher limits for contamination on accessible surfaces and even higher limits for non-accessible surfaces. Finally, SCO-III applies to large solid objects that cannot be transported in the types of package described in the IAEA (2018) regulations. For SCO-III there are limits on the contamination on external and inaccessible surfaces plus a requirement to seal all openings to prevent the release of radioactive material.

More highly radioactive materials are transported in Type A, B and C packages. For Type A packages, the primary constraint is on the amount of activity in the package. However, because the hazard associated with a radionuclide depends on the half-life, emissions, chemistry and physico-chemical form of that radionuclide, an index rule is used to define the maximum permissible activity to be included in a package. That rule is:

$$\sum_i B(i)/A_1(i) + \sum_j C(j)/A_2(j) \leqslant 1 \qquad (18.1)$$

where $B(i)$ is the activity (Bq) of radionuclide i as special form radioactive material, whereas $C(j)$ is the activity (Bq) of radionuclide j as other than special form radioactive material. The term 'special form radioactive material' means either an indispersible solid radioactive material or a sealed capsule containing radioactive material. Values of A_1 and A_2 are given in table 2 of IAEA (2018), as are activity limits and activity concentration limits for exempt material. In general, A_1 values are equal to or larger than A_2 values, reflecting the constraints imposed on special form radioactive material. In addition to setting limits on the activity contents of packages, A_1 and A_2 values are used for other purposes, e.g. multiples of A_2 values are used to define the maximum permissible releases from packages under accident conditions. Testing of Type A packages includes water spray, free drop, stacking and penetration tests.

Type B packages are distinguished into Type B(U) and B(M). Here attention is focused on Type B(U) because Type B(M) packages are restricted to transport within a specified country or between specified countries. Thus, B(M) packages are required to meet the same conditions as B(U) packages, except that they may be varied by the competent authorities in the specified countries.

Type B(U) packages may contain radioactive materials that generate significant amounts of heat. Therefore, there is a requirement to consider the effects of such heat on the containment and shielding offered by the package if it was left unattended for one week. Effects of particular concern include the degradation of the radioactive material or any can or other receptacle in which it is contained, lessening of the efficiency of the packaging through differential thermal expansion, or cracking or melting of the radiation shielding material, and acceleration of corrosion. The testing of Type B(U) packages is more rigorous than for Type A packages. It includes a mechanical, thermal and water-immersion tests applied sequentially to the same package. The mechanical tests comprise a 9 m drop onto a flat target, a 1 m drop onto a perpendicular cylindrical bar and a dynamic crush test in which a 500 kg mass is dropped onto the package from a height of 9 m. The thermal test involves exposure for 30 min at 800 °C followed by an extended period at an ambient temperature of 38 °C and with the maximum permitted rate of internal heat generation until temperatures in all parts of the package are either decreasing or approaching steady state conditions. Finally, Type B(U) packages are tested by water immersion under a head of water of at least 15 m for at least eight hours (with a further test under a head of 200 m for at least one hour for packages containing particularly large amounts of activity). Type C packages must be able to satisfy similar requirements to Type B(U) packages, but with some variations to make them suitable for transport by air (e.g. in respect of the velocity of impacts that they must be able to survive without loss of integrity).

Finally, there are two special categories. The first of these relates to materials that must be transported under a special arrangement. This provides an option for the competent authority to license the transport of consignments for which conformity with the IAEA (2018) regulations is impracticable, but for which the requisite standards of safety set out in the regulations have been demonstrated through alternative means. The second relates to bulk quantities of uranium hexafluoride. Uranium hexafluoride is the chemical form of uranium that is used during the uranium enrichment process. Within a reasonable range of temperature and pressure, it can be a solid, liquid, or gas. Solid UF_6 is a white, dense, crystalline material that resembles rock salt. Uranium hexafluoride does not react with oxygen, nitrogen, carbon dioxide, or dry air, but it does react with water or water vapour to form corrosive hydrogen fluoride (HF) and uranyl fluoride (UO_2F_2). For this reason, UF_6 is always handled in leak-tight containers and processing equipment. Because of these considerations, transport requirements for uranium hexafluoride relate primarily to the avoidance of over-pressurisation of the transport containers. Specifically, the UF_6 should be in solid form and the internal pressure of the container should not be above atmospheric when it is presented for transport. Additionally, the ullage within the container should be at least 5% at the maximum temperature of the package as specified for the plant systems where the package might be used.

Holden (2015) provides a useful summary of the various types of package and their uses:

- Excepted packages are used to carry very low levels of radioactive materials, e.g. small sources and samples;
- Industrial packages are generally used for waste materials, e.g. coveralls/wipes, decommissioning debris and contaminated, redundant plant;
- Type A packages are used to carry sources of various types (both sealed and unsealed), including radiopharmaceuticals, and irradiated reactor components;
- Type B packages are used to carry spent nuclear fuel, materials containing substantial amounts of plutonium and high-activity sources;
- Type C packages are for the air transport of highly active, easily dispersible materials, e.g. plutonium oxide powders.

18.3 Package designs

The international regulations (IAEA 2018) set out what packages must achieve to comply with those regulations, but do not prescribe those designs. Therefore, there are a variety of packages available that comply with the regulations. The package may be anything from a taped plastic bag (e.g. for an excepted package containing a sample of slightly radioactive material) through to a massive, highly engineered structure (such as a cask for the transport of spent nuclear fuel). Industrial packages (IP-1, IP-2 and IP-3) may comprise soft-sided bags or 200 litre drums (sometimes with an inner plastic bag because the joint between the body and lid of the drum may leak and then reseal as a consequence of being subject to the standard drop test) (Holden 2015).

For Type A packages, containment may be intrinsic to the source (if it is a sealed source). In other cases, a container with an elastomeric O-ring seal between the lid and body is typically appropriate. Because some gases can permeate conventional O-ring materials, metal O-seals may be required, with implications on the surface finish required and on the seal face clamping load to be applied. Where an elastomeric seal is used, consideration needs to be given as to whether it could be degraded by exposure to extremes of temperature or to the radiation field arising from the transported source. Shielding of Type A packages will generally be required. However, for a small source, the dose rate at the surface of the package may be limited simply by placing the source in the centre of the package (e.g. in the centre of a large box). However, it must be demonstrated that the distance between the source and the surface of the package could not decrease during transport through the provision of suitable filling material or in-package furniture. Where shielding is required, dense materials such as steel, lead, depleted uranium, tungsten and concrete are generally used for beta–gamma emitters (Holden 2015, Issard 2015). However, for neutron emitters, shield materials should be high in light elements, specifically hydrogen and boron. Thus,

Figure 18.1. Illustrative type A package. Images courtesy of Lemer Pax (https://www.lemerpax.com/en/products/posisafe-20–30-en/).

water, polythene, various polymers, and boronated steel or aluminium may be used (Holden 2015, Issard 2015). Impact mitigation is typically achieved by deformation of the outer part of the packaging. This may be distinct from the shielding component of the package to ensure that the degree of shielding is not compromised by the deformation (Holden 2015). An illustrative Type A package is shown in figure 18.1.

Type B packages are typically massive, engineered structures. In terms of containment, they require that the sealing integrity should be verified prior to each shipment. For lids and closures of other penetrations, twin seals are usual with the space between the seals pressurised. This means that a failure in either or both seals is signalled by a drop in pressure. Shielding thicknesses are generally substantial and large amounts of heat may be generated by the radioactive material requiring the packaging to have fins to facilitate heat transfer (Holden 2015). An example of a Type B package for the transport of nuclear fuel is shown in figure 18.2.

18.4 Dose rates in the vicinity of packages, transport indices and criticality indices

IAEA (2018) regulations permit a relatively high dose rate of 2 mSv h^{-1} on the outer surface of a package. Indeed, if a package is transported under an exclusive use arrangement, the surface dose rate can be up to 10 mSv h^{-1}. However, optimisation in the context of package loading and handling may result in substantially lower surface dose rates. The acceptability of a package for off-site transport is determined

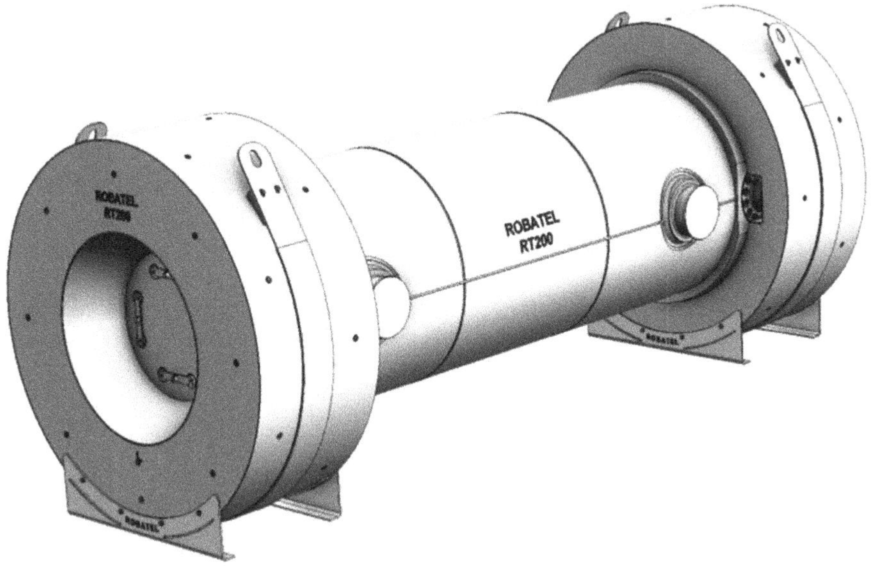

Figure 18.2. A Type B transport cask for activated hardware, including control rod blades. Image courtesy of Robatel Technologies (from https://robateltech.com/transportation-and-storage-casks/).

by calculating its Transport Index (TI). For a package, the TI is calculated as follows.

- Determine the maximum dose rate (mSv h^{-1}) at a distance of 1 m from the external surface of the package, overpack, freight container or unpackaged waste (unpackaged LSA-I, SCO-I and SCO-III may be transported) or use pre-defined dose rates for uranium and thorium ores and their concentrates.
- In the cases of tanks, freight containers and unpackaged wastes, multiply by a factor that depends on the size of the load. The size of the load is defined as its largest cross-section and the multiplication factor is 1 for a load of less than 1 m^2, 2 for 1–5 m^2, 3 for 5–20 m^2 and 10 for more than 20 m^2.
- Round the resultant value up to the first decimal place, except that values of 0.05 or less may be rounded down to zero.

For a rigid overpack, freight container or conveyance, the TI may be determined by direct measurement using the above procedure. Alternatively, the TI may be determined as the sum of the TI values for all the packages included in the overpack, freight container or conveyance. However, for a non-rigid overpack, freight container or conveyance, the TI may only be determined as the sum of the individual package TI values. Except under exclusive use arrangements, the TI of any package or overpack may not exceed 10 (larger values of 50 or 200 are permitted for freight containers or conveyances, see table 10 of IAEA (2018)). Thus, in principle, for transport of a single package, the dose rate at 1 m from the package could be up to 10 mSv h^{-1}. However, in those circumstances, the dose rate somewhere on the surface of the package would exceed 2 mSv h^{-1} and it would have to be transported under an exclusive use arrangement. There is also a requirement that dose rates at 2 m from the external surface of a vehicle or freight container, unless employed for exclusive use, should not exceed 0.1 mSv h^{-1} (see paragraph 566 of IAEA (2018)).

These relatively high permitted dose rates in the vicinity of packages make it important to undertake assessments of the radiological impacts arising from the transport of radioactive materials. This is particularly the case for practices that involve the shipment of large numbers of packages, e.g. routine dispatches of radiopharmaceutical products. Specifically, consideration needs to be given to locations where significant hold-up and storage of packages could arise, because this could result in a limited number of individuals being exposed to multiple packages and therefore being exposed for a significant total period each year.

In evaluating the radiological impact of transport of packages, the two main situations that have to be addressed are where a package is stationary, and an individual is present in its vicinity and where a package is in motion and passes a stationary individual. The first of these cases is addressed by multiplying the dose rate at the location of the individual by their occupancy of that location. The second requires a more formal mathematical treatment. The relevant geometry is shown in figure 18.3.

A package passes the location of an individual with closest distance of approach h m. At time t seconds, the distance of approach is $d = (x^2 + h^2)^{½}$ m where $x = vt$ and v m s^{-1} is the speed at which the package passes the individual. If the dose rate at the individual is a function of d, it can be written as $D(d)$ Sv s^{-1}. Therefore, the total dose received by the individual, D_{tot} Sv, is given by:

$$D_{tot} = \int_{-\infty}^{\infty} dtD(d) = \int_{-\infty}^{\infty} dxD(d)/v \tag{18.2}$$

where the substitution $dt = dx/v$ has been made.

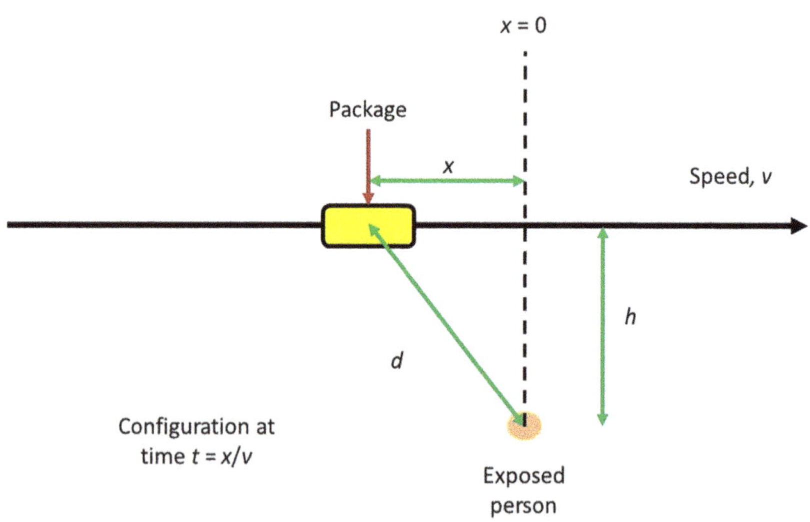

Figure 18.3. Geometry of a package being transported close to an exposed person.

A simple approximation that is often made is that the package can be treated as a point source. Neglecting attenuation of the flux in air between the package and the exposed individual, this gives $D(d) = D_0/(x^2 + h^2)$, where D_0 m^2 Sv s^{-1} is the dose rate at 1 m from the point source (hence the inclusion of m^2 in the units of the dose rate). This gives:

$$D_{\text{tot}} = \int_{-\infty}^{\infty} dx\, D_0/(x^2 + h^2)/v = \pi\, D_0/\{hv \sin(\pi/2)\} = \pi\, D_0/hv \qquad (18.3)$$

Note that the inverse square law that applies to a stationary point source is replaced by an inverse law because the source, when integrated over time, is equivalent to an infinitely long line source present for an infinitesimal period.

Limiting values of the TI ensures that the accumulation of packages does not give rise to excessively high external dose rates. Control on the accumulation of packages at a single location is even more important when those packages contain fissile material, because accumulation of a large amount of fissile material and moderator at a single location could result in a criticality incident. Control, in this case, is achieved by placing limits on the Criticality Safety Index (CSI) of a package, overpack, freight container, consignment or conveyance. The CSI of a package is defined as follows:

$$\text{CSI} = 50 \times S \times M_{U\text{-}235}/Z + M_{\text{OTHER}}/280 \qquad (18.4)$$

where S (2 or 5) is a scale factor determined by the dimensions of the package, $M_{\text{U-235}}$ (g) is the mass of U-235 in the package, and M_{OTHER} (g) is the mass of other fissile radionuclides present in the package. Z is a modifying factor that allows for the consideration that criticality becomes of decreasing concern the more that U-235 is diluted with U-238. Thus, values of Z decrease with increasing enrichment (expressed as the mass percentage of U-235). These values are 2200 (up to 1.5% enrichment), 850 (1.5%–5%), 660 (5%–10%), 580 (10%–20%), 450 (20%–100%). The limiting value of the CSI for freight containers and conveyances not under exclusive use is generally 50. A higher limit of 200 applies to the total cargo of a sea-going vessel, but each hold on such a vessel will have a limit of 50. Under exclusive use arrangements, the limit on CSI, where applicable, is 100.

18.5 Conditions of transport

Some general conditions of transport apply to all types of packages. Before the first shipment, there is a requirement to confirm that the packaging has been manufactured in conformity with the design specifications, complies with the relevant provisions of the regulations and has been approved for use. For each shipment, there is a requirement to ensure that the radionuclide contents, and the physico-chemical form of the contents are in conformance with the design specification. Other specifications relate to the removal of temporary lifting attachments, holding of packages until thermal conditions have stabilised, ensuring that closures and valves are properly closed and sealed, and that the neutron multiplication factor has been estimated, as appropriate, for packages containing fissile material.

Other general provisions prohibit the transport of any items other than those that are necessary for the use of the radioactive material and require that any other dangerous properties of the contents of the package, e.g. explosiveness, have been considered and that the package complies with the dangerous goods regulations in respect of those properties.

Of relevance to radiological protection, there is a specific requirement for monitoring fixed and non-fixed contamination on surfaces of packages and of conveyances used to transport those packages. Contamination levels are required to be below specified limits, but there is also a more general requirement that they should be as low as practicable.

Specific requirements for individual package types relate to aspects such as limiting TI and CSI, the marking, labelling and placarding of packages, overpacks and freight containers, the responsibilities of the consigner, including the required contents of the transport documentation, the responsibilities of the carrier, and the notification of the competent authorities. Specific guidance is also given on segregation during transport from workers, members of the public and undeveloped photographic film, and for storage in transit (see section V of IAEA (2018) for details).

The various regulations on the transport of dangerous goods, which include radioactive materials, are quite specific on the need for training of all those involved in the transport process. These include those who classify, pack, mark and label dangerous goods, prepare transport documents, offer or accept dangerous goods for transport, mark, placard, load or unload dangerous goods into or from transport vehicles, bulk packaging or freight containers, or are otherwise involved in the transport of dangerous goods. These requirements include general awareness training, function-specific training, safety training (including emergency response training) and security training. Details on issues relating to training are given by McNamara (2015).

18.6 Security considerations

In general, the security of nuclear installations, and other security related issues, such as the safe keeping of high-activity sealed sources, are outside the remit of this book. However, it is relevant to note that substantial releases of radionuclides from radioactive materials in transit or in storage during transport, could be due either to accidents or from deliberate actions intended to give rise to a radiation incident. The security of transport of radioactive materials is addressed by the IAEA who have issued guidance to states and their competent authorities on how to establish and maintain the elements of the nuclear security regime relating to the transport of radioactive material (IAEA 2020). The IAEA considers that this publication may also assist shippers, carriers and others with transport security responsibilities in designing their security systems. It includes a discussion of security measures that can be implemented to prevent threats to the security of transport and of measures to locate or recover radioactive material lost or stolen in transport.

18.7 Accidents and incidents

More than 1×10^8 packages containing radioactive materials have been transported, without there having been any transport accidents resulting in a casualty confirmed to be associated with the radioactive nature of the material being transported. Nevertheless, a substantial number of incidents associated with the transport of radioactive materials occur, e.g. in France about 100 such incidents are declared each year (Sert 2015). Furthermore, some of these incidents had the potential to have significant radiological consequences, e.g. improper shielding of an Ir-192 source being transported by air from Washington D.C. to Atlanta, Georgia in 1974, and various train and highway accidents, including train derailments and collisions, shipwrecks involving cylinders of UF_6 and large sealed sources of Cs-137, and a collision between a truck transporting dairy products and a van transporting a type B(U) package containing sealed sources of caesium. These incidents, and others, are summarised by Sert (2015), who discusses similarities and differences between incidents and accidents that may occur during transport and those that may occur at fixed installations. Some of the main differences are that although potential areas of contamination are likely to be smaller with transport accidents, the accidents can happen anywhere, including densely populated urban areas. Furthermore, such accidents will tend to occur rapidly, e.g. due to collisions or fires, requiring a rapid response. However, the driver and accompanying crew, and the resources available to local authorities, may have limited competence in handling incidents involving radioactive materials. These considerations make it desirable to put in place national arrangements to respond to serious transport accidents involving radioactive materials. Examples of such national arrangements are given by Sert (2015). Preparedness for and response to radiation emergencies are discussed further in chapter 21.

References

Holden G V 2015 Functional requirements for the design of transport packages *Safe and Secure Transport and Storage of Radioactive Materials* ed K B Sorenson (Sawston: Woodhead Publishing) pp 7–25

Issard H 2015 Radiation protection by shielding in packages for radioactive materials *Safe and Secure Transport and Storage of Radioactive Materials* ed K B Sorenson (Sawstone: Woodhead Publishing) pp 123–40

IAEA 2018 *Regulations for the Safe Transport of Radioactive Material* 2018 edn (Vienna: International Atomic Energy Agency) IAEA Safety Standards Series No. SSR-6 (Rev. 1)

IAEA 2020 *Security of Radioactive Material in Transport: Implementing Guide* (Vienna: International Atomic Energy Agency) IAEA Nuclear Security Series No. 9-G (Rev. 1)

McNamara P 2015 Training in the nuclear transport industry *Safe and Secure Transport and Storage of Radioactive Materials* ed K B Sorenson (Sawston: Woodhead Publishing) pp 27–41

Sert G 2015 Responding to emergencies associated with the transport of radioactive material *Safe and Secure Transport and Storage of Radioactive Materials* ed K B Sorenson (Sawston: Woodhead Publishing) pp 65–92

IOP Publishing

Principles and Techniques of Radiological Protection

Michael Charles Thorne

Chapter 19

Radiation accidents leading to large-scale releases of radionuclides to the environment

M C Thorne and R Wakeford

19.1 Introduction

In the period since 1950, military and commercial uses of nuclear power have expanded enormously. Although, in general, research, military and power-production reactors have operated safely, there have been several large accidents at reactors or other nuclear facilities that have resulted in significant releases of radionuclides to the environment. This chapter presents a brief account of those accidents and draws some general conclusions relevant to emergency preparedness, planning and responses to potential future major accidents. The accidents discussed comprise (UNSCEAR 1993, 2011, 2014):

- The Windscale reactor fire at Sellafield, England, in 1957;
- The 'Kyshtym Accident'—a waste tank explosion at Mayak, Russia, in 1957;
- The Three Mile Island reactor core meltdown, at Harrisburg, USA, in 1979;
- The Chornobyl reactor explosion in Ukraine, in 1986;
- The Fukushima Dai-ichi reactor core meltdowns in Japan, in 2011.

For each of these accidents. a brief account is given of its causes and consequences (section 19.2). This brief account is followed by selected comments on the implications of these accidents in terms of radiological impacts and health effects in workers (section 19.3), and radiological impacts and health effects in members of the public (section 19.4). However, before presenting this information, it may be helpful to consider how the severity of these five accidents has been assessed by the International Atomic Energy Agency (IAEA).

In characterising incidents of all sizes at nuclear installations, use is often made of the International Nuclear and Radiological Event Scale (INES). This is a tool that IAEA Member States use for communicating the safety significance of nuclear and radiological events to the public. It is used for the rating of events that result in a

doi:10.1088/978-0-7503-2424-3ch19

release of radioactive material into the environment and in the radiation exposure of workers and the public. In addition, it is used for events that have no such consequences, but where the measures put in place to prevent them did not function as intended. The scale is also applied to events involving the loss or theft of radioactive sources and the discovery of uncontrolled radioactive sources in scrap metal. The scale comprises seven levels, as illustrated in figure 19.1.

The lowest three levels are characterised as incidents, whereas the upper four levels are characterised as accidents. In broad terms, each successive level is a factor of ten more significant than the one immediately below. Brief descriptions of incidents or accidents at each level are given in table 19.1.

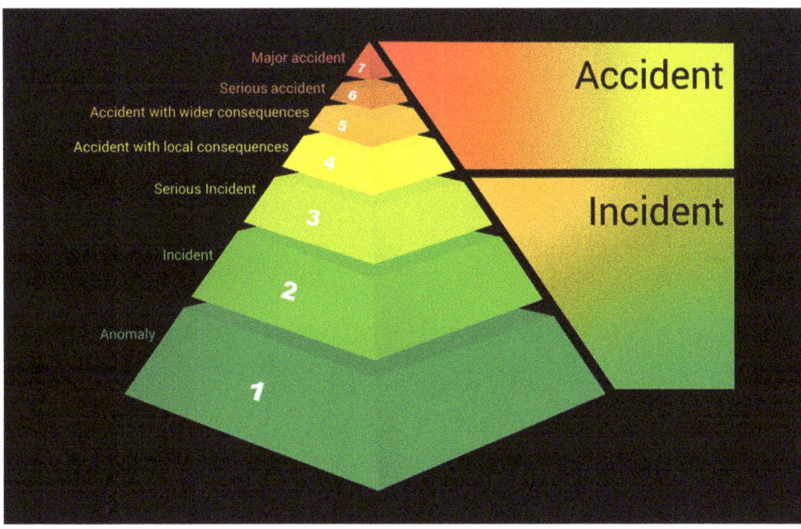

Figure 19.1. The INES International Nuclear and Radiological Event Scale. Reproduced from https://www.iaea.org/resources/databases/international-nuclear-and-radiological-event-scale. © IAEA

Table 19.1. Characteristics of incidents and accidents (from https://www.iaea.org/resources/databases/international-nuclear-and-radiological-event-scale).

Level	Description	Characteristics
1	Anomaly	Has an impact on defence-in-depth resulting in: • Exposure of a member of the public above statutory annual limits • Minor problems with safety components, but with significant defence-in-depth preserved • Low activity lost or stolen radioactive source, device or transport package
2	Incident	Has an impact on people and the environment, radiological barriers and control, or on defence-in-depth.

		Impact on people and environment: • Exposure of a member of the public to more than 10 mSv • Exposure of a worker above the statutory annual limits *Impact on radiological barriers and control:* • Radiation levels in an operating area of more than 50 $mSv\,h^{-1}$ • Significant contamination within the facility in an area where this is not expected by design *Impact on defence-in-depth:* • Significant failures in safety provisions but with no actual consequences • Found highly radioactive sealed orphan source, device or transport package with intact safety provisions • Inadequate packaging of a highly radioactive sealed source
3	Serious incident	Has an impact on people and the environment, radiological barriers and control, or on defence-in-depth. *Impact on people and environment:* • Exposure to more than ten times the statutory annual limit for workers • Non-lethal deterministic health effect (e.g., burns) from radiation *Impact on radiological barriers and control:* • Exposure rates of more than 1 $Sv\,h^{-1}$ in an operating area • Severe contamination in an area where this is not expected by design, but with a low probability of significant public exposure *Impact on defence-in-depth:* • Near-accident at a nuclear power plant with no safety provisions remaining • Lost or stolen highly radioactive sealed source • Misdelivered highly radioactive sealed source without adequate procedures in place to handle it
4	Accident with local consequences	Has an impact on people and the environment, or on radiological barriers and control. *Impact on people and environment:* • Minor release of radioactive material unlikely to result in implementation of planned countermeasures other than local food controls • At least one death from radiation *Impact on radiological barriers and control:* • Fuel melt or damage to fuel resulting in more than 0.1% release of the core inventory

(*Continued*)

Table 19.1. (*Continued*)

Level	Description	Characteristics
		• Release of significant quantities of radioactive material within an installation with a high probability of significant public exposure
5	Accident with wider consequences	Has an impact on people and the environment, or on radiological barriers and control. *Impact on people and environment:* • Limited release of radioactive material likely to require implementation of some planned countermeasures • Several deaths from radiation *Impact on radiological barriers and control:* • Severe damage to reactor core • Release of large quantities of radioactive material within an installation with a high probability of significant public exposure
6	Serious accident	Has an impact primarily on people and the environment and results from a significant release of radioactive material likely to require implementation of planned countermeasures.
7	Major accident	Has an impact primarily on people and the environment and results from a major release of radioactive material with widespread health and environmental effects requiring implementation of planned and extended countermeasures.

Of the accidents considered in this chapter, Windscale and Three Mile Island were classified as Level 5, Kyshtym as Level 6, and Chornobyl and Fukushima Dai-ichi as Level 7; these are the only accidents to have been assigned INES Levels 5–7.

Not surprisingly, vast amounts of information, comment and speculation have been published on each of these accidents. The material included in this chapter is, necessarily, only a summary, with an emphasis mainly on the radiological significance of each accident.

19.2 Summaries of the five major accidents

19.2.1 The Windscale Reactor Fire (INES level 5)

The accident at Windscale Works, Sellafield, in North-west England occurred in October 1957 and arose during a routine release of the Wigner energy stored in the graphite moderator of Windscale No. 1 Pile, one of the two air-cooled, open-circuit reactors at the installation. The accident has been described in detail in an authoritative book by Lorna Arnold (2007). Wigner energy arises because the flux of neutrons in the moderator displaces carbon atoms from their positions within the graphite lattice and results in some of them ending up in interstitial positions, leading to potential energy building up in the moderator. This potential energy is referred to as Wigner energy and if released in an uncontrolled manner can result in localised temperature increases and the possibility of a fire. However, Wigner energy can be released in a controlled way by heating the graphite to allow the interstitial atoms and vacancies to recombine, but because heating increases the rate of release of Wigner energy, a positive feedback can occur. In later graphite-moderated reactors, such as the Magnox reactors that formed the first generation of nuclear power stations in the UK, the operating temperature was sufficiently high that Wigner energy storage was not a problem because it was continually released, but the operating temperature of the Windscale Piles was low enough for Wigner energy to be stored in the graphite. In the Windscale accident, the planned release of Wigner energy led to a localised high temperature in the core, which produced overheating of the uranium-metal fuel and a reactor fire that gave rise to releases of radioactive materials to the atmosphere during 10–11 October 1957. Radionuclide releases occurred mainly over two periods: the first was when air flow was restarted through the core on the evening of 10 October and the second was when water was pumped into the reactor on the morning of 11 October to extinguish the fire (Garland and Wakeford 2007).

Radionuclide releases from this accident were dominated by ^{131}I (1.8 10^{15} Bq), with smaller quantities of ^{137}Cs (1.8 10^{14} Bq) and ^{106}Ru (3.0 10^{12} Bq). A large amount of the noble gas ^{133}Xe was released (2.6 10^{16} Bq), but this is of low radiotoxicity compared with the other radionuclides released. These radionuclides are fission products and arose from the irradiated fuel. In addition, there were releases of 4.2×10^{13} Bq of ^{210}Po that was being produced by neutron irradiation of bismuth ingots in the core, and 5.0 10^{15} Bq of ^{3}H produced by the irradiation of lithium cartridges in the core (Garland and Wakeford 2007). The plume from the fire initially travelled in a north easterly direction, but then a north-westerly wind carried the radioactive releases southeast over England and then continental Europe (Johnson *et al* 2007).

The predominance of the release of ^{131}I in the Windscale accident meant that the focus of concern related to deposition on pasture, with subsequent transfer of the ^{131}I to humans in milk from cattle grazing that pasture, particularly in respect of intakes of radioiodine by children. Early recognition of the potential significance of this pathway and the prompt imposition of a ban on milk supplies in the locally affected area had the effect of reducing doses. Because ingested and inhaled iodine is efficiently taken up by the thyroid gland where it is used in the production of thyroid hormones, radiation doses to the thyroids of children were of particular concern following the Windscale accident. Maximum absorbed doses to the thyroid for persons close to the site were estimated as 10 mGy in adults and 100 mGy in children. Further from the accident, in Leeds and London, absorbed thyroid doses in adults were less than 1 mGy, with young children incurring doses that were about a factor of two larger (Crick and Linsley 1984, UNSCEAR 1993).

Although ^{131}I was the radionuclide of greatest concern in the immediate aftermath of the accident, it was only one of several contributors to the collective effective dose. The total collective effective dose received in the UK and Europe has been estimated as 2000 person-Sv (1900 person-Sv by the population of the UK), of which about 900 person-Sv was due to inhalation, 800 person-Sv was due to the ingestion of milk and other foodstuffs, and 300 person-Sv was due to external irradiation from ground deposits. The main contributors to the collective effective dose were ^{131}I (37%), ^{210}Po (37%) and ^{137}Cs (15%) (Crick and Linsley 1984, UNSCEAR 1993).

19.2.2 The Kyshtym Waste Tank Explosion (INES level 6)

In the early-1950s high-level radioactive wastes from the Chelyabinsk-40 (later Chelyabinsk-65) plutonium production facility near Kyshtym in the southern Ural Mountains (the installation is now called Mayak Production Association) were stored in water-cooled tanks encased in concrete. The corrosion and failure of process-monitoring equipment led to a breakdown in the cooling system of a 300 m^3 tank, allowing 70 to 80 tonnes of waste, mainly in the form of nitrate and acetate compounds, to heat up, resulting in the drying out of the waste until it reached a temperature of 330 °C–350 °C. On 29th September 1957 the contents of the tank exploded, creating a crater 10 m deep and 20 m in diameter. The explosion ejected 740 PBq (20 MCi) of radioactive material into the atmosphere. About 90% of the activity released from the tank was deposited locally within the Mayak site, whereas the remaining 10% (about 1 10^{17} Bq) travelled as a plume in a north easterly direction beyond the Mayak site. The main radionuclides released were, expressed as a percentage of activity, ^{144}Ce + ^{144}Pr (66%), ^{95}Zr + ^{95}Nb (24.9%), ^{90}Sr + ^{90}Y (5.4%) and ^{106}Ru + ^{106}Rh (3.7%). In addition, traces of ^{89}Sr, ^{147}Pm, ^{155}Eu and $^{239/240}$Pu were released. The composition is as expected for one-year-cooled fission products, except for a deficit in ^{137}Cs, which represented 0.35% of the total activity. This deficit arose because the waste was produced by precipitation with sodium hydroxide, which left the caesium predominantly in solution from which it was later concentrated separately (UNSCEAR 1993, Akleyev *et al* 2017, Batorshin and Mokrov 2018, Vasilenko *et al* 2020).

The radioactive cloud reached a height of about 1 km and dispersed under relatively stable meteorological conditions resulting in a long (~300 km) and narrow (~50 km) area of deposition known as the East Urals Radioactive Trace (EURT), as illustrated in figure 19.2.

Redistribution of deposited radionuclides occurred to some degree, mainly in the first few days following the accident, but the contaminated area changed little after 1958. Defining the contaminated area as a deposition of ^{90}Sr of more than 4 10^3 Bq m^{-2}, i.e., twice the level of global fallout, gave a contaminated area of between 15 000 and 23 000 km^2, with a total population of about 270 000. Of these, ~1200 were within an area with a deposition density of more than 4×10^7 Bq m^{-2} and ~11 000 were in areas with a deposition density of more than 7×10^4 Bq m^{-2} (UNSCEAR 1993, Akleyev *et al* 2017, Batorshin and Mokrov 2018). The ~1200 persons in the area with the highest deposition density were evacuated within ten days

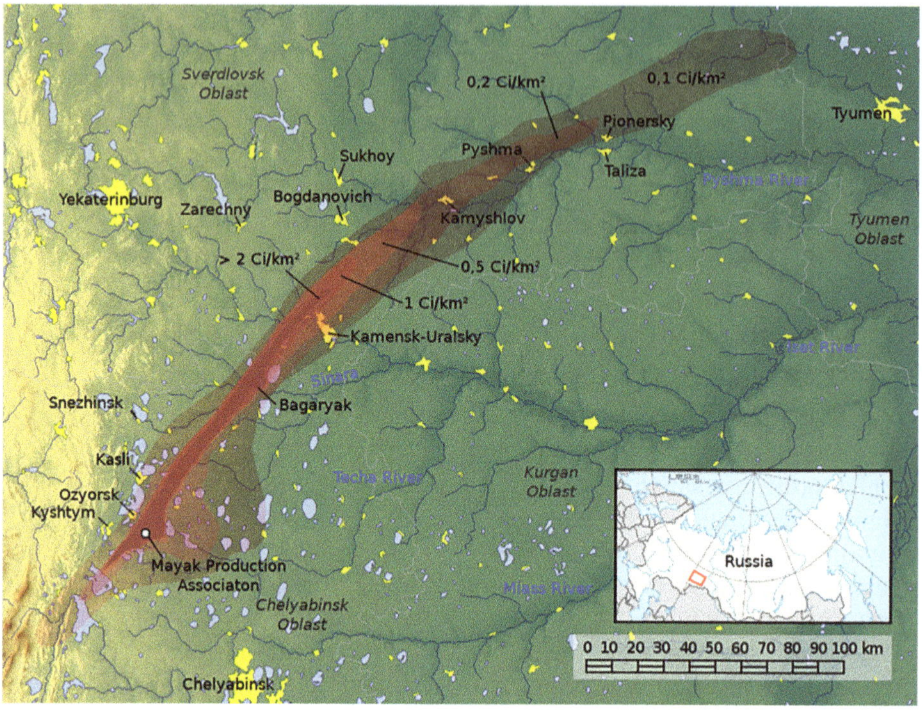

Figure 19.2. The East Urals Radioactive Trace (EURT). This Ostural-Spur.xcf image has been obtained by the author(s) from the Wikimedia website where it was made available by User:JanRieke under a CC BY-SA 3.0 licence. It is included within this article on that basis. It is attributed to User:JanRieke.

of the accident. A ban on consumption of local, contaminated food was also imposed. From 8 to 18 months after the accident, evacuation and resettlement was undertaken for the population resident in areas with a ^{90}Sr deposition density of more than 1.5×10^5 Bq m^{-2} (150 kBq m^{-2}). The average effective dose to the population group evacuated within 10 days of the accident was 170 mSv from external radiation and 350 mSv from internal irradiation (UNSCEAR 1993). For the non-evacuated population, the effective dose over the thirty years following the accident was estimated as 0.32 mSv per kBq m^{-2} of ^{90}Sr. For a deposition density of 150 kBq m^{-2}, this corresponds to an effective dose of 48 mSv.

19.2.3 The Three Mile Island accident (INES level 5)

The accident at Three Mile Island (TMI) occurred in Unit 2 of a two-unit Pressurised Water Reactor power station near Harrisburg, Pennsylvania, USA (Kemeny 1979, Anonymous 1981, UNSCEAR 1993). Unit 2 of the TMI station had a power output of 906 MW of electricity and had only been operating at power for a short time before the accident happened at 4 am on 28th March 1979, when the reactor was operating at 97% power. It involved a relatively minor malfunction in the secondary cooling circuit that caused the temperature in the primary coolant to rise. This in turn caused the reactor to shut down automatically.

Within seconds of the shutdown, the pilot-operated relief valve on the reactor cooling system opened. About 10 seconds later it should have closed. However, it remained open, leaking reactor coolant water to the reactor coolant drain tank. The operators believed the relief valve had shut because instruments showed them that a 'close' signal was sent to the valve, but they did not have an instrument indicating the actual state of the valve. Responding to the loss of cooling water, high-pressure injection pumps automatically pushed replacement water into the reactor system. As water and steam escaped through the relief valve, cooling water surged into the pressuriser, raising the water level in it. Operators responded by reducing the flow of replacement water. Because the pressuriser level was increasing, they thought the reactor system was too full of water. Also, pumping a mixture of steam and water caused the reactor cooling pumps to vibrate. Because the severe vibrations could have damaged the pumps and made them unusable, the operators shut them down. This ended forced cooling of the reactor core and as reactor coolant water boiled away, the core became uncovered and heated further.

Subsequently, at 6:22 am, operators closed a valve between the relief valve and the pressuriser. This action stopped the loss of coolant water through the relief valve. However, superheated steam and gases blocked the flow of water through the core cooling system. Throughout the morning, operators attempted to force more water into the reactor system to condense steam bubbles that they believed were blocking the flow of cooling water. Then, during the afternoon, operators attempted to decrease the pressure in the reactor system to allow a low-pressure cooling system to be used and emergency water supplies to be put into the system.

In the late afternoon, operators began high-pressure injection of water into the reactor cooling system to increase pressure and to collapse steam bubbles. By 7:50 pm on 28th March, they restored forced cooling of the reactor core when they were able to restart one reactor coolant pump. This brought the main phase of the accident to an end.

During the accident, radioactive gases from the reactor cooling system built up in the makeup tank in the auxiliary building. During 29th and 30th March, operators used a system of pipes and compressors to move the gas to waste gas decay tanks. However, the compressors leaked, and some radioactive gases were released to the environment. These went through high-efficiency particulate air (HEPA) filters and charcoal filters that removed most of the radionuclides, except for the noble gases, the estimated total of which was about 3.7×10^{17} Bq, mainly ^{133}Xe, and a smaller amount of ^{131}I (5.5×10^{11} Bq).

Because of the predominance of noble gases in the release, effective doses were small. Individual doses averaged 0.015 mSv within 80 km of the plant and the maximum effective dose that any member of the public could have received is estimated to have been 0.85 mSv (UNSCEAR 1993). These doses were primarily due to external irradiation. The collective effective dose to the population within 80 km of the plant was 20 person-Sv and the collective effective dose to the population beyond 80 km was similar in magnitude, giving 40 person-Sv in total (UNSCEAR 1993).

19.2.4 The Chornobyl accident

The accident on 26 April 1986 at the Chornobyl Nuclear Power Plant occurred during a low-power engineering test of the Unit 4 reactor, a Soviet RBMK reactor. Safety systems had been switched off and improper operation of an inherently unstable design of reactor allowed an uncontrollable power surge to occur that resulted in successive steam explosions that destroyed the reactor. The radionuclide releases from the destroyed reactor occurred mainly over a 10-day period at varying rates. An initial high release rate on the first day was caused by mechanical discharge because of the explosions in the reactor. There followed a 5-day period of declining releases associated with the hot air and fumes from the red-hot core material. During the next few days, the release rate increased until day 10, when the releases dropped abruptly (UNSCEAR 2011).

Releases of ^{131}I and ^{137}Cs were of greatest radiological significance in terms of impacts on the public. These releases were estimated to have been $\sim 1.76 \times 10^{18}$ and $\sim 8.5 \times 10^{16}$ Bq, respectively (UNSCEAR 2011).

Radioactive deposition on the ground was found to some extent in practically every country of the northern hemisphere. The most contaminated areas, where the average deposition density of ^{137}Cs exceeded 3.7 10^4 Bq m^{-2}, were mainly in Belarus, the Russian Federation and Ukraine. These areas were widely distributed around the site (figure 19.3),

Figure 19.3. Map of levels of ^{137}Cs deposition in 1989 in Belarus, the Russian Federation and Ukraine. Reprinted from UNSCEAR (2008), with permission of UNSCEAR.

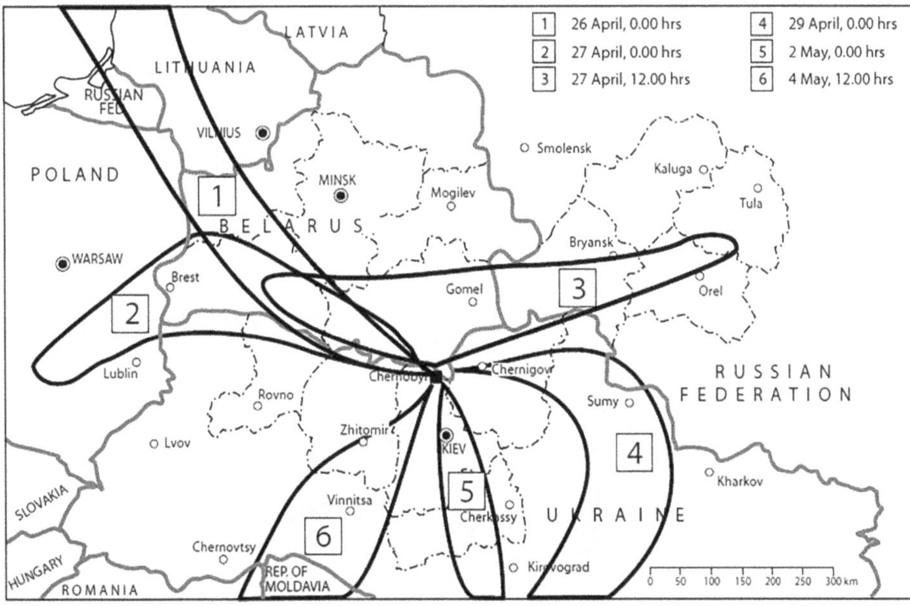

Figure 19.4. Estimated plumes for instantaneous releases on the dates and at the times (UTC) indicated considering the prevailing meteorological conditions. Reprinted from UNSCEAR (2008), with permission of UNSCEAR.

due to substantial changes in wind direction during the ten-day period over which most of the release occurred (figure 19.4).

Average individual and collective doses received by the population groups that were exposed because of the Chornobyl accident are given by UNSCEAR (2011) and are summarized in table 19.2. Absorbed doses to the thyroid over the first few weeks after the accident for those members of the population drinking fresh milk containing ^{131}I were much higher than the dose to the thyroid due to natural sources of radiation. This was especially true for infants and children who consumed proportionally more milk than adults—in excess of 10 000 children received thyroid doses from radioiodine that were greater than 1 Gy.

The deposited material consisted of hot particles in addition to more homogeneously distributed radioactive material, with the hot particles of greatest relative importance close to the site of the accident. These hot particles have been classified into two broad categories: (a) fuel fragments with a mixture of fission products bound to a matrix of uranium oxide, similar in composition to that of the fuel in the core, but sometimes very much depleted in caesium, iodine and ruthenium, and (b) particles consisting of one dominant element (ruthenium or barium) but sometimes having traces of other elements. Typical activities were $1\ 10^{2}$ to $1\ 10^{3}$ Bq for a fuel fragment hot particle and $5\ 10^{2}$ to $1\ 10^{4}$ Bq for a ruthenium hot particle. A typical aerodynamic effective diameter was about 10 μm, to be compared with 0.4–0.7 μm for particles associated with ^{131}I and ^{137}Cs.

Table 19.2. Average individual and collective doses from the Chornobyl accident (UNSCEAR 2011).

Population Group	Population Size	Average thyroid dose in 1986 (mGy)	Average effective dose in 1986–2005 (mSv)	Collective thyroid dose in 1986 (person-Gy)	Collective effective dose in 1986–2005 (person-Sv)
Recovery operation workers	5.30×10^5	Not estimated[a]	117[b]	Not estimated	6.12×10^4
Evacuees	1.15×10^5	490	31[c]	5.70E + 04	3.60×10^3
Inhabitants of contaminated areas[d] of Belarus, Russia and Ukraine	6.40×10^6	102	9[c,e]	6.50E + 05	5.89×10^4
Inhabitants of Belarus, the Russian Federation and Ukraine	9.80×10^7	16	1.3[c,e]	1.60E + 06	1.25×10^{5}[e]
Inhabitants of distant countries[f]	5.00×10^8	1.3	0.3[c,e]	6.60E + 05	1.30×10^{5}[e]

Notes

[a] Thyroid doses only exist for a small number of workers and it is not possible to give a valid average value for the whole group.

[b] Effective dose estimates for the workers include only the doses from external irradiation, delivered essentially from 1986 to the end of 1990. It is assumed that the recorded dose in mGy is numerically equal to the effective dose in mSv.

[c] Effective dose estimates are the sum of the contributions from external and internal irradiation, excluding the thyroid dose.

[d] The contaminated areas were defined arbitrarily in the former Soviet Union as areas where the ^{137}Cs levels in soil were greater than 3.7×10^4 Bq m^{-2}.

[e] The total dose will continue to accumulate to be perhaps 25% higher for the whole lifetime.

[f] All the European countries except the three republics, Turkey, countries of the Caucasus, Andorra and San Marino.

19.2.5 The Fukushima Dai-ichi accident (INES level 7)

The Fukushima Dai-ichi Nuclear Power Station is in the Fukushima Prefecture of the Tōhoku region in Japan and is about 230 km north-east of Tokyo, facing onto the Pacific Ocean. The total power-generating capacity of the six Boiling Water Reactors on site was 4700 MW of electricity. On 11th March 2011, an earthquake of magnitude 9.0 occurred along the Japan Trench at 14:46 Japan Standard Time (JST). The earthquake and the following tsunami triggered a severe nuclear accident at the power station. The accident has been described in detail by UNSCEAR (2014) and the IAEA (2015).

When the earthquake occurred, Units 1–3 of FDNPS were in normal operation, whereas Units 4–6 were undergoing periodic maintenance and refuelling operations. As designed, the emergency shutdown feature went into operation at Units 1–3 immediately after seismic activity started. The seismic tremors damaged electricity transmission facilities inside and outside the site, resulting in total loss of off-site electricity. In response, the emergency diesel generators automatically activated, as designed, to provide backup power for the reactor cooling systems and other plant safety systems.

However, the earthquake also caused a tsunami to hit the Japanese coastline at 15:35 JST. This had an estimated maximum wave height of about 15 m at Fukushima Dai-ichi, much higher than the 6 m protective seawall and above the elevation of approximately 10 m at which key buildings were constructed. The tsunami damaged or destroyed the emergency diesel generators, the seawater cooling pumps, the electric wiring system and the DC power supply for Units 1, 2 and 4, resulting in the loss of all on-site power, except for Unit 6 that was supplied with electricity from an air-cooled emergency diesel generator.

With no cooling to remove heat, damage to the reactor core may have begun at Unit 1 on 11th March. Injection pumps (driven by steam generated by the reactors) were used to provide cooling water to the reactors on Units 2 and 3, but these pumps eventually stopped working, and all cooling to the reactors was lost until fire engines were used to restore water injection. Without adequate cooling, pressure inside the reactor vessels increased, but was relieved to some degree for Units 2 and 3 by venting through the safety relief valves. In addition, water or steam in direct contact with the over-heated fuel assemblies reacted with the zirconium of the fuel cladding to produce hydrogen gas. This accumulated in the upper portion of the reactor buildings and ignited, producing explosions in the Unit 1 and Unit 3 reactor buildings on 12th and 14th March, respectively. Hydrogen generated in Unit 3 seems to have migrated into the Unit 4 reactor building, resulting in a subsequent explosion and damage there on 15th March. Severe damage, including meltdown, occurred in the cores of the three reactors (Units 1, 2 and 3). In all three units, melted fuel fell to, and subsequently penetrated, the bottom of the reactor pressure vessels, resulting in molten-fuel–concrete interactions that further increased the pressure within the containments.

Operations to reduce pressure in, or leaks from, the reactor vessels resulted in releases of volatile radionuclides into the containment vessel, the reactor buildings and the outside environment. These volatile radionuclides were present in both gaseous and aerosol form, although a significant fraction of the aerosols was trapped in the water in the reactor containment and in the turbine buildings.

At 20:50 JST on 11th March 2011, the Governor of Fukushima Prefecture issued instructions to evacuate settlements within 2 km of the facility. Shortly afterwards, the Director-General of the Nuclear Emergency Response Headquarters ordered the evacuation of residents and others within 3 km of the facility and the sheltering indoors of all residents and others within 10 km. At 05:44 JST the next morning, the people within 10 km were ordered to be evacuated. At 18:25 JST on same day, the evacuation radius was expanded to 20 km. Following the hydrogen explosion between about 06:00 and 06:12 JST on 15th March, an instruction was issued ordering all people living between 20 and 30 km to shelter indoors. In addition, on 16th March, an instruction was issued that anyone remaining within 20 km should take stable iodine. This instruction was not implemented, because the area was considered to have already been evacuated.

Monitoring of food and drinking water by Japanese national and prefectural governments began on 16th March 2011. Selected foodstuffs with radionuclide concentrations that exceeded the provisional regulation values were prohibited from distribution on 21st March 2011 and from consumption on 23rd March 2011. On 25th March, the residents in the area between 20-km and 30-km radius of the site, who had been sheltering since 15th March, were advised to begin voluntary evacuation and instructed to be prepared to evacuate depending on future developments at the power station. This instruction was terminated on 30th September.

Significant releases of radionuclides to the atmosphere began on 12th March and the rate of release varied considerably over the following week, with marked increases associated with specific events at each unit. After the first week, the rates of release gradually declined and by the beginning of April, release rates had fallen to a thousandth or less of the release rates that occurred during the first week of the accident. The releases occurred from different locations, at different heights and with quite different characteristics, all of which affected their subsequent dispersion in, and deposition from, the atmosphere.

Radionuclides entered the marine environment by direct and indirect routes. Direct release into the ocean resulted from leakage of highly contaminated water from a trench outside Unit 2 and the deliberate discharge of slightly contaminated water from storage tanks; the latter were emptied to create capacity for the storage of more highly contaminated water remaining in the trench. Radioactive material entered the ocean indirectly, mainly from deposition onto the ocean surface of material released to the atmosphere, but also from run-off into rivers of material deposited on land and transported downstream into the ocean.

In total, about 1×10^{17} to 5×10^{17} Bq of ^{131}I and 6×10^{15} to 2×10^{16} Bq of ^{137}Cs were released to the atmosphere, and about 7×10^{16} to 1.2×10^{17} Bq of ^{131}I and 8×10^{15} to 1.4×10^{16} Bq of ^{137}Cs entered the ocean, either directly or indirectly.

Because radionuclide releases were protracted and varied with time under changing meteorological conditions, the pattern of deposition was spatially complex. Material released on 12th March initially went towards the Pacific Ocean, but the release in the afternoon of 12th March, spread northwards along the eastern coast of the main island and later shifted to a north-north-easterly direction, over the coastal area of Miyagi. Subsequently, material released from late at night on 14th March was dispersed toward the south and, on the morning of 15th March, resulted in dry deposition of radionuclides in the prefectures of Tokyo, Saitama and Kanagawa. By the afternoon of 15th March, this dispersing material encountered precipitation, which resulted in enhanced levels of wet deposition in areas of the prefectures of Gunma, Tochigi and Fukushima. A further major release occurred on the morning of 15th March, with the released material moved towards the south then progressively to the north-west. Finally, material released from 20th to 23rd March was subject to wet deposition, particularly in the prefectures of Iwate, Miyagi, Ibaraki and Chiba.

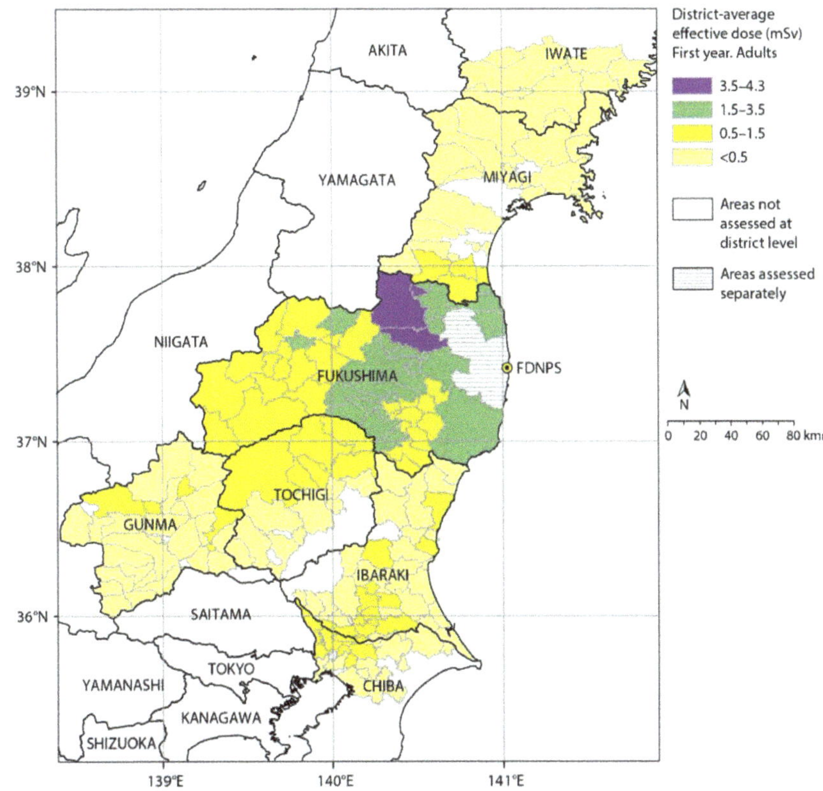

Figure 19.5. Estimated district-average effective doses in the first year following the accident to adults living in districts of Fukushima Prefecture and some districts of other prefectures that were not evacuated. Reprinted from UNSCEAR (2014), with permission of UNSCEAR. © United Nations, 2008. All rights reserved.

Figure 19.5 shows estimated effective doses in the year following the accident for individuals not evacuated from the area. Note that these effective doses are relatively small, being up to a few mSv expressed as district-averaged values. Effective doses to evacuees were of a similar magnitude, ranging up to about 10 mSv, and included a significant contribution from dose received at the evacuation destination. UNSCEAR (2014) estimated that the evacuation of settlements within the 20-km zone averted effective doses to adults of up to about 50 mSv.

19.3 Radiological and health effects impacts on workers

19.3.1 The Windscale reactor fire

Of the workers at Windscale Works, Sellafield, in October 1957, 471 have been identified as involved in either fighting the fire or involved in clean-up operations afterwards. The relatively small number of workers together with the low average dose received during October 1957 of 6 mSv implies that any study of health effects that might be attributable to the accident will have very low power. The 83 deaths among these workers that occurred in the 50 years after the accident was statistically non-significantly less than the number expected from the experience of those workers employed at Sellafield in 1957 but not directly involved in fighting the fire or recovery operations (McGeoghegan *et al* 2010). No effect of the fire upon the mortality or cancer morbidity experience of the Windscale fire workers could be detected (McGeoghegan *et al* 2010).

19.3.2 The Kyshtym accident at Mayak, Russia

During 1957–59, around 38 500 individuals were involved in emergency and clean-up operations following the chemical explosion in a liquid radioactive waste tank at the Mayak nuclear complex. About 10 500 of these people received doses of around 250 mSv because of the release, and some doses exceeded 1 Sv (Vasilenko *et al* 2020). The collective dose received during the accident and in the immediate aftermath was approximately 7300 person-Sv and over the following three months, 4500 person-Sv; in 1958 and 1959 the collective doses were 2250 and 480 person-Sv, respectively (Vasilenko *et al* 2020). Although epidemiological studies of Mayak workers have been conducted, no specific studies of those individuals involved in the 'Kyshtym Accident' have been reported.

19.3.3 The Three Mile Island accident

Any additional exposure of TMI workers in the immediate aftermath of the accident was very small, as was that of workers involved in clean-up operations over the following 14 years. No study specific to these workers has been conducted.

19.3.4 The Chornobyl accident

Some doses received by plant and emergency workers soon after the explosion at Chornobyl Unit 4 were several Gy and sufficiently high to produce tissue reactions (deterministic effects). A total of 134 of these workers suffered from acute radiation syndrome (ARS) and the radiation-induced injuries of 28 people were sufficiently severe that they died within weeks of their exposure (UNSCEAR 2011).

Clean-up operations during 1986–90 involved about 530 000 'liquidators' who are estimated to have received a collective effective dose of 60 000 person-Sv, with an average effective dose of $\sim$120 mSv and a range extending from $<$10 mSv to $>$1000 mSv (UNSCEAR 2011). Various groups of these liquidators have been the subject of several epidemiological studies.

Nearly 70 000 Russian liquidators have been followed up for health effects and significantly raised rates of solid cancer incidence and mortality have been reported (Ivanov *et al* 2020). Kesminiene *et al* (2008) conducted a case-control study of 70 cases of lympho-haematological malignancies in liquidators from Belarus, Russia and the Baltic countries, and found non-significantly raised risks both of leukaemia excluding chronic lymphocytic leukaemia (CLL) and of CLL, and a marginally significant raised risk of non-Hodgkin lymphoma; these findings are difficult to interpret. Kesminiene *et al* (2012) also reported a significant increase in the radiation-related risk of thyroid cancer in this group of liquidators, which is surprising given that the exposure is of adults and there is limited evidence of a raised risk of thyroid cancer following adult exposure to radiation. In a case-control study of 137 cases of leukaemia in Ukrainian liquidators, Zablotska *et al* (2013) found a marginally non-significant radiation-related increased risk of leukaemia excluding CLL, and an indication of a raised risk for CLL. Other health effects, such as an increase of diseases of the circulatory system and of ocular cataracts, have been reported in Chornobyl liquidators (Little *et al* 2021a), although the interpretation of some of these results is problematical (Wakeford 2019).

19.3.5 The Fukushima Dai-ichi accident

Doses on the scale of those received by workers dealing with the Chornobyl emergency were avoided during the Fukushima Dai-ichi accident, where an initial effective dose limit of 100 mSv was initially put in place but was subsequently raised to 250 mSv. Six workers received effective doses that exceeded the 250 mSv limit, with the largest effective dose assessed as 679 mSv (UNSCEAR 2021). However, the effective doses received by these six workers were dominated by doses to the thyroid received from iodine-131, so the range of thyroid doses was 4.8–32 Gy (UNSCEAR 2021). It is possible that this level of thyroid dose (and the level of dose of seven other workers who received thyroid doses of more than 2 Gy) could lead to hypothyroidism (UNSCEAR 2021), so health monitoring is advisable. Otherwise, discernible health effects in emergency workers are unlikely. Clean-up workers are receiving low levels of dose during extensive operations. Emergency and clean-up workers are the subject of epidemiological studies, but little information is available at present.

19.4 Radiological and health effects impacts on members of the public

19.4.1 The Windscale Reactor fire

The principal concern arising from the radioactive releases during the Windscale Fire was the impact of iodine-131 upon the risk of thyroid cancer, particularly for children drinking contaminated cow's milk. Largely for this reason, a milk distribution ban was implemented in those areas around Sellafield that were most heavily contaminated (Dunster *et al* 2007). From the (relatively few) measurements of thyroid activities that were carried out in the immediate aftermath of the accident, this milk ban seems to have been broadly successful, with the largest thyroid dose estimated as 160 mGy in a child (possibly due to the milk distribution ban not being 100% effective, especially immediately after the start of the releases), but generally in the range 10–20 mGy (Dunster *et al* 2007, Environment CoMAoRit 2016). If these thyroid doses were representative of the exposed population, then an excess of thyroid cancer cases would not be expected.

A study of thyroid cancer incidence during 1974–2012 in Cumbria, where Sellafield is located, and the neighbouring county of Lancashire, which was affected by contamination to a lesser extent, found generally raised rates in Cumbria and generally lowered rates in Lancashire when compared with rates in the rest of

England (McNally *et al* 2016). The pattern of incidence is not easy to interpret because, for example, raised incidence rates of thyroid cancer in Cumbria extended to those born during 1959–63, who would not have been affected by iodine-131 released during the Fire. This geographical study did not provide evidence of an effect of the Fire upon the risk of thyroid cancer, and further studies are planned (Environment CoMAoRit 2016).

19.4.2 The Kyshtym accident at Mayak, Russia

The major release of radionuclides from the explosion in a liquid waste tank at the Mayak nuclear complex in the Southern Urals of Russia in 1957 led to communities for a few 100 km to the north-east of Mayak being contaminated, some to a sufficient extent to be evacuated—around 10 000 people were eventually evacuated. Unfortunately, the Kyshtym Accident was kept secret by the USSR until 1989, and details of health studies that had been conducted in the Soviet era are sparse—those studies that have been reported have been summarised by Akleyev *et al* (2017) who suggested that an excess of solid cancer incidence and mortality in the exposed population could be detected. Further studies are planned (Tolstykh *et al* 2017).

19.4.3 The Three Mile Island accident

Given the limited quantities of noble gases released during the accident and the consequent low doses, any health effects produced by the radioactive releases during the TMI accident will be far below the level that could be detected against variations in background rates. Epidemiological studies have not found evidence of any effect of the accident (Hatch *et al* 1990), although some authors have claimed otherwise (Wing *et al* 1997), but these claims are highly controversial (Hatch *et al* 1997).

Recently, it has been claimed that a chromosome aberration study around TMI showed that doses were much larger than previously believed (Datesman 2020). This claim has been demonstrated to be without foundation (Little *et al* 2021b).

19.4.4 The Chornobyl accident

Of the radionuclides released during the Chornobyl accident of 1986, of greatest concern were the radioisotopes of iodine, because these concentrate in the thyroid gland giving rise to an increased risk of thyroid cancer, especially among those exposed as children. A combination of delaying evacuation of the worst-affected areas for 36 h and the failure to halt the consumption of heavily contaminated milk led to more than 10 000 children receiving thyroid doses greater than 1 Gy, so it is unsurprising that an excess of thyroid cancer cases began to be apparent in the areas with the highest levels of contamination from 1990 onwards (UNSCEAR 2011). UNSCEAR (2018) assessed that of the nearly 20 000 cases of thyroid cancer among those less than 18 years of age at the time of the Chornobyl accident and diagnosed during 1991–2015 in Belarus, Ukraine and the four most affected regions of Russia, about 5000 cases were attributable to exposure to the releases.

There is no reasonable doubt that intakes of radioiodine released during the Chornobyl accident led to a large excess incidence of thyroid cancer in the heavily affected areas of the former USSR, but of interest is how the size of this excess compares with what would be predicted by standard thyroid cancer risk models. These risk models are derived from children exposed to external sources of radiation exposure (the Japanese atomic bombings and radiotherapy) (Ron *et al* 1995, Lubin *et al* 2017, Veiga *et al* 2016), so a comparison with the risk experienced by children exposed to iodine-131 from Chornobyl is important. Several studies of thyroid cancer among those exposed as children around Chornobyl have been carried out, and the results are summarised in table 19.3.

It will be seen from table 19.3 that the risk estimates derived from the Chornobyl studies are compatible with those obtained from studies of external exposure, although there is a suggestion that the risks in the iodine-131-exposed children are somewhat less than those in the externally exposed children. Caution in interpretation is required because several factors could be affecting the radiation-related risk, such as the age-at-exposure and time-since-exposure in the comparison groups, and whether other factors such as iodine deficiency in the Chornobyl population or

Table 19.3. Thyroid cancer incidence (<15 years of age at exposure)

Exposure	Study	ERR/Gy (95% CI)
External	Pooled analysis (Ron *et al* 1995)	7.7 (2.1, 29)
External	Pooled analysis (Veiga *et al* 2016)	5.5 (4.1, 7.5)
Chornobyl	Case-control (Belarus & Russia) (Cardis *et al* 2005)	4.5 (1.2, 7.8)
Chornobyl	Case-control* (Bryansk, Russia) (Kopecky *et al* 2006) * <20 years of age at exposure	48.7 (4.8, 1151)
Chornobyl	Cohort* (Ukraine) (Tronko *et al* 2006) *<18 years of age at exposure	5.2 (1.7, 27)
Chornobyl	Cohort* (Belarus) (Zablotska *et al* 2011) *<18 years of age at exposure	2.2 (0.8, 5.5)
Chornobyl	Cohort* (Ukraine) (Brenner *et al* 2011) *<18 years of age at exposure	1.9 (0.4, 6.3)

intakes of stable iodine used to 'block' the uptake of radioiodine could be playing a role. Nonetheless, it does not appear that standard thyroid cancer risk models derived from those exposed to external radiation seriously underestimate the risk of thyroid cancer from iodine-131 released during the Chornobyl accident. There is little convincing evidence for an excess of thyroid cancer among those exposed as children outside the heavily contaminated areas (UNSCEAR 2011). A substantial amount of work is being expended on 'radiation signatures' of exposure following the Chornobyl accident, and it appears that DNA double-strand breaks play an early role in the radiation-related thyroid cancers (Morton *et al* 2021).

Doses from Chornobyl radionuclides to other organs/tissues have been much less than the thyroid doses received from intakes of radioiodine, so excess levels of cancers other than thyroid cancer would not be anticipated, at least at the level found for thyroid cancer. A large case-control study of childhood leukaemia among those exposed in utero or <6 years of age while present in the most heavily contaminated areas around Chornobyl found little evidence for a raised risk in Belarus or Russia, but a raised risk in Ukraine (Davis *et al* 2006). However, there were concerns over the accuracy of the Ukrainian component of the study (Davis *et al* 2006, Moysich *et al* 2011), and a subsequent study from Ukraine reported a much lower level of risk (Noshchenko *et al* 2010). There are indications of raised levels of breast cancer in a contaminated region of Russia (Rivkind *et al* 2020), but not generally (Zupunski *et al* 2021), and further studies are required to clarify the situation.

19.4.5 The Fukushima Dai-ichi accident

The releases of radioactive materials from the Fukushima Dai-ichi nuclear power station were less than those from Chornobyl, and the majority of the releases were carried over the Pacific Ocean, but an area to the north-west of the reactor site was contaminated to a level that required evacuation of over 100 000 people (UNSCEAR 2014, 2021). Following the dramatic rise in thyroid cancer incidence that occurred among those exposed at a young age to radioiodine in the heavily contaminated areas around Chornobyl, the principal concern over the releases from Fukushima must be the risk of thyroid cancer in the exposed children.

The thyroid doses assessed to have been received by the most exposed children were far less than those received in the heavily contaminated communities around Chornobyl (Shinkarev 2021), so a discernible increase in thyroid cancer incidence is not anticipated (UNSCEAR 2014, 2021). The Japanese authorities established a prefecture-wide health surveillance programme in Fukushima Prefecture that included a sensitive ultrasound screening of the thyroids of around one-third of a million people who were less than 19 years of age in March 2011 (Kamiya 2021). In the first round of thyroid screening conducted within three years of the accident, 116 confirmed or suspected cases of thyroid cancer were found, which was around ten times greater than the number of incident cases expected from Japanese background rates. This led to speculation that excess cases consequent to the releases during the accident were being found (Tsuda *et al* 2016). However, it was pointed out (e.g. by

Wakeford *et al* 2016) that the prevalent cases detected by sensitive screening could not be compared with incident cases found by conventional medical investigations, that the geographical distribution of the cases did not follow the levels of contamination in Fukushima Prefecture, and that those affected had a notably older distribution of age-at-exposure than those at Chornobyl. Subsequent screening rounds found successively fewer cases, so that 21 cases were detected in the fourth round (Kamiya 2021), which could indicate that early screening was detecting thyroid cancers that were largely already present in March 2011 and new cases were being found in later screening. What is clear is that sensitive ultrasound thyroid screening can lead to a marked rise in cases of thyroid cancer, as was found, for example, in South Korea in the early-2000s following the government making screening available to the public (Ahn *et al* 2014). Such is the concern about 'overdiagnosis' of thyroid cancer that an expert group convened by the International Agency for Research on Cancer recommended that mass thyroid screening should not take place after a nuclear accident (Togawa *et al* 2018).

References

Anonymous 1981 The three mile island nuclear accident: lessons and implications *Ann. N. Y. Acad. Sci.* **365** 1–343

Ahn H S, Kim H J and Welch H G 2014 Korea's thyroid-cancer 'epidemic'—screening and overdiagnosis *N. Engl. J. Med.* **371** 1765–7

Akleyev A, Krestinina L Y, Degteva M and Tolstykh E 2017 Consequences of the radiation accident at the Mayak production association in 1957 (the'Kyshtym Accident') *J. Radiol. Prot.* **37** R19

Arnold L 2007 *Windscale 1957: Anatomy of a Nuclear Accident* (Berlin: Springer)

Batorshin G S and Mokrov Y G 2018 Experience and the results of emergency management of the 1957 accident at the Mayak production association *J. Radiol. Prot.* **38** R1

Brenner A V, Tronko M D, Hatch M, Bogdanova T I, Oliynik V A, Lubin J H *et al* 2011 I-131 dose response for incident thyroid cancers in Ukraine related to the Chornobyl accident *Environ. Health Perspect.* **119** 933–9

Cardis E, Kesminiene A, Ivanov V, Malakhova I, Shibata Y, Khrouch V *et al* 2005 Risk of thyroid cancer after exposure to ^{131}I in childhood *J. Natl. Cancer Inst.* **97** 724–32

Crick M and Linsley G 1984 An assessment of the radiological impact of the Windscale reactor fire, October 1957 *Int. J. Radiat. Biol. Relat. Stud. Phys., Chem. Med.* **46** 479–506

Datesman A M 2020 Radiobiological shot noise explains Three-Mile Island biodosimetry indicating nearly 1,000 msv exposures *Sci. Rep.* **10** 10933

Davis S, Day R W, Kopecky K J, Mahoney M C, McCarthy P L, Michalek A M *et al* 2006 Childhood leukaemia in Belarus, Russia, and Ukraine following the Chornobyl power station accident: results from an international collaborative population-based case-control study *Int. J. Epidemiol.* **35** 386–96

Dunster H J, Howells H and Templeton W L 2007 District surveys following the Windscale incident, October 1957–1958 *J. Radiol. Prot.* **27** 217–30

Environment CoMAoRit 2016 *COMARE Seventeenth Report. Further Consideration of the Incidence of Cancers around the Nuclear Installations at Sellafield and Dounreay* (Chilton: Public Health England)

Garland J and Wakeford R 2007 Atmospheric emissions from the Windscale accident of October 1957 *Atmos. Environ.* **41** 3904–20

Hatch M, Susser M and Beyea J 1997 Comments on 'a reevaluation of cancer incidence near the Three-Mile Island nuclear plant *Environ. Health Perspect.* **105** 12

Hatch M C, Beyea J, Nieves J W and Susser M 1990 Cancer near the Three-Mile Iisland nuclear plant: radiation emissions *Am. J. Epidemiol.* **132** 397–412 discussion 413-397

IAEA 2015 *The Fukushima Daiichi Accident* (Vienna, Austria: International Atomic Energy Agency)

Ivanov V K, Karpenko S V, Kashcheev V V, Lovachev S S, Kashcheeva P V, Shchukina N V *et al* 2020 Relationship between follow-up periods and the low-dose ranges with statistically significant radiation-induced risk of all solid cancers in the Russian cohort of Chornobyl emergency workers *Radiat. Environ. Biophys.* **59** 415–21

Johnson C, Kitchen K and Nelson N 2007 A study of the movement of radioactive material released during the Windscale fire in October 1957 using era40 data *Atmos. Environ.* **41** 3921–37

Kamiya K 2021 Health management and care following the Fukushima nuclear power plant accident: overview of Fukushima health management survey *Ann. ICRP* **50** 82–9

Kemeny J G 1979 Report of the President's Commission on the Accident at Three-mile Island: The Need for Change: The Legacy of TMI: The Commission

Kesminiene A, Evrard A S, Ivanov V K, Malakhova I V, Kurtinaitis J, Stengrevics A *et al* 2008 Risk of hematological malignancies among Chornobyl liquidators *Radiat. Res.* **170** 721–35

Kesminiene A, Evrard A S, Ivanov V K, Malakhova I V, Kurtinaitise J, Stengrevics A *et al* 2012 Risk of thyroid cancer among Chornobyl liquidators *Radiat. Res.* **178** 425–36

Kopecky K J, Stepanenko V, Rivkind N, Voillequé P, Onstad L, Shakhtarin V *et al* 2006 Childhood thyroid cancer, radiation dose from Chornobyl, and dose uncertainties in Bryansk Oblast, Russia: a population-based case-control study *Radiat. Res.* **166** 367–74

Little M P, Azizova T V and Hamada N 2021a Low- and moderate-dose non-cancer effects of ionizing radiation in directly exposed individuals, especially circulatory and ocular diseases: a review of the epidemiology *Int. J. Radiat. Biol.* **97** 782–803

Little M P, Wakeford R, Hatch M, Ainsbury E A and Tawn E J 2021b Chromosome aberrations in a group of people exposed to radioactive releases from the Three-mile Island nuclear accident and inferences for radiation effects *Radiat. Res.* **195** 584–9

Lubin J H, Adams M J, Shore R, Holmberg E, Schneider A B, Hawkins M M *et al* 2017 Thyroid cancer following childhood low-dose radiation exposure: a pooled analysis of nine cohorts *J. Clin. Endocrinol. Metab.* **102** 2575–83

McGeoghegan D, Whaley S, Binks K, Gillies M, Thompson K and McElvenny D M 2010 Mortality and cancer registration experience of the Sellafield workers known to have been involved in the 1957 Windscale accident: 50 year follow-up *J. Radiol. Prot.* **30** 407–31

McNally R J, Wakeford R, James P W, Basta N O, Alston R D, Pearce M S *et al* 2016 A geographical study of thyroid cancer incidence in north-west England following the Windscale nuclear reactor fire of 1957 *J. Radiol. Prot.* **36** 934–52

Morton L M, Karyadi D M, Stewart C, Bogdanova T I, Dawson E T, Steinberg M K *et al* 2021 Radiation-related genomic profile of papillary thyroid carcinoma after the Chornobyl accident *Science* **372**

Moysich K B, McCarthy P and Hall P 2011 25 years after Chornobyl: lessons for Japan? *Lancet Oncol.* **12** 416–8

Noshchenko A G, Bondar O Y and Drozdova V D 2010 Radiation-induced leukemia among children aged 0–5 years at the time of the Chornobyl accident *Int. J. Cancer* **127** 412–26

Rivkind N, Stepanenko V, Belukha I, Guenthoer J, Kopecky K J, Kulikov S *et al* 2020 Female breast cancer risk in Bryansk Oblast, Russia, following prolonged low dose rate exposure to radiation from the Chornobyl power station accident *Int. J. Epidemiol.* **49** 448–56

Ron E, Lubin J H, Shore R E, Mabuchi K, Modan B, Pottern L M *et al* 1995 Thyroid cancer after exposure to external radiation: a pooled analysis of seven studies *Radiat. Res.* **141** 259–77

Shinkarev S M 2021 Comparison of thyroid doses to the public from radioiodine following the Chornobyl and Fukushima accidents *Ann. ICRP* **50** 174–80

Togawa K, Ahn H S, Auvinen A, Bauer A J, Brito J P, Davies L *et al* 2018 Long-term strategies for thyroid health monitoring after nuclear accidents: recommendations from an expert group convened by IARC *Lancet Oncol.* **19** 1280–3

Tolstykh E I, Peremyslova L M, Degteva M O and Napier B A 2017 Reconstruction of radionuclide intakes for the residents of East Urals Radioactive Trace (1957–2011) *Radiat. Environ. Biophys.* **56** 27–45

Tronko M D, Howe G R, Bogdanova T I, Bouville A C, Epstein O V, Brill A B *et al* 2006 A cohort study of thyroid cancer and other thyroid diseases after the Chornobyl accident: thyroid cancer in Ukraine detected during first screening *J. Natl. Cancer Inst.* **98** 897–903

Tsuda T, Tokinobu A, Yamamoto E and Suzuki E 2016 Thyroid cancer detection by ultrasound among residents ages 18 years and younger in Fukushima, Japan: 2011 to 2014 *Epidemiology* **27** 316–22

UNSCEAR 1993 *Sources and Effects of Ionizing Radiation: UNSCEAR 1993 Report to the General Assembly with Scientific Annexes* Annex B Exposures from Man-Made Sources of Radiation http://unscear.org/docs/publications/1993/UNSCEAR_1993_Annex-B.pdf

UNSCEAR 2008 *UNSCEAR Sources and Effects of Ionizing Radiations, 2008 Report to the General Assembly with Scientific Annexes* (New York: United Nations)

UNSCEAR 2011 *Annex D: Health Effects Due to Radiation from the Chornobyl Accident, vol II* UNSCEAR 2008 Report (New York: United Nations)

UNSCEAR 2014 *UNSCEAR 2013 Report Annex A: Levels and Effects of Radiation Exposure Due to the Nuclear Accident after the 2011 Great East-Japan Earthquake and Tsunami* (New York: United Nations)

UNSCEAR 2018 *Evaluation of Data on Thyroid Cancer in Regions Affected by the Chornobyl Accident* (New York: United Nations)

UNSCEAR 2021 *UNSCEAR 2020/2021 Report, Sources, Effects and Risks of Ionizing Radiations* **vol 4** (New York: United Nations)

Vasilenko E K, Aladova E E, Gorelov M V, Knyazev V A, Kolupaev D V and Romanov S A 2020 The radiological environment at the Mayak PA site and radiation doses to individuals involved in emergency and remediation operations after the 'Kyshtym accident' in 1957 *J. Radiol. Prot.* **40** R23–R45.

Veiga L H, Holmberg E, Anderson H, Pottern L, Sadetzki S, Adams M J *et al* 2016 Thyroid cancer after childhood exposure to external radiation: an updated pooled analysis of 12 studies *Radiat. Res.* **185** 473–84

Wakeford R, Auvinen A, Gent R N, Jacob P, Kesminiene A, Laurier D *et al* 2016 Re: thyroid cancer among young people in Fukushima *Epidemiology* **27** e20–21

Wakeford R 2019 Does low-level exposure to ionizing radiation increase the risk of cardiovascular disease? *Hypertension* **73** 1170–1

Wing S, Richardson D, Armstrong D and Crawford-Brown D 1997 A reevaluation of cancer incidence near the Three-mile Island nuclear plant: the collision of evidence and assumptions *Environ. Health Perspect.* **105** 52–7

Zablotska L B, Ron E, Rozhko A V, Hatch M, Polyanskaya O N, Brenner A V *et al* 2011 Thyroid cancer risk in Belarus among children and adolescents exposed to radioiodine after the Chornobyl accident *Br. J. Cancer* **104** 181–7

Zablotska L B, Bazyka D, Lubin J H, Gudzenko N, Little M P, Hatch M *et al* 2013 Radiation and the risk of chronic lymphocytic and other leukemias among Chornobyl cleanup workers *Environ. Health Perspect.* **121** 59–65

Zupunski L, Yaumenenka A, Ryzhov A, Veyalkin I, Drozdovitch V, Masiuk S *et al* 2021 Breast cancer incidence in the regions of Belarus and Ukraine most contaminated by the Chornobyl accident: 1978 to 2016 *Int. J. Cancer* **148** 1839–49

IOP Publishing

Principles and Techniques of Radiological Protection

Michael Charles Thorne

Chapter 20

Radiation incidents and accidents involving localised sources of exposure

M C Thorne

20.1 Introduction

Although major accidents, such as those at Chornobyl and Fukushima, tend to be the focus of media and public attention, it is important to recognise that smaller-scale accidents and incidents involving radioactive materials have the potential to give rise to significant radiation exposures to small numbers of individuals (typically ranging from exposure of a single individual to up to several tens of individuals). This chapter reviews the types of accident and incident that have occurred in the past and the lessons that can be learned from them. These accidents and incidents are addressed in the context of industrial uses of sealed radioactive sources, notably in industrial radiology (section 20.2), uses of sealed and unsealed radioactive sources in diagnostic radiology, nuclear medicine and radiotherapy (section 20.3), criticality accidents (section 20.4) and other types of incidents and accidents (section 20.5). Overall, Nénot (2009) has estimated that since the end of World War II there have been at least 600 events that caused significant radiation exposures of about 6000 individuals, with about 70 accidents resulting in one or more fatalities each. Due to these accidents, there have been about 200 deaths due to acute radiation syndrome.

20.2 Industrial uses of sealed radioactive sources

The IAEA (1998a) compiled details of a substantial number of accidents and incidents that had occurred in industrial radiology, with a view to identifying the principal and contributing causes of such accidents and incidents. The various cases identified by the IAEA (1998a) are summarised in table 20.1 arranged by primary cause as identified by the IAEA. Case numbers are as given in IAEA (1998a). This compilation illustrates that many such incidents involve an equipment failure compounded by an inappropriate response often due to a lack of training, experience and/or supervision. In some cases, a lack of appreciation of safety issues

doi:10.1088/978-0-7503-2424-3ch20

Table 20.1. Accidents and incidents that have occurred in the context of operations involving industrial radiology.

Case	Description	Initiating event	Contributing factors and prevention
Primary cause: inadequate regulatory control			
1	Source not transferred from the source changer to the disposal container prior to the return of the source changer to the manufacturer.	The worker forgot to remove the source from the source changer before transportation	The regulatory authority had not established adequate requirements for the disposal, tracking and transport of radiographic sources.
2	A cable-drive exposure device malfunctioned, and the source could not be retracted into the shielded position. The worker reported the problem to his supervisor. However, because it was still possible to perform radiographic exposures, he was instructed to continue work.	Malfunction of the exposure device.	Because of lack of regulatory control, radiography took place without any licensing or regulatory requirements.
3	A pipeline rolled onto the exposure device crushing the lead collimator onto the guide tube and preventing retraction of the source. The operator used a welding torch to melt the collimator and guide tube to free the source, which was of higher activity than was suitable for the exposure device.	Rolling of a pipeline onto the collimator and guide tube of the exposure device.	Absence of regulatory control and a lack of safety culture meant that the operator performed an improper source recovery procedure. He was not trained in source recovery and the operation was undertaken to satisfy production requirements.
4	A previous accident involving the same operator and source assembly as Case 3. The guide tube was not securely fastened and disconnected from the exposure device as the source assembly was projected. In trying to recover from this, the drive cable disconnected from the source assembly, which fell into the sea. Recovery of the source assembly by a diver could have resulted in a large extremity dose.	Improper guide tube connection to the exposure device.	Absence of regulatory control and a lack of safety culture meant that the operator performed an improper source retrieval procedure.

5	Repeat radiography of a weld at the end of a 12-h shift was undertaken by two workers who undertook the operation in a hurry without using a survey meter and personal radiation alarms, because these had been left behind. At the end of the operation, the source was not withdrawn to a shielded position and the unshielded source was placed in the back of their small van.	The source assembly was not properly connected to the drive cable.	The individuals were untrained, unlicensed and had been on an extended work shift. The assistant was not familiar with the equipment being used. The operating organisation deliberately violated regulations in permitting unlicensed and untrained individuals to perform radiography. To prevent such wilful violations the regulatory authority must perform adequate inspections and have an effective enforcement policy.
Primary cause: failure to follow operational procedures			
6	When an x-ray unit was replaced, the interlock on the room door was disconnected and not reconnected. A year later, a radiographer turned on the x-ray set to warm up, then entered the radiography room to make final adjustments to the piece to be radiographed. This involved locating the beam centre with a plumb bob held over the beam port with his thumb. There were no indicators in the room to show that the x-ray unit was activated.	Commissioning of a new x-ray unit did not ensure reconnection of the interlock system.	Procedures need to be in place to ensure that all systems are functional after repair or replacement. A daily check of the operability of the interlock would have revealed the issue and a radiation survey during operations would have detected the high radiation levels and prevented exposure.
7	An exposure device was suspended 6 m above the floor. After exposure, the radiographer used an aerial platform to approach the device and performed a radiation survey to confirm that the source was in the shielded	The source shifted when the exposure device was moved.	A complete radiation survey must be performed after locking the source to confirm that it is in the shielded condition. Also, the radiographer failed to follow instructions because his personal alarming

(Continued)

Table 20.1. (*Continued*)

Case	Description	Initiating event	Contributing factors and prevention
	position. The radiographer then moved the device to lock the source in the shielded position. However, this movement shifted the source such that it was locked in the exposed position. The radiographer was wearing a personal alarming device but had turned it off to conserve battery power.		device had been turned off before the operation was complete.
8	During a piping inspection, a worker failed to connect the source assembly to the drive cable. After the first exposure, the source assembly remained in the exposed position unknown to the radiographer for over 100 exposures. No survey meter or direct reading dosimeter was used in the course of this work.	Failure to make a connection of the drive cable to the source assembly.	Failure to verify the connection at the start of operations. Also, no radiation survey was performed to confirm that the source had been returned to the shielded position after each exposure and direct reading dosimeters were not consulted during the work.
9	When winding a source assembly back into the exposure device, a break in the source assembly resulted in the source remaining exposed. No radiation surveys were undertaken to determine that the source had been returned. The exposure device with the exposed source was sent back to the base and from there to the supplier. However, it was not adequately packaged to compensate for the lost shielding.	Breakage of the source assembly in use.	The radiographer failed to follow instructions that required verification of return of the source to its shielded position. There was also a failure to properly package the damaged source assembly for transport. Furthermore, the required radiation survey was not performed before transporting the source.
10			

	A radiographer propped open the door of a shielded enclosure to allow air to circulate as he set up the next exposure. When he did this for the first time, he switched off the door alarm. This also defeated the enclosure radiation alarm. He and a colleague entered the enclosure with the source in an exposed position without using a survey meter and when the alarms were defeated. Neither of them was wearing a personal dosimeter.	Interlock and radiation alarms for the enclosure were deliberately defeated.	Defeating the door alarm should not have also defeated the radiation alarm. Operational procedures should have been followed to verify that the source had returned to the shielded position and to ensure that appropriate dosimeters were worn.
11	While having difficulties locking an exposure device, a radiographer failed to respond correctly to the warnings given by his alarming device and to an off-scale reading of the radiation survey meter. He failed to follow the operating organisation's emergency procedures by attempting to correct the problem instead of leaving the area immediately.	Difficulties in locking the exposure device.	The operator failed to follow established operating and emergency procedures.
12	In radiography operations to perform 35 exposures of welds on a waste-water storage tank, the source tube and attached collimator were clamped to a stand that was attached magnetically to the tank wall. After cranking out the source for the sixth exposure, the stand fell from the side of the tank. The radiographer straightened out a loop in the guide tube and was then able to	The magnetic holder was not adequate for the application. Also, it was damaged in its fall from the tank, such that the source assembly was disconnected from the drive cable.	No survey was performed to verify that the source had returned to its shielded position. The radiographer based his further actions on the belief that the source was in the shielded position. The deliberate removal of his dosimeter indicates a lack of safety culture in the organisation.

(*Continued*)

Table 20.1. (*Continued*)

Case	Description	Initiating event	Contributing factors and prevention
	fully retract the guide cable and consequently thought that the source was in the camera. He also removed his personal dosimeter to conceal the radiation exposure that he thought that he might have received. He then disassembled the collimator and end cap from the guide tube. When he removed the end cap, the source assembly fell out.		
13	At the end of a session of radiography of a pipeline located in a trench and after retracting the source, the radiographer approached the exposure device from the rear with his survey meter. He surveyed only the rear (drive cable) end of the device before locking it and carrying it to his truck. He then went back to the pipeline to collect other materials, but then noticed a high dose rate on his survey meter as he approached the truck. This was because the source was not in its fully shielded position. As he approached the truck, he became aware that his alarming device was emitting a signal. He had not heard this earlier because of the noise from equipment operating in the area.	The source assembly did not retract fully.	The radiographer did not follow appropriate procedures and carry out a complete radiation survey. Also, the locking mechanism on the exposure device failed due to excessive wear. Daily inspection prior to use would have detected the worn lock and prevented its use.

14	A radiographer and film placer were performing radiography on the insert plates of the hull of a ship. Several hours after starting the work, the radiographer thought he heard the 'go ahead' signal from the film placer and exposed the source. However, the film placer had not sent a message and was still working in the area. He observed that his survey meter was off-scale and exited the area.	Lack of communication procedures.	Production pressures meant that the radiographer was in a hurry to expose the source and did not verify that the film placer was out of the controlled area. Also, the company used unapproved procedures for communication, because the regulations required visual contact or equivalent control.
15	Under adverse weather conditions a radiographer's survey meter ceased to function and the lock on the exposure device became jammed with ice and dirt and could not be locked into the shielded position. He continued to perform radiography and during movement the source shifted out of its shielded position.	Adverse weather conditions resulting in failure of a survey meter and the locking mechanism of an exposure device.	Continuing of radiography operations without correctly operating equipment. Also, the equipment was being used beyond its design capabilities and was not adequate for the environment of use. There was also a failure to use an alarming device or to check a direct reading dosimeter. Either would have made him aware of the high dose rate.
16	A radiographer had difficulties in locking his exposure device. Also, he saw that his direct reading dosimeter was off-scale. However, his survey meter was malfunctioning and did not show any radiation. He struck the locking assembly with a hammer blow to achieve the locked position and left the device unattended on site while he went to fetch a new survey meter. On return, he found that he had the	Difficulties with the locking of an exposure device.	Failure to follow operating procedures when the equipment malfunctioned.

(*Continued*)

Table 20.1. (*Continued*)

Case	Description	Initiating event	Contributing factors and prevention
	same problems with the locking assembly, his direct reading dosimeter was still off-scale, and his new survey meter was not working properly. Also, he had left his thermoluminescent dosimeter behind and so he continued working on the site without it.		
17	A radiographer had a permit to carry out x-ray radiography on a pipeline. A barrier clearly identified the extent of the controlled area and warning signals were given before the work commenced. Several exposures had been made when two men emerged from a hole further along the pipeline. They also had a permit to work and had been inspecting the pipeline internally. They had crawled through the x-ray beam twice while performing their inspection.	Lack of co-ordination of work.	Failure of the radiographer to maintain the required control of the area.
18	An individual received more than 100 Sv to his right hand and more than 10 Sv to his whole body. Although it has not been possible to establish how the exposures occurred, it is hypothesised that they arose when removing source capsules from source assemblies using shielding and a hacksaw.	Performance of unsafe operations with radiographic sources.	Lack of detection of the high doses, indicates that the dosimeter was not worn properly.
19	While a radiographer and his assistant were performing radiography at a temporary	The source assembly became disconnected from the drive cable.	The key factor was the lack of radiation surveys that led to over-exposure of the

	work site, the source became disconnected and remained in the guide tube. After completing two exposures, the radiographer went to develop the film while the assistant moved the exposure device to the next location. While doing this he placed the guide tube round his neck. When he set the equipment down, the source assembly fell from the guide tube. During radiography, no radiation surveys were undertaken, and the radiographer was not wearing his personal dosimeter.		assistant during transport of the exposure device.
20	A source became disconnected from its drive cable and was not properly returned to its shielded container. Subsequently the guide tube was disconnected from the exposure device and the source eventually dropped to the ground where it was picked up by a passer-by. Eventually, eight persons died because of this accident.	The source assembly became disconnected from the drive cable and was left at the work site.	No radiation surveys were performed to ensure that the source had returned to its shielded position. Had there been warning signs on the source these might have been recognised by the passer-by and the consequences of the accident could have been mitigated.
Primary cause: inadequate training			
21	An individual received radiography training over a two-week period and was then allowed to work without supervision. On completion of work he placed the source container on the front passenger seat of his car when returning to the facility. Upon arrival, a fellow worker noticed that the source container shutter was open with the	The radiographer did not have adequate training to work without supervision.	Use of equipment that allowed the source to be inadvertently removed from its container may have contributed to this event.

(*Continued*)

Table 20.1. (*Continued*)

Case	Description	Initiating event	Contributing factors and prevention
	radiation beam directed towards the driver's seat. In addition, burns on the chest of the individual that developed later suggest that the source may have been removed from its container and placed in his shirt pocket for a short time.		
22	A source shifted during the movement of an exposure device. No survey meter seems to have been used. The operator was wearing a personal alarming device which did send out a signal. However, the operator decided that it was malfunctioning and so turned it off for the duration of the work.	The source shifted during movement of the exposure device.	The radiographer did not believe or have trust in the radiation detection equipment. This indicates that he did not understand the radiation hazard or the function of the detection equipment. These failures of understanding can be traced back to inadequate training.
23	An unskilled local labourer was used to undertake radiography of weld joints in gas pipelines. He was neither trained in radiation protection nor was he aware of the hazards associated with ionising radiation. During radiography, the source remained continuously in its exposed condition. As there was no provision made for monitoring, the radiographer was unaware of the hazard.	An untrained individual was allowed to operate the exposure device.	No personnel dosimetry or radiation survey meters were provided, and the untrained individual was allowed to operate without supervision.
24	A source became disconnected in the guide tube. The radiographer covered it with a	The source became disconnected from the drive cable.	The radiographer performed procedures for which he was not adequately trained.

	2.5 cm thick lead sheet, but when he disconnected the guide tube to recover the source assembly, it fell into the mud at the bottom of a ditch. In picking up the source with channel-lock pliers, the source assembly slipped, and the radiographer apparently touched the source while attempting to align the source assembly with the exposure device. He then pushed the source assembly back into the exposure device, but the wrong way round, such that the source capsule was located outside the exposure device. This was indicated by survey meter readings. The radiographer subsequently rectified this under the lead sheet such that the source was secured in the shielded condition. Inspection of the exposure device was performed by the radiation safety officer the day after the incident and it was found that a similar, but incorrect, source assembly had been provided and used in the exposure device.		Specifically, he did not know the proper techniques or the correct tools for source recovery and used an inadequate amount of shielding. Also, daily inspection would have detected an incorrect source assembly prior to use.
Primary cause: inadequate maintenance			
25	A camera-locking mechanism came apart from the exposure device allowing the source to be pulled from the exposure device. On moving the exposure device, the drive cable fell to the ground. Thinking he had a disconnect, the radiographer picked	The roll pins that secure the lock insert were missing.	The radiographer assumed a source disconnect and did not confirm the actual situation with a survey meter. Daily inspection may have detected the looseness of the lock insert prior to performing radiography. The radiographer had

(*Continued*)

Table 20.1. (*Continued*)

Case	Description	Initiating event	Contributing factors and prevention
	up the crank-out and moved his hand quickly to the connector end. He grabbed what he thought was the connector, brought it to within 0.15 m of his face and realised that it was the source. He dropped it, alerted his partner and ran from the area.		removed his film badge prior to the retakes that were the occasion of the incident and this is indicative of a lack of safety culture.
26	The source assembly of an exposure device became wedged in a cut in the guide tube. In attempting to retrieve the source assembly, the radiographer disconnected the guide tube and shook it violently with both hands.	The cut in the guide tube caused the source assembly to become wedged and remain exposed.	Daily inspection would have detected the cut in the guide tube prior to use. Also, source recovery was not performed properly.
27	When a radiographer's assistant was returning an exposure device to store, he tilted the shelf and the source assembly fell to the floor, triggering the radiation alarm. Subsequent investigation showed that the exposure device had not been properly maintained, the spring-loaded latch was not working and was jammed in the unlocked position by dirt. In addition, the radiographer had not placed the shutter control in the off position, nor had he installed the dust cover cap on the front of the device.	Jamming of the lock in the open position.	Failure to fulfil secondary security requirements of placing the shutter control in the off position and installing the dust cover cap.

Primary cause: human error			
28	A longer source assembly was used than that for which the exposure device was designed. This led to high radiation levels at the front of the device when the source assembly was retracted, because the source was not as fully shielded as with a standard-length assembly. Observing this, the two workers using the source panicked, attempted to shield the source with their van, dropped and damaged a radiation alarm, and then attempted to shield the source by dropping the exposure device, guide tube and controls down a hole in the ground by pulling it using the drive cable. In attempting this, they dropped and broke their survey meter. Eventually, they used a personal alarming dosimeter at the front of the exposure device to verify that the source was in the shielded position.	Survey meter off-scale because of the source assembly being used.	The personnel did not attempt a proper radiation survey to confirm the radiation levels. They could have measured the dose rate further from the source and estimated dose rates close to it using the inverse square law. Based on their inappropriate responses, it is apparent that they did not have the proper training to handle such an emergency.
29	An exposure device was placed on the rear bumper of a truck. It fell off when the truck was driven back to the office. The driver recovered it by retracing his route with a survey meter on the dashboard. The exposure device was damaged when it fell on the road and the unshielded source assembly was found close to the exposure device.	Driver forgetting that he had placed an exposure device on the bumper of his truck.	A system of double-checking, such a written checklist, might have prevented this incident.

(*Continued*)

Table 20.1. (*Continued*)

Case	Description	Initiating event	Contributing factors and prevention
30	While performing radiography, the guide tube came in contact with an unshielded welder's high-voltage cable. The guide tube melted, and the source assembly could not be returned to the exposure device by normal means. The faulty welding cable had been noted earlier, but nothing had been done to correct the situation.	Faulty welding cables coming into contact with a guide tube.	Production pressures to continue work with faulty equipment, and two tasks (welding and radiography) being performed in the same area without co-ordination.
31	Two radiography teams were working at opposite ends of a large manufacturing workshop. The x-ray control panels, and the warning systems were in the centre of the workshop out of direct sight of the teams. Therefore, as an added safety measure during the preparation phase the cables from the x-ray tube heads were disconnected from their respective control panels. When the first team had completed preparations, they connected the cable and initiated a 7-minute exposure. However, on returning to the area after the exposure they noticed that the wrong cable had been connected activating the second x-ray set where preparations were still ongoing.	Connection of the wrong control cable and hence activation of the wrong x-ray apparatus.	The radiographer should have verified that the correct connection had been made by his assistant. Also, the control panels should have been designed such that interchangeability was not possible.

Primary cause: equipment malfunction or defect			
32	A radiographer was unable to retract a source because of separation of the inner liner between the control housing and the gear-box assembly.	The inner liner separated from the control housing because of a defect.	Appropriate recovery operations were performed. Malfunctions and defects should be reported to the manufacturer, so that this causes can be determined and appropriate actions taken.
33	A source assembly fell out of an exposure device because of a defective locking mechanism. The radiographer failed to notice the loss of the source, but two young canteen workers noticed the source and picked it up. They saw the danger marking on it and reported their discovery.	Defective locking mechanism for a source assembly.	An advisory note was issued to alert other users to potential problems with the lock.
34	A source failed to retract because the male connector of the drive cable was broken. The radiation protection officer was informed, and the source was retrieved correctly.	Breakage of the male connector of a drive cable causing disconnection of the source assembly.	The routine radiation survey immediately detected the problem and appropriate emergency procedures were followed.
35	A leaking source caused contamination of the guide tube, drive cable and radiographer's hand and forearm.	A faulty weld had remained undetected during manufacture of the source.	Inadequate quality control by the manufacturer. Also, use of double encapsulation significantly reduces the likelihood of leakage.
Primary cause: design flaws			
36	During a routine survey, a narrow beam of radiation was found emerging from an exposure device. This arose because the two halves of the uranium shield had separated because of over-torqueing of the retaining bolts.	Over-torqueing of shield retaining bolts during refurbishment.	The manufacturer was notified, and the design was changed to a one-piece uranium shield. Operators were notified of the potential problem with the previous shield design.

(Continued)

Table 20.1. (*Continued*)

Case	Description	Initiating event	Contributing factors and prevention
37	A source fell out of a pneumatically operated exposure device and was not detected by the radiographer because of a faulty survey meter. A construction supervisor picked up the source, assumed it to be a component of a mobile crane and took it home with him. Loss of the source was discovered next day, with subsequent identification and recovery.	A design fault in a pneumatically operated exposure device.	Failure to perform a proper survey because of use of a faulty survey meter.
38	A source assembly became disconnected from the source drive cable in the guide tube, because a hook mechanism was used that could easily become disconnected.	Disconnection of a source assembly from the source drive cable because of a poorly designed connector.	The regulatory authority recommended changing the design to a ball-type connection.
Primary cause: wilful violation			
39	An attempt was made to interchange sources between two pneumatically operated exposure devices projecting the sources into the guide tubes and exchanging the tubes. However, control of both sources was lost and attempts at recovery resulted in one of the sources falling to the ground and not being discovered for some time. Two of the radiographers involved removed their personal dosimeters when they knew that a problem had arisen.	Performance of a source exchange that was not carried out according to approved procedures.	Confusion during the source exchange and recovery operation added to the doses received. Removal of required personal dosimeters was a wilful violation of established requirements.

40	A source in a lead shield and an empty exposure device were stolen from a storage pit and the theft went unnoticed for several days. The source assembly changed hands many times and was ultimately recovered at the shop of a scrap dealer.	Theft of radioactive material.	Adequate physical security should be provided.
41	An operating organisation allowed untrained individuals to perform radiography. As a result, two untrained workers, not supervised by a radiographer, received over-exposures from a source disconnect. The individuals were not familiar with the operation of the equipment or survey instruments.	Use of untrained individuals to operate radiographic equipment.	Wilfully negligent actions by the operating organisation in using untrained individuals were compounded by a lack of supervision.
42	A company allowed an untrained individual to perform radiographic work at night. During the work, he was unable to retract the source. Recognising the problem, he tried to communicate with the radiation protection officer of the facility but was unable to reach him. He finally contacted personnel of the regulatory authority and the event was resolved.	Use of an untrained individual to operate radiographic equipment.	Although the individual was not trained, he recognised the problem and took appropriate actions to resolve it, thus avoiding over-exposure.
43	A radiographer was performing a 1-h panoramic exposure of a joint of a pipe. At 10:30 pm he posted a watchman to keep a watch on the source while he went to check whether a second joint was ready for radiography. When he returned after thirty	Deliberate movement of a source.	Inadequate security of the source was compounded by a delay in reporting its loss that contributed to the excessive dose incurred by the construction worker.

(Continued)

Table 20.1. (*Continued*)

Case	Description	Initiating event	Contributing factors and prevention
	minutes, he found the source had gone. He did not report the matter to his supervisor immediately and he did not monitor the area because he did not know how to use the survey meter. The matter was reported to the supervisor next morning and the source was found 4 h later about 8 m from its original location, i.e. 16 m above the ground. Subsequently, a construction worker who had slept on the mesh platform from which the source had been found hanging developed skin lesions.		

and a poorly developed safety culture were primary issues leading, for example, to individuals deliberately defeating safety interlocks or other safety systems, or going beyond their competence, e.g. in attempting source recovery actions. Subsequent accidents of this type are described in more recent IAEA reports (IAEA 2004a, 2009).

The radiological consequences of the various incidents summarised in table 20.1 differed enormously depending upon whether exposure arose from an x-ray unit or a source, the activity of the source, and the duration and other aspects of the exposure situation. In some cases, the doses received were as small as a few millisieverts, whereas whole-body doses of more than 1 Sv were received in some accidents, leading to the induction of acute radiation effects in some exposed individuals (cases 2, 6, 8, 20, 21, 24, 37 and 43) and leukaemia in others (case 18). Acute effects were typically skin burns, either of the extremities or of the chest where source assemblies had been placed in shirt pockets.

20.3 Incidents associated with radiotherapy and nuclear medicine

Some examples of the types of incident that can arise in radiotherapy are described below. These incidents illustrate that errors in equipment calibration or operation are of particular concern, because they have the potential to adversely impact substantial numbers of patients treated sequentially using the incorrectly calibrated or malfunctioning equipment.

A detailed report on the accidental over-exposure of radiotherapy patients in San José, Costa Rica has been published by the IAEA (1998b) and a summary has been provided by Nénot (2009). Errors in the calibration of a new Co-60 radiotherapy source resulted in exposures of 114 patients that were 50%–60% higher than prescribed. The error resulted from a confusion in the time units adopted between the second and 1/100 of a minute. The error was recognised only after about one month when a large series of patients exhibited abnormal signs of radiation injuries. The over-exposures resulted in severe consequences, sometimes worsened by sub-optimal treatment procedures. Severe effects in four patients included quadriplegia, paraplegia, spinal-cord demyelination, and severe digestive and cutaneous injuries. Marked effects occurred in a further 16 patients. Nénot (2009) estimates that among 61 deaths within two years of the accident, 13 could be directly attributed to their over-exposures and four to radiation-induced complications. In addition, among the 51 patients still alive two years later, two were suffering from late severe complications and 12 exhibited marked and disabling effects.

In its review of this accident, the IAEA (1998b) drew some important conclusions relating to radiotherapy.

- Radiation accidents with severe and even fatal consequences can and do occur in medical facilities.
- Human error is the most common cause of radiation accidents.
- Prior to the accident, external auditing had detected the poor quality of record keeping, the lack of redundancy in procedures, and inadequate

education and training; had actions been taken on these findings, the accident might have been prevented.
- Investigation of radiation accidents generally reveals faults that should be corrected.
- When there is high incidence/severity of acute effects during radiotherapy treatment, the treatment should be stopped, and the source calibration checked immediately.
- In radiotherapy accidents, the tumour dose may not be the parameter of primary interest. Often, the biologically equivalent 2 Gy per fraction dose (the dose that would be biologically equivalent had it been delivered in fractions of 2 Gy) to sensitive structures such as the spinal cord, heart and intestine is more important.
- Accepted radiotherapy protocols have little margin for error, since both normal and malignant cells are killed; significant overdoses (errors much greater than 10%) will result in an unacceptable incidence of severe consequences.
- Radiotherapy administered in fewer than the normal number of treatments with higher doses per treatment results in an excessive number of early and, particularly, late complications.
- When radiation therapy sources are replaced, the calibration should be done by appropriately trained persons and independently checked.
- A properly operating machine does not guarantee good radiotherapy treatment; adequate ancillary equipment, education and training, staffing and management are essential.
- Regulations should cover the level of training and competence required to deal with potentially hazardous radiation sources.
- Specific training should be given after an individual working in a radiotherapy unit has received a thorough basic education and should not consist of simply attending occasional short courses.
- Radiotherapy records should be uniform, clear, consistent and complete.
- Early and reliable information and communication are crucial for the good management of radiation accidents.
- Radiation accidents can have major short- and long-term psychosocial consequences.

In an earlier incident in Columbus, Ohio between 1974 and 1976, an incorrect calibration of a Co-60 source led to 426 patients receiving doses that were 15%–45% higher than prescribed. Among the 183 patients still alive one year after treatment, more than one third had severe complications of the central nervous and gastrointestinal systems. Conversely, in Stoke-on-Trent, UK between 1982 and 1991, 1045 patients received doses 5%–35% lower than expected, but the consequences of these under-exposures are difficult to evaluate. Various other accidents have occurred in hospital contexts arising from errors in programming, from disposal of a catheter containing an Ir-192 brachytherapy source, from adaptation of a computerised treatment to incorporate additional shielding and from malfunctioning of a linear accelerator used for the treatment of breast cancer (Nénot 2009). This malfunctioning resulted from a transitory loss of electrical power that caused an automatic shutdown of the accelerator. Following the restoration of electrical power, the machine was restarted after its controls had been checked. However, subsequent dosimetry measurements revealed that the output of the accelerator was significantly higher than expected. Further checks revealed that the dose monitoring system was not functioning properly, and that one of the electronic components of the safety interlock system was damaged. Thus, the over-exposures resulted from faults developed in two separate electronic circuits (IAEA 2004b).

20.4 Criticality incidents and accidents

Criticality accidents were relatively common in the first two decades after World War II, but only six process-related accidents have occurred since 1966. In total, 22 accidents have occurred in process facilities, whereas 38 have occurred during critical experiments or in operations with research reactors. Details of all 60 accidents are presented in McLaughlin *et al* (2000). The characteristics of process accidents are different from those arising in critical experiments or in work with research reactors. Process facilities aim to avoid criticality accidents through physical and administrative controls. Thus, for example, controls on degree of enrichment or the geometry of vessels containing fissile solutions can be used to avoid critical or near-critical configurations. Nevertheless, in process facilities and under normal working conditions, operatives, who are not generally specialists in criticality, can be close to potentially critical configurations. Therefore, if physical and/or administrative controls fail and a criticality accident occurs, large acute radiation doses may be received by persons local to the accident. In contrast, research reactors and experimental configurations studying issues related to criticality are usually staffed by operating personnel with expertise in criticality physics. Furthermore, although the staff carry out hands-on work with fissile materials, operations in which critical or near-critical configurations are planned to arise are generally undertaken remotely. Thus, in the event of criticality accidents the staff will often incur only small radiation doses.

When a criticality accident occurs, the primary consequence is the release of one or more pulses of neutrons. Multiple pulses may occur, for example, if the first pulse changes the geometry of the system decreasing its reactivity but the geometry changes back after the pulse with an increase in reactivity. For example, in an aqueous system the heat generated by the first pulse of neutrons may evaporate some of the liquid present, decreasing the moderation and hence the reactivity of the system. However, if further liquid then enters the system, the reactivity will increase, and a second pulse of neutrons may be generated. The generation of a second pulse is illustrated by an accident at the Oak Ridge Y-12 plant in 1958 (Process Accident 4 in McLaughlin *et al* (2000)). In this accident, uranyl nitrate solution was drained from small-diameter vessels into a 55-gallon drum, which is a compact geometry that is more conducive to criticality. After about 15 min, an initial criticality arose, contributing about 6×10^{16} fissions. It is thought that this first pulse generated enough radiolytic gas bubbles in the solution to reduce the reactivity. However, about 15 s later, when these bubbles had left the system, a second and larger pulse occurred. Subsequent smaller pulses occurred over the next 2.6 min, until the system started to boil, which reduced its reactivity. The total number of fissions in this accident was about 1.3×10^{18} fissions, i.e. a factor of 20 larger than in the first pulse.

McLaughlin *et al* (2000) summarise the main characteristics of the 22 process accidents that have occurred. Of these, 21 occurred with fissile material in solutions or slurries, whereas only one occurred with metal ingots and none with powders. Eighteen of the accidents occurred in manned, unshielded facilities, but no accidents occurred during the transport or storage of fissile material. Adverse health effects were mainly incurred by process operatives and included nine fatalities plus three survivors who required limb amputations. In contrast, only one accident resulted in measurable fission-product contamination and only one resulted in measurable (but small) doses to members of the public. Total fissile yields for these accidents ranged from 1×10^{15} fissions up to 4×10^{19} fissions (see Table 10 of McLaughlin *et al* (2000)).

Reactor and critical experiment accidents have been much more diverse than process accidents, though with total fission yields spanning much the same range (see Table 11 of McLaughlin *et al* (2000)). The numbers of accidents that have occurred in different types of system are:

- Fissile solution systems: 5;
- Bare and reflected metal systems: 15;
- Moderated metal and oxide systems: 13;
- Miscellaneous systems: 5.

A total of 12 fatalities have occurred in accidents involving bare and reflected metal systems and in moderated metal and oxide systems. These fatalities typically occurred when near-critical assemblies were being constructed or modified by hand by individuals with expertise in criticality. They typically reflected errors of judgement by the operatives often combined with violations of operating procedures. In addition, in one accident at a research reactor, saturation of neutron-detection instruments led to misinterpretation of the state of the reactor.

In practice, criticality safety is generally assured by performing Monte Carlo calculations of neutron transport for geometrical configurations of fissile material, moderator and other materials appropriate to the application. These calculations are used to determine the effective neutron multiplication factor, k_{eff}, taking both prompt and delayed neutron emissions into account. Sub-critical systems have a k_{eff} value of <1.0 and there is generally a requirement to ensure that the k_{eff} value is below some limiting value of less than 1.0 to take account of uncertainties in the calculations. The Monte Carlo codes MONK and WIMS are examples of the types of code available (see https://www.answerssoftwareservice.com/expertise/criticality.html). Experimental studies of critical and sub-critical assemblies are used to provide data to validate these codes and provide estimates of their accuracy in various contexts.

Criticality safety assessment is a specialist discipline. It requires not only an ability to use the available software, but also a deep understanding of the physics involved, because an important aspect is the identification of the geometrical configurations of fissile material, moderator and other materials that could arise under normal and accident conditions, and that would result in the highest reactivity of the system under investigation.

20.5 Larger-scale accidents

Major accidents at nuclear power plants and reprocessing facilities are discussed in chapter 19, whereas accidents involving one or a few individuals are described in the earlier subsections of this chapter. However, this leaves still to be considered various accidents that do not readily fit into the above categories. Such accidents often involve exposure of a significant number of individuals and/or relatively extensive contamination of the environment. Because of their significance, each such accident is usually the subject of a report published by the International Atomic Energy Agency (IAEA) and developed in conjunction with other relevant agencies that reviews both the characteristics of the accident and the response to it. Here, information on some of these various accidents is collated in summary form to illustrate the various contexts in which they may occur.

20.5.1 Loss or abandonment of radiotherapy sources

Radiotherapy sources typically contain large amounts of gamma-emitting radionuclides and are susceptible to being lost, abandoned or stolen. These sources may subsequently be disassembled. Thus, in Juarez, Mexico in 1983 a teletherapy machine containing 15.6 TBq of Co-60 in 6000 pellets was dismantled. During disassembly, 800–1000 pellets were spilt into a truck bed and recycled via a scrap yard and iron foundries. Consequently, about 4000 members of the public were exposed to non-trivial doses, with 800 receiving more than 50 mSv and eight between 1 and 7 Gy. Because of the protracted nature of the exposures no deaths occurred. However, rehabilitation of the environment took four months and necessitated control of 17 000 houses of which 800 were destroyed because they were judged impossible to decontaminate (Nénot 2009).

In 1987, in Goiânia, Brazil, a similar accident resulted in much more severe consequences because the 51 TBq source was made of highly soluble Cs-137 chloride. Early symptoms of acute radiation syndrome in two scrap dealers were attributed to tropical disease, before the radiological nature of the incident was suspected by an oil prospector. Because of the wide distribution of the radioactive material large numbers of people had to be monitored for contamination. In total, four victims died and 28 had to undergo surgery, comprising grafts or amputations. Internal doses were of significance, particularly amongst children who had played with the ‘magic’ luminescent powder. Large quantities of radioactive waste were generated, and it took ten years to find a viable solution for dealing with this material (IAEA 1988, Nénot 2009).

A serious radiological accident occurred in Istanbul, Turkey, in December 1998 and January 1999 when two packages used to transport Co-60 teletherapy sources were sold as scrap metal. The persons who purchased the two packages opened them and broke open the shielded containers. Consequently, they developed acute radiation syndrome. In total, 18 persons (including seven children) were admitted to hospital. Of these, ten adults exhibited clinical signs and symptoms of acute radiation exposure (IAEA 2000a).

In respect of medical treatment, the IAEA (1988) concluded that the treatment of casualties of radiological accidents is extremely varied and complex. They must be cared for in hospitals by staff who are routinely engaged in the treatment of patients at risk from cancer, immunosuppression and blood dyscrasias. Therefore, provision should be made in radiological emergency plans for immediate assistance from medical specialists trained to handle such patients. Recognition of the nature of radiation injury, however, depends on the general education of workers as well as on trained health professionals.

20.5.2 Theft of sources

The serious effects that can arise from the theft of a large source are illustrated by an incident in Samut Prakarn, Thailand in 2000. A 15.7 TBq Co-60 teletherapy machine, purchased in 1974 but never used, was stolen by four scrap collectors and rapidly resold. Over the next three weeks, 13 people were exposed. Ten of them were hospitalised for haemorrhages and burns, and three died during the second month following the theft. Overall, about ten people required intensive medical care and some required amputations. Also, a further 44 individuals exhibited symptoms that might have been due to over-exposure (Nénot 2009). An important contributing cause of this accident was that there was no clear or effective end of life provision for the source (IAEA 2002). Because there were no provisions for returning the disused source to the manufacturer, the hospital contracted with a local supplier of similar equipment to handle the disused device (including the disused source) and sold it without notifying the regulatory authority. The local supplier did not have satisfactory arrangements for prompt disposal of the source, which was kept in insecure storage (IAEA 2002). In respect of medical treatment, it is notable that exposed individuals reported to hospitals and outpatient clinics, but their original symptoms (diarrhoea, nausea and vomiting) were not initially diagnosed as being caused by radiation exposure (IAEA 2002). Also, there was apparently no adequate biological dosimetry available to assess the probable range of the radiation doses received by the individuals involved (IAEA 2002).

A theft with more limited consequences occurred in Tammiku, Estonia in 1994, when an abandoned, military 1.6 TBq Cs-137 source was stolen from a waste repository and kept in a house for 27 days. During this period, the death of a young man who had kept the source in his pocket before taking it home was attributed to toxaemia. However, when the victim's nephew was found to be suffering from a haematological disorder and burns, the origins of the diseases was discovered (IAEA 1998c, Nénot 2009). Biological dosimetry, based on scoring dicentric chromosomes, was found to underestimate the dose received by the victim's nephew, indicating its limitations in cases of inhomogeneous irradiation (IAEA 1998c).

20.5.3 Accidental discovery of sources

Abandoned or inadequately secured sources may be discovered by individuals who fail to recognise their nature and the associated hazard. For example, near the village of Lia, Georgia, three woodcutters found two metallic cylinders around which the snow was melting. During the following cold night, they used these cylinders (which were thermoelectric generators each containing 1.3PBq of Sr-90) for heating. After a few hours, all three presented prodromal signs of the acute radiation syndrome and extensive burns appeared within one to two weeks. Nevertheless, the radiological cause of their injuries was not recognised until three weeks later (Nénot 2009). Similarly, in Xinzhou, Shenxi, China, a worker found a 100 GBq Co-60 source in a deep hole in the ground on a site formerly used for the storage of radioactive wastes. He, his father and his brother died after two or three weeks, with their deaths attributed to an infectious disease. It was only after their deaths that the real cause of their injuries was determined, and the source was not recovered until more than two months after its original discovery (Nénot 2009). In Lilo, Georgia, nine servicemen of the Lilo Training Detachment of Frontier Troops developed local radiation-induced skin diseases on various parts of their bodies and a subsequent inquiry established that this was due to the improper and unauthorized abandonment of twelve Cs-137 sources at the Lilo site (IAEA 2000b). The lack of documentation relating to the plant suggested that there had been only limited contact between the former owners of the sources and the current operating organization (IAEA 2000b).

20.5.4 Entry into high-dose-rate areas

In some contexts, individuals may decide to enter high dose rate areas to remediate fault conditions. For example, at an industrial irradiation facility near San Salvador, repackaged medical products were sterilized by irradiation using an intensely radioactive Co-60 source in a movable source rack. The accident happened when this source rack became stuck in the irradiation position. The operator bypassed the already degraded safety systems and entered the radiation room with two other workers to free the source rack manually. All three workers were exposed to high radiation doses and developed the acute radiation syndrome. Although their initial hospital treatment was effective in countering the acute effects, the legs and feet of two of the three men were so seriously injured that amputation was required. Moreover, the worker who had been most exposed died six and a half months after the accident, his death being attributed to residual lung damage due to irradiation, exacerbated by injury sustained during treatment (IAEA 1990). This accident illustrates the potential significance of allowing both physical protection and safety procedures to progressively degrade (IAEA 1990). This accident happened in 1989 and the following year a similar accident occurred at a commercial irradiation facility at Soreq, Israel. In this accident, only one operative entered the irradiation room. He received a whole-body dose of between 10 and 20 Gy, and died, because of haemopoietic and gastrointestinal effects, 36 days after the exposure (IAEA 1993). A similar accident occurred in an irradiation facility in the town of Nesvizh, Belarus in October 1991 (IAEA 1996a). An accident with more limited radiological consequences occurred in Hanoi in 1992, in which an individual adjusted the position of a sample at an electron accelerator without being aware that the beam was switched on for part of the time that he was in close proximity to the sample. During the sample repositioning, his hands were within 0.03–0.3 m of the target. Consequently, he experienced severe injuries to both of his hands, and one had to be amputated (IAEA 1996b).

20.6 Malicious activities

It should also be kept in mind that both sealed and unsealed sources may be used maliciously to cause injury or death. Two attempted murders in the 1970s involved the use of radioactive sources. In one, a man deliberately exposed his son to a 37 GBq Cs-137 source that he was holding for oil prospecting and in the second a reprocessing plant employee, intending to cause severe injuries to a colleague, placed under a car seat a radioactive bar that he had stolen from a workshop. A Russian publication reports at least four criminal actions involving gamma sources (Krasniouk 2004, Nénot 2009). More recently, Po-210 was used in the poisoning of Mr Alexander Litvinenko (Harrison *et al* 2017a), which led also to collateral contamination of other individuals, including eight that received effective doses in the range of 10 to 100 mSv (Harrison *et al* 2017b).

References

Harrison J, Fell T, Leggett R, Lloyd D, Puncher M and Youngman M 2017a The polonium-210 poisoning of Mr Alexander Litvinenko *J. Radiol. Prot.* **37** 266–78

Harrison J, Smith T, Fell T, Smith J, Ham G, Haylock R, Hodgson A and Etherington G 2017b Collateral contamination concomitant to the polonium-210 poisoning of Mr Alexander Litvinenko *J. Radiol. Prot.* **37** 837–51

IAEA 1988 *The Radiological Accident in Goiânia* (Vienna: International Atomic Energy Agency)

IAEA 1990 *The Radiological Accident in San Salvador* (Vienna: International Atomic Energy Agency)

IAEA 1993 *The Radiological Accident in Soreq* (Vienna: International Atomic Energy Agency)

IAEA 1996a *The Radiological Accident at the Irradiation Facility in Nesvizh* (Vienna: International Atomic Energy Agency)

IAEA 1996b *An Electron Accelerator Accident in Hanoi, Viet Nam* (Vienna: International Atomic Energy Agency)

IAEA 1998a *Lessons Learned from Accidents in Industrial Radiography, Safety Reports Series No. 7* (Vienna: International Atomic Energy Agency)

IAEA 1998b *Accidental Overexposure of Radiotherapy Patients in San José, Costa Rica* (Vienna: International Atomic Energy Agency)

IAEA 1998c *The Radiological Accident in Tammiku, Estonia* (Vienna: International Atomic Energy Agency)

IAEA 2000a *The Radiological Accident in Istanbul* (Vienna: International Atomic Energy Agency)

IAEA 2000b *The Radiological Accident in Lilo* (Vienna: International Atomic Energy Agency)

IAEA 2002 *The Radiological Accident in Samut Prakarn* (Vienna: International Atomic Energy Agency)

IAEA 2004a *The Radiological Accident in Cochabamba* (Vienna: International Atomic Energy Agency)

IAEA 2004b *Accidental Exposure of Radiotherapy Patients in Bialystok* (Vienna: International Atomic Energy Agency)

IAEA 2009 *The Radiological Accident in Nueva Aldea* (Vienna: International Atomic Energy Agency)

Krasniouk V I 2004 Radiation injuries in man after criminal exposure *IRPA 11* (Madrid: International Congress of the International Radiation Protection Association (IRPA))

McLaughlin T P, Monahan S P, Pruvost N L, Frolov V V, Ryazanov B G and Sviridov V I 2000 *A Review of Criticality Accidents, 2000 Revision, Los Alamos National Laboratory Report LA-13638* (New Mexico: Los Alamos National Laboratory)

Nénot J-C 2009 Radiation accidents over the last 60 years *J. Radiol. Prot.* **29** 301–20

IOP Publishing

Principles and Techniques of Radiological Protection

Michael Charles Thorne

Chapter 21

Emergency preparedness, planning and response

K I Pearce

21.1 Introduction

The requirement for emergency preparedness is ubiquitous in legislation and international guidance across the world. Much of this has been informed by the bitter experience of sub-optimal responses to previous events and by site safety cases that identify a range of possible events that cannot be totally engineered away.

The early development of nuclear power and nuclear armaments involved numerous experiments with fissile material, reflectors and moderators. In this field, and in others, there was intense competition between the major powers both for bragging rights and for strategic advantage at a time when war-damaged economies struggled to provide resources. A substantial number of critical accidents, with fatalities resulted (see section 21.4).

21.2 Learning from experience

Through the late 1940s and the 1950s, the development of nuclear reactors was rapid with new designs and more powerful reactors being designed and built in great haste. In these early stages, the main plan for the protection of the public was to place nuclear facilities in areas with low population densities. In 1950, the WASH-3 report suggested an exclusion area around a nuclear reactor given by the following rule of thumb (US NRC Undated):

$$R = 0.016 \quad P^{0.5} \tag{21.1}$$

Where R (km) is the radius of exclusion area (no residents permitted) and P (kW) is the reactor power (defined as its thermal output).

For Sizewell B (P = 3479 MW), in Suffolk in the UK and 2.5 km from Leiston (population 5500) this gives an exclusion zone of radius about 30 km.

This arrangement was thought to provide adequate protection of the public even when the USA authorities deliberately took an uncontained reactor prompt-critical in March 1955 (the BORAX reactor).

doi:10.1088/978-0-7503-2424-3ch21

Caution ensured that early nuclear reactors had emergency shutdown systems that would quickly insert neutron absorbers into the assembly if there were any signs of an unplanned rise in neutron population, but otherwise safety standards were much less stringent than those now adopted. As operating experience was gained, accidents and near misses were recorded and it was clear that ever larger exclusion zones were not going to be acceptable as reactor power increased. Containment buildings were introduced to reduce the impact of potential accidents and allow more powerful reactors to be built in areas that did not satisfy the above criterion. Greater emphasis was also placed on prevention of accidents by conservative design and the introduction of more safety systems. An important driver for safety by design, rather than by isolation, was the US nuclear submarine programme where there are obviously highly constrained limits on the distances between the crew and the nuclear reactor that can achieved. Later, less restrictive, criteria placed limits on population numbers in a series of concentric circles centred on the proposed site.

Despite developments in reactor safety, in 1954 the influential US congressional oversight Joint Committee on atomic energy was told that '*no matter how careful anyone in the atomic energy business may try to be, it is possible that accidents may occur*'. In 1956 they were told that because of technical uncertainties and limited operating experience, '*the determination that the hazard is acceptably low is a matter of competent judgment if safeguards are included in facility design against all possible accidents having unacceptable consequences, then it could be argued that any site would be acceptable assuming that safeguards would not fail and that some dangerous accidents had not been overlooked.*'

Chapters 19 and 20 discuss various accidents that occurred in the past. These can inform our view of the need for emergency preparedness and help us formulate a functional specification for an emergency scheme.

On 12 December 1952 at the NRX reactor core at Chalk River, Ontario a series of operator errors and communication failures led to control rods being removed and a power excursion followed by a hydrogen–oxygen explosion. Part of the core melted, fission products were released to the atmosphere and contaminated water was dumped. This is possibly the earliest recorded reactor accident.

In 1956 the draft statute establishing the IAEA was discussed at the United Nations Headquarters in New York. It came into force on 29 July 1957.

The 1957 Kyshtym disaster, an explosion at an inadequately cooled tank containing 60–70 tons of heat-generating liquid radioactive waste, released an estimated 800 PBq (20 MCi) of radioactivity making it the third worst nuclear accident to date behind Chornobyl and Fukushima. This heavily contaminated the local area and the Techna River (already contaminated by previous discharges) and produced a plume that deposited radionuclides, notable Cs-137 and Sr-90, over a wide area up to 350 km downwind, an area now called the East-Urals Radioactive Trace (EURT). There were mass evacuations and crude clean-up attempts following the accident, but any potential learning for the west was delayed by decades of cover-up from the Soviet authorities, including the designation of the area as a wild life reserve with very limited access.

The Windscale works in Cumbria, UK was the site of two primitive graphite-moderated, once-through air-cooled nuclear reactors used for research purposes but mainly to produce material for nuclear weapons. In October 1957, when an annealing process to release Wigner energy led to one of the cores catching fire, there was little in place that we would understand as emergency preparedness, but the site personnel improvised a response, put the fire out and put in place activities to reduce radiation doses to the public; notably the destruction of milk produced in the area contaminated by the fall out. It was an extremely uncomfortable experience. Tom Tuohy, the deputy general manager at the site, is quoted as saying '*Mankind had never faced a situation like this; there's no-one to give you any advice*'. You can read a detailed first-hand account of the off-site health physics response in (Dunster *et al* 1958).

In the aftermath of the fire, the 1959 the Nuclear Installations Act (HMG 1965) was passed to enhance the licensing and safety management of the UK civil nuclear industry. This established the Nuclear Installations Inspectorate (NII), forerunner of the current Office for Nuclear Regulation (ONR) and gave them the duty to 'attach to [the site licence] such conditions as the authority considers necessary or desirable in the interests of safety'. All licensed sites in the UK are subject to the same site license conditions (ONR 2007).

During the Three Mile Island (TMI) incident in March 1979, against the background of technical debates about the size of hydrogen bubbles in the containment and the potential consequences of their ignition, schools were closed, residents were advised to stay indoors and farmers told to keep animals under cover and feed them with stored feed. Two days after the onset of core damage, the local Governor advised evacuation of pregnant women and pre-school children within 5 km of the site. Later this distance was increased to 20 km. Instead, many more people (about 140 000) left the area with most returning over the next three weeks. Among other lessons, TMI taught the industry to pay more careful attention to the messages given out to the public, considering the content, tone and consistency of those messages. This led to mechanisms to coordinate the media teams of the responding organisations and to organise media briefings. It was also shown that it was necessary to improve the coordination of responders, and that this should be done off-site (one reason being that the site's telephone exchange could not cope with the communications demand). In the UK, this led to the establishment of Local Emergency Centres where information and understanding could be exchanged and joint decisions about the off-site response reached. These were forerunners of the current Strategic Coordination Centres (SCCs).

The Chornobyl accident on 26 April 1986 was a shock to the nuclear industry. This accident refocussed emergency planning thought on the possibility of a large accident although it was still rather complacent as it turned out. It was thought very unlikely that such an accident could occur in the western world because of better reactor design and construction, better operator training and better compliance.

The Piper Alpha accident on a North Sea oil rig on 6th July 1988 resulted in 165 deaths out of the 226 on board. The enquiry after the event concluded that there was poor response management in the habitation module and critical delays in switching off pumps on neighbouring rigs while someone with authority to order them stopped was found. Following this finding, the nuclear industry strengthened the authorisations of their on-duty emergency responders to ensure they had the authority needed to make rapid decisions on-site.

The UK experience of fuel protests (2000), flooding (2000), foot and mouth (2001) and the September 11th terrorist attacks in the USA in 2001 led, in the UK, to a realignment of the defence-based Civil Defence Act with the broader Civil Contingencies Act 2004 placing revised duties on local authorities, emergency services and other organisations.

Fukushima reinforced the message that poor site, local and national coordination can make the response look inept and it finally brought home the potential for large accidents. In the aftermath there was a significant review of the safety of nuclear facilities in extreme events (ENSREG Stress reports) and the new UK Radiation (Emergency Preparedness and Public Information) Regulations (REPPIR-19) have moved outline planning, that is the consideration about how the response to a severe accident might be managed, from 'good practice' to a regulatory requirement. The UK government has also reorganised the committee structure managing its preparations for a nuclear emergency with severe accidents in mind.

In its analysis of the lessons learned from the responses to radiation emergencies from 1945 to 2010, the IAEA (2012) concluded that the lessons demonstrate the importance of establishing, maintaining and exercising appropriate arrangements for dealing with radiation emergencies at national, local and site level. Arrangements should include the management of orphan sources and sources in the scrap metal economy and medical personnel should be able to identify radiation injuries and should report them to national bodies. Arrangements should also allow for clear and informed messages to be provided to the public.

Current siting criteria for the USA can be found in chapter 100 of the NRC regulations (NRC 2020). This determines an exclusion area and a low population zone based on limiting the dose that might be received by a person at the outer edge of these areas to 25 rem (250 mSv) whole body or 300 rem (3 Sv) thyroid dose as a result of '*a major accident, hypothesized for purposes of site analysis or postulated from considerations of possible accidental events, that would result in potential hazards not exceeded by those from any accident considered credible. Such accidents have generally been assumed to result in substantial meltdown of the core with subsequent release of appreciable quantities of fission products*'. For the exclusion zone the exposure time is two hours, for the low population zone the exposure time is the full duration of the release.

21.3 The protection of the public in the event of an atmospheric release

The protection of the public in the event of a nuclear or radiological emergency is usually taken to mean the reduction in their additional radiation exposure as a result of the accident although in the wider sense the protection of their mental and economic well-being and their way of life are all important.

The IAEA defines an urgent protective action as '*a protective action in the event of an emergency which must be taken promptly (normally within hours) in order to be effective, and the effectiveness of which will be markedly reduced if it is delayed*'. In off-site plans these can be 'automatic protective actions', ones that are implemented without discussion on declaration of an emergency, and 'deliberated protective actions', those that are implemented after discussion.

Understanding of the available protective actions comes from investigation of the pathways from release to effective dose (see chapter 10). The dose routes of interest are generally (figure 21.1):

- Inhalation, which requires that the person is within the cloud of radioactive material;
- Cloud beta/gamma, which requires that the person is close enough to the cloud to be exposed to ionising radiations emanating from it;
- Ground gamma dose, which requires the exposed person to be within the range of the ionising radiations emanating from surfaces upon which the radioactive material has deposited or to which it has migrated;
- Ingestion dose, from the consumption of foods contaminated by radioactive material.

In the short-term, ingestion dose is dominated by deposition onto green vegetables and fruit (which can be significantly reduced by rinsing the food or discarding the outer layers). Within a few days the radioactivity can be found in milk and eggs, a particular issue with iodine and caesium isotopes. In subsequent growing seasons, the root uptake of radionuclides can be important. Direct consumption of contaminated soil can also be important.

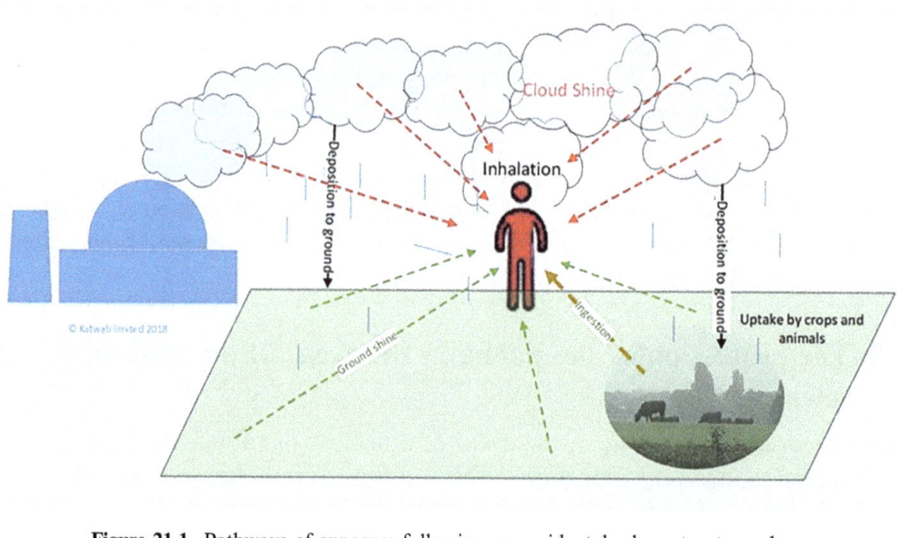

Figure 21.1. Pathways of exposure following an accidental release to atmosphere.

Prospective estimates of these doses, for emergency planning purposes, require an estimate of the release of radioactivity, including its time structure and release characteristics (chemical and aerodynamic properties, temperature and kinetic energy can all feature in models, see chapter 10), the dispersion in the atmosphere (which depends on the assumed weather conditions and the effects of buildings, slopes and ground cover), the assumed location of the exposed individuals and their characteristics (age, gender and activity levels) and a model of their dose uptake.

Real-time estimates, for emergency response purposes, require the same set of parameters but in this case some of the assumptions can be replaced with measurements or estimates based on the situation and the real-time weather data can be utilised. Model predictions can be compared with actual measurements to improve the parameterisation. To be useful for setting protective actions, it is necessary to project these estimates forward in time thus requiring a release prognosis from the on-site emergency response structure as well as weather forecasts. Combining weather data and forecasts, monitoring data and the release prognosis, where all can constantly change and there are never enough data or enough time, is a non-trivial problem.

Whether for emergency planning or emergency response purposes, the estimates of avertable dose (the amount by which radiation exposure could be reduced by applying a protective action) are compared in a cost-benefit framework to evaluate the determent resulting from the potential exposure against the cost and inconvenience of planning for and implementing each available protective action (see below) to inform the protective action strategy.

In terms of dose reduction, the tools available in the early stages are shelter-in-place, evacuation, the provision of stable iodine and controls on the consumption of potentially contaminated foods. Later actions might include decontamination and the restriction of access to, or usage of, contaminated land.

In the UK, shelter in place is associated with the phase 'go in, stay in, tune in'. In the US, the phrase used is 'get inside, stay inside, stay tuned', the intent is the same; to get people into shelter and to get them to tune into the local radio or TV station or other media for updated information. There is significant literature discussing the effectiveness of this protective action. PHE report dose-reduction factors (DRFs) for external gamma dose to be in the range 0.15 for typical residential brick-built homes and 0.05 for multi-storey buildings and recommend a default DRF 0.6 for 'inhalation dose to an individual sheltering during the entire passage of the plume, until both the indoor and outdoor air concentrations fall back down to (or close to) zero, with no opening of windows and doors to the external environment (and under such circumstances the DRF remains constant irrespective of the release duration)'. This does depend on the duration of the release, the airtightness of the building in question and any attempts by the occupants to improve this by improvising a shelter or taping up gaps around windows and doors (figure 21.2).

The dose to the thyroid gland can be reduced significantly by the timely consumption of relatively large amounts of stable iodine which saturate the gland for several hours (see figure 21.3). In the UK, this is a common prompt protective

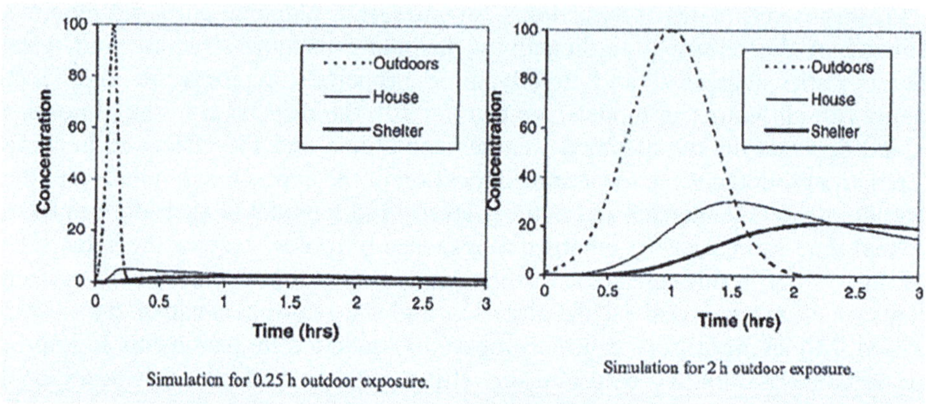

Figure 21.2. Indoor air concentrations relative to outdoor concentrations for two different exposure durations. Reprinted from Jetter and Whitfield (2005), Copyright (2005), with permission from Elsevier.

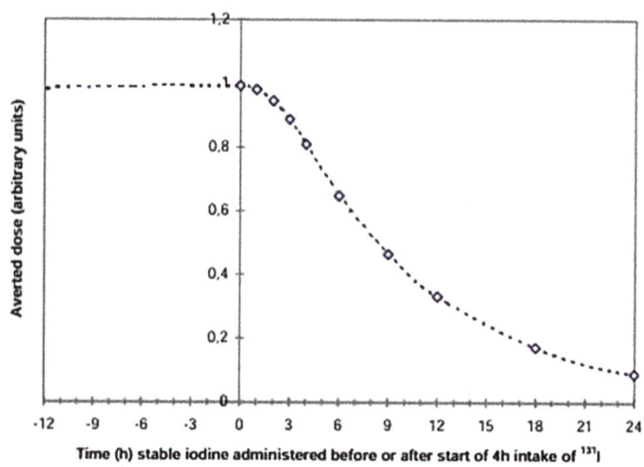

Figure 21.3. Protective effect of the administration of stable iodine before and after inhalation of I-131. Reproduced from WHO (1999), Copyright (1999).

action adopted around operating nuclear sites and nuclear submarine berths to reduce inhalation dose but is not considered acceptable for reducing ingestion dose where food controls can be implemented to achieve the same end. It should be noted that stable iodine has some protective value even if it is taken shortly after inhalation of radioiodine.

Evacuation, if completed before the radioactive cloud arrives can avert all the short-term dose (the sum of inhalation and cloud doses) but is the most disruptive protective action considered and can be distressing (see Box 21.1). In situations where the transport systems do not allow for rapid evacuation, shelter in place may give lower doses than being caught in a traffic jam under a plume. Evacuations can be temporary, where people are allowed back to their homes and businesses within hours or days of the event, or long term, where people are relocated permanently or for several years.

Box 21.1: Adverse effects of evacuation.

It is widely reported that several vulnerable people died in the Fukushima evacuation; deaths that would probably not have happened if these people had stayed in their homes or hospitals. Hypothermia, deterioration of underlying medical problems and dehydration were suspected as the causes of death. The lack of medical support before, during and after the evacuation was regarded as the major reason for the loss of life during the evacuation. The lesson here is, if there is going to be a total evacuation of an area there is a need to take care of those with compromised health.

Deciding to evacuate the able bodied from an area and leave behind those who an evacuation is likely to harm, hopefully with suitable care provided by volunteers, may be a rational decision, but it could also be a public relations disaster.

After Fukushima, excess deaths among those relocated continued in the weeks, months and even years after the evacuation. 'The leading causes of the majority of those early deaths were disruption to the smooth operation of hospitals, the exacerbation of pre-existing health problems, and the general 'mental fatigue' from dramatic changes in life situation'.

A repeat issue from Chornobyl and Fukushima is that of multiple moves for those evacuated as the expanding evacuation zone repeatedly reaches the places to which they have been sent for shelter. This is not conducive to mental or physical well-being.

See: Hasegawa et al. (2016)

The IAEA provides generic criteria for protective actions to reduce the risk of stochastic effects in emergency exposure situations in table 3 of GSG-2 (IAEA 2011). These include:

- Iodine thyroid blocking for $H_{\mathrm{Thyroid}} > 50$ mSv in the first seven days;
- Sheltering, evacuation, decontamination, food controls for E or $H_{\mathrm{Fetus}} > 100$ mSv in the first seven days;
- Temporary relocation, decontamination, replacement of contaminated foods for $E > 100$ mSv per annum;
- Screening based on equivalent doses to specific radiosensitive organs (as a basis for medical follow-up); counselling > 100 mSv in a month.

In the UK, Public Health England provides national recommendations called Emergency Reference Levels (ERLs) which come in pairs (CRCE 2019). If the avertable dose is below the lower ERL the protective action is not indicated. If the avertable dose is above the upper ERL then the protective action should be recommended. Between the two ERLs, the responders are to use their judgement based on the ease of application (see table 21.1).

In the US similar material can be found in the Protective Action Guides (PAGs) (EPA 2020). This gives the 'early phase' protective action recommendations shown in table 21.2.

Difficult decisions will have to be made about relaxing protective actions. There is one school of thought that this should be done cautiously and that, for example, shelter advice should not be withdrawn until the area has been adequately surveyed to assess the risk. However, this may take a considerable amount of time and there will be a lot of scope for scientists within the decision-making process to disagree with each other about the effectiveness of the monitoring programme, the interpretation of the data and the conclusions with regard to public protection. Meanwhile, the affected members of the public may be getting impatient and alarmed with their activities and plans disrupted, and the media will be flooded with learned and social opinion. The other school of thought is that when you reach the stage at which you would not recommend the protective action i.e. the remaining projected dose (the dose that will be received between now and the end of the event) is below the lower ERL you should withdraw the protective action advice.

Table 21.1. Emergency reference levels adopted in the UK (CRCE 2019).

		Emergency reference level (ERL) averted dose (mSv)	
Countermeasure	Body or organ dose	Lower	Upper
Sheltering	Whole body	3	30
Evacuation	Whole body	30	300
Stable iodine	Thyroid	30	100

Table 21.2. Protective actions for the early phase of an accident given in the US Protective Action Guides (PAGs) (EPA 2020).

Protective action recommendation	PAG, or guideline
Shelter-in-place or evacuation	PAG: 10–50 mSv projected dose over four days.
Administration of stable iodine	PAG: 50 mSv projected child thyroid dose from exposure to radioactive iodine.
Limit emergency worker exposure (total dose incurred over entire response)	Guideline: 50 mSv year^{-1} (or greater under exceptional circumstances)

Note: When comparing these numbers notice that ERLs are in averted dose and PAGs in projected dose.

Certainly, over the days and months following any significant release to the environment there will have to be assessments of the potential doses from consumption of contaminated food and water, external radiation and resuspension of radioactive dusts and gases. This may lead to food controls, as well as controls over access to, and activities in, contaminated areas. An impression of the potential complexity of this effort can be gleaned from the experiences following Fukushima. See, for example IAEA (2015b) volumes 4 and 5.

21.4 International advice

There is a great deal of advice aimed at those responsible for preparedness for, and response to, a nuclear emergency. This advice is from international, national and local players. National and local advice naturally varies from place to place, but generally most countries follow ICRP and IAEA advice to some extent in their regulations, standards and advice.

21.4.1 ICRP

Most countries base the radiological aspects of their emergency preparedness and response on the advice of the ICRP.

In line with ICRP Publication 103 (ICRP 2007), emergency plans are designed to keep projected doses below an agreed reference level (the ICRP recommendation of 100 mSv is commonly used) so far as is reasonably practical and to use protective actions when the avertable dose is estimated to be above an agreed intervention level.

21.4.2 IAEA

At the top of the hierarchy for international advice on nuclear reactor safety sits the IAEA. The IAEA Safety Standards are not legally binding on nations but are the basis of EU Directives and regulations around the world.

Over the years the IAEA has published many documents related to emergency preparedness and response. The current key documents are:

- Preparedness and Response for a Nuclear or Radiological Emergency (Series No. GSR Part 7) (IAEA 2015);
- Arrangements for Preparedness for a Nuclear or Radiological Emergency (Series No. GS-G-2.1) (IAEA 2007);
- Criteria for Use in Preparedness and Response for a Nuclear or Radiological Emergency (Series No. GSG-2) (IAEA 2011);
- Planning and Preparing for Emergency Response to Transport Accidents involving Radioactive Material, Safety Guide No. TS-G-1.2 (ST-3) (IAEA 2002).

There are also an increasing number of useful documents in the Emergency Preparedness and Response series to be found at https://www.iaea.org/publications/search/type/emergency-preparedness-and-response.

Emergency planning is deeply embedded in IAEA advice and is not a peripheral add-on.

Principle 9 from IAEA Safety Fundamentals Number SF-1 (IAEA 2006) (the keystone of IAEA publications) states that '*Arrangements must be made for emergency preparedness and response for nuclear or radiation incidents.*' It defines the primary goals of emergency preparedness to include:

- *To ensure that arrangements are in place for an effective response at the scene and, as appropriate, at the local, regional, national and international levels, to a nuclear or radiation emergency,*
- *For any incidents that do occur, to take practical measures to mitigate any consequences for human life and health and the environment.*

The document goes on to discuss the need for local, regional and national arrangements that reflect the likelihood and possible consequences of a nuclear or radiation emergency and the scope and extent of arrangements.

IAEA GSR Part 7 Preparedness and Responses for a Nuclear or Radiological Emergency (2015) (IAEA 2015), in the document layer below SF-1, '*establishes the requirements for an adequate level of preparedness and response for a nuclear or radiological emergency. The application of these requirements is also intended to mitigate the consequences of a nuclear or radiological emergency if such an emergency arises despite all efforts made to prevent it.*' It is largely talking to national governments and setting the standards that they should achieve.

The document suggests a list of 'requirements' covering general requirements (1–5), functional requirements (6–19) and infrastructure (20–26). It also recommends limits on the doses that responders should be exposed to and the ranges of doses that would lead to the recommendation to impose off-site protective actions.

The document presents a rather crude hazard assessment that recognises five 'threat categories': (i) large operating commercial reactors; (ii) smaller research reactors; (iii) facilities such as irradiation facilities with very strong sources; (iv) processes such as the transport of radioactive material and radiography; and (v) areas within the planning zones of sites in neighbouring countries. It then uses the site's threat category to suggest appropriate capabilities of the site's emergency plan including the readiness of:

(a) Precautionary urgent protective actions to avoid or to minimize severe deterministic effects;
(b) Urgent protective actions and other response actions to avoid or to minimize severe deterministic effects and to reduce the risk of stochastic effects;
(c) Early protective actions and other response actions;
(d) Other emergency response actions such as longer term medical actions and emergency response actions aimed at enabling the termination of the emergency; or
(e) Protection of emergency workers.

In the UK, this duty is discharged, in more detail, by the Local Resilience Forum Risk Register and the REPPIR-19 consequence reports.

IAEA Safety Guide GS-G-2.1 Arrangements for Preparedness for a Nuclear or Radiological Emergency (2007) (IAEA 2007) drills further down into the detail, suggesting functional requirements (what should be achievable in a given time-frame), infrastructure needs, and a concept of operations outlining an ideal emergency response at each threat level.

The bulk of the IAEA radiological advice for emergency preparedness is contained in GSG-2 (IAEA 2011). This concerns itself with preventing or reducing:

- Development of severe deterministic effects (by the implementation of urgent protective actions and immediate medical examinations, consultation and treatment as indicated, contamination control, decorporation where applicable, registration for long-term health monitoring, and comprehensive psychological counselling);
- An increase in stochastic effects (by the implementation of urgent protective actions and initiation of early protective actions as far as reasonably possible and screening for possible medical effects based on dose assessments);
- Adverse effects on the environment and property;
- Other adverse effects (e.g. psychological effects, social disorder, economic disruption).

21.4.3 EU basic safety standards

The EU sets out standards for radiation protection in the Member States in the revised Basic Safety Standard that was adopted in 2013 by the Council (Council Directive 2013/59/Euratom of 5 December 2013 laying down basic safety standards for the protection against the dangers arising from exposure to ionising radiation) (EU 2013).

This document defines an 'emergency' as '*a non-routine situation or event involving a radiation source that necessitates prompt action to mitigate serious adverse consequences for human health and safety, quality of life, property or the environment, or a hazard that could give rise to such serious adverse consequences*'.

It proposes that '*With regard to the management of emergency exposure situations, the current approach based on intervention levels should be replaced by a more comprehensive system comprising an assessment of potential emergency exposure situations, an overall emergency management system, emergency response plans, and pre-planned strategies for the management of each postulated event*'.

The EU BSS also introduces the concept of reference levels in emergency and existing exposure situations from ICRP Publication 103 (ICRP 2007). They define a reference level as '*in an emergency exposure situation or in an existing exposure situation, the level of effective dose or equivalent dose or activity concentration above which it is judged inappropriate to allow exposures to occur as a result of that exposure situation, even though it is not a limit that may not be exceeded*'.

The relationship between the phase of an emergency and the dose control regime is shown in figure 21.4 (IAEA 2018).

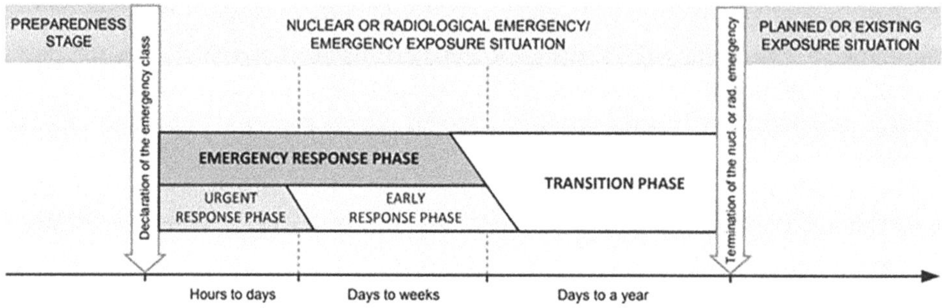

Figure 21.4. Dose control regime in different phases of an emergency. IAEA (2018), © IAEA, 2018, p. 13, ISBN 978–92–0–108017–2

Reference levels are helpful concepts giving useful guidance about the levels of dose that would be unacceptable as residual doses (the dose that remains after protective actions have been implemented) when planning protective action strategies. This was previously missing.

The BSS mandates that, in general, the reference level for a radiation worker should be below an effective dose of 100 mSv although, in exceptional situations, in order to save life, prevent severe radiation-induced health effects, or prevent the development of catastrophic conditions, a reference level for an effective dose from external radiation of emergency workers may be set above 100 mSv, but not exceeding 500 mSv. At these levels of dose, deterministic effects are expected to be absent or minor and the risk of induction of fatal cancer is about 0.006 to 0.03, i.e. small but not negligible.

21.5 National guidance

Most countries with a nuclear power programme produce copious regulations and guidance for the nuclear industry and this includes defining the expectations for emergency preparedness and response.

Current UK guidance for nuclear emergency planning is contained within the approved code of practice and guidance to REPPIR-19 and in the *National Nuclear Emergency Planning and Response Guidance* (BEIS 2015) published by the Department for Business, Energy & Industrial Strategy. The advice is divided into five parts:

- Concept of Operations;
- Part 1—Preparedness;
- Part 2—Response;
- Part 3—Recovery;
- Annexes.

In the USA, the Nuclear/Radiological Incident Annex to the Response and Recovery Federal Interagency Operational Plans (FEMA 2016) fulfils a similar function and '*provides guidance and serves as a reference for federal agency planning efforts involving nuclear/radiological incidents. Other stakeholders (*e.g.*, local, state, tribal, territorial, and insular area governments; nongovernmental organizations; voluntary agencies; and the private sector) engaged in their own planning will find this document useful in enhancing their understanding of how the Nuclear/ Radiological Incident Annex will be implemented and how their planning efforts can be complementary*'.

21.6 The development of an off-site emergency plan

21.6.1 Purpose

In GSG Part 7 the IAEA states that the goals of emergency response are:

(a) to regain control of the situation and to mitigate consequences;
(b) to save lives;
(c) to avoid or to minimize severe deterministic effects;
(d) to render first aid, to provide critical medical treatment and to manage the treatment of radiation injuries;
(e) to reduce the risk of stochastic effects;
(f) to keep the public informed and to maintain public trust;
(g) to mitigate, to the extent practicable, non-radiological consequences;
(h) to protect, to the extent practicable, property and the environment; and
(i) to prepare, to the extent practicable, for the resumption of normal social and economic activity.

The site operator will be focussed on (a) to (e) producing an on-site plan, whereas the local authority will focus on items (f) to (i), producing an off-site plan. It is important that the two plans should dovetail and provide the appropriate support for each other.

21.6.2 Contents of an on-site plan

It would be expected that an on-site plan would detail how an event would be detected, the conditions under which, and how, a declaration of emergency would be made, how the alert would be promulgated (on-site and off-site), the initial actions taken on alert including the initiation of automatic protective actions and the alerting of the public, the arrangements to keep others on the site safe, communications with off-site organisations, and the team structures, roles and responsibilities for the on-site responders.

Each response role would have a person specification and training needs analysis to ensure that the right people were available, and the preparedness plan might have a rota to ensure that someone is available to fill each role.

Each role might have an aide-memoir briefing reminding them of their role and interfaces with the rest of the emergency scheme, an action list, and any data and information they need to fulfil their role.

Health physicists would be deployed in the front line surveying the radiological situation, giving advice on stay-times and PPE/RPE; in the access control point ensuring that dispatched teams were suitably prepared and returning teams monitored and, if necessary, decontaminated; in the emergency control centre providing situational awareness and radiological advice to the controller and maintaining maps of monitoring data.

Health physicists would also be managing any off-site radiation surveys, collating the results and running dispersion models to support public protective action strategy formulation. This is a complex process requiring close working with the technical teams supporting the accident scene to ensure as full an understanding of the uncertainties in the release prognosis and trying to integrate disparate monitoring results into the model.

21.6.3 Contents of an off-site plan

In the UK schedule 6 Part 2 of REPPIR-19 mandates content for the off-site plan and has associated guidance. To be included are:

- Arrangements to set the emergency plan in motion. This includes how the site will raise the alarm and promulgate initial information and how the responding organisations will react to hearing the alarm. This is usually to mobilise their response teams and to send personnel to the various control centres.
- Arrangements to coordinate the off-site protective actions. This involves the setting up of the Strategic Coordination Centre (SCC) and Tactical Support Centres (TSCs). These centres will have an agreed *modus operandi* detailing their intent, which organisations attend and how they coordinate. This may be an arrangement of main and sub-groups, each with their own terms of reference, an outline 'battle rhythm' and default attendance lists and agenda for meetings.
- The arrangements for off-site protective actions. The off-site plan is expected to provide all the procedures required to initiate the automatic protective actions, if they are included in the plan, and the procedures to consider 'the particular characteristics' of the event facing the responders and the decision-making process to determine the protective action strategy, the communication strategy and the recovery strategy going forward.

The off-site plan is also expected to contain helpful information such as maps, demographics, contact details for the Detailed Emergency Planning Zone (DEPZ) and in less detail for the Outline Planning Zone (OPZ) (see the following subsection).

21.6.4 Emergency planning zones

The IAEA defines several planning zones around major nuclear facilities (see figure 21.5):

- The on-site area is the area under the control of the operator or first responders,
- The off-site area is the area beyond that area under the control of the facility, operator or first responders which is divided into:

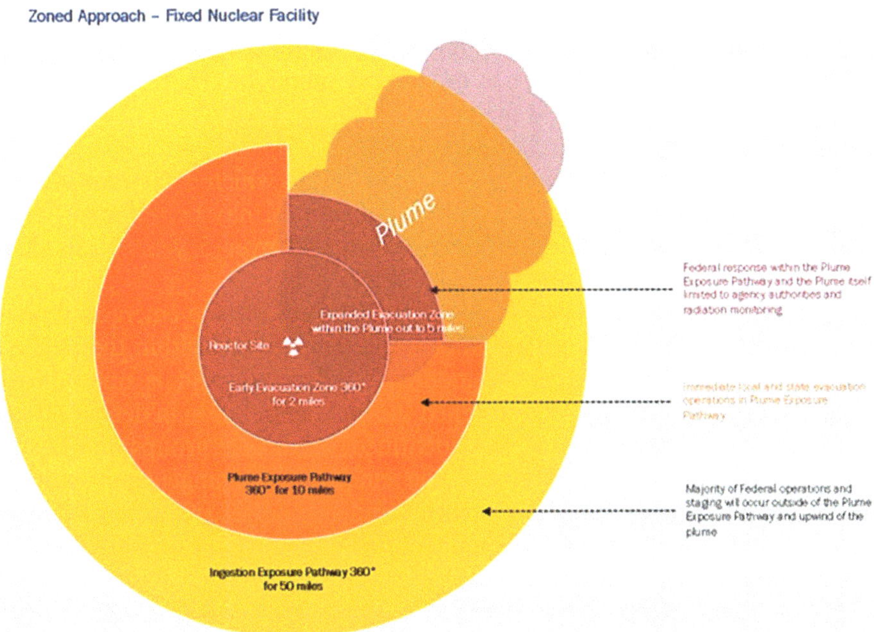

Figure 21.5. Emergency planning zones around nuclear facilities. Reproduced from https://www.nrc.gov/about-nrc/emerg-preparedness/about-emerg-preparedness/planning-zones. Public domain.

- ○ A Precautionary Action Zone (PAZ), for facilities in threat category I, for which arrangements shall be made with the goal of taking precautionary urgent protective action, before a release of radioactive material occurs or shortly after a release of radioactive material begins, on the basis of conditions at the facility (such as the emergency classification) in order to reduce substantially the risk of severe deterministic health effects,
- ○ An urgent protective action planning zone (UPZ), for facilities in threat category I or II, for which arrangements shall be made for urgent protective action to be taken promptly, in order to avert dose off the site in accordance with international standards.

The IAEA suggests sizes for the PAZ and UPZ depending on the nature of a site including PAZs of 3–5 km and UPZs of 5–30 km for reactors with power ratings of 1000 MW_{th} or more but it is important to note that the IAEA states that '*a different distance should be used when this is substantiated by a detailed safety analysis*' (IAEA 2007).

In the USA, the NRC defines two emergency planning zones (NRC 2018):

- the plume exposure EPZ, about 10 miles in radius for which predetermined protective action plans are in place designed to avoid or reduce dose from potential exposure to radioactive materials. These actions include sheltering, evacuation, and the use of potassium iodide where appropriate; and

- the ingestion exposure pathway EPZ, about 50 miles in radius for which predetermined protective action plans are in place designed to avoid or reduce dose from potential ingestion of radioactive materials. These actions include a ban on contaminated food and water.

In the UK, the PAZ and UPZ are replaced by a DEPZ which is the zone in which it is estimated that prompt and urgent protective actions may be required for faults with a frequency estimate greater than 1 in 20 000 over 5 years. Less detailed planning is required for a larger zone, the OPZ.

Local authorities are expected to produce 'detailed' off-site emergency plans to manage protective actions (both automatic and deliberated) within the DEPZ and 'outline plans' for the OPZ. In this context, a detailed plan is one where, for example, the local authority has researched the local population in detail and knows the numbers with some accuracy, has identified vulnerable groups and has made contact with them and has discussed the emergency arrangements with them. Databases of telephone numbers of homes and businesses in the DEPZ may be maintained. In outline planning, the population has been quantified in larger sectors and some thought has been given to distribution of stable iodine (reactor sites only) and how to advise shelter-in-place to the community. Outline planning can also encompass the evacuation of people near the site for whom detailed planning involves shelter in place. Such evacuation might be implemented in the event of a more extreme accident than those considered in detailed planning.

21.6.5 Risk assessment

The first stage of emergency planning is to determine the capabilities required in the response. The IAEA states that '*the emergency management system shall be designed to be commensurate with the results of the hazard assessment and shall enable an effective emergency response to reasonably foreseeable events (including very low probability events)*' (IAEA 2015). The IAEA gives guidance on the scope of plans and the times within which important activities should be starting following an alarm based on a crude risk assessment in which facilities are put into five different 'Threat Categories' (IAEA 2007).

In the UK, regulation 4 of REPPIR-19 requires the operator of sites containing more than threshold levels of radioactivity to make a written evaluation '*sufficient to identify all hazards arising from the work undertaken which have the potential to cause a radiation emergency*'. This requires a trawl through the site's safety case looking at the estimated frequency and consequences of each fault identified and analysed.

Regulation 5 then requires the operator to '*consider and evaluate a full range of possible consequences of the identified radiation emergencies*' and regulation 7 requires the preparation of a report setting out those consequences which should be sent to the local authority.

Guidance requires that the assessment consider a range of potential source terms and weather conditions, the different persons that may be exposed, the effective and equivalent doses they are likely to receive, the different pathways involved (classically inhalation of plume, cloud dose, ground dose, ingestion dose and resuspension dose) and the distances within which urgent protective action may be warranted when assessed against the relevant Emergency Reference Level (ERL). CRCE (2019b) outlines how Public Health England (PHE) would do these assessments.

Various operators' consequence reports can be found on the internet (signposting to these can be found at http://www.katmal.co.uk/reppir2019progress.html). They generally provide the information mandated by Schedule 4 of the regulations but little else. Generally, the consequence reports state that shelter is required out to between 1 and 5 km depending on the nature of the site, stable iodine is needed to between 1 and 2 km for the reactor sites and evacuation is generally not needed beyond a few hundred metres of the site boundary. The advice is usually to implement the protective actions 'as soon as possible', as most sites admit to the potential for a no-notice event (nuclear submarine sites are an exception to this, claiming a warning time of about three hours for core damage to progress to the point of increasing radiation doses). These equip the local authority with a statement about how far downwind the operator thinks that protective actions will be needed if the accidents considered occur.

21.6.6 Functional specification

An early step in the design of an off-site plan is the determination of the required functional capabilities. This is likely to be based on the operator's risk assessment. The questions that must be answered at this stage are:

- Do we need to be able to evacuate areas, if so, how quickly to be useful ?
- Do we need to be able to shelter areas, if so, how quickly to be useful ?
- Do we need to be able to enable stable iodine tablet consumption, if so, how quickly to be useful ?
- Do we need to be able to control food and water sources at short notice, if so, how quickly to be useful ?
- How will we know the area over which these protective actions will be needed following an alert ?
- Might we be required to extend any of the above protective actions beyond the distances assumed?
- Will action be required to keep individual doses below the local reference level (by default 100 mSv)?

At this stage, the off-site authorities would benefit from knowing the outlines of a number of different faults including their frequencies, any warning time between initiation and the start of the release of radioactive material, the potential duration of the release, the potential off-site doses and what factors could affect any of these. The off-site authority would then know how quickly it might have to put protective actions into place, how long they might have to support local populations under restrictions and what information they would need to be able to collect and process on the day to understand the developing situation.

There may, for example, be a warning period of several hours between initiation and release for all reasonably foreseeable faults in which case the public could be given notice that shelter is likely to be imposed and given time to prepare or the option of self-evacuation. If the warning period relates to reactor faults, then stable iodine may be stored locally and distributed during the warning period.

On the other hand, there may be reasonably foreseeable faults that do not have any significant warning period. If this were the case, the planners might consider pre-distribution of stable iodine tablets to homes and businesses in the area and plans to quickly alert the public (by sirens or automatic telephone message systems) and automatically impose protective actions on declaration rather than after a confirmation and discussion period.

Complications arise if there are a number of more frequent faults with a usable warning time and a few that are significantly less likely but give no notice. If the no-notice events are smaller than the ones with notice, the local authority might consider a fast-track application of protective actions in a smaller area and a more deliberative process for further out. But if the no-warning low frequency faults are also among the more severe, the local authority might feel compelled to design the plan to deliver protective actions over a wider area recognising that this increases the cost of preparations and may impose protective actions over a wider than necessary area in the event of a smaller event.

Clearly the scoping of an emergency plan is as much an art as it is a science. The fuller understanding of the range of events that could occur and the variability of consequences the better founded the resulting functional specification will be.

Understanding the time available to implement protective actions is important as the thresholds (the ERLs in the UK) are based on avertable dose and this reduces with time as the release progresses, as shown in figure 21.6.

Figure 21.6 shows four different release profiles in terms of the dose rate at some point downwind (left plot) and the avertable dose (remaining dose) (right plot). Also shown is the 7.5 mSv avertable dose that represents the threshold for sheltering in the UK, adopting the accepted assumption for the effectiveness of shelter of 40% and the lower ERL for shelter of 3 mSv. These plots show that if you cannot achieve shelter in just over an hour for the 2-h uniform release, about 2.5 h for the 4-h release and exponentially reducing release and about 5.5 h for the release with a 4-h warning period, then it is not worth doing.

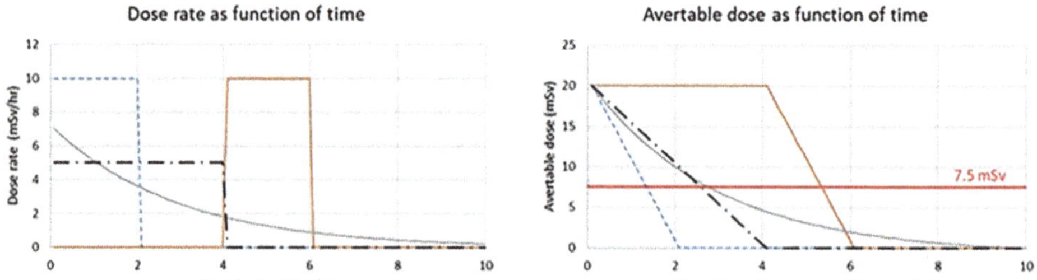

Figure 21.6. Avertable dose as a function of time during an accident.

For the three no-notice releases, the plan will aim to establish shelter as quickly as possible, whereas the release with the warning period may use that time to make the shelter more bearable by providing time to get animals under cover and unite families, for example.

Similarly, the potential for a no-notice reactor accident would suggest pre-distribution of stable iodine tablets to those near the site and a plan to get the message to them to take those tablets without undue delay (telephone warning or sirens initiated on declaration rather than after evidence gathering and decision process). For processes where a delay before the release is expected in all but the most unlikely events. it may be decided to have nearby stocks of tablets coupled with on-the-day distribution of the tablets and advice notes.

It is worth considering what an emergency plan should look like for an 'explosive dispersion' event which gives no notice and is over almost as soon as it begins. A radioactive cloud produced by such an explosion will disperse downwind at the speed the wind is moving giving people times from a few seconds to a few tens of minutes to gain shelter if they are to receive any protection from the plume.

If it is possible that the protective action strategy may need to be revised on the day to respond to the characteristics of the event, then the emergency plan needs to consider what information would be needed, where it would come from, how it would be communicated, how decisions based on it would be made, and how real-time advice would be communicated to those affected.

In this respect, sites that deal with radioactive materials of fixed isotopic composition in discrete and uniform quantities with a simple set of accident initiators, such as containers being breached or ignited can readily give a reasonable upper bound fault against which to scale the emergency plan and the protective action strategy.

For reactor sites, where the release depends on the core history and the exact details of the fault, there may be a wide range of faults and for each fault a wide range of potential releases depending on the details. Selecting the default automatic protective action strategy is then more debatable and considering how the characteristics of an event in train will be determined on the day and how the protective actions strategy will be updated is more important. Reactor sites also have a potential, albeit claimed to be at extremely low frequencies, for severe accidents, of significantly greater consequence than the 'design basis accidents' upon which detailed planning is based. This is the regime of outline planning.

IAEA GS-G-2.1 (IAEA 2007) provides an appendix giving recommended response time objectives. These include times for announcing who is the on-site director of the emergency response (<15 min), activating site and local emergency operations facilities (<1 h for category 1 and 2 sites, <3 h for category 3), these should be fully functional no more than an hour later, emergencies should be classified in < 15 min, and the local authorities notified on the same time scale.

The output of the risk assessment and scoping phase of the emergency planning process might be a functional specification that details actions and target times such as:

- Declare emergency, identify on-site controller and initiate alert call-down chain within 15 min of detection of fault;
- Activate on-site damage control within 30 min;
- Activate urgent protective advice to the public within 30 min of declaration;
- Activate an on-site command post and an off-site multi-agency coordination centre within 1 h;
- Discuss the need to extend protective action advice within 2 h of declaration;
- Dispatch first mobile off-site survey vehicle within 30 min of declaration;
- Put in place public help lines and other community support within 2 h of declaration.

Note: the times here are not taken from a specific plan. The target times would depend on the distribution of fault consequences that the plan is designed to protect against.

The capabilities of the facilities will need to be carefully defined, including when they should first become available and when they would become fully operational.

21.6.7 Concept of operations

Following on from the functional specification the emergency planners would have to consider how best to deliver the requirements, who would do what, how it would be coordinated and how they would respond to the characteristics of the real event.

The output of this stage can be a 'concept of operations' paper which can be a short narrative explaining the promulgation of the alert, the initial response of those alerted including all automatic actions and then how the situation will be reviewed and how strategic, tactical and operational coordination and decision making will proceed. The narrative should continue to the end of the emergency and the beginning of the recovery stage. An example of a concept of operations for an operating reactor site is given on page 48 of GS-G-2.1 (IAEA 2007).

The US concept of operations for a nuclear or radiological event can be found in FEMA (2016). The UK nuclear emergency response concept of operations can be found in HMG (2015). What these have in common is that the operator is responsible for the on-site response but not the off-site response.

The UK concept of operations gives the default objectives of the response strategy as:

- Protect human life and, as far as possible, property and the environment;
- Alleviate suffering;
- Support the continuity of everyday activity and the restoration of disrupted services at the earliest opportunity; and
- Uphold the rule of law and the democratic process.

The effort is coordinated by a Strategic Coordination Group (SCG) based at the local police headquarters and usually chaired by a senior police officer. This body takes overall responsibility for the multi-agency management of the emergency and establish the policy and strategic framework within which tactical and operational tiers will operate. The SCG comprises senior members of the emergency services, local authority, government bodies, health bodies and the operator, among others, who are authorised to commit the resources of their organisations. They share information and coordinate the response.

The response structure in England and Wales is shown in figure 21.7.

The response will 'grow upwards' from the site's emergency control centre as far as it needs to go given the severity of the accident and the public and media response.

Scientific advice is provided at the Scientific and Technical Advice Cell (STAC) and, if initiated, the Scientific Advisory Group for Emergencies (SAGE) for central government. These groups review the science and data available and provide their client strategic groups with situation reports and options for public protection, public communications and recovery strategies.

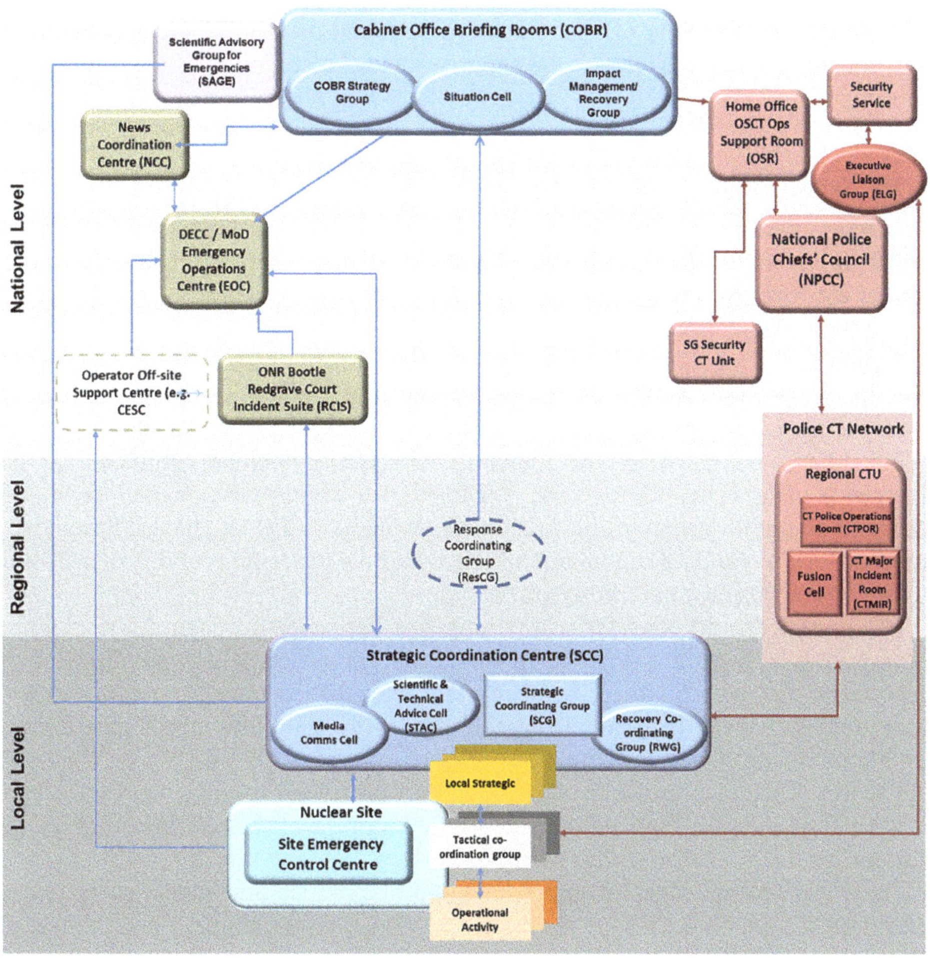

Figure 21.7. The emergency response structure applicable in England and Wales. Reproduced with permission from https://assets.publishing.service.gov.uk/media/5a8014d2ed915d74e33f8550/NEPRG00_-_Concept_of_Operations.pdf. © Crown copyright 2015. Open Government license.

21.6.8 Consultation and plan writing

The emergency planner must consult with all of those who will be expected to deliver something in the execution of an off-site plan. Obvious consultees are the operator, the local emergency services, the local social services, the local medical services and government bodies. Less obvious are any charity or other organisation that might be involved in setting up and running evacuation centres, transport system managers, utilities.

The purpose of consultation is to engage with and take account of the views of relevant parties during the preparation of the emergency plan. Ideally the result will be an agreed plan with all of the key parties confirming that they will maintain the processes and equipment they require to manage their part of the response and will ensure that they have an adequate number of trained personnel to respond at any time.

Various local authority off-site plans can be found on their emergency preparedness and response websites (see http://www.katmal.co.uk/reppir2019progress.html for signposting).

21.6.9 Reviewing and testing of emergency plans

Emergency plans are complex entities that need the coordination of numerous organisations and require difficult decisions to be made about public safety in the light of elevated levels of radioactivity in the atmosphere. It is valuable to test and review such plans periodically.

Regulation 12 of REPPIR-19 requires that each UK operator or local authority who has prepared a REPPIR emergency plan must review, and where necessary revise the plan and must test the plan 'at suitable intervals not exceeding three years unless otherwise agreed by the regulator'. Exercises may be conducted to test emergency plans, where an exercise is defined as a simulation designed to validate the capability of an organisation to manage incidents and emergencies. Specifically, exercises seek to validate training undertaken and the procedures and systems within emergency or business continuity plans.

The review should consider if anything has happened to change the planning assumptions, if there are any changes in the organisation and capabilities of the responders, and whether there has been any relevant learning from experience or from previous tests of the plan. This generally involves a multi-agency meeting that is quite often combined with the first exercise planning meeting.

The test of the plan 'must be adequate to test the ability to implement the plan in question' and a report must be written within 3 months of the conclusion of the test and sent to the regulator within the next 28 days.

There are several pages of approved code of practice and guidance related to regulation 12, which is likely to dominate the discussions about review and testing, but prior to REPPIR-19 the main source of advice was *Nuclear Emergency Planning and Response Guidance—Part 1* (BEIS 2015).

21.6.10 Prior and real-time information to the public

Experience and common sense both suggest that if you may require members of the public in a given area to undertake some actions without delay at some as yet unknown time in the future you would be well advised to give them prior notice so that they can think about it and can understand what is expected of them and why.

In the UK, such information is mandated by REPPIR-19 regulation 21 which puts a duty on the local authority to 'ensure that members of the public are made aware of the relevant information, and where appropriate, provided with it. The information required for those living in the areas subject to urgent protective actions (those that are imposed on declaration of the emergency without waiting for any assessments) is:

1. Basic facts about ionising radiation and its effects on persons and on the environment;
2. The various types of radiation emergency identified and their consequences for the general public and the environment;
3. Protective actions envisaged to alert, protect and assist the general public in the event of a radiation emergency;
4. Appropriate information on protective actions to be taken by the general public in the event of a radiation emergency;
5. The authority or authorities responsible for implementing the protective actions referred to in items 3 and 4 above;
6. The extent of the detailed emergency planning zone.

This information is provided in the form of leaflets and calendars and is often repeated on the websites of the local authority and the local resilience forum. A poster containing the generic information for the 2001 REPPIR regulations can be found at https://webarchive.nationalarchives.gov.uk/20131102061245/http://www.hpa.org.uk/webc/HPAwebFile/HPAweb_C/1272031414086

The IAEA suggests that during an emergency the public will want to know 'is the situation safe?[1]' and 'How do I keep my family and myself safe and how do I protect my interests'. They give useful advice and communications materials in EPR-Communicate.

21.7 Emergency response

21.7.1 Event initiation

The initiating event could be many things: a human error such as an operator pressing the wrong button or a technician disconnecting the wrong system, a valve

[1] Of course, the answer the public is looking for is 'yes, you are safe' but health physicists can't help diving into complex epidemiology and comparative risk statistics. This approach seldom reassures. A more recent trend of claiming that anything under 100 mSv once in a lifetime is fine works better until you assess someone's dose as 105 mSv.

failing on demand, a pressure circuit or tank failure, the outbreak of a fire, the ignition of a build-up of explosive gases. The initial result might be automatic alarms triggering, someone hitting an alarm switch or an eye-witness report to a control centre.

The cause of the alarm would be investigated and, if necessary, an appropriate formal declaration would be made. This triggers several responses: the alert is promulgated, the on-site emergency controller assumes responsibility, response teams muster at their duty stations and others on-site muster in safe areas. The processes of understanding the event, assisting casualties, ensuring that all personnel are accounted for, and deploying damage repair teams and monitoring teams are all initiated as soon as resources allow.

Emergency services might be summoned to the site or, more likely, would respond with a pre-agreed level of response automatically, with the responders briefed on route.

The off-site centres would be opened and made ready and responders would start to arrive and take up their posts. A senior person would take charge (in the UK this falls to the police service).

Information would start to flow from the site to the off-site centres.

Automatic protective actions would be triggered, and the operator would immediately start considering if these might need extending.

21.7.2 Decision making

In the off-site centre (Strategic Coordination Centre in the UK) the strategic group would meet to exchange information and start to make sense of the situation. Situation information and scientific advice would start to flow and a 'battle rhythm' would be established. This has to allow the strategic coordination group time to exchange information, discuss and make decisions without hampering the members' ability to communicate with their teams beyond the room, to give them actions and be briefed on the state of their organisation.

The UK Nuclear Emergency Planning and Response Guidance Concept of Operations document (BEIS 2015) '*sets out how national and local emergency responders may operate together to deliver an effective and consistent response to any radiation emergency at any nuclear civil or military operating site anywhere in the UK*'.

It identifies the core characteristics of an effective response as:

- **Anticipation:** Risk assessment and risk communication.
- **Preparedness:** Clarity of roles and responsibilities, supported by training, exercising and the availability of equipment and agreed procedures.
- **Continuity:** It is helpful if individuals and teams are asked to carry out similar roles, using the same tools as they employ in their normal work rather than expect them to do something very different or use unfamiliar systems with occasional training.
- **Subsidiarity:** Decisions should be taken at the lowest appropriate level, with coordination at the highest necessary level.

- **Direction:** Clarity of purpose should be delivered through a common awareness of the strategic aims and supporting objectives for the response agreed at the Strategic Co-ordinating Group (SCG).
- **Integration:** Effective coordination should be exercised between and within organisations both at the local and national levels of a response. For the emergency services this is second nature based on the JESIP principles (Joint Emergency Services Interoperability Principles), (JESIP 2020).
- **Communication:** Good two-way communications are critical to an effective response. This is both about physical communications technologies and about situational awareness and shared understanding.
- **Co-operation:** Positive engagement based on mutual trust and understanding will facilitate information-sharing and deliver effective solutions to arising issues.

21.7.3 Strategic, tactical and operational considerations

Off-site command and control are exercised at three levels: strategic, tactical and operational. Each of these may have its own control room or improved control area near the scene (the emergency services are equipped with control vehicles that can be quite complex). In the UK, the terms Gold, Silver and Bronze are used by the police service to denote strategic, tactical and operational, respectively, and these labels can also be found in other settings.

The strategy (gold) commander is in overall control of their organisation's resources at the incident but will not be at the scene. The role is to determine the strategy and provide resources to enable the strategy to be realised. The gold commander sets the strategic direction, coordinates responders and prioritises resources. In the UK, the Civil Contingencies Act nominates the police to coordinate the strategic level of a multi-agency response.

The tactical (silver) commander coordinates the efforts at the scene determining the tactics required to meet the strategy. The silver commander can be in a remotely located control room, a command vehicle near the scene or at the scene, depending on the service and the situation. The silver commander interprets strategic direction, develops a tactical plan, and coordinates activities and assets.

The operational (bronze) commander(s) (there can be several for each organisation working on different aspects of the response) will be at the scene and is in direct control of their organisation's resources at the incident. The bronze commander executes the tactical plan, commands the single-service response, and coordinates actions.

For example, it might be a strategic decision to reduce the number of people in the open in an area. The tactical level might be to decide to close the inward lanes of roads, close public footpaths and put messages on broadcast and social media and would assign teams to each task. Operational teams would manage roadblocks, close footpaths and promulgate the message through media lines and progress would be reported back up the line.

The UK emergency services have a formalised and practiced method for working together (figure 21.8). The Joint Decision Model is a process to help bring together the available information, reconcile objectives and make effective decisions as a collective. It is based on agreeing the situation, aims, objectives, overarching values, priorities and actions and then reviewing and repeating. See the Joint Emergency Services Interoperability Principles (JESIP) website for more details (JESIP 2020).

21.7.4 Command and control

'Command and Control' has been an important phrase in the nuclear industry since the aftermath of the Piper Alpha accident. There have been several attempts to give it a precise definition, but this is hampered by it meaning different things to different organisations. The definitions in the following box were taken from an MOD publication published in 2017.

Box 21.2: Definition of the term 'Command and Control'.

'Command and control capability' is defined as: A dynamic and adaptive socio-technical system configured to design and execute joint action.

'Command and control' is defined as: To provide focus for individuals and organisations so that they may integrate and maximise their resources and activities to achieve desired outcomes.

JCN 2/17, Future of Command and Control

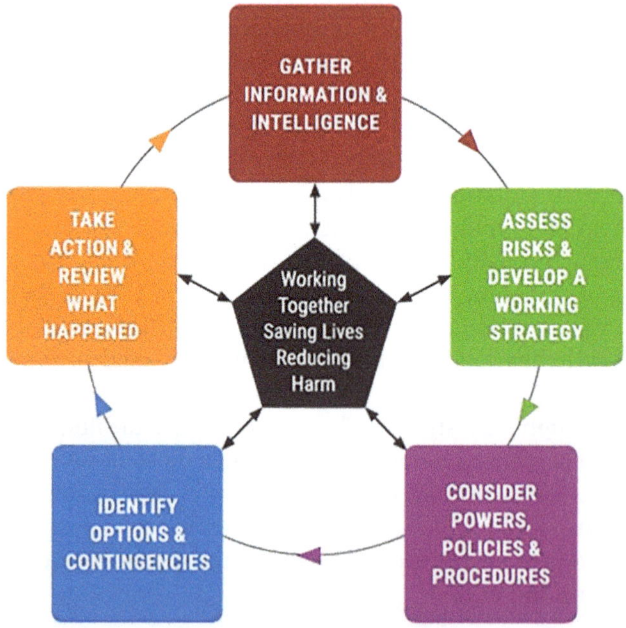

Figure 21.8. UK joint decision model. Image: Courtesy of JESIP.

From the point of view of operators of nuclear power stations, the introduction of 'Command and Control' made clear the fact that a duly appointed emergency controller in a formally declared emergency has full control over any resources on the site; the normal management chain and formal decision-making processes being suspended. The purpose of the emergency response organisation is to ensure that the person in the command role has the fullest picture of what is happening (situational awareness), the best technical advice available under the time constraints, and command over the resources to enact their will.

On-site command rooms vary in size, complexity and function but, across the nuclear industry there tend to be a few common factors.

In front of the emergency controller will be a bank of information boards or screens showing the history of the event and the current situation. Based on this, and on advice received, the Emergency Control produces a 'focus list' which are the areas to which they want priority given. These might include, for example 'Complete roll call and determine if anyone is missing', 'Aid and recover all casualties', 'Assess situation at <scene of event>', 'Determine options for action'. Importantly, the focus list is not a shopping list of everything that has to be achieved; it is the current priorities that the organisation should focus on at the expense of everything else. Progress on these would be reviewed periodically and they should change as the event progresses. These foci would be displayed on one of the information boards.

A group of people, with suitable qualifications and experience, would work on turning the foci into a list of actions, each with a person responsible for managing them through to the end and each with a target completion time. Actions would be displayed on another board with expected completion times and people can expect to be pressed hard to achieve or beat these targets.

Periodic briefs would be given to ensure that everyone in the room (and beyond) is aligned in thoughts and actions and to drive progress.

This is an exhausting process that can be unrelenting in the acute phases of a response. For an excellent review of the controller's role and the stresses it imposes see Larken (1995).

21.7.5 Radiological advice

The emergency services, local authorities and health organisations are very experienced at working together to respond to natural events such as storms, floods, animal disease and to events such as major fires and industrial accidents. What makes nuclear and radiological events more difficult for them is a lack of familiarity with ionising radiation and radiological contamination both within the responding organisations and by the general public.

Providing clear radiological advice is an important component of managing a nuclear or radiological event.

Initially, this advice would come from the operator, but many countries have government- or industry-backed independent expertise that will be bought into the discussions at an early stage.

The decision makers are hoping to get a clear situation report. For example:

The reactor containment is breached, and the core is damaged and is getting hotter resulting in a release of radioactive material to the environment. The operator, with help from the Fire and Rescue Service, is trying to (a) seal the breach to stem the release and (b) provide additional cooling to prevent further core damage. If either of these succeeds, we expect the release rate to drop significantly over the next two—three hours and the current protective action advice to prove adequate. However, if both are delayed or fail then the release rate will continue to rise to the extent that the current protective action strategy is inadequate and we would have wished to have extended shelter advice and considered evacuation of those on-site and members of the public close to the site.

The damaged container continues to release uranium hexafluoride posing a chemo-toxic as well as a radiological hazard to those downwind. We estimate that somewhat less than half the content was lost in the initial event and that at current release rates the container will be empty within 4 h. Attempts are being made to seal the leak. The current protective action advice should prove adequate unless an unforeseen circumstance causes a significant increase in the current release rate.

The fire is now confirmed to be out, and the release rate has dropped significantly. There is still a quantity of radioactive material 'at risk' in the damaged building and attempts are being made to prevent that being dispersed.

Preparing these briefs in a timely manner requires the integration of the efforts of several experts. Figure 21.9 shows, in outline, the system for the operating nuclear reactors in the UK. Similar systems are in place across the nuclear industry in other countries.

Information is gathered from the site including a description of the event and a discussion about how it is likely to develop and the worst reasonable extrapolation. This would include an estimated source term and any radiation monitoring results from fixed detectors on or in the plant and on the site boundary and beyond and from any mobile monitoring units deployed in the surrounding area.

Most of the measurements will be external dose rates which can only be used indirectly to estimate the dose to the public, since this is usually dominated by inhalation dose which correlates poorly with external dose rate.

Off-site survey vehicles would move into the plume and take external dose rate measures to confirm the position of the plume and airborne concentration measurements to help assess inhalation dose rates. Exposed filters can be sent to laboratories for high-resolution gamma spectroscopy or alpha spectroscopy. Vehicles are increasingly equipped with gamma-spectroscopy systems that operate at room temperature rather than liquid nitrogen temperature and are able to give a faster

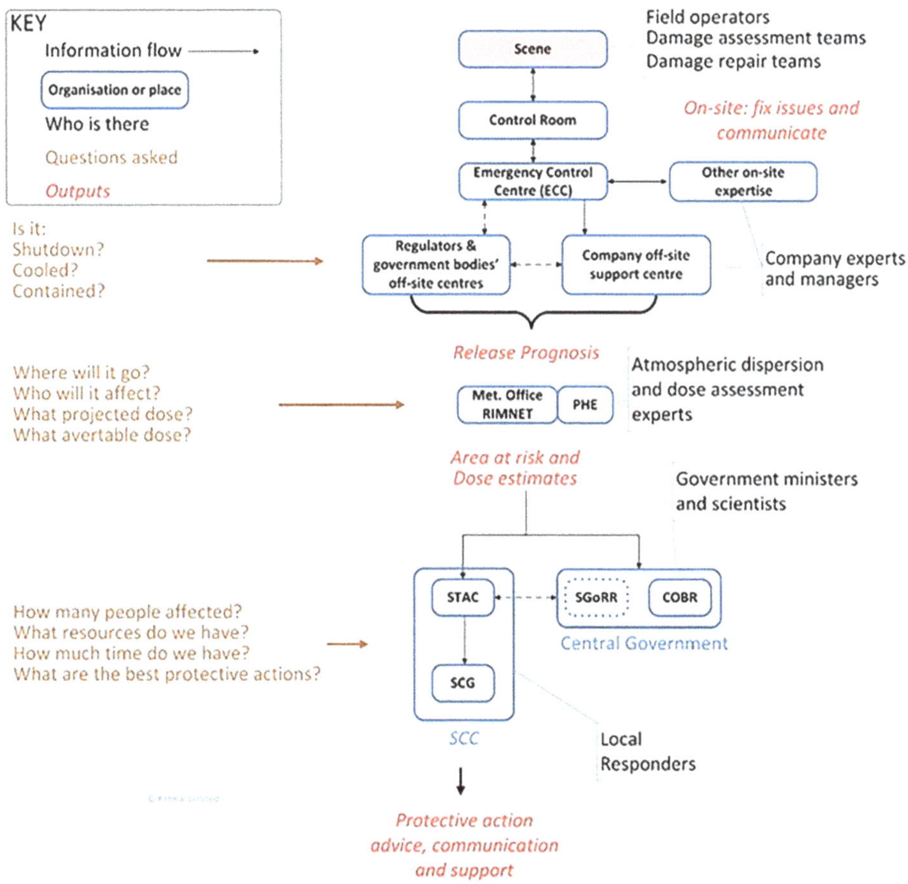

Figure 21.9. Inputs to the preparation of a briefing report.

indication of the isotopic composition of the release, aiding dose assessment (Pearce 2019).

A quantity discussed within the dose assessment teams is 'time to ERL'. This usually follows the gamma spectroscopy of an air sample and is the lower ERL for shelter[2] (3 mSv) divided by the estimated inhalation dose rate (mSv h^{-1}). This is a useful measure. A 'time to ERL of two hours', for example, means that if the situation (release rate and dispersion) does not change over the next two hours plus the time it would take to get someone into shelter, then it is worth advising shelter at the point the measurement was made. The time to ERL together with an assessment from the reactor team about whether the release rate will rise, stay the same or fall, and an estimate of the variation of airborne concentration with location, provides a basis for making a judgement about the need to advise protective actions.

[2] Time to ERL can, of course, be applied to any of the ERLs.

21.7.6 Detection of oversea accidents

One clear outcome of the Chornobyl accident was the building of systems in many countries to detect any clouds of radioactive material from nuclear accidents in other countries. In the UK, RIMNET collects radiation dose rate readings (gamma plus cosmic) from 96 fixed Geiger counter sets at sites spread around the UK to monitor for foreign radioactivity and a number of newer rings of detectors around nuclear power stations to check for any indication of abnormal increases (HMG 2013).

21.7.7 Integration of weather forecast information into dose assessments

All countries with nuclear power have processes to integrate weather forecasts into their dose assessments. In the UK, the Procedures and Communications in the event of a release of Radioactive Materials (PACRAM) Service provides meteorological advice on the atmospheric dispersion of radioactive materials that enables resulting contamination in the air and on the ground to be assessed.

Once alerted by the operator, the Meteorological Office will run computer codes to determine the ‘area at risk’, i.e. the area likely to be under the plume based on their understanding of the release conditions (location, height and buoyancy) and the real-time weather conditions. The software can also provide a reasonable estimate of the population within the area at risk.

These calculations are refined as the event progresses and greater clarity about the source term and prognosis is obtained, see https://www.metoffice.gov.uk/services/government/environmental-hazard-resilience/cbrn-incident-management

21.7.8 Protecting the environment

There is a requirement running through IAEA, EU BSS and UK regulation to consider the protection of the environment in nuclear emergency planning and response but little by way of practical advice on how that might be achieved.

REPPIR-19 recognises a range of environmental harm (the risk framework) from catastrophic with ‘exclusion zones increase and heavy restrictions extended to further distances’, through ‘restricted or temporary loss of environmental growth or produce’ to ‘reluctance to use environmental areas or produce’. Following Fukushima there were highly emotional debates about the acceptable level of radiation contamination and radiation dose in school playing fields.

REPPIR guidance recognises actions such as preventing or minimising contamination of land, water, air, plants and animals as environmental protection actions and claims that, if appropriate levels of attention are given to the reduction of human doses through all significant dose pathways then other types of serious consequences (e.g. to human life, health and safety, quality of life, property and the environment) do not require further explicit assessment. Importantly the environment agencies are involved in the planning and response phases with their focus on environmental protection.

Environmental protection issues would be very much to the fore in any discussions about remediation work in contaminated environments post-accident.

As experiences at Chornobyl and Fukushima have shown, if plans are not thought through, the results of expensive remediation can be large volumes of low-level waste to manage and an environment that does not provide a significantly enhanced quality of life or is poorer than it was before the measures were taken.

The best way to protect the environment is to reduce any release. Real-time debates on environmental harm may result from discussions about potential protective actions that deliberately move radioactivity from one route to another such as using water sprays to limit atmospheric dispersion at the cost of producing contaminated water that can get into the ground and watercourses.

21.7.9 Public information

The IAEA states a requirement to provide information to members of the public within the emergency planning zones around category I and II sites with information on the response to a nuclear or radiological emergency. '*This information shall include information on the potential for a nuclear or radiological emergency, on the nature of the hazards, on how people would be warned or notified, and on the actions to be taken in such an emergency*'.

In GSG-14, the IAEA identifies the following objectives of public communication for a nuclear or radiological accident:

- To protect the public;
- To inform the public, both at the preparedness stage and during the response, of protective actions and other response actions, and of the nature of any hazards, and to facilitate emergency response actions;
- To gain and maintain public trust in the emergency response by means of transparent, timely, clear and accurate public communication;
- To address public concerns regarding potential adverse consequences for human life, health, property and the environment;
- To prevent undue concern, to mitigate anxiety and long-term psychological effects, and to help to ensure that actions taken do more good than harm;
- To respond to misinformation and rumours.

It then outlines the requirements for a public communication plan in some detail and discusses the thorny issue of putting radiological health hazards into perspective suggesting a traffic light system:

- Red: Dangerous to health;
- Amber: Possible health effects resulting from radiation exposure;
- Green: No observable health effects resulting from radiation exposure.

IAEA EPR-Public Communications 2012 provides advice about the nature and delivery of this information. This document provides resources and 'plain language explanations' to help in this endeavour.

In the UK, REPPIR-19 regulation 21 requires that the local authority ensure that members of the public within the DEPZ and the OPZ are made aware of information mandated in schedules attached to the regulations.

This includes the basic facts about ionising radiation, how they would be alerted in the event of a radiation accident and the protective actions they might be advised to adopt. For those in the OPZ the information includes the factors that might lead to a triggering of the outline plan. Examples of local authority REPPIR information can be found on their emergency preparedness websites. For example, the AWE leaflet can be found at https://info.westberks.gov.uk/CHttpHandler.ashx?id=39234&p=0.

A good example of prior information, in this case earthquake response in Tokyo, can be found at https://www.bousai.metro.tokyo.lg.jp/book/pdf/en/01_Simulation_of_a_Major_Earthquake.pdf.

Regulation 22 requires the local authority to prepare and keep up to date arrangements to supply information to the public in their area in the event of an emergency however that emergency might arise. This regulation applies to all local authorities, not just those that host nuclear sites. This information should include the basic details of the incident, advice and guidance about protective actions, the impact on other services, particularly childcare and schools and reassurance (including a helpline).

21.7.10 Care for the public

It is not always entirely clear how long the public may be asked to shelter in the event of a nuclear emergency. Estimates range from a few hours (while a short duration plume transits) to several days (in the event of a prolonged release or pending environmental monitoring indicating that advice to shelter can be lifted). Government advice suggests up to 72 h (HMG 2014). Although normal activities such as waste collection, road works and grass cutting would be suspended in an area subject to advice to shelter, more difficult decisions have to be made about the support given to vulnerable people in their homes and the response to medical emergencies or to such things as reports of gas leaks and fires.

People affected by the emergency may require some, or all, of (CO 2016):

- emotional support;
- first aid;
- shelter, food, clothing;
- information updates about the incident and individuals directly involved;
- advice and support on financial, legal and insurance issues;
- support to restore social networks;
- opportunities for remembrance/memorialisation;
- input into any evaluation process following the emergency (for example a public inquiry).

If people are displaced from their homes, they may prefer to go to friends or relatives, but some might need short-term help to find shelter. In the UK, local authorities are responsible for the setting up Reception/Rest Centres to provide shelter, advice and food to those that need it. These are often managed by voluntary organisations.

A Reception/Rest Centre is a designated building or buildings for the temporary accommodation of evacuees or uninjured survivors. It should offer safe and secure shelter, at a safe distance from the incident site, with facilities for the provision of refreshments, access to welfare services and short-term sleeping arrangements, if necessary, as a minimum.

If there are widespread concerns about personal contamination, there may be the need to set up a Radiation Monitoring Unit (RMU). The intent of an RMU is to provide reassurance monitoring to people who fear that they may have been contaminated by a release as well as provide measurements and records of any actual contamination. RMUs will be managed and staffed by people with suitable medical physics/health physics backgrounds. UK planning and operational guidance can be found in CRCE (2011).

The flow of people through an RMU is shown in figure 21.10 (taken from CRCE (2011). At initial registration, a preliminary risk assessment will be undertaken to

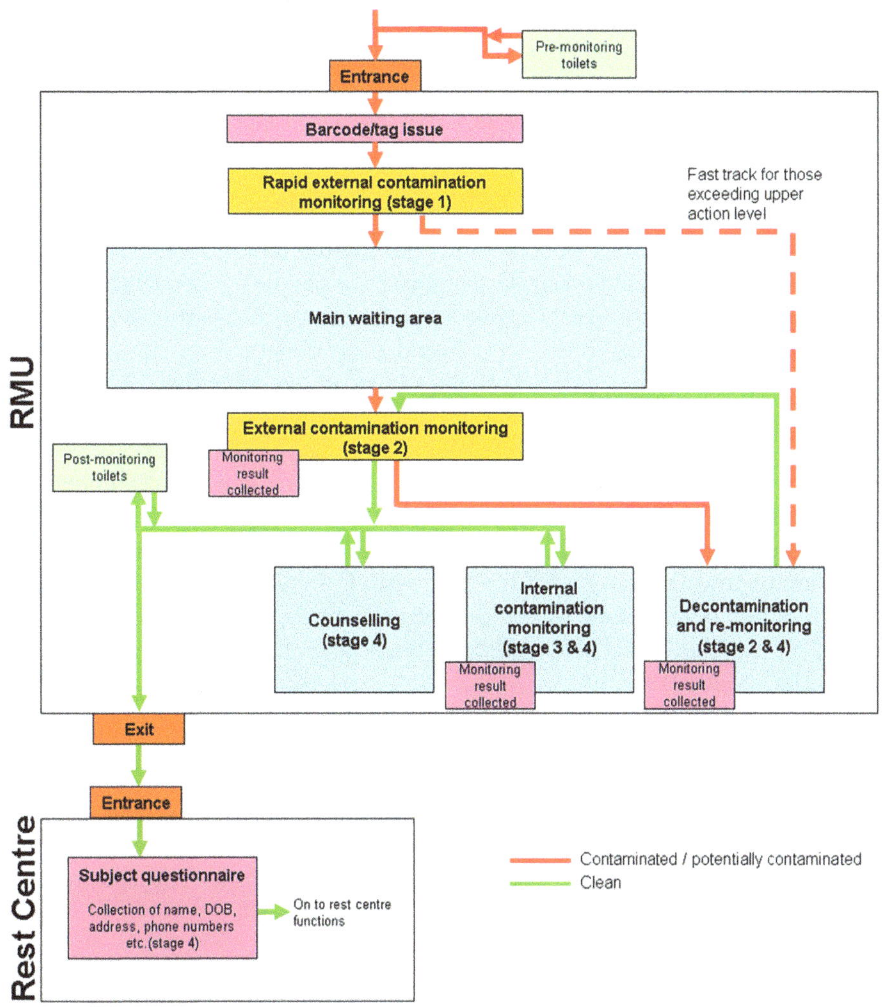

Figure 21.10. Flow of people through a radiological monitoring unit. Reproduced with permission from CRCE (2011). © Health Protection Agency, 2011. Open Government license.

determine the likelihood of contamination and those thought to be more at risk will be fast-tracked onwards. Initial monitoring for contamination by gamma-emitting radionuclides will be by portal monitors with advisors on hand to ensure that people stay in the sensitive area for the correct amount of time[3]. An initial screening action level will be used to remove those more heavily contaminated from the main flow of people and fast-track them into decontamination and re-monitoring. Slower monitoring will then be used on the main flow of people to monitor to lower levels of contamination to identify those that require decontamination. Any members of the public giving results above the second action level will be monitored using handheld instruments and, if necessary, decontaminated and monitored again. Monitoring for internal contamination will be undertaken on those more heavily contaminated and, if resources allow, on random samples of those shown to be contaminated.

For the system to work it needs to be well laid out, there is a need for good facility managers, competent Health Physicists and Health Physics Monitors, good record keeping and good communication with the users coming through. Leaflets that explain the process should be available to all users and further leaflets that explain the consequence of a positive report available, along with counselling, to those who are identified as contaminated.

Fukushima provided experience of using RMUs. Their initial action level to trigger decontamination was 13 000 counts per minute (cpm). However, they found this impractical as it selected a very large number of people for decontamination and was difficult to measure in areas with elevated dose rates. The action level was raised to 100 000 cpm to identify those who would most benefit from decontamination. Fukushima Prefecture reported performing radiation screening for 253 844 residents including those who had evacuated from within 20 km from the NPP from 13 March to 6 June 2012. Of these, 102 people were found to have levels in excess of 100 000 cpm. After removing clothes, the counts decreased to less than 100 000 cpm; nobody had contamination levels that would lead to health problems. Changing clothes was one of the recommended protective actions for decontamination. However, because of the Tsunami and the manner of the evacuation, there was a shortage of clothing and the reception centres were too cold to remove clothing without replacing it. The reception centres and RMU also suffered from a lack of water, again because of the Tsunami (IAEA 2014).

21.7.11 Recovery

An important feature of emergency plans is a strategy for getting back to normal. This requires that the emergency response organisation recognises when it has fulfilled its mandate and that there are better mechanisms for continuing the response.

[3] Different instruments are appropriate for sites where the releases may be dominated by alpha emitters.

Transition to recovery is a process rather than an event. It is likely that the police will want to hand over the role of chair of the SCG once the site has been made safe and the urgent protective actions are being withdrawn. The site will stand down their emergency declaration once the timeframe for decision making allows for important decisions to be made by a collaborative process. At this stage, there may well be considerable disruption to the lives of the people in the vicinity; some may have been evacuated, some may be in shelter, many will want information, reassurance and support. Decisions will have to be made about any decontamination and any lasting protective actions. The local authority may stay at major incident level for considerably longer as it grapples with recovery issues. In general, the command, control and coordination will change in response to the changing situation punctuated by a few formal handover processes.

An issue with this is that regulation talks about the acute phase and the recovery phase as though there is a clear distinction and time of transition between them. This impacts dose control, because, at the transition, the reference level of dose moves from that applicable to an emergency situation to that applicable to an existing situation.

Post-accident a UK local authority can set new reference levels to guide the recovery process and so to can the Secretary of State.

When approaching such transition points, the organisations involved are likely to take stock of their current situation, foci and outstanding actions and then produce a transition plan. This plan will determine the actions that need to be completed and the situation that needs to be achieved before handover can take place.

At the handover process the current situation will be articulated clearly and the inheriting organisation will draw up their focus list and start to manage the actions required to progress the recovery. It is important that at this stage any incomplete actions are reviewed and either cancelled or handed over to new supervisors.

The immediate challenges for the multi-agency recovery effort will be to (HMG 2015b):

- Support sectors of the population temporarily evacuated or permanently relocated because of the event;
- Understand the extent and level of contamination based on available monitoring and sampling capability and what this means for long-term radiation exposures;
- Understand the potential long-term impact on infrastructure, business activities, agriculture, food production, drinking water and leisure;
- Identify where access to contaminated areas needs to be restricted, where remediation is necessary and where 'doing nothing' is the best long-term option;
- Coordinate remediation activities on the ground—these will include monitoring, hazard assessment, clean-up, waste disposal and clearance monitoring of public and private assets;
- Manage communications with the public, media and stakeholders;
- Manage the expectations of the public and stakeholders;
- Establish a decision-making process that is open, transparent and involves representatives of the affected community;

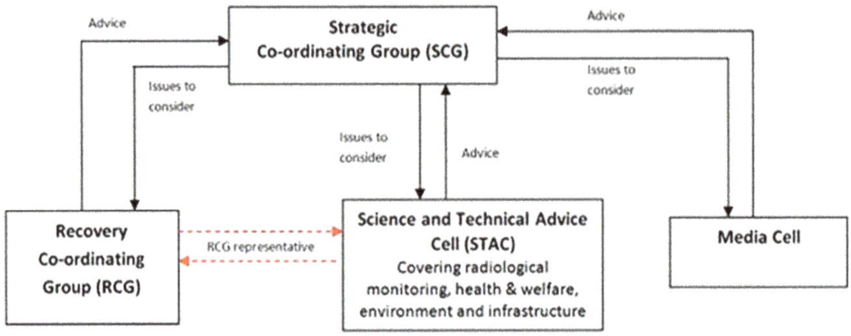

Figure 21.11. UK arrangements for the organisation of recovery effort in the response phase. Reproduced with permission from https://assets.publishing.service.gov.uk/media/5a807dbce5274a2e87dba1c8/NEPRG03_-_Recovery.pdf. © Crown copyright 2015. Open Government license.

- Establish effective processes for compensation claims; and
- Manage the long-term impacts on the local economy.

In the UK, this is managed by the recovery co-ordinating Group (RCG) which fits into the organisation of the Strategic Coordination Centre as shown in figure 21.11.

Technical advice for this Group can be found in the *UK recovery handbooks for radiation incidents* (HMG 2015c). These handbooks discuss different decontamination methods and outline their costs, strengths and weaknesses and show how a recovery strategy can be developed.

21.8 Non-reactor faults

21.8.1 Criticality accidents

The IAEA records 27 criticality faults in ERP-LESSONS LEARNED 2012. These include many people being exposed to large doses of radiation (up to 100 Gy), several deaths and a number of life changing injuries (see also section 20.4).

Although the frequency of criticality excursions has reduced as nuclear research has matured there is still the possibility of process faults at fuel preparation or reprocessing plants and nuclear weapons establishments leading to a criticality event.

A well-known example of a criticality accident occurred at Tokai Mura in Japan in 1999. Safety standards at a small fuel fabrication plant had been slipping for several years and several badly thought through and unapproved changes to the processes made. On the day of the accident, they were handling higher than usually enriched uranium oxides and did not follow the process which was designed to ensure a criticality margin through geometry. The result was a cyclic criticality event lasting intermittently for 20 h. Two operators received fatal doses of radiation and over 100 people received additional doses of more than 1 mSv.

On-site plans for criticality events generally involve detecting the radiation pulse, sounding local alarms and evacuating the area. Teams will then be formed to try to understand the situation, and determine if there is a possibility of further criticalities, then make the area safe and start the clean-up.

Personnel close to the event will be identified and a process to determine their potential doses will be initiated. This involves several steps including taking beta dose rate measurements from criticality lockets worn by the workforce, which contain gold which is activated by neutrons through the reaction $^{197}Au(n,\gamma)^{198}Au$. ($^{198}Au$ is beta active with a half-life of 2.697 days) and an indium foil activated by $^{115}In(n,gamma)^{116}In$ (the ^{116}In has a 54-minute half-life.)

Other elements in the locket give different information, including the neutron flux at different energies and beta/gamma dose. However, this information takes more time to extract.

An alternative screening approach is to measure body activity due to neutron activation of the sodium in the blood via the $^{23}Na(n, gamma)^{24}Na$ reaction. Sodium-24 has a 15-h half-life and emits a 1.4-MeV beta (maximum energy) and two gammas (1.37 and 2.75 MeV).

Other rapidly available information comes from the person's location at the time of the event.

Those with high doses of radiation will identify themselves with the onset of Acute Radiation Syndrome and it is important that the emergency response records the onset of symptoms as that gives clues about the exposure which in turn informs medical treatment.

See, for example table 21.3, taken from IAEA Safety Reports Series No. 101.

Table 21.3. Initial decisions on medical treatment based on observations of vomiting and erythema (IAEA Safety Reports Series No. 101).

Clinical manifestations		Estimated dose		
Whole body exposure	Local exposure	Whole body exposure	Local exposure	Initial decision
No vomiting	No erythema	<1 Gy	<3 Gy	Outpatient with five-week surveillance (blood, skin).
Vomiting 2–3 h after exposure	Primary erythema* 12–24 h after exposure	1–2 Gy	>3–8 Gy	Monitoring in a general hospital.
Vomiting 1–2 h after exposure	Primary erythema 8–15 h after exposure	2–4 Gy	15 Gy < 25 Gy	Hospitalisation in a haematological or surgical (burn) department) or specialized surgical department.
Vomiting earlier that 1 h after exposure	Primary erythema within 3–6 h (or less) associated with itching, oedema and pain	> 4 Gy	> 25 Gy	Hospitalisation in a haematological or specialized surgical department.

21.8.2 Industrial accidents

The IAEA record 68 industrial accidents with radioactive sources, with industrial radiography and irradiation facilities both featuring regularly (see sections 20.2 and 20.5). Generally, a failure of process, equipment or of interlocks, often combined with lax use of safety measures, such as dosimeters, leads to one or more people being exposed to high doses of radiation. Quite often the dose, and the subsequent damage, is localised, but the impact can be gruesome with several lingering deaths but also with fingers and hands lost in the months following exposure.

The IAEA has published a number of reports of radiation accidents of this nature (https://www.iaea.org/publications/search/topics/accident-reports). These reports detail the accident, the investigation, the medical treatment of the victims and the outcomes (see also chapter 20).

The Health Physics role in these accidents is prevention. Post-accident, the scene has to be made safe, by recovering any sources into shielded containers or switching the power off on x-ray sets, and medical aid given to those hurt.

21.8.3 Orphan sources and malicious source placement

An orphan source is one that has escaped control, as has happened in the past with end-of-life industrial radiography, radiotherapy and irradiation sources. Such sources often end up in recycling facilities, where they should be detected, but sometimes turn up in inhabited areas giving large doses, sometimes to many people.

The Goiânia incident in Brazil in 1987, where a caesium source was dismantled led to 129 people being exposed to significant doses and 4 people dying within four weeks of admission to hospital. The first signs of the event were people developing prodromal manifestations of acute radiation syndrome and local radiation injuries (cutaneous radiation syndrome). Several people had presented at medical facilities with gastrointestinal symptoms, but the cause was not understood until part of the source was handed in to the authorities. https://www.iaea.org/publications/3684/the-radiological-accident-in-goiania.

One lesson from Goiânia is that health personnel must be aware of the clinical manifestations of whole body and local radiation exposures ('acute' radiation and cutaneous radiation syndromes).

The authorities were faced with a situation where many high specific activity caesium beads had been loose in the environment for a number of days being deliberately and accidentally spread.

In the response 112 000 persons were monitored. Over a period of two days all the more than 67 km^2 urban areas of Goiânia were monitored. Action levels in this initial response were set for the control of access (10 $\mu Sv\ h^{-1}$), for evacuation and prohibited access (2.5 $\mu Sv\ h^{-1}$) and later 10 $\mu Sv\ h^{-1}$ for houses, and 150 $\mu Sv\ h^{-1}$ for unoccupied areas. The investigation levels selected for considering the various actions corresponded to an effective dose of 5 mSv in the first year and a long-term projected effective dose of 1 mSv per year in subsequent years.

A lesson that recurs is that, in real events, where clean-up targets are determined in part by politicians in the light of media scrutiny, the targets set are much more

restrictive than organisations such as the IAEA suggest as international defaults. This has consequences.

The decontamination necessitated the demolition of seven residences and various other buildings, and the removal of the topsoil from large areas. The packaging of wastes required 3800 metal drums (200 l), 1400 metal boxes (5 tonnes), 10 shipping containers (32 m^3) and 6 sets of concrete packaging. The final total volume of waste stored was 3500 m^3. The best estimate of the radioactivity accounted for in contamination is around 44 TBq, compared with the known radioactivity of the caesium chloride source before the accident of 50.9 TBq.

This type of accident could, in principle, happen anywhere and does not have an operator with responsibilities to prepare and resource an adequate emergency plan. Metal recycling operations are expected to have radiation monitors to ensure that large sources are not allowed into their processes https://www.hse.gov.uk/waste/radioactive-contamination.htm#application-of-health.

There is a recognised risk that powerful radiation sources could be deliberately placed in public places. Many organisations, such as the London Underground, have taken steps to protect themselves and their users by placing radiation detectors in passenger barriers. The impact of a deliberately placed source would be similar to that of an orphan source but if left in a mass transit system could lead to presentations at hospitals and doctors over a wide area.

21.8.4 Dirty bombs

A dirty bomb or radiological dispersal device (RDD) is one that uses the primary explosion to disperse a material hazardous to health. This could, in principle include radioactive material. From the terrorist point of view, the addition of radioactive material would cause more alarm, may deny access to an area for considerably longer and may entail more significant clean-up costs. If no warning or explanation were given this would look like a standard terrorist explosive detonation.

The immediate priority would be to assist those injured in the blast and it may not be recognised as a radiation incident at first. In the UK, and many other countries, the emergency services are provided with radiation monitors and basic training and would be expected to detect radioactive material in the blast area if it were β–γ in nature and there in significant quantities.

The IAEA advice to first responders (IAEA 2006b) is that the radiation monitoring should locate and mark areas where ambient dose rates are >100 mSv h^{-1} in which only lifesaving activities should be performed with a stay time of less than 30 min. The inner cordon should be established where the ambient dose rate is no greater than 0.1 mSv h^{-1}. This advice applies to all radiological events not just dirty bombs.

For casualties without life-threatening injuries some at-scene decontamination may be appropriate. The bulk of any contamination is likely to be removed by removing the outer clothes. Bare skin can be decontaminated by brushing, showering or washing.

In the longer term, the local authority is likely to have to manage a cordoned off area for an extended period of time with people displaced from their homes and businesses. The response may be similar to the Salisbury nerve agent incident (2018). See the government summary of the first two months of this event (Novichok nerve agent use in Salisbury: UK government response, March to April 2018 (HMG 2018)).

The US Department of Homeland Security has published a planning document showing what an emergency plan and response for this type of event could look like (DHS 2017).

References

BEIS 2015 National Nuclear Emergency Planning and Response Guidance https://gov.uk/government/publications/national-nuclear-emergency-planning-and-response-guidance)

CO 2016 Cabinet Office, Human Aspects in Emergency Management Guidance on Supporting Individuals Affected by Emergencies https://assets.publishing.service.gov.uk/government/uploads/system/uploads/attachment_data/file/564306/human_aspects_guidance_2016_final.pdf

CRCE 2011 HPA-CRCE-017 RMU Planning and Operational Guidance https://gov.uk/government/uploads/system/uploads/attachment_data/file/340153/HPA-CRCE-017_for_website.pdf

CRCE 2019a PHE-CRCE-049 (2019), Public Health Protection in Radiation Emergencies

CRCE 2019b PHE-CRCE-050 REPPIR 2019 consequence assessment methodology

Dunster H J, Howells H and Templeton W L 1958 District surveys following the windscale incident. In *October 1957 Proceedings of the Second United Nations International Conference on the Peaceful Uses of Atomic Energy* (Geneva, 1 September–13 September 1958). *Volume 18: Waste Treatment and Environmental Aspects of Atomic Energy* (Geneva: United Nations) pp 296–308 reproduced as Dunster H J *et al* 2007 *J. Radiol. Prot.* **27** 217

DHS 2017 Radiological Dispersal Device (RDD), Response Guidance, Planning for the First 100 min https://dhs.gov/sites/default/files/publications/NUSTL_RDD-ResponsePlanningGuidance-Public_171215-508.pdf

EPA 2020 Protective Action Guides (PAGs)

EU 2013 Council Directive 2013/59/EURATOM http://eur-lex.europa.eu/legal-content/EN/TXT/PDF/?uri=CELEX:02013L0059–20140117&from=EN]

FEMA 2016 Nuclear/Radiological Incident Annex to the Response and Recovery Federal Interagency Operational Plans https://fema.gov/media-library-data/1488825883577-fb52b6b7fd784dfb64aaee1fd886393a/NRIA_FINAL_110216.pdf

Hasegawa A *et al* 2016 Emergency Responses and Health Consequences after the Fukushima Accident; Evacuation and Relocation *Clin. Oncol.* **28** 237–44

HMG 1965 Nuclear Industry Act https://legislation.gov.uk/ukpga/1965/57

HMG 2013 Radioactive Incident Monitoring https://gov.uk/government/collections/radioactive-incident-monitoring

HMG 2014 Evacuation and Shelter Guidance, Non-statutory guidance to complement Emergency preparedness and Emergency response and recovery. https://assets.publishing.service.gov.uk/government/uploads/system/uploads/attachment_data/file/274615/Evacuation_and_Shelter_Guidance_2014.pdf

HMG 2015 Nuclear Emergency Planning and Response Guidance, Concept of Operations https://assets.publishing.service.gov.uk/government/uploads/system/uploads/attachment_data/file/472419/NEPRG00_-_Concept_of_Operations.pdf

HMG 2015b National Nuclear Emergency Planning and Response Guidance Part 3—Recovery https://assets.publishing.service.gov.uk/government/uploads/system/uploads/attachment_-data/file/472424/NEPRG03_-_Recovery.pdf

HMG 2015c UK Recovery Handbooks for Radiation Incidents https://gov.uk/government/publications/uk-recovery-handbooks-for-radiation-incidents-2015

HMG 2018 Novichok Nerve Agent Use in Salisbury: UK Government Response, March to April 2018 https://gov.uk/government/news/novichok-nerve-agent-use-in-salisbury-uk-government-response

IAEA 2002 Planning and Preparing for Emergency Response to Transport Accidents Involving Radioactive Material, Safety Guide No. TS-G-1.2 (ST-3)

IAEA 2006a IAEA Safety Fundamentals Number SF-1

IAEA 2006b EPR-First Responders 2006, Manual for First Responders to a Radiological Emergency https://www-pub.iaea.org/MTCD/Publications/PDF/EPR_FirstResponder_web.pdf)

IAEA 2007 IAEA Safety Guide *GS-G-2.1 Arrangements for Preparedness for a Nuclear or Radiological Emergency (2007)*

IAEA 2011 Criteria for Use in Preparedness and Response for a Nuclear or Radiological Emergency (Series No. GSG-2) https://www-pub.iaea.org/MTCD/Publications/PDF/Pub1467_web.pdf

IAEA 2012 EPR Lessons Learned, Lessons Learned from the Response to Radiation Emergencies (1945–2010) http://www-pub.iaea.org/books/IAEABooks/8920/Lessons-Learned-from-the-Response-to-Radiation-Emergencies-1945–2010

IAEA 2014 IAEA EPR Medical/T 2014 Medical Preparedness and Response for a Nuclear or Radiological Emergency, Lecture 44

IAEA 2015 Preparedness and Response for a Nuclear or Radiological Emergency, GSR Part 7 2015

IAEA 2015b The Fukushima Daiichi Accident, Non-serial Publications, STI/PUB/1710 978-92-0-107015-9 https://iaea.org/publications/10962/the-fukushima-daiichi-accident

IAEA 2018 Arrangements for the Termination of a Nuclear or Radiological Emergency, GSG-11 https://www-pub.iaea.org/MTCD/Publications/PDF/PUB1796_web.pdf

ICRP 2007 The 2007 Recommendations of the International Commission on Radiological Protection, ICRP Publication 103 *Ann. ICRP* **37(2–4)**

Jetter J and Whitfield C 2005 Effectiveness of expedient sheltering in place in a residence *J. Hazard. Mater.* **119** 31–40

JESIP 2020 Joint Emergency Services Interoperability Principles https://jesip.org.uk/home

Larken E S J 1995 Practical Emergency Management, ICHEME Symposium Series No. 139, Paper 27

NRC 2018 Emergency Planning Zones https://nrc.gov/about-nrc/emerg-preparedness/about-emerg-preparedness/planning-zones.html

NRC 2020 NRC Regulations Title 10, Code of Federal Regulations https://nrc.gov/reading-rm/doc-collections/cfr/

NRC Undated Perspectives on Reactor Safety, NUREG/CR-6042, Rev. 2 SAND 93–0971 https://nrc.gov/docs/ML0210/ML021080026.pdf

ONR 2007 *Licence Condition Handbook* http://onr.org.uk/documents/licence-condition-handbook.pdf

Pearce K I 2019 *Mobile Measurements* Nuclear Engineering International https://neimagazine.com/features/featuremobile-measurements-7539961/

WHO 1999 *Guidelines for Iodine Prophylaxis following Nuclear Accidents* Report WHO/SDE/PHE/99.6

www.ingramcontent.com/pod-product-compliance
Ingram Content Group UK Ltd.
Pitfield, Milton Keynes, MK11 3LW, UK
UKHW051909100926
8556IPUK00005B/818

9 780750 32422